Lecture Notes in Computer Science 7675

Commenced Publication in 1973
Founding and Former Series Editors:
Gerhard Goos, Juris Hartmanis, and Jan van Leeuwen

Lecture Notes in Computer Science 7875

Commenced Publication in 1973
Founding and Former Series Editors:
Gerhard Goos, Juris Hartmanis, and Jan van Leeuwen

Editorial Board

David Hutchison
    Lancaster University, UK
Takeo Kanade
    Carnegie Mellon University, Pittsburgh, PA, USA
Josef Kittler
    University of Surrey, Guildford, UK
Jon M. Kleinberg
    Cornell University, Ithaca, NY, USA
Alfred Kobsa
    University of California, Irvine, CA, USA
Friedemann Mattern
    ETH Zurich, Switzerland
John C. Mitchell
    Stanford University, CA, USA
Moni Naor
    Weizmann Institute of Science, Rehovot, Israel
Oscar Nierstrasz
    University of Bern, Switzerland
C. Pandu Rangan
    Indian Institute of Technology, Madras, India
Bernhard Steffen
    TU Dortmund University, Germany
Madhu Sudan
    Microsoft Research, Cambridge, MA, USA
Demetri Terzopoulos
    University of California, Los Angeles, CA, USA
Doug Tygar
    University of California, Berkeley, CA, USA
Gerhard Weikum
    Max Planck Institute for Informatics, Saarbruecken, Germany

Yuexian Hou   Jian-Yun Nie   Le Sun
Bo Wang   Peng Zhang (Eds.)

Information Retrieval Technology

8th Asia Information Retrieval
Societies Conference, AIRS 2012
Tianjin, China, December 17-19, 2012
Proceedings

 Springer

Volume Editors

Yuexian Hou
Tianjin University, School of Computer Science and Technology
300072 Tianjin, China
E-mail: yxhou@tju.edu.cn

Jian-Yun Nie
University of Montreal, DIRO
CP. 6128, succursale Centre-ville, Montreal, H3C 3J7, QC, Canada
E-mail: nie@iro.umontreal.ca

Le Sun
Chinese Academy of Sciences
Institute of Software, Storage & Information Retrieval Laboratory
100190 Beijing, China
E-mail: sunle@iscas.ac.cn

Bo Wang
Tianjin University, School of Computer Science and Technology
300072 Tianjin, China
E-mail: bo.wang.1979@gmail.com

Peng Zhang
Robert Gordon University, School of Computing
St Andrew Street, Aberdeen, AB25 1HG, UK
E-mail: p.zhang1@rgu.ac.uk

ISSN 0302-9743   e-ISSN 1611-3349
ISBN 978-3-642-35340-6   e-ISBN 978-3-642-35341-3
DOI 10.1007/978-3-642-35341-3
Springer Heidelberg Dordrecht London New York

Library of Congress Control Number: 2012953221

CR Subject Classification (1998): H.3, H.4, F.2.2, I.4-5, E.1, H.2.8

LNCS Sublibrary: SL 3 – Information Systems and Application, incl. Internet/Web
and HCI

Typesetting: Camera-ready by author, data conversion by Scientific Publishing Services, Chennai, India

Printed on acid-free paper

Springer is part of Springer Science+Business Media (www.springer.com)

Preface

These proceedings contain the refereed papers and posters presented at the 8th Asia Information Retrieval Societies Conference (AIRS 2012). The conference was held during December 17–19, 2012, in Tianjin University (TJU), which is the oldest university in China, founded in 1895 as Peiyang University.

The annual AIRS conference has established its reputation as the main information retrieval (IR) conference for the Asia-Pacific region, where the research community and industry in IR have been growing rapidly. The conference aims to bring together researchers and developers to exchange new ideas and the latest achievements in the broad area of IR, covering topics on theories, systems, technologies, and applications of IR on both text and multimedia data.

This year, we received 77 submissions from Asian countries and all over the world; 22 (28.5%) of them were accepted as regular papers for oral presentations and 26 (33.7%) for poster presentations. All the papers went through a rigorous review process. Each paper was reviewed by three reviewers and a meta-reviewer.

The conference also featured two keynote speeches. Norbert Fuhr, the winner of the Salton Award of SIGIR in 2012, gave a keynote presentation on the calibrated values of probability of relevance and its application on vertical search, optimum clustering framework, and interactive IR. Haifeng Wang, a senior director of Baidu, gave a keynote speech on Web-based machine translation and cross-language IR. In addition, an industrial track was organized, in which several researchers from industry shared their technologies and experiences with the conference participants.

We would like to thank all the authors for submitting their research work to the conference. We are particularly grateful to the members of the Program Committee for their efforts in providing timely and high-quality reviews. We thank Tianjin University and Microsoft Research Asia for their financial support. We also thank the editorial staff of Springer for their assistance in publishing the conference proceedings as part of the *Lecture Notes in Computer Science* (LNCS). Finally, a special thanks goes to all the participants and student volunteers for their contributions and help in making AIRS 2012 a success.

December 2012

Yuexian Hou
Jian-Yun Nie
Le Sun
Bo Wang
Peng Zhang

Organization

AIRS 2012 was organized by Tianjin University, University of Montreal, Institute of Software and Institute of Computing Technology at the Chinese Academy of Sciences, and Robert Gordon University.

Conference Organizing Committee

General Co-chairs

Yuexian Hou   Tianjin University, China
Le Sun   Institute of Software, Chinese Academy of
                        Sciences, China

Program Chair

Jian Yun Nie   University of Montreal, Canada

Local Chair

Ruiguo Yu   Tianjin University, China

Industry Track Chair

Bin Wang   Institute of Computing Technology, Chinese
                        Academy of Sciences, China

Publicity Chair

Bo Wang   Tianjin University, China

Proceedings Chair

Peng Zhang   Robert Gordon University, UK

Senior Program Committee

Wanxiang Che   Harbin Institute of Technology, China
Hsin-Hsi Chen   National Taiwan University, Taiwan
Jianfeng Gao   Microsoft Research, USA
Jimmy Huang   York University, USA
Joemon Jose   University of Glasgow, UK
Noriko Kando   National Institute of Informatics, Japan
Wai Lam   The Chinese University of Hong Kong,
                        Hong Kong

Hang Li   Huawei Noah's Ark Lab, Hong Kong
Massimo Melucci   University of Padua, Italy
Ian Soboroff   National Institute of Standards and
                                    Technology, USA
Dawei Song   Tianjin University, China
Bin Wang   Institute of Computing Technology, China
Shiqi Zhao   Baidu Inc., China
Guodong Zhou   Soochow University, China

Program Committee

Lidong Bing   The Chinese University of Hong Kong,
                                    Hong Kong
Fan Bu   Tsinghua University, China
Ben Carterette   University of Delaware, USA
Michael Chau   The University of Hong Kong, Hong Kong
Pu-Jen Cheng   National Taiwan University, Taiwan
Kevyn Collins-Thompson   Microsoft Research, USA
Alberto Costa   LIX, Ecole Polytechnique, France
Gerard De Melo   ICSI Berkeley, USA
Emanuele Di Buccio   University of Padua, Italy
Kevin Duh   Nara Institute of Science and Technology,
                                    Japan
Georges Dupret   yahoo! Labs, USA
Ingo Frommholz   University of Bedfordshire, UK
Wei Gao   Qatar Computing Research Institute, Qatar
Shlomo Geva   Queensland University of Technology, Australia
Jiafeng Guo   Institute of Computing Technology, China
Cathal Gurrin   Dublin City University, Ireland
Martin Halvey   University of Glasgow, UK
Xianpei Han   ISCAS, China
Ben He   Graduate University of Chinese Academy of
                                    Sciences, China
Ruifang He   Tianjin University, China
Xiaodong He   Microsoft Research, USA
Yulan He   The Open University, UK
Yu Hong   Soochow University, China
Yuexian Hou   Tianjin University, China
Xuanjing Huang   Fudan University, China
Jing Jiang   Singapore Management University, Singapore
Long Jiang   Microsoft Research Asia, China
Hideo Joho   University of Tsukuba, Japan
Gareth Jones   Dublin City University, Ireland
Leszek Kaliciak   Robert Gordon University, UK
Jaap Kamps   University of Amsterdam, The Netherlands
Kazuaki Kishida   Keio University, Japan

Kazuko Kuriyama	Shirayuri College, Japan
Oren Kurland	Technion, Israel Institute of Technology, Israel
Yanyan Lan	Chinese Academy of Sciences, China
Jun Lang	National University of Singapore, Singapore
Raymond Y.K. Lau	City University of Hong Kong, Hong Kong
Gary Lee	POSTECH, South Korea
Teerapong Leelanupab	University of Glasgow, UK
Shoushan Li	Soochow University, China
Wenjie Li	The Hong Kong Polytechnic University, Hong Kong
Zhenghua Li	Harbin Institute of Technology, China
Shasha Liao	New York University, USA
Fei Liu	University of Texas at Dallas, USA
Feifan Liu	University of Wisconsin-Milwaukee, USA
Jingjing Liu	MIT CSAIL, USA
Kang Liu	National Laboratory of Pattern Recognition, Institute of Automation, China
Xiaohua Liu	Microsoft Research Asia, China
Yang Liu	Tsinghua University, China
Yiqun Liu	Tsinghua University, China
Zhiyuan Liu	Tsinghua University, China
Hao Ma	Microsoft Research, USA
Andrew Macfarlane	City University London, UK
David McClosky	Stanford University, USA
Mandar Mitra	Indian Statistical Institute, India
Alistair Moffat	The University of Melbourne, Australia
Jian-Yun Nie	University of Montreal, Canada
Zhengyu Niu	Baidu Inc., China
Manabu Okumura	Tokyo Institute of Technology, Japan
Iadh Ounis	University of Glasgow, UK
Kristen Parton	Columbia University, USA
Tao Qin	Microsoft Research Asia, China
Vijay V. Raghavan	University of Louisiana at Lafayette, USA
Reede Ren	University of Glasgow, UK
Tetsuya Sakai	Microsoft Research Asia, China
Yohei Seki	University of Tsukuba, Japan
Jialie Shen	Singapore Management University, Singapore
Ruihua Song	Microsoft Research Asia, China
Yang Song	Microsoft Research, USA
Valentin Spitkovsky	Stanford University, USA
Aixin Sun	Nanyang Technological University, Singapore
Mihai Surdeanu	Yahoo! Research Barcelona, Spain
Hisami Suzuki	Microsoft Research, USA
Paul Thomas	CSIRO, Australia
Andrew Trotman	University of Otago, New Zealand
Ming-Feng Tsai	National Chengchi University, Taiwan

Yuen-Hsien Tseng	National Taiwan Normal University, Taiwan
Stephen Wan	CSIRO, Australia
Xiaojun Wan	Peking University, China
Bo Wang	Tianjin University, China
Lei Wang	Robert Gordon University, UK
Mengqiu Wang	Stanford University, USA
Quan Wang	Peking University, China
William Webber	University of Maryland, USA
Tak-Lam Wong	The Hong Kong Institute of Education, Hong Kong
Mingfang Wu	RMIT University, Australia
Yunqing Xia	Tsinghua University, China
Wei Xiang	Baidu Inc., China
Deyi Xiong	Institute for Infocomm Research, Singapore
Jun Xu	Microsoft Research Asia, China
Xiaobing Xue	University of Massachusetts, USA
Zheng Ye	York University, USA
Xing Yi	University of Massachussets, Amherst, USA
Meishan Zhang	Harbin Institute of Technology, China
Min Zhang	Tsinghua University, China
Peng Zhang	Robert Gordon University, UK
Qi Zhang	Fudan University, China
Zhichang Zhang	Northwest Normal University, China
Hai Zhao	Shanghai Jiao Tong University, China
Jun Zhao	Institute of Automation, China

Sponsors

Tianjin University
University of Montreal
Microsoft Research Asia
Chinese Information Processing Society of China
ACM SIGIR
Springer

The Return of the Probability of Relevance
(Invited Talk)

Norbert Fuhr

University of Duisburg-Essen, Germany
norbert.fuhr@uni-due.de

The probability ranking principle (PRP) [6] proves that ranking documents by decreasing probability of relevance yields optimum retrieval quality. Most research on probabilistic models has focused only on producing a probabilitic ranking, without estimating the actual probabilities. For classical IR applications, however, knowing these parameters would e.g. allow for estimating the quality of a query result, or for determining the best cutoff point for document filtering.

This talk discusses models for three types of modern IR applications which rely on calibrated values of the probability of relevance.

In vertical search, there are several collections of different type or media (like e.g. Web pages, news, tweets, videos, images). Given a query, the system has to select the collections containing potentially relevant items, process the query on these collections and then merge the results. The model presented in [2] describes the optimum solution to the selection problem, which is based on the probabilistic estimation of the number of relevant documents per resource; the subsequent merging step is straightforward in case the probabilities of relevance are available.

The optimum clustering framework [4] provides not only the first theoretic foundation for document clustering, it also proves the clustering hypothesis. Its key idea is to base cluster analysis and evalutation on a set of queries, by defining documents as being similar if they are relevant to the same queries. Given a set of queries, a probabilistic retrieval method estimates of the relevance probability for all possible query-document pairs, after which a document similarity metric is applied. It can be shown that more or less all clustering methods are implicitly based on these three components, but that they use heuristic design decisions for some of them.

The interactive PRP [3] generalizes the classical PRP for interactive retrieval. It characterizes interactive retrieval as a sequence of situations, where, in each situation, the user is confronted with a list of choices. Each choice is described by three parameters, namely the effort for evaluating it, the probability that the user will accept it, and the benefit resulting from acceptance. While the user effort and expected benefit can be derived from appropriate user studies involving e.g. gaze tracking [7], the acceptance probabilities depend on the quality of the choices presented by the system. The probabilities of relevance of the documents in the result list are just one example of these parameters, other choice lists for which the acceptance parameters have to be estimated are e.g. lists of query expansion terms or clickthrough rates for result list items. Once these parameters are available, the expected benefit of a choice can e.g. be estimated as the time saved for reaching the search goal.

For estimating the actual probabilities of relevance required for applying these models, there are three general methods.

1. Direct estimation of the relevance probabilities from the underlying model is very difficult, since it requires the estimation of other parameters (like e.g. the generality of the current query) for which no appropriate observation data exists.
2. Some 'learning to rank' methods [5] (like e.g. logistic regression) aim at a direct estimation of the probability of relevance.
3. Score distribution models [1] regard the distribution of retrieval scores in relevant and nonrelevant documents, from which appropriate transformations onto the probability of relevance can be derived.

By combining these estimation methods with the models described above, it becomes possible to implement approaches based on solid theoretic foundations, which are more transparent than heuristic approaches, thus allowing for theory-guided adaptation and tuning.

References

1. Arampatzis, A., Robertson, S.: Modeling score distributions in information retrieval. Inf. Retr. 14(1), 26–46 (2011)
2. Fuhr, N.: A decision-theoretic approach to database selection in networked IR. ACM Transactions on Information Systems 17(3), 229–249 (1999)
3. Fuhr, N.: A probability ranking principle for interactive information retrieval. Information Retrieval 11(3), 251–265 (2008), http://dx.doi.org/10.1007/s10791-008-9045-0
4. Fuhr, N., Lechtenfeld, M., Stein, B., Gollub, T.: The optimum clustering framework: Implementing the cluster hypothesis. Information Retrieval 15, 93–115 (2012), doi:10.1007/s10791-011-9173-9
5. Liu, T.-Y.: Learning to Rank for Information Retrieval. Springer (2011)
6. Robertson, S.E.: The probability ranking principle in IR. Journal of Documentation 33, 294–304 (1977)
7. Tran, V.T., Fuhr, N.: Using eye-tracking with dynamic areas of interest for analyzing interactive information retrieval. In: Hersh, W.R., Callan, J., Maarek, Y., Sanderson, M. (eds.) Proceedings of the 35th International ACM SIGIR Conference on Research and Development in Information Retrieval, pp. 1165–1166. ACM (2012)

Web-Based Machine Translation and Cross-Language Information Retrieval
(Invited Talk)

Haifeng Wang

Baidu Campus,
No. 10 Shangdi 10th Street,
Beijing, 100085, China
wanghaifeng@baidu.com

Machine translation (MT) aims to translate text or speech from one language to another, while Cross-Language Information Retrieval (CLIR) tries to retrieve documents in a language different from that of the query. These two are well connected. A CLIR system has the potential to extend the searchable information from a single language to multiple languages. MT system enables users to understand the essential contents of a retrieved document in a foreign language and it makes it easier to implement CLIR by adding an MT system for query translation on top of an IR system. There are many potential web-based applications for MT and CLIR, such as translation of web-page, translation of instant messages, cross-language search, etc. On the one hand, more web data can be crawled to train theses systems. Both the scale and quality of the web data poses major challenges to web-based MT and CLIR applications. In order to build better web-based applications, it is essential to develop techniques to adapt state-of-the-art MT and CLIR technologies to the web scenario. In this talk, I will introduce our work on web-based machine translation and cross-language information retrieval.

Table of Contents

Session 1: IR Models

Session 2: Evaluation and User Studies

Session 3: NLP for IR

Session 4: Machine Learning and Data Mining

Session 5: Social Media

Session 6: IR Applications

Session 7: Multimedia IR and Indexing

Session 8: Collaborative and Federated Search

Poster Session

Exploring and Exploiting Proximity Statistic for Information Retrieval Model

Yadong Zhu, Yuanhai Xue, Jiafeng Guo, Yanyan Lan,
Xueqi Cheng, and Xiaoming Yu

Institute of Computing Technology, Chinese Academy of Sciences
{zhuyadong,xueyuanhai,yuxiaoming}@software.ict.ac.cn
{guojiafeng,lanyanyan,cxq}@ict.ac.cn

Abstract. Proximity among query terms has been recognized to be useful for boosting retrieval performance. However, how to model proximity effectively and efficiently remains a challenging research problem. In this paper, we propose a novel proximity statistic, namely Phrase Frequency, to model term proximity systematically. Then we propose a new proximity-enhanced retrieval model named BM25PF that combines the phrase frequency information with the basic BM25 model to rank the documents. Extensive experiments on four standard TREC collections illustrate the effectiveness of the BM25PF model, and also shows the significant influence of the phrase frequency on retrieval performance.

Keywords: Phrase Frequency, Proximity Statistic, BM25.

1 Introduction

In the past decades, many different retrieval models have been proposed and successfully applied in document ranking, including vector space models [19], classical probabilistic models [16, 18], and statistical language models [14]. By assuming full independence between terms, most of them represent query and document as bag of words, and mainly exploit term based statistics such as within-document term frequency (tf), inverse document frequency (idf) and document length in ranking. However, the existing methods often lack a way to explicitly model the proximity among query terms within a document, which has been widely recognized to be useful for boosting retrieval performance [24].

Recently, many studies have been conducted to capture proximity in retrieval. One is to index phrases instead of terms which can capture proximity indirectly [5]. However, it is shown that the index methods may not perform consistently well across a variety of collections [12]. Another way is the dependency modeling [20, 22, 6, 11, 8], which considers the use of term dependency such as high-order n-grams, and integrates them into the existing retrieval models. The main problem of those dependency models lies in that the parameter space becomes very large, which will lead to difficulty in parameter estimation and sensitivity to data sparse and noise [26].

A simple but effective way is direct proximity modeling. Some early studies on direct proximity modeling try to add proximity factors into probabilistic ranking

Y. Hou et al. (Eds.): AIRS 2012, LNCS 7675, pp. 1–13, 2012.

models in a heuristic manner, but their experiment results are not conclusive and show limited success [3, 7, 15, 2]. Recent work [24] and [25] obtains a promising performance improvement essentially by a pairwise term distance measure. While considering only word pairs may not be appropriate for long queries, and many redundant proximity information may be involved which may be harmful to the final performance. For example, given a user query "second hand car in Detroit", word pairs such as "second car" or "hand car" can not convey the underlying concept of "second hand car" correctly. Therefore, how to model proximity effectively and efficiently remains a challenging research problem.

In this paper, we propose a novel proximity statistic named Phrase Frequency (pf). It reflects the statistical frequency information of a query phrase provided by a document, beyond the traditional term statistics such as tf and idf. It is defined based on the definition of Span_Cover. A Span_Cover is a basic document segment that covers each query term at least once within a limited window size, and no overlap between each other. It is different from the existing two span-based definitions: Span (or FullCover) and MinCover [24, 4]. Four proximity-based density functions are considered to produce proximity score of a span_cover instance, which is small when query terms within the span_cover are far away, and large when query terms are close to each other. The phrase frequency is then computed by accumulating the proximity scores within a document.

Based on the concept of phrase frequency, we propose a proximity-enhanced model referred as BM25PF. Through extensive experiments on different sizes and genres of TREC collections, we show that the BM25PF outperforms the BM25 baseline and is also comparable to the state-of-art approaches in general.

The rest of this paper is organized as follows. In Section 2, we introduce the previous related work. In Section 3, we introduce the concept of pf, corresponding calculating algorithm, and the practical application strategy. In Section 4, we propose a proximity-enhanced model named BM25PF based on pf. Then, in Section 5, we test the proposed model BM25PF on four TREC collections, and compare their performance with state-of-the-art approaches. Finally in Section 6, we conclude the paper with a discussion of our work and future work.

2   Related Work

Term proximity has been previously studied in several approaches. One approach tries to index phrases instead of terms. In such a way, dependence between words can be captured indirectly [5, 12]. However, the index method does not perform consistently well across a variety of collections.

Another way to capture proximity is the dependency modeling. Several work has been conducted to integrate term dependence into the language model. Song et al. propose a general model that combines bigram language model with unigram language model under several smoothing techniques [20]. Srikanth et al. introduce a biterm language model that takes into account dependency between unordered adjacent word pairs [22]. [13, 6] put forward dependency language models that consider the relation between terms existed in a dependency tree.

Metzler and Croft propose a general framework for model term dependencies based on Markov Random Field [11], in which term structures with different levels of proximity can be defined and captured. He et al. [8] extend the traditional BM25 model to a combination of a series of n-gram BM25 models that will take the proximity between query terms into account.

Direct proximity modeling has been considered widely recently. [3] and [7] appear to be the first to evaluate proximity on TREC data sets. Both of them try to measure proximity by using the shortest interval containing a match set. Recent work has attempted to heuristically incorporate proximity factor into probabilistic ranking functions [15, 2], but their experiments are not conclusive and show limited success. Song et al. represent term proximity by co-occurrences of terms in non-overlapping phrases [21]. Zhao and Yun's work views the query term proximity as the Dirichlet hyper-parameter that weights the parameters of the unigram document language model [26]. Lv and Zhai apply a series of kernel functions to estimate a language model for every position in a document [10]. Overall, direct proximity modeling is an economic and effective way to go beyond the "bag of words" assumption in retrieval. In our paper, we will also follow the way by introducing a novel proximity statistic.

Density functions based on proximity have been used to propagate term influence [10, 25]. Lv and Zhai [10] incorporate the term proximity by using four kernel functions: Gaussian, Triangle (Linear), Cosine, and Circle. [25] additionally introduces three more kernel functions: Quartic, Epanechnikov, Triweight. In our work, we utilize two of them: Gaussian and Linear, and introduce two more kernel functions containing Exponential and Negative Power.

3 A Novel Proximity Statistic: Phrase Frequency

In this section, we introduce the concept of Phrase Frequency and the associated calculating algorithm. Before giving the formal definition of Phrase Frequency, we will first define a new concept: *Span_Cover*, which is different from the previous notions of Span (or FullCover) and MinCover [24, 4].

3.1 Span_Cover

The traditional span-based notions: FullCover and MinCover, which can be viewed as two kinds of "boundary" definitions, while the Span_Cover can be viewed as a basic unit between them.

Definition 1. *(Span_Cover) Span_Cover is defined as a document segment that will have following properties:*

(1) Starting from a query term and ending with a query term;
(2) Covering each query phrase term at least once;
(3) Its length is no more than a limited maximal window size, which is usually adaptively set as integral times of the number of unique query terms;

(4) Only the shortest of several span_covers sharing a common starting points is counted as an instance

(5) There is no overlapped terms between any two span_cover instances under a certain scanning way;

Given a short document d as an example to explain our definitions [24].

$$d = t_1, t_2, t_1, t_3, t_5, t_4, t_2, t_3, t_4$$

Given a query: $\{t_1, t_2\}$, and setting the maximal windows size as 4 times the number of query terms, its corresponding Span_Cover instances would be $\{t_1, t_2\}$ and $\{t_1, t_3, t_5, t_4, t_2\}$ by a sequentially scanning way. The alert reader will notice that, the segment $\{t_2, t_1\}$ will be also a proper span_cover satisfying all the above properties. Usually, we will detect all the Span_Cover instances provided by a relevant document in a sequential scanning way. Certainly there may exist some other scanning ways, we will study them in our future work.

3.2 Phrase Frequency

The Phrase Frequency is a novel proximity statistic which reflects the statistical frequency information of a query phrase within a relevant document. Given a user query phrase that contains at least two terms and a candidate document, there exists several factors that may affect the relevance of a document:

1. Density: More tighter or shorter a Span_Cover instance, the greater the like-lihood that the document is relevant.
2. Number: The more Span_Cover instances provided by a document, the greater the likelihood that the document is relevant.
3. Order: The order in a certain Span_Cover instance may also have some influences on query proximity.

Obviously, the traditional method of proximity modeling based on Mincover, which only considers the Density factor, may not be appropriate. Similarly, the method based on FullCover can not explicitly detect the number of Span_Cover instances or the density of a specific instance. In fact, the experiment results in [24] also show that they are not effective in general. Meanwhile, many previous work has shown that the order of appearances of query term is not important [11, 8]. Therefore, instead of strict order constraint, rather, we expect the query terms to appear ordered or unordered within a Span_Cover instance. And the definition of phrase frequency must systematically consider the density and number of Span_Cover instances.

Given a query phrase Q with K unique query terms and the allowed maximal windows size as wK (w usually set as positive integer). Supposing m Span_Cover instances have been detected in a candidate document D as follows: $\{Span\_Cover_1, Span\_Cover_2, ..., Span\_Cover_m\}$

Definition 2. *The **Phrase Frequency** (pf) of a given query phrase Q in D will be calculated as the following.*

$$pf(Q, D) = \sum_{i=1}^{m} pf_i = \sum_{i=1}^{m} Density(|Span\_Cover_i| - K) \qquad (1)$$

where $Density(\cdot)$ is a type of proximity-based density function that transforms a $Span\_Cover_i$ instance to a proximity score belonging to (0, 1], which is notated as pf_i. $|Span\_Cover_i|$ is the length of the ith Span_Cover: $|Span\_Cover_i| = pos_{end} - pos_{start} + 1$. The more tighter $Span\_Cover_i$, the higher pf_i will be.

While if there does not exist a legal Span_Cover instance in document D, we will assign a uniform minimal value for pf in the following way:

$$pf(Q, D) = Density(wK) \qquad (2)$$

Obviously, $Density(\cdot)$ plays an important role in pf. It must follow several properties such as non-negative, monotonic [10, 25]. We will analyze and explore a number of proximity-based density functions in following subsection.

3.3 Proximity-Based Density Functions

A major technical challenge in the definition of pf is how to define the density functions $Density(\cdot)$. Following some previous work [24, 10, 25], we study four representative kernel functions: Gaussian, Linear, Exponential, and Negative Power. They are all non-increasing functions, and represent different ways in which the proximity information (pf_i) decreases when $|Span\_Cover_i|$ increases.

1. Gaussian Kernel
$$Kernel(x) = e^{-\frac{x^2}{2a^2}} \qquad (3)$$

2. Linear Kernel
$$Kernel(x) = ax + 1, \ a < 0 \qquad (4)$$

3. Exponential Kernel
$$Kernel(x) = e^{-ax}, \ a > 0 \qquad (5)$$

4. Negative Power Kernel
$$Kernel(x) = (ax + 1)^k, \ k < 0, a > 0 \qquad (6)$$

The first two kernels have been used in [10, 25], and the last two kernel functions are firstly introduced in our work. The performance of using all the kernel functions will be investigated in the experiments. Obviously, when $|Span\_Cover_i| = K$, we will get the maximal value: $pf_i = 1$, which means a consecutive occurrence of a query phrase. a is a normalization parameter to make sure the output belonging to (0, 1]. It is usually determined by query length and the maximal windows size together.

3.4   Algorithm and Time Complexity

In this section, we present an effective and efficient calculating algorithm of pf by a sequential scanning way. Then we give an analysis of its time complexity. We show that its time complexity is close to linear in terms of N, which is the total number of occurrences of all the query terms within a document. The corresponding calculating algorithm is summarized in Algorithm 1.

Given a query phrase $Q = \{t_1, t_2, ..., t_K\}$, supposing that a candidate document D matches the K unique query terms, and the total number of occurrences

---

Algorithm 1. pf **Calculating using a Sequential Scanning**

Input:
  K- query length in terms of unique words
  N- the total number of query term occurrence
  w- the parameter controlling the maximal windows size
  $Hit[]$- a ordered chain of every query term occurrence
  $List[]$- a list of length K
  $kernel()$- a certain kernel function
  $existpf$- a flag indicating whether pf exists or not
Output: pf score provided by a relevant document
1: for each $i \in [0, K-1]$ do
2:     $List[i] = -1$
3: end for
4: $pf = 0$
5: $existpf = false$
6: for $i = 0$ to $N - 1$ do
7:     for $j = 0$ to $K - 1$ do
8:         if $Hit[i].term == t_j$ then
9:             $List[j] = Hit[i].pos$
10:             break
11:         end if
12:     end for
13:     $start = min(List[0], List[1], ..., List[K-1])$
14:     $end = Hit[i].pos$
15:     if $start \neq -1$ then
16:         $cover\_len = end - start + 1$
17:         if $cover\_len \leq w * K$ then
18:             $pf+ = kernel(cover\_len - K)$
19:             $existpf = true$
20:             for $m = 0$ to $K - 1$ do
21:                 $List[m] = -1$
22:             end for
23:         end if
24:     end if
25: end for
26: if $!existpf$ then
27:     $pf = kernel(w * K)$
28: end if

---

of these K query terms is N. All the query term occurrences compose a chain of ordered hits: $\{t_{P_1}, t_{P_2}, ..., t_{P_N}\}$, and $t_{P_i} \in Q$, $\forall i \in \{1, 2..., N\}$. We record the position and the term information of every query term occurrence in a Hit set. While scanning, we maintain a list of length K: $List[0, ..., K-1]$, in which we store the temporary position of each seen query term. We set the allowed maximal windows size as $w * K$, w is a positive integer.

The first 5 steps are initialization. Specially we assign every position in $List[]$ a negative value of -1. For steps 7 to 12, we update the corresponding position information of a certain query term when scanning an occurrence sequentially. In steps 13 and 14, we record the *start* and *end* position information after every position updating. Then we will judge whether every position in $List$ has been updated in step 15, which corresponds with the second property of Span_Cover. In steps 16 and 17, we calculate the length of the segment, and then judge whether it satisfies the length constraint property. In steps 18 and 19, we accumulate the *pf* score provided by a certain Span_Cover instance and set the *existpf* flag as true. From step 20 to 22, we reinitialize the position information in $List[]$ to make sure there is no overlapping between any two Span_Cover instances. Finally we will assign a uniform minimal value for *pf* from step 26 to 28 if there does not exist a legal Span_Cover instance. The algorithm for calculating *pf* has the time complexity: $O(N * K)$. Since K is often very small, the time complexity is close to linear in terms of N: $O(N)$, and N is the total number of query term occurrence within a document.

3.5   A Practical Application Strategy

In practical retrieval environment, there exists many long user queries (5 or more terms). A Span_Cover instance of a user query requires to cover each query term at least once within a limited windows size. This condition may be too strong for these long user queries. On the other hand, when user constructs a long query, he will often aggregate additional terms to the concepts which will make them less meaningful [1]. Ideally, we should select the meaningful term sequences, and consider the proximity information of them only. Therefore, we naturally consider to combine proximity modeling with the query segmentation techniques together. Query segmentation is an important task toward understanding queries accurately[9, 23], which is essential for improving search results in modern information retrieval. It aims to separate query words into segments so that each segment maps to a semantic component or sub-query phrases.

For a user query, if its query length is less than 5 terms, we calculate the *pf* score as Algorithm 1 normally, or else we need to do a user query structure analysis first by query segmentation techniques, and then calculate as follows:

$$pf(Q, D) = \sum_{sub_i \subseteq Q} wei_i \times pf(sub_i, D), \quad |sub_i| \geq 2, \; wei_i \in (0, 1] \quad (7)$$

where sub_i is a sub-query phrase, which contains at least two query terms, wei_i is a normalization weight factor assigned for sub_i, which can reflect how meaningful the sub-query phrase is. $pf(sub_i, D)$ denotes the pf score of a specific sub-query phrase, which will also be calculated by Algorithm 1.

There exists several methods for query segmentation, and the result sub-query phrases can be weighted correspondingly. In our paper, we use the method presented in [9], and define the wei_i as follows:

$$wei_i = \frac{connexity(sub_i)}{\sum_{sub_i \subseteq Q} connexity(sub_i)} \tag{8}$$

$$connexity(sub_i) = freq(sub_i) \times I(w_{i_1}...w_{i_{n-1}}, w_{i_2}...w_{i_n}) \tag{9}$$

where $connexity(sub_i)$ denotes the *connexity* score of the *ith* sub-query phrase (or segment)[9], which is defined as a product of the global frequency of the segment ($freq(sub_i)$) and the mutual information (I) between longest but complete subsequence of sub_i.

4   Proximity-Enhanced Retrieval Model

In this section, we will test whether our proposed proximity statistic can really boost the retrieval performance of the existing retrieval models, which are usually derived from bag-of-words assumption (e.g. Okapi BM25, KL-divergence language model). In this paper, we choose to combine the pf score with the representative state-of-the-art retrieval models: Okapi BM25 model [17, 18]. Previous extensive experiments show that BM25 can provide a robust and effective retrieval performance. The BM25 retrieval model can be expressed as follows:

$$BM25(Q, D) = \sum_{q_i \in q \cap d} (w_i \times \frac{(k_1 + 1) \times c(q_i, d)}{k_1((1 - b) + b\frac{|d|}{avdl}) + c(q_i, d)} \times \frac{(k_3 + 1) \times c(q_i, q)}{k_3 + c(q_i, q)})$$

where $c(q_i, d)$ is within-document term frequency, and $c(q_i, q)$ is within-query term frequency. Additionally, $|d|$ is document length, and $avdl$ is average document length. k_1, k_3 and b are tuning parameters. $w_i = ln\frac{N-n+0.5}{n+0.5}$ is q_i's term weight. N is the number of documents within a collection, n is the document frequency of term q_i.

And then we can define a new combined proximity-enhanced retrieval model as follows:

$$BM25PF(Q, D) = \lambda BM25(Q, D) + (1 - \lambda)pf(Q, D) \tag{10}$$

where λ is a trade-off parameter reflecting the influence of the proximity statistic. When λ equals to 1, the BM25PF model degenerates to the basic BM25 model.

5   Experiments

We conduct a series of experiments on four standard TREC collections: AP88-89, FR88-89, WT10G and ClueWebB. They represent different sizes and genres of

Table 1. Basic Data Set Statistics

	AP88-89	FR88-89	WT10G	ClueWeb B
Topics	51-100	51-100	451-550	rf.01-rf.50
# of Topics	50	21	100	50
# of Docs	164,597	45,820	1,692,096	49,375,681

Table 2. Performance comparison between BM25 and BM25PF with different kernels

	WT10G			ClueWeb B			AP88-89			FR88-89		
	MAP	P@5	P@10	MAP	P@5	P@10	MAP	P@5	P@10	MAP	P@5	P@10
BM25	0.2131	0.3864	0.291	0.2109	0.3780	0.2806	0.2670	0.4358	0.3970	0.2892	0.1670	0.1450
BM25PF_E	0.2210	0.4006	0.3019	0.2134	0.4034	0.2897	0.2725	0.4480	0.4153	0.2971	0.1741	0.1520
BM25PF_N	0.2226	0.4017	0.3023	0.2176	0.4053	0.2894	0.2730	0.4479	0.4200	0.2952	0.1743	0.1515
BM25PF_L	0.2231	0.4093	0.3051	0.2201	**0.4073***	0.2976*	0.2743	**0.4621***	0.4200	0.2973	0.1780*	0.1530
BM25PF_G	**0.226***	**0.4128***	**0.3057***	**0.2213**	0.4071*	**0.2981***	**0.2780**	0.4610*	**0.4208***	**0.30**	**0.1780***	**0.1539***

text collections. AP88-89 and FR88-89 are chosen as small homogeneous collections with little noise. The WT10G is a medium size crawl of Web documents. ClueWeb collection is the largest TREC test collection currently with a very large crawl of the Web. We use the category B of ClueWeb which contains about 50 million English Web pages. Queries are taken from the title field of the TREC topics. The basic statistics of the collections are illustrated in Table 1. We preform stemming with the Porter stemmer. And no stop words removing is done in both documents and queries to make sure the minimum preprocessing of them.

On each collection, we conduct a 2-fold cross-validation. The associated test topics are split into two equal subsets, referred as odd-number and even-number topics. In each fold, we use one subset of topics for training, and use the remaining subset for testing. The overall retrieval performance is averaged over the two test subsets of topics. Both top precision and average precision are used to evaluate the experiment result which include MAP/statMAP, P@5, P@10. All statistical tests are based on Wilcoxon matched-pairs signed-rank test at the 0.05 level.

5.1 Effectiveness of BM25PF

In this section, we evaluate the effectiveness of the proximity-enhanced model BM25PF. We use the optimal BM25 as our baseline. The BM25 has three main parameters. We set $k_1 = 1.2$ and $k_3 = 1,000$ suggested in [24] and tune b to be optimal. And the parameter b is set to be 0.3.

The related experimental results are presented in Table 2. We apply four kinds of kernel functions for instantiating BM25PF model, and notate them correspondingly as: BM25PF_G (BM25PF Gaussian), BM25PF_L (BM25PF Linear), BM25PF_E (BM25PF Exponential), BM25PF_N (BM25PF Negative Power). Additionally, the parameter a of them is set as: $w * K$, $-1/((w + 1) * K)$, $w * K$ and 1, respectively. The parameter k in Equation 6 is set as -1 (in fact any negative value is allowed). All results are evaluated in terms of MAP, P@5, P@10.

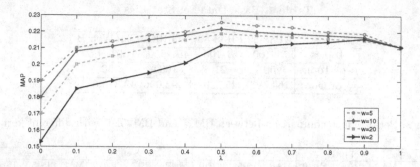

Fig. 1. Sensitivity to BM25PF parameter λ on ClueWeb B

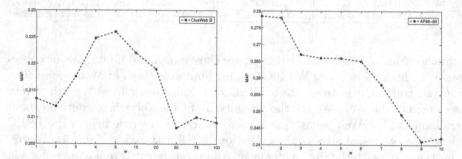

Fig. 2. Sensitivity to BM25PF proximity parameter w on ClueWeb B and AP88-89

Results with "$*$" indicate the improvement is significant at 0.05 level. The best result obtained on each collection is highlighted.

The results show that the proposed BM25PF model outperforms BM25 on all four data collections. Notablely, the proposed model has significant performance improvement especially on top retrieved documents in terms of P@5, P@10. And each kernel function shows a stable performance improvement especially Gaussian kernel, which is also consistent with the conclusion of the previous work [10]. The Gaussian kernel exhibits the following property: the density measurement would drop slowly when the Span_Cover length is small, but drop quickly as the length is in a middle range, and then drop slowly again when its length become large. Such an "S-shape" trend is reasonable for that: the dependent query terms are not always adjacent in documents, but can be a little far from each other, thus we would not like to make the density measurement of a Span_Cover instance so sensitive to the length when its length is small. However, when the Span_Cover length is just around the boundary of strong semantic associations, the density measurement should be more sensitive to the length change. Then as the Span_Cover length increases further, all query terms are presumably only loosely associated, and thus the density measurement again should not be so sensitive to the change of Span_Cover length.

We further compare the retrieval performance of the BM25PF with two state-of-the-art approaches proposed in [24] and [25] respectively. For the first approach, to be fair, we use the best retrieval formula suggested in [24] (BM25+MinDist), and tune parameter α to be optimal. We label this approach as 'BM25MD'. For the second approach, the CRTER model proposed in [25] which measures the association of every two query terms, we set their free parameters by using cross-validation, which is also the same way using for our approach. We conduct the performance comparison on all four data collections and report the comparison results in Table 3. Overall, the proposed BM25PF model shows a steady performance improvement on all four data collections. And it is at least comparable to, if no better than, the two state-of-art approaches.

Table 3. Performance comparison of different ranking models in terms of MAP

	WT10G	ClueWeb B	AP88-89	FR88-89
BM25	0.2131	0.2109	0.267	0.2892
BM25MD	0.2197	0.213	0.2713	0.2918
CRTER	0.2207	0.2138	0.2751	0.2926
BM25PF	0.2261*	0.2213	0.2780	0.3000

5.2 Parameter Sensitivity of BM25PF

The BM25PF model introduces two parameters λ and w which will affect the retrieval performance. The parameter λ balances the influence of basic BM25 model and the proximity statistic. w is a proximity parameter for modeling different level proximity information. We choose the Gaussian for instantiating the BM25PF model based on the results shown in Table 2.

Figure 1 plot the sensitivity curves over λ values ranging from 0 to 1 in terms of MAP on ClueWeb data. And a group of value settings of parameter w are applied such as $w = 2, 5, 10, 20$. Overall, a reliable λ value is relatively insensitive to different w values, and $\lambda = 0.5$ seems to be best.

Figure 2 plot the sensitivity tendency over proximity parameter w on ClueWeb and AP88-89 respectively, in terms of MAP. These two data collections represent different data genres. Additionally, the λ is fixed at 0.5. From the first subfigure (ClueWeb B), we find that the BM25PF achieves its best performance when w value is around 5, which means the model capturing proximity information at a sentence-level. The second subfigure shows that when w is set a small value such as 1 or 2, the overall performance tend to be best. A larger windows size will bring noise, and the retrieval performance will decrease and be not stable. This may be due to the homogeneous, clean nature of the documents in these small newswire collections. And a strictly tighter matching Span_cover instances in the calculating of pf will capture high quality proximity information. While for the Web collections, which is heterogeneous and noisy collections, a larger windows size of sentence-level is appropriate because they can deal better with the noise inherent in Web documents.

6   Conclusions and Future Work

In this paper, we propose a novel proximity statistic, namely Phrase Frequency (pf), to model term proximity for boosting retrieval performance. It reflects the statistical frequency information of a query phrase within a document, and is approximated by a density function based on the definition of Span_Cover. pf systematically considers the density of a span_ cover instance and the numbers of span_cover within a document. In addition, we also present an efficient calculating algorithm of pf. Based on the concept of pf, we propose a proximity-enhanced retrieval model: BM25PF. Through extensive experiments on different sizes and genre of TREC collections, we show that the BM25PF outperforms the BM25 baseline and is also comparable to the state-of-art approaches in general.

The proposed proximity statistic pf has shown significant influence on retrieval performance. It is a novel and important proximity statistic as the same as the traditional term statistics such as tf and idf. In the future work, we can study how to eventually obtain a unified retrieval model with the incorporation of pf in a more soundly theoretical way.

Acknowledgements. This research work was funded by the National Natural Science Foundation of China under Grant No. 60933005, No. 61173008, No. 61003166, 973 Program of China under Grants No. 2012CB316303, and National Key Technology R&D program under Grant No. 2011BAH11B02.

References

[1] Bai, J., Chang, Y., Cui, H., Zheng, Z., Sun, G., Li, X.: Investigation of partial query proximity in web search. In: Proc. of the 17th WWW, pp. 1183–1184 (2008)

[2] Büttcher, S., Clarke, C.L.A., Lushman, B.: Term proximity scoring for ad-hoc retrieval on very large text collections. In: Proc. of the 29th SIGIR, pp. 621–622 (2006)

[3] Clarke, C.L.A., Cormack, G.V., Burkowski, F.J.: Shortest substring ranking (multitext experiments for trec-4). In: TREC, pp. 295–304 (1995)

[4] Cummins, R., O'Riordan, C.: Learning in a pairwise term-term proximity framework for information retrieval. In: Proc. of the 32nd SIGIR, pp. 251–258 (2009)

[5] Fagan, J.: Automatic phrase indexing for document retrieval. In: Proc. of the 10th ACM SIGIR, pp. 91–101 (1987)

[6] Gao, J., Nie, J.-Y., Wu, G., Cao, G.: Dependence language model for information retrieval. In: Proc. of the 27th ACM SIGIR, pp. 170–177 (2004)

[7] Hawking, D., Thistlewaite, P.B.: Proximity operators - so near and yet so far. In: TREC (1995)

[8] He, B., Huang, J.X., Zhou, X.: Modeling term proximity for probabilistic information retrieval models. Inf. Sci. 181, 3017–3031 (2011)

[9] Risvik, K.M., Mikolajewski, T., Boros, P.: Query segmentation for web search. In: Proc. of the 12th WWW (2003)

[10] Lv, Y., Zhai, C.: Positional language models for information retrieval. In: Proc. of the 32nd ACM SIGIR, pp. 299–306 (2009)

[11] Metzler, D., Croft, W.B.: A markov random field model for term dependencies. In: Proc. of the 28th ACM SIGIR, pp. 472–479 (2005)

[12] Mitra, M., Buckley, C., Singhal, A., Cardie, C.: An analysis of statistical and syntactic phrases. In: Proc. of 5th RIAO, pp. 200–214 (1997)

[13] Nallapati, R., Allan, J.: Capturing term dependencies using a language model based on sentence trees. In: Proc. of the 11th ACM CIKM, pp. 383–390 (2002)

[14] Ponte, J.M., Croft, W.B.: A language modeling approach to information retrieval. In: Proc. of the 21st ACM SIGIR, pp. 275–281 (1998)

[15] Rasolofo, Y., Savoy, J.: Term Proximity Scoring for Keyword-Based Retrieval Systems. In: Sebastiani, F. (ed.) ECIR 2003. LNCS, vol. 2633, pp. 207–218. Springer, Heidelberg (2003)

[16] Robertson, S.E., Sparck Jones, K.: Relevance weighting of search terms. Journal of the American Society for Information Science 27(3), 129–146 (1976)

[17] Robertson, S.E., Walker, S., Hancock-Beaulieu, M.: Okapi at trec-7: Automatic ad hoc, filtering, vlc and interactive. In: TREC, pp. 199–210 (1998)

[18] Robertson, S.E., Zaragoza, H.: The probabilistic relevance framework: Bm25 and beyond. Foundations and Trends in Information Retrieval 3(4), 333–389 (2009)

[19] Salton, G., Wong, A., Yang, C.S.: A vector space model for automatic indexing. Commun. ACM 18, 613–620 (1975)

[20] Song, F., Croft, W.B.: A general language model for information retrieval. In: Proc. of the 8th ACM CIKM, pp. 316–321 (1999)

[21] Song, R., Taylor, M.J., Wen, J.-R., Hon, H.-W., Yu, Y.: Viewing Term Proximity from a Different Perspective. In: Macdonald, C., Ounis, I., Plachouras, V., Ruthven, I., White, R.W. (eds.) ECIR 2008. LNCS, vol. 4956, pp. 346–357. Springer, Heidelberg (2008)

[22] Srikanth, M., Srihari, R.: Biterm language models for document retrieval. In: Proc. of the 25th ACM SIGIR, pp. 425–426 (2002)

[23] Tan, B., Peng, F.: Unsupervised query segmentation using generative language models and wikipedia. In: Proc. of the 17th WWW, pp. 347–356 (2008)

[24] Tao, T., Zhai, C.: An exploration of proximity measures in information retrieval. In: Proc. of the 30th ACM SIGIR, pp. 295–302 (2007)

[25] Zhao, J., Huang, J.X., He, B.: Crter: using cross terms to enhance probabilistic information retrieval. In: Proc. of the 34th ACM SIGIR, pp. 155–164 (2011)

[26] Zhao, J., Yun, Y.: A proximity language model for information retrieval. In: Proc. of the 32nd ACM SIGIR, pp. 291–298 (2009)

A Category-integrated Language Model for Question Retrieval in Community Question Answering

Zongcheng Ji[1,2], Fei Xu[1,2], and Bin Wang[1]

[1] Institute of Computing Technology, Chinese Academy of Sciences, Beijing, China
[2] Graduate University of Chinese Academy of Sciences, Beijing, China
{jizongcheng, feixu1966}@gmail.com
wangbin@ict.ac.cn

Abstract. Community Question Answering (CQA) services have accumulated large archives of question-answer pairs, which are usually organized into a hierarchy of categories. To reuse the invaluable resources, it's essential to develop effective Question Retrieval (QR) models to retrieve similar questions from CQA archives given a queried question. This paper studies the integration of category information of questions into the unigram Language Model (LM). Specifically, a novel Category-integrated Language Model (CLM) is proposed which views category-specific term saliency as the Dirichlet hyper-parameter that weights the parameters of LM. A point-wise divergence based measure is introduced to compute a term's category-specific term saliency. Experiments conducted on a real world dataset from Yahoo! Answers show that the proposed CLM which integrates the category information into LM *internally at the word level* can significantly outperform the previous work that incorporates the category information into LM *externally at the word level* or *at the document level*.

Keywords: Community Question Answering, Question Retrieval, Category, Category-integrated Language Model.

1 Introduction

Community Question Answering (CQA) has recently become a popular type of web service where users ask and answer questions and access historical question-answer pairs from the large scale CQA archives. Examples of such CQA services include Yahoo! Answers[1] and Baidu Zhidao[2], etc.

To effectively share the knowledge in the large scale CQA archives, it is essential to develop effective question retrieval models to retrieve historical question-answer pairs that are semantically equivalent or relevant to the queried questions.

As a specific application of the traditional Information Retrieval, Question Retrieval in CQA archives is distinct from the search of web pages in that historical questions are organized into a hierarchy of categories. That is to say, each question in

---

[1] http://answers.yahoo.com/
[2] http://zhidao.baidu.com/

Y. Hou et al. (Eds.): AIRS 2012, LNCS 7675, pp. 14–25, 2012.

the CQA archives has a category label. The questions in the same category or subcategory usually relate to the same topic. For example, the questions in the subcategory "Travel.Asia Pacific.China" mainly relate to travel to or in the country of China. This makes it possible to exploit category information to enhance the question retrieval performance. To exemplify how a categorization of questions can be exploited, consider the following queried question (q): "Can you recommend sightseeing opportunities for senior citizens in China?" The user is interested in sightseeing specifically in China, not in other countries. Hence, the historical question (Q) "Can you recommend sightseeing opportunities for senior citizens in Korea?" and its answers are not relevant to the queried question although the two questions are syntactically very similar, making it likely that existing question retrieval methods will rank question Q highly among the returned ranking list. If we can establish a connection between q and the category "Travel.Asia Pacific.China", the ranking of the questions in that category could be promoted, thus perhaps improving the question retrieval performance.

Several work [3, 2] has been done to incorporate the category information into existing retrieval models. These work shows that exploiting the category information for each question properly can improve the retrieval performance effectively. However, there is one common shortcoming in both previous work: they just incorporate the category information into the existing retrieval models in a very heuristic way that they just linearly combine the category information related models with the existing retrieval models *externally*, no matter *at the word level* [2] or *at the document level* [3].

Language modeling has become a very promising direction for information retrieval because of its empirical good performance. In this paper, we try to integrate category information into the unigram language modeling approach. Our intuition is that a term may be more important in some category than other categories. For example, "China" is an important or salient word in the category "Travel.Asia Pacific.China", but not in the category "Travel.Asia Pacific.Korea". We model the individual term's saliency in a specific category as Dirichlet hyper-parameter that weights the corresponding term emission parameter of the multinomial question language model. Thus, we attain an *internal* formula that effectively boosts the score contribution from terms when the term is more salient in the category of the historical question. This incorporation of category information *at the word level* is mathematically grounded. Furthermore, it performs empirically better than previous heuristic incorporation of category information *at the word level* [2] or *at the document level* [3] when we conduct experiments on a large scale real world CQA dataset from Yahoo! Answers.

The rest of this paper is organized as follows. Section 2 reviews the related work on question retrieval and especially gives detail of the existing two most effective methods on how to exploit the category information to enhance the question retrieval performance. Section 3 details our proposed category-integrated language model. Section 4 gives a method of measuring a term's saliency in a specific category. Section 5 describes the experimental study on a large scale real world CQA dataset from Yahoo! Answers. Finally, we will conclude our work and offer the future work in section 6.

2   Related Work

Recently, question retrieval has been widely investigated in CQA data. Most of previous work focuses on the translation-based methods [4, 7, 10, 12] to alleviate the lexical gap problem between the queried question and the historical questions. Beside using translation-based methods, Wang et al. [9] propose a syntactic tree matching method to find similar questions. Ming et al. [8] explore domain-specific term weight to improve question retrieval within a certain category. Cao et al. [2] and Cao et al. [3] propose different strategies to exploiting the category information to enhance the question retrieval performance in the whole collection with the whole categories. Cai et al. [1] propose to learn the latent topics from the historical questions to alleviate the lexical gap problem, while Ji et al. [5] propose to learn the latent topics aligned across the historical question-answer pairs to solve the lexical gap problem. Moreover, Cai et al. [1] also incorporate the category information of the historical questions into learning the latent topics for further improvements.

Because our work in this paper mainly focuses on investigating a new method to incorporate the category information into the ranking function, thus in what follows we will review in detail the two most effective methods [3, 2] on how to exploit the category information to enhance the question retrieval performance.

2.1   Leaf Category Smoothing Enhancement

As the first attempt to exploit the category information, Cao et al. [2] propose a two-level smoothing model to enhance the performance of language model based question retrieval. A category language model is computed and then smoothed with the whole question collection; the question language model is subsequently smoothed with the category language model. The experimental results showed that it can improve the performance of the language model significantly. Following [2], given a queried question $\mathbf{q}$ and a historical question Q with the leaf category $Cat(Q)$, the ranking function of the two-level smoothing model can be written as:

$$P(\mathbf{q}|Q) = \prod_{w \in \mathbf{q}} (1 - \lambda) P_{ml}(w|Q) + \lambda[(1 - \beta) P_{ml}(w|Cat(Q)) + \beta P_{ml}(w|Coll)] \quad (1)$$

where λ and β are the two different smoothing parameters. If we set $\beta = 1$, the ranking function will reduce to the unigram language model with Jelinek-Mercer (JM) smoothing [11] without any category information. $P_{ml}(w|Q), P_{ml}(w|Cat(Q))$, $P_{ml}(w|Coll)$ are the maximum likelihood estimation of word w in the historical question Q, the leaf category $Cat(Q)$ and the whole question collection $Coll$, respectively.

From Equation (1), we can see that this approach combines the smoothed category language model with the original question language model linearly only in an *external* way *at the word level*. However, in our work we will integrate the category information into the original language model in an *internal* way *at the word level*, which seems to be more promising.

2.2   Category Enhancement

Realizing that the leaf category smoothing enhancement method is tightly coupled with the language model and is not applicable to other question retrieval models, Cao et al. [3] further propose a method called the category enhanced retrieval model, which is to compute the ranking score of a historical question Q based on an interpolation of two relevance scores: a global relevance score between the queried question $\mathbf{q}$ and the category $Cat(Q)$ containing Q, and a local relevance score between the queried question $\mathbf{q}$ and the historical question Q. In this case, various retrieval models can be used to compute the two scores, which may have very different ranges. Then they normalize them into the same range before linear combining. Following [3], the final ranking function can be written as:

$$Score(\mathbf{q}, Q) = (1 - \alpha)N(S_{\mathbf{q},Q}) + \alpha N(S_{\mathbf{q},Cat(Q)}) \qquad (2)$$

where $S_{\mathbf{q},Q}$, $S_{\mathbf{q},Cat(Q)}$ are the local and global relevance scores, α is the parameter to control the linear interpolation, and $N()$ is the normalization function for the local and global relevance scores.

From Equation (2), we can see that this approach combines the two relevance scores linearly only in an *external* way *at the document level*. The most promising idea of this method is that it can be easily applied to any existing question retrieval models. However, in our work we will focus on the language modeling framework to integrate the category information into the original question language model in an *internal* way *at the word level*.

Overall, leaf category smoothing enhancement and category enhancement methods seem to be heuristic but effective to exploit the category information for question retrieval. Our work tries to improve the previous work of exploiting the category information by the following aspects. We integrate the category information measured by the category-specific term saliency into the unigram language modeling approach in a more systematic and internal way at the word level which is more effective than linear combination in an external way at the word level or at the document level. We proceed to present our method in the following sections.

3   Category-integrated Language Model

3.1   The Unigram Language Model

The unigram language model has been widely used in previous work [3, 4, 10] for question retrieval in CQA archives. The basic idea of the language model is to estimate a model for each historical question, and then rank the questions by the likelihood of the queried question according to the estimated models.

Formally, let $\mathbf{q} = w_{q_1} \dots w_{q_{|q|}}$ be a queried question and $Q = w_{Q_1} \dots w_{Q_{|Q|}}$ be a historical question in a question collection $Coll$. Let θ_Q be the most popular multinomial generation model estimated for the historical question Q, where θ_Q has the same number of parameters (i.e. word emission probabilities) as the number of words in the vocabulary set V, i.e., $\{P(w_i|\theta_Q)\}_{i=1}^{|V|}$. Thus, the ranking function of the query likelihood would be

$$P(\mathbf{q}|Q) = \prod_{i=1}^{|\mathbf{q}|} P(w_i|\theta_Q) = \prod_{w \in V} P(w|\theta_Q)^{c(w,\mathbf{q})} \tag{3}$$

where $c(w, Q)$ is the count of word w in the queried question $\mathbf{q}$.

With such a model, the retrieval problem is reduced to the problem of estimating the multinomial parameters $\{P(w_i|\theta_Q)\}_{i=1}^{|V|}$ for a given historical question Q. The simplest way is the maximum likelihood estimation:

$$P_{ml}(w|\theta_Q) = \frac{c(w,Q)}{\sum_{w \in V} c(w,Q)} = \frac{c(w,Q)}{|Q|} \tag{4}$$

To avoid zero probability, various smoothing strategies are usually applied to the estimation, which commonly interpolate the estimated document language model with the collection language model [11].

3.2 Integration with Category Information

Now, we discuss how to integrate the category information of each historical question into the unigram language model.

Our intuition of the category information is that a term may be more important in some category than other categories. That is to say, a term of a historical question in one specific category should have different term weighting than the term in other categories, we call this different term weighting as *category-specific term saliency*. For example, "virus" in the category "Computers & Internet.Security" is more salient than the term in the category "Computers & Internet.Software".

By empirical Bayesian analysis, we could express our knowledge or belief on the uncertainty of the parameters by some prior distribution on it. The conjugate of the distribution where the parameters come from is usually exploited to express the prior belief. The natural conjugate of multinomial distribution is Dirichlet distribution.

Specifically, supposing that Sal is the category-specific term saliency computing model (which will be introduced in section 4) and $Sal(w, Cat(Q))$ is the computed term saliency of term w in category $Cat(Q)$ containing Q, we use a Dirichlet prior on θ_Q with hyper-parameters $u = (u_1, u_2, \dots, u_{|V|})$, given by

$$P(\theta_Q|u) = Z_u \prod_{i=1}^{|V|} P(w_i|\theta_Q)^{u_i-1} \equiv Dir(\theta_Q|u) \tag{5}$$

where $u_i = \lambda_{Cat} Sal(w_i, Cat(Q))$, λ_{Cat} is the parameter, and $Z_u = \frac{\Gamma(\sum_{i=1}^{|V|} u_i)}{\prod_{i=1}^{|V|} \Gamma(u_i)}$ does not depend on the parameter θ_Q.

Then, with the Dirichlet prior, the posterior distribution of θ_Q is given by

$$P(\theta_Q|Q,u) = \frac{P(Q|\theta_Q)P(\theta_Q|u)}{P(Q|u)} = Z_{u'} \prod_{i=1}^{|V|} P(w_i|\theta_Q)^{c(w_i,Q)+u_i-1} \equiv Dir(\theta_Q|u') \quad (6)$$

By the property of natural conjugate distributions, $Z_{u'} \prod_{i=1}^{|V|} P(w_i|\theta_Q)^{c(w_i,Q)+u_i-1}$ in the above equation is also a Dirichlet distribution denoted as $Dir(\theta_Q|u')$ with parameters $u' = (c(w_1,Q) + u_1, c(w_2,Q) + .u_2, ..., c(w_{|V|},Q) + u_{|V|})$. The prior distribution reflects our prior beliefs about the weight of θ_Q, while the posterior distribution of θ_Q reflects the updated beliefs about θ_Q.

Given the posterior distribution, the estimation of the word emission probability can be noted as

$$P(w_i|\theta_Q^{Cat}) = \int_{\theta_Q} Dir(\theta_Q|u') \, \theta_Q \, d\theta_Q - \frac{c(w_i,Q) + \lambda_{Cat}Sal(w_i,Cat(Q))}{|Q| + \sum_{i=1}^{|V|} \lambda_{Cat}Sal(w_i,Cat(Q))} \quad (7)$$

In the above estimated question model, we can see that the category information of a historical question is integrated into the estimated unigram language model. In this way, the category-specific term saliency could be seen as transformed to word count information which is the primary element that unigram language model has the ability to model. From another point of view, we could consider that the "bag-of-words" representation of the question Q is transformed into a pseudo "bag-of-words" representation Q^{Cat} given the question's category-specific term saliency computing model. In Q^{Cat}, the matching term's frequency is transformed from $c(w_i,Q)$ to $c(w_i,Q) + \lambda_{Cat}Sal(w_i,Cat(Q))$ and the question length is transformed from $|Q|$ to $|Q| + \sum_{i=1}^{|V|} \lambda_{Cat}Sal(w_i,Cat(Q))$. Then, the problem of ranking the historical question Q given a queried question $\mathbf{q}$ is changed to ranking Q^{Cat}. On such ground, any "bag-of-words" based language model, e.g. the query likelihood model or the KL divergence model [6], could work on Q^{Cat}, and thus integrate the category information of the historical question in an internal way.

Next, we further smooth the category-integrated language model by a collection language model $P(w|Coll)$ to account for unseen words as in [11]. Specifically, the collection-based Dirichlet prior $(\mu P(w_1|Coll), \mu P(w_2|Coll), ..., \mu P(w_{|V|}|Coll))$ is used for smoothing. Thus, we can get the smoothed category-integrated estimation:

$$P(w_i|\hat{\theta}_Q^{Cat}) = \frac{c(w_i,Q) + \lambda_{Cat}Sal(w_i,Cat(Q)) + \mu P(w_i|Coll)}{|Q| + \sum_{i=1}^{|V|} \lambda_{Cat}Sal(w_i,Cat(Q)) + \mu} \quad (8)$$

Note that, the final word count for the smoothed category-integrated language model $\hat{\theta}_Q^{Cat}$ for historical question Q consists of the document-level count $c(w_i,Q)$, the category-level pseudo count $\lambda_{Cat}Sal(w_i,Cat(Q))$ and the collection-level pseudo count $\mu P(w_i|Coll)$.

If we set $\lambda_{Cat} = 0$, the ranking function will ignore the category information, thus will reduce to the unigram language model with Dirichlet smoothing [11].

Finally, the smoothed category-integrated ranking function in the query likelihood language modeling framework can be written as:

$$P(\mathbf{q}|Q) = \prod_{i=1}^{|q|} P(w_i|\hat{\theta}_Q^{Cat}) = \prod_{w \in V} P(w|\hat{\theta}_Q^{Cat})^{c(w,\mathbf{q})} \tag{9}$$

4   Category-specific Term Saliency Measure

A key notion in our proposed category-integrated language model is a term's category-specific term saliency $Sal(w, Cat(Q))$ in category $Cat(Q)$ containing question Q, which represents a term's importance or saliency in the category that the question belongs to. Term distribution in a specific category is biased as compared to that in a whole collection. In this section, we will introduce a measure of term saliency based on the divergence of term distribution in a specific category from the general whole collection.

4.1   Divergence-based Feature

The divergence of term distribution in a specific category from the whole collection reveals the significance of terms globally in its category. We employ *Jensen-Shannon* (JS) divergence, which is a well adopted distance measure between two probability distributions, to capture the difference of term distribution in two collections. It is defined as the mean of the relative entropy of each distribution to the mean distribution. As we evaluate the divergence at term level rather than the whole sample set, thus we examine the point-wise function for each individual term as follows:

$$d_{JS}(w, S||G) = \frac{p_s(w)\log\frac{p_s(w)}{\frac{1}{2}\big(p_s(w) + p_g(w)\big)} + p_g(w)\log\frac{p_g(w)}{\frac{1}{2}\big(p_s(w) + p_g(w)\big)}}{2} \tag{10}$$

Here, S and G denote the category-specific and general vocabularies, and $p_s(w)$ and $p_g(w)$ denote their corresponding probability distribution obtained by maximum likelihood estimation. $d_{JS}(w, S||G)$ means the JS divergence of term w between the specific category and general whole collection. The higher $d_{JS}(w, S||G)$ is, the more important or salient term w is.

4.2   Estimating Term Saliency from Divergence-based Feature

Having the point-wise divergence feature at hand, now we will define the function needed to transform the JS divergence feature into the term saliency score. The transformation function plays a very important role in setting the scale of proportional ratio between the JS divergence feature score of different terms. The following exponential form is used and tested, of which x is the JS divergence feature score, *para* is the parameter to control the scale of the transformed score and ϵ is a smoothing parameter which is set as 1e-8 in our experiments.

$$f(x) = para^{\log(x+\epsilon)} \qquad (11)$$

Then, the category-specific term saliency computing model can be written as:

$$Sal(w, Cat(Q)) = f(d_{JS}(w, S||G)) = para^{\log(d_{JS}(w,S||G)+\epsilon)} \qquad (12)$$

5   Experiments and Discussion

5.1   Experimental Setup

Question Retrieval Methods. To evaluate the performance of the proposed category-integrated language model, we use the following four types of baseline methods:

— LM: Unigram Language Model without considering the category information, we will give the results of unigram language model using JM smoothing and Dirichlet smoothing together, denoted as LM_{JM} and LM_{Dir} respectively[3].
— LM@OptC: This method performs retrieval only in the queried question's category specified by users of Yahoo! Answers using LM, where JM smoothing is used.
— LM@LS: Leaf Category Smoothing Enhancement method. (in section 2.1)
— LM+LM: Category Enhancement method that the global relevance and local relevance are computed using LM, where we also use JM smoothing. (in Section 2.2)
— CLM: This is our proposed Category-integrated Language Model.

Dataset. We use an open dataset[4] which is used in the pioneer work [3, 2] of exploiting the category information to enhance the question retrieval performance in a real world CQA dataset from Yahoo! Answers. The open dataset contains 252 *queried questions* with totally 1,373 relevant historical questions as the *ground-truth*. In our experiments, we perform preprocess as follows: all the questions are lowercased and stopwords are removed using a total of 418 stopwords from the standard stoplist in Lemur toolkit (version 4.10)[5]. Thus, after preprocessing, we get our final *questions repository* (3,111,219 questions), which contains 26 categories at the first level and 1262 categories at the leaf level. Each question belongs to a unique leaf category.

Metrics. We evaluate the performance of all the ranking methods using the following metrics: **Mean Average Precision** (MAP) and **Precision@n** (P@n). MAP rewards methods that return relevant questions early and also rewards correct ranking of the results. P@n reports the fraction of the top-n questions retrieved that are relevant. We also perform a significance test using a paired t-test with a significant level of 0.05.

---

[3] Note that the baseline models LM_{JM} and LM_{Dir} are reported as references, and the main purpose of this work is to investigate a new method to exploit the category information for enhancing the question retrieval performance.
[4] http://homepages.inf.ed.ac.uk/gcong/qa/
[5] http://www.lemurproject.org/

5.2   Parameter Setting

There are several parameters to be set in our experiments. Following the literature [3, 2], we set $\beta = 1, \lambda = 0.2$ in Equation (1) for getting the unigram language model using JM smoothing and $\beta = 0.2, \lambda = 0.2$ in Equation (1) for getting LM@LS. In LM+LM, we set $\alpha = 0.1$ in Equation (2) [3].

In LM_{Dir}, we set $\lambda_{Cat} = 0$ in Equation (8) and then tune the Dirichlet smoothing parameter μ by setting it as $1, 2, 3, ..., 20, 30, 40, 50$. From Figure 1, we can see that when $\mu > 5$ the performance on MAP metric reaches high and is almost stable; when $3 \leq \mu \leq 11$ the performance on P@10 metric is relative high, however when $\mu > 11$, the performance on P@10 drops quickly. Thus, the best Dirichlet smoothing parameter μ is around 5~11, which is different from the common setting (around 500~2500) on the retrieval tasks on TREC data [11]. The main reason may be that the historical questions in question retrieval tasks are very short, which is not the case on the traditional retrieval tasks on TREC data. Specifically, we will fix $\mu = 10$ in all the following experiments.

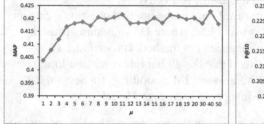 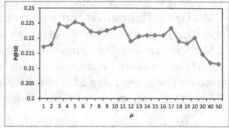

Fig. 1. Sensitivity to the Dirichlet smoothing parameter μ of the unigram language model on the dataset for question retrieval.

In CLM, the most two important parameters are λ_{Cat} and $para$. λ_{Cat} is the parameter that controls proportional weight of the prior factor which is the category-specific term saliency. $para$ is the parameter that controls the scale of the transformation function in Equation (11) in section 4. In our experiments, we set these two parameters by maximizing the empirical performance on the collections through exhaustive searches in the following parameter space showed in Table 1.

Table 1. Parameters and their coresponding search space in CLM

Parameters	Search Space
λ_{Cat}	$1, 2, 3, ..., 20, 30, 40, 50$
$para$	$2.1, 2.2, 2.3, ..., 3.0$
μ	$1, 2, 3, ..., 20, 30, 40, 50$

Figure 2 reports the parameter sensitivity of CLM on the test collection. From Figure 2, we can see that the performance is relative stable when $1 \leq \lambda_{Cat} \leq 10$, but drops quickly when $\lambda_{Cat} > 10$. The best performance is achieved when λ_{Cat} is set to about 7. Moreover, when $para \geq 2.3$ the best value of λ_{Cat} is relative insensitive to different values of $para$ in the category-specific term saliency measure. However, the exponential weight $para$ which controls the scale of the transformation function does have some influence on the ranking performance. From the figure, a relatively bigger value of $para$ about 2.7 is preferred. Thus, the parameter setting of the best performance of CLM reported in the next subsection is: $\lambda_{Cat} = 7, para = 2.7$.

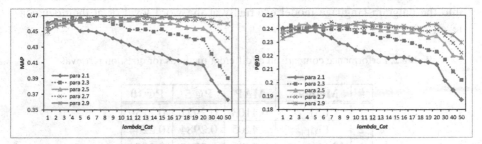

Fig. 2. Sensitivity to the parameters λ_{Cat} and $para$ for CLM

5.3 Performance Comparison

Table 2 presents the comparison of different methods for question retrieval. Row 1 and Row 2 are the Unigram Language Model without considering the category information, using JM smoothing and Dirichlet smoothing respectively. Row 3 is the method which performs retrieval only in the category that specified to the queried question. Row 4 and Row 5 are the Leaf Category Smoothing Enhancement and Category Enhancement methods, which are the proposed methods in the pioneer work [3, 2]. Row 6 is our proposed method. There are some clear trends in the results of Table 2:

1. Unigram language model using Dirichlet Smoothing significantly outperforms that using JM smoothing in the setting of our experiments (Row 1 vs. Row 2).
2. The method retrieving in the optimal category that specified to the queried question by users of Yahoo! Answers does not improve the methods without considering the category information (Row 3 vs. Row1, 2). The results show that the queried question's category information does not help the retrieval performance when used directly and this is consistent with the former work [3, 2]. This maybe because relevant questions are also contained in the other categories but not only in the category of the queried question.
3. When using the methods proposed in [3, 2], which incorporate the category information into the retrieval model externally at the word level (LM@LS) or at the document level (LM+LM), the performance can significantly outperform the unigram language models that do not consider the category information (Row 4, 5 vs. Row 1, 2). These results are also consistent with the form work [3, 2]. However, our experiments of LM+LM does not outperforms LM@LS, which is not the case in the results of [3]. We think that there are mainly two possible reasons: First, the preprocessing of the dataset maybe different. Second, LM+LM combines two relevant scores externally at the document level, which is not natural than that of LM@LS at the word level.

4. Our proposed CLM significantly outperforms LM@LS and LM+LM (Row 6 vs. Row 4, 5). We conduct a significant test (a paired t-test with a significant level of 0.05) on the improvements of our approach over LM@LS and LM+LM. The results indicate that the improvements are statistically significant in terms of all the evaluation metrics. This demonstrates that our proposed CLM which integrate the category information into the unigram language model at the word level internally is more effective than the methods which incorporate the category information into the unigram language model externally at the word level or at the document level.

Table 2. Performance comparison of different methods for question retrieval

#	Methods	MAP	P@5	P@10
1	LM_{JM}	0.4103	0.2921	0.2159
2	LM_{Dir}	0.4205	0.2984	0.2234
3	LM@OptC	0.3496	0.2762	0.2056
4	LM@LS	0.4638	0.3270	0.2425
5	LM+LM	0.4424	0.3159	0.2341
6	**CLM**	**0.4684**	**0.3294**	**0.2452**

6   Conclusions

In this paper, we propose a novel category-integrated language model (CLM) to integrate the category information into the unigram language model internally at the word level for improving the performance of question retrieval in CQA archives, which views category-specific term saliency as Dirichlet hyper-parameters to weight the parameters of the multinomial language model. This integration method has solid mathematical foundation, and experiments conducted on a large scale real world CQA dataset from Yahoo! Answers demonstrate that our proposed CLM can significantly outperforms the previous work that incorporats the category information into the unigram language model externally at the word level or at the document level.

Some interesting future work should be continued. First, it is of relevance to apply and evaluate other question retrieval methods, e.g., translation-based methods, in the proposed framework. Second, we believe it is interesting to include answers into our proposed framework. Finally, the hierarchical category structures may perhaps be exploited to further improve the performance of the CLM model.

Acknowledgements. This work is supported by the National Science Foundation of China under Grant No. 61070111 and the Strategic Priority Research Program of the Chinese Academy of Sciences under Grant No. XDA06030200.

References

1. Cai, L., Zhou, G., Liu, K., Zhao, J.: Learning the latent topics for question retrieval in community qa. In: IJCNLP, pp. 273–281 (2011)
2. Cao, X., Cong, G., Cui, B., Jensen, C.S., Zhang, C.: The use of categorization information in language models for question retrieval. In: CIKM, pp. 265–274 (2009)
3. Cao, X., Cong, G., Cui, B., Jensen, C.S.: A generalized framework of exploring category information for question retrieval in community question answer archives. In: WWW, pp. 201–210 (2010)
4. Jeon, J., Bruce Croft, W., Lee, J.H.: Finding similar questions in large question and answer archives. In: CIKM, pp. 84–90 (2005)
5. Ji, Z., Xu, F., Wang, B., He, B.: Question-answer topic model for question retrieval in community question answering. In: CIKM (2012)
6. Lafferty, J., Zhai, C.: Document language models, query models, and risk minimization for information retrieval. In: SIGIR, pp. 111–119 (2001)
7. Lee, J.-T., Kim, S.-B., Song, Y.-I., Rim, H.-C.: Bridging lexical gaps between queries and questions on large online q&a collections with compact translation models. In: EMNLP, pp. 410–418 (2008)
8. Ming, Z.-Y., Chua, T.-S., Cong, G.: Exploring domain-specific term weight in archived question search. In: CIKM, pp. 1605–1608 (2010)
9. Wang, K., Ming, Z., Chua, T.-S.: A syntactic tree matching approach to finding similar questions in community-based qa services. In: SIGIR, pp. 187–194 (2009)
10. Xue, X., Jeon, J., Bruce Croft, W.: Retrieval models for question and answer archives. In: SIGIR, pp. 475–482 (2008)
11. Zhai, C., Lafferty, J.: A study of smoothing methods for language models applied to ad hoc information retrieval. In: SIGIR, pp. 334–342 (2001)
12. Zhou, G., Cai, L., Zhao, J., Liu, K.: Phrase-based translation model for question retrieval in community question answer archives. In: ACL-HLT, pp. 653–662 (2011)

The Reusability of a Diversified Search Test Collection

Tetsuya Sakai[1], Zhicheng Dou[1], Ruihua Song[1], and Noriko Kando[2]

[1] Microsoft Research Asia, P.R. China
tetsuyasakai@acm.org, {zhichdou,Song.Ruihua}@microsoft.com
[2] National Institute of Informatics, Japan
kando@nii.ac.jp

Abstract. Traditional ad hoc IR test collections were built using a relatively large pool depth (e.g. 100), and are usually assumed to be reusable. Moreover, when they are reused to compare a new system with another or with systems that contributed to the pools ("contributors"), an even larger measurement depth (e.g. 1,000) is often used for computing evaluation metrics. In contrast, the web diversity test collections that have been created in the past few years at TREC and NTCIR use a much smaller pool depth (e.g. 20). The measurement depth is also small (e.g. 10-30), as search result diversification is primarily intended for the first result page. In this study, we examine the reusability of a typical web diversity test collection, namely, one from the NTCIR-9 INTENT-1 Chinese Document Ranking task, which used a pool depth of 20 and official measurement depths of 10, 20 and 30. First, we conducted additional relevance assessments to expand the official INTENT-1 collection to achieve a pool depth of 40. Using the expanded relevance assessments, we show that run rankings at the measurement depth of 30 are too unreliable, given that the pool depth is 20. Second, we conduct a leave-one-out experiment for every participating team of the INTENT-1 Chinese task, to examine how (un)fairly new runs are evaluated with the INTENT-1 collection. We show that, for the purpose of comparing new systems with the contributors of the test collection being used, condensed-list versions of existing diversity evaluation metrics are more reliable than the raw metrics. However, even the condensed-list metrics may be unreliable if the new systems are not competitive compared to the contributors.

1 Introduction

Traditional ad hoc IR test collections were built using a large *pool depth*: typically, top 100 documents were collected from every run that was submitted to an evaluation task, and these pooled documents were assessed for relevance (pool depth $pd = 100$). Although the target document collection is usually much larger than the pooled document sets, these IR test collections are often assumed to be *reusable*: they are used for comparing a new system with another and with systems that contributed to the pools ("contributors"). Moreover, in ad hoc IR, a *measurement depth* of 1,000 is often used: that is, top $l = 1000$ documents

Y. Hou et al. (Eds.): AIRS 2012, LNCS 7675, pp. 26–38, 2012.
© Springer-Verlag Berlin Heidelberg 2012

returned by the system are used for computing evaluation metrics (e.g. [22]). While more intricate techniques for efficiently and reliably obtaining relevance assessments exist (e.g. [7]), the traditional method of using a static pool depth is still widely used, due to its simplicity, its convenience for assessment cost estimation, and its independence to the choice of evaluation metrics.

In contrast to the above practices in ad hoc (particularly non-web) IR, the web *diversity* test collections that have been created in the past few years at TREC and NTCIR use a much smaller pool depth, typically $pd = 20$ [8,17] or $pd = 25$ [9]. These collections are used specifically for evaluating *search result diversification*, which aims to produce a single Search Engine Result Page (SERP) that satisfies different users or user intents that share the same search query (e.g. [8,15]). In web diversity evaluation, the measurement depth is also very small (e.g. $l = 10, 30$), as the target of diversification is typically the *first* SERP (i.e. URLs ranked at the very top). Given the small pool depth, what is the appropriate measurement depth for diversity evaluation? Are existing web diversity test collections reusable to any degree? If they are, what are the appropriate ways to reuse them?

To address the above questions, we examine the reusability of a typical web diversity test collection, namely, one from the NTCIR-9 INTENT-1 Chinese Document Ranking task, which used $pd = 20$ and official measurement depths of $l = 10, 20, 30$ for ranking the submitted runs [17]. First, we conducted additional relevance assessments to expand the official relevance assessments of the INTENT-1 collection to achieve $pd = 40$. Using the expanded data, we show that run rankings at $l = 30$ are too unreliable, given that $pd = 20$. Second, we conduct a *Leave-One-Out* experiment (e.g. [3,12,22]) for every participating team of the INTENT-1 Chinese task, to examine how (un)fairly new runs are evaluated with the INTENT-1 collection. In addition to a set of state-of-the-art diversity evaluation metrics, we experiment with *condensed-list* versions of these metrics, which remove all *unjudged* documents from runs prior to computation [10]. We show that, for the purpose of comparing new systems with the contributors of the test collection being used, condensed-list diversity metrics are more reliable than the raw metrics. However, even the condensed-list metrics may be unreliable if the new systems are not competitive compared to the contributors.

2   Related Work

It is well-known that IR test collections built through pooling are *incomplete* and may be *biased* [2,3]. Their relevance assessments are said to be incomplete if some relevant documents exist among the *unjudged* documents in the collection. Furthermore, the relevance assessments are said to be biased if they represent some limited aspects of the complete set of relevant documents. For example, *shallow pooling* (i.e. using a small *pd*) may cause a *pool depth bias*: the relevant document sets thus obtained may contain only documents that are very easy to retrieve, for example, by keyword matching [1]. Moreover, if the systems that participate in the pooling process all use similar search strategies, this will cause a *system bias*: the relevant documents thus obtained are the ones that are retrievable by that particular class of systems. Hence relevant documents that can be retrieved by a "novel" system (which we really want to build) may be outside the relevance assessments.

In the context of ad hoc IR, several studies addressed the problem of test collection incompleteness by randomly sampling documents from the original relevance assessments (e.g. [2,10,14,21]). But random sampling does not directly address the bias problem. Zobel's seminal work [22] examined the effect of pool depth bias and system bias; in particular, his Leave-One-Out (LOO) methodology for studying system bias was later adopted by TREC for validating their test collections. This method removes *unique contributions* of a particular team from the original relevance assessments, where a unique contribution is a document contributed to the pool by that particular team only[1]. Thus, the team that has been left out can be seen as a "new" team that did not contribute to the pool. If the outcome of the evaluation for the "new" team based on the LOO relevance assessments is similar to that based on the original relevance assessments, then that suggests that the test collection may be reusable: many of the relevant documents retrieved by the "new" team are already covered by the test collection, even if this team did not contribute to the pool. More recent system bias and pool depth bias studies for ad hoc IR include the work by Büttcher *et al.* [3] and that by Sakai [11,12].

Some evaluation metrics have been designed specifically for the purpose of coping with incompleteness and bias [2,10,14,21]: among them, Sakai's simple approach of using *condensed lists* obtained by removing all unjudged documents from the runs is applicable to *any* existing evaluation metrics, including those that handle graded relevance. However, in his subsequnt study on handling system bias, Sakai [12] reported that *"condensed-list metrics overestimate new systems while traditional metrics underestimate them."* When a new run is evaluated with a condensed-list metric, many relevant documents go up the ranks as the unjudged documents in the run are removed. The above work of Sakai generalises an earlier finding by Büttcher *et al.* who focussed on binary relevance metrics such as Average Precision (AP) and Binary Preference (bpref): *"Where AP underestimates the performance of a [new] system, bpref overestimates it"* [3]. However, these studies were about *traditional* IR, where typically $pd = 100$ and $l = 1000$.

More intricate approaches to handling incompleteness and bias exist. For example, Webber and Park [19] have proposed to adjust the evaluation metric values computed for new systems, but this methodology requires some new relevance assessments for the new systems. Carterette *et al.* [5] propose to quantify the reusability of test collections, but this requires several kinds of computation, such as estimating the relevance probability of each unjudged document using several features. Carterette *et al.* [6] propose an approach to conducting

---

[1] Zobel's original method removed documents contributed by a particular *run*, but it is now common practice to conduct a more stringent test by removing documents contributed by a particular *team*, as one team typically submits multiple runs [18].

relevance assessments while monitoring reusability. While the approach is interesting, the focus of the present study is to examine the reliability of an existing web diversity test collection that was constructed in a traditional manner.

3   NTCIR-9 INTENT-1 Task

3.1   Task and Data

NTCIR (NII Testbeds and Community for Information access Research) is a sesquiannual series of international evaluation workshops hosted by National Institute of Informatics (NII), Japan[2]. The first INTENT task (INTENT-1), introduced at the ninth NTCIR workshop (NTCIR-9), had two subtasks: *Subtopic Mining* (SM) and *Document Ranking* (DR). Both subtasks covered two languages: Chinese and Japanese [17].

The SM subtask was defined as: given a query, return a ranked list of *subtopic strings*, which represent diverse *intents* behind the query. By pooling subtopics submitted by the SM participants and manually clustering them, the INTENT-1 task organisers identified a set of *intents* for each query[3]. The organisers also estimated the probability of each intent given the query based on assessor voting.

The DR subtask is similar to the TREC web track diversity task [8,9]: given a query, each participating system returns a diversified ranked list of web pages. The main differences between the evaluation practices at TREC and INTENT-1 are: (a) While TREC treats each intent for a given topic as equally likely, INTENT-1 leveraged the intent probabilities, to prioritise documents relevant to popular intents; (b) While TREC evaluates runs based on per-intent *binary* relevance, INTENT-1 leveraged per-intent *graded* relevance[4]. Also, the evaluation metrics used in these two forums are different, as we shall discuss in Section 3.2.

In this study, we examine the reusability of the INTENT-1 *Chinese DR* test collection, because we were able to hire Chinese assessors and obtain additional relevance assessments for this collection, *and* because the INTENT-1 *Japanese* DR collection is highly unlikely to be reusable: it only involved three participating teams. Table 1 shows some statistics of the Chinese DR test collection.

The additional relevance assessments were done in exactly the same way as the official relevance assessments of $pd = 20$. The same assessor interface was used, which let assessors to view each pooled document and to select a relevance grade for each intent: "highly relevant", "relevant" and "nonrelevant." Two assessors were assigned to each topic, and the relevance grades were aggregated to form a five-point relevance scale, from $L0$ (judged nonrelevant) to $L4$ (highest relevance level) [17][5]. The only difference is how the document pools were obtained: at

---

[2] http://research.nii.ac.jp/ntcir/

[3] In the TREC web diversity tasks, the intents are referred to as "subtopics."

[4] The TREC 2011 web diversity test collection actually contains graded relevance assessments, but they were treated as binary relevance assessments in the evaluation.

[5] We assume that the disagreements between the old and the new assessors are negligible. Ideally, this assumption should be verified by letting the new assessors re-judge some of the "old" documents and computing kappa statistics.

Table 1. Statistics of the INTENT-1 Chinese DR test collection.

#web pages	138 million (SogouT collection)
#topics	100
#intents	917 across 100 topics (max.16; min 4)
#teams (#runs)	7 (24)
relevance levels	5-point ($L0$: judged nonrelevant – $L4$: highest level)
pool depth	20
#relevant docs	12,144 across 100 topics (max.182; min 9)
#relevant intents/doc	mean across 12,144: 1.94 (max. 10; min 1)
#$L0$ (judged nonrelevant)	6,335

Table 2. Statistics of the expanded relevance assessments

pool depth	40
#relevant docs	21,596 across 100 topics (max. 343; min 13)
#relevant intents/doc	mean across 21,596: 1.89 (max. 10; min 1)
#$L0$ (judged nonrelevant)	15,176

Table 3. Statistics of each team at the INTENT-1 Chinese DR subtask ($pd = 20$)

Team	#runs	Unique contributions per topic	Unique relevant per topic	Best run	Official Mean D♯-nDCG @10
THUIR	5	29.92	16.18	THUIR-D-C-5	.5717
uogTr	5	19.51	15.28	uogTr-D-C-5	.5499
MSINT	5	18.10	7.87	MSINT-D-C-1	.5461
HIT2jointNLPlab	2	23.04	16.30	HIT2jointNLPLab-D-C-2	.4749
NTU	1	10.90	7.70	NTU-D-C-1	.4747
SJTUBCMI	5	34.19	22.70	SJTUBCMI-D-C-2	.4663
III_CYUT_NTHU	1	17.10	8.59	III_CYUT_NTHU-D-C-1	.3335

INTENT-1, the top 20 documents from every run was included in the pool; in our study, we first obtained the top 40 documents from every run, and then removed the aforementioned depth-20 documents[6]. Table 2 shows the statistics of the expanded relevance assessments thus obtained.

The left half of Table 3 shows some statistics for each of the seven teams that participated in INTENT-1. The *unique contributions* of a team are documents that were contributed to the pool by *this team only*. The *unique relevant* documents of a team are the relevant documents among its unique contributions. We will discuss the right half of Table 3 in Section 4.1.

3.2   Evaluation Metrics

In this study, we use the evaluation metrics that were used officially for ranking the INTENT-1 runs, namely, *I-rec* (intent recall), *D-nDCG* and *D♯-nDCG* [16]. I-rec is simply the proportion of intents covered by a ranked list; D-nDCG is a ranked retrieval metric for diversity evaluation that takes into account the popularity of intents and per-intent graded relevance. Intuitively, it encourages systems to retrieve documents that are highly relevant to many popular intents

---

[6] In both cases, the pooled documents are sorted by "popularity" prior to assessments [13]. This practice is not used at TREC.

before those that are marginally relevant to a few minor intents. D♯-nDCG is simply a linear combination of I-rec and D-nDCG, and has been shown to have several advantages over other diversity metrics [15,16].

In addition to the three official metrics, our LOO experiments consider their condensed-list versions, which we call I-rec′, D-nDCG′ and D♯-nDCG′. While standard metrics treat both judged nonrelevant documents (L0 documents) and *unjudged* documents (i.e. those that were never included in the pool) as nonrelevant, condensed-list metrics *remove* all unjudged documents from the ranked list before computation. Sakai [10] and Sakai and Kando [14] showed that some condensed-list metrics are more robust to the *incompleteness* of test collections than others such as bpref. As was mentioned in Section 2, however, Sakai [12] showed in his study on test collection *bias* that "*condensed-list metrics overestimate new systems while traditional metrics underestimate them*" and that "*the overestimation [by the condensed-list metrics] tends to be larger than the underestimation [by the raw metrics]*." The overestimation occurs because, when a new run is evaluated with a condensed-list metric, many relevant documents go up the ranks as the unjudged documents in the run are removed. However, while his study was about traditional IR where the measurement depth was $l = 1000$, our present study concerns diversified search where the measurement depth is very small, e.g. $l = 10$. We thought it possible that the overestimation effect of condensed-list metrics may be small in our case, as the number of unjuded documents that are removed will be small.

4   Experiments

4.1   Pool Bias: Pool Depth and Measurement Depth

The INTENT-1 task used $pd = 20$ and officially reported Mean I-rec, D-nDCG and D♯-nDCG values at the measurement depths of $l = 10, 20, 30$. In this section, we examine the effect on the evaluation outcome at $l = 10, 30$ when relevance assessments based on $pd = 40$ are used instead[7]. Figure 1 shows how run rankings change if $pd = 40$ is used instead of $pd = 20$, for all three official metrics. Kendall's τ rank correlation and *symmetric* τ_{ap} values are also shown [20]. τ_{ap} is similar to τ but is more sensitive to the rank changes near the top. The left half of the figure shows that expanding the relevance assessments has very little effect on the system ranking for $l = 10$; while the right half of the figure shows that the effect is not negligible for $l = 30$. For example, the graphs for Mean D♯-nDCG@30 show that top runs actually change if we add more relevance assessments. This shows that the system rankings with $l = 30$, given the pool depth of $pd = 20$, should not be trusted.

We also examined the effect of adding more relevance assessments to statistical significance testing, in particular, significant differences across different teams, as this is important at evaluation forums like NTCIR and TREC. From each of the

---

[7] Note that this section does not discuss condensed-list metrics, as there are no unjudged documents in the top 20 of any of the runs.

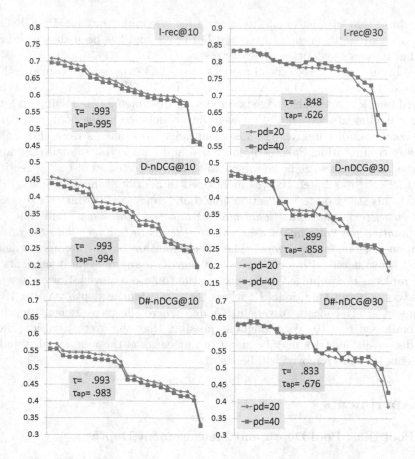

Fig. 1. How run rankings change if $pd = 40$ is used instead of $pd = 20$. The x-axis represents runs sorted by the performance based on relevance assessments with $pd = 20$. The y-axis represents the performance values.

seven teams, we selected the official "best" run in terms of Mean D♯-nDCG@10, as shown in the right half of Table 3. Then we applied a randomised version of two-sided Tukey's Honestly Significant Differences (HSD) test as described by Carterette [4] for the three evaluation metrics with different settings. Table 4 shows the results. For example, from Table 4(I), we can see that even though the difference between THUIR and MSINT was officially significant in terms of D-nDCG@10 (but not in terms of I-rec@10 or D♯-nDCG@10), the difference is *not* statistically significant when less incomplete (i.e. $pd = 40$) relevance assessments are used. Thus, assuming that less incomplete assessments provide more reliable conclusions, the significant difference in terms of D-nDCG@10 is a *false alarm*. Such discrepancies between the $pd = 20$ and $pd = 40$ results are indicated by $\star$'s in Table 4, and it can be observed by comparing Parts (I) and (II) of the table that the false alarms occur more frequently when $l = 30$. These significance test results also show that results with $l = 30$ are not to be trusted when

Table 4. Statistically significant differences across teams according to a randomised Tukey's HSD test at $\alpha = 0.05$. Significant differences by I-rec, D-nDCG and D♯-nDCG are indicated by I, D and D♯, respectively. If none of the differences is significant, this is indicated by a "no." Discrepancies across the pool depths are indicated by ⋆'s..

(I) Measurement depth $l = 10$		(b)	(c)	(d)	(e)	(f)	(g)
(a) THUIR	pd = 20	no	D⋆	I, D, D♯	D, D♯	I, D, D♯	I, D, D♯
	pd = 40	no	no	I, D, D♯	D, D♯	I, D, D♯	I, D, D♯
(b) uogTr	pd = 20	–	no	I, D♯	D, D♯	D, D♯	I, D, D♯
	pd = 40	–	no	I, D♯	D, D♯	D, D♯	I, D, D♯
(c) MSINT	pd = 20	–	–	I, D♯	I, D♯	I, D♯	I, D, D♯
	pd = 40	–	–	I, D♯	I, D♯	I, D♯	I, D, D♯
(d) HIT2jointNLPLab	pd = 20	–	–	–	no	no	I, D, D♯
	pd = 40	–	–	–	no	no	I, D, D♯
(e) NTU	pd = 20	–	–	–	–	no	I, D, D♯
	pd = 40	–	–	–	–	no	I, D, D♯
(f) SJTUBCMI	pd = 20	–	–	–	–	–	I, D, D♯
	pd = 40	–	–	–	–	–	I, D, D♯
(g) III_CYUT_NTHU	pd = 20	–	–	–	–	–	–
	pd = 40	–	–	–	–	–	–
(II) Measurement depth $l = 30$		(b)	(c)	(d)	(e)	(f)	(g)
(a) THUIR	pd = 20	no	D	I, D, D♯	I, D, D♯	D, D♯	I, D, D♯
	pd = 40	no	D	I, D, D♯	I, D, D♯	D, D♯	I, D, D♯
(b) uogTr	pd = 20	–	D	I⋆, D, D♯	I⋆, D, D♯	D, D♯	I, D, D♯
	pd = 40	–	D	D, D♯	D, D♯	D, D♯	I, D, D♯
(c) MSINT	pd = 20	–	–	I, D♯⋆	I, D♯⋆	no	I, D, D♯
	pd = 40	–	–	I	I	no	I, D, D♯
(d) HIT2jointNLPLab	pd = 20	–	–	–	no	no	I, D, D♯
	pd = 40	–	–	–	no	no	I, D, D♯
(e) NTU	pd = 20	–	–	–	–	no	I, D, D♯
	pd = 40	–	–	–	–	no	I, D, D♯
(f) SJTUBCMI	pd = 20	–	–	–	–	–	I, D, D♯
	pd = 40	–	–	–	–	–	I, D, D♯
(g) III_CYUT_NTHU	pd = 20	–	–	–	–	–	–
	pd = 40	–	–	–	–	–	–

$pd = 20$. Thus we recommend that the INTENT task organisers focus on $l = 10$ measurements when officially announcing the participants's performances in the future. In Section 4.2 where we discuss the effect of system biases by means of LOO tests, we will focus on $l = 10$ measurements only.

4.2   System Bias: Evaluating New Systems

To examine how the INTENT-1 Chinese DR test collection evaluates a "new" system that did not contribute to the pool, we conducted a Leave-One-Out (LOO) experiment for each of the seven participating teams at the INTENT-1 Chinese DR task. For example, as Table 1 shows, Team THUIR, the official top performer of the task in terms of Mean D♯-nDCG, had a total of 2,992 unique contributions across the 100 topics when $pd = 20$. The LOO relevance assessment set for this team at $pd = 20$ ("pd20loo-THUIR") was constructed by removing all of these unique contributions from the original relevance assessments with $pd = 20$. Then, all of the 24 runs were evaluated using pd20loo-THUIR. Note that, when the runs from THUIR are evaluated using pd20loo-THUIR, the evaluation relies entirely on contributions from teams *other than* THUIR. The above

Table 5. Leave-One-Out results for each participating team at the INTENT-1 Chinese Document Ranking subtask. For each team, the first row shows the difference between the performance with the original relevance assessments and the performance with that teams' LOO relevance assessments; the second row show the ranks before and after leaving out that team. In Part (b), the results that are more *effective* than those in Part (a) are shown in **bold**.

	(a) Raw-list metrics			(b) Condensed-list metrics		
	I-rec	D-nDCG	D♯-nDCG	I-rec'	D-nDCG'	D♯-nDCG'
	(I) $pd = 20$ ($l = 10$)					
THUIR	−.0720	−.1015	−.0867	**+.0183**	**+.0279**	**+.0231**
	6↓12	1↓12	1↓10	**6↓7**	**1→1**	**1↓2**
uogTr	−.0708	−.1123	−.0915	**+.0212**	**+.0233**	.0223
	7↓16	5↓16	3↓14	**7↓8**	**5→5**	3↓6
MSINT	−.0774	−.0843	−.0809	**+.0318**	**+.0695**	**+.0507**
	2↓9	8↓14	4↓12	**2↓3**	**8↑5**	**4↑2**
HIT2jointNLPlab	−.1221	−.1396	−.1308	**+.0418**	**+.0515**	**+.0466**
	22↓23	13↓22	13↓22	**22↑14**	**13↑9**	**13→13**
NTU	−.1527	−.1260	−.1394	−.0090	**+.0353**	**+.0131**
	14↓23	16↓23	14↓23	14↓15	**16↑14**	**14↑13**
SJTUBCMI	−.2081	−.1718	−.1900	−.0390	**+.0071**	−.0160
	17↓20	15↓20	15↓20	**17↓18**	**15→15**	15↓16
III_CYUT_NTHU	−.2123	−.1398	−.1760	**+.0928**	**+.0648**	**+.0788**
	24→24	24→24	24→24	**24↑23**	**24↑21**	24→24
	(II) $pd = 40$ ($l = 10$)					
THUIR	−.0558	−.0671	−.0615	**+.0045**	**+.0170**	**+.0107**
	6↓10	1↓8	1↓10	**6↓8**	**1→1**	**1→1**
uogTr	−.0442	−.0670	−.0556	**+.0058**	**+.0146**	**+.0102**
	7↓10	5↓11	3↓9	**7→7**	**5↓6**	**3↓4**
MSINT	−.0461	−.0511	−.0485	**+.0339**	**+.0636**	**+.0488**
	2↓5	8↓14	4↓9	**2↑1**	**8↑4**	**4↑1**
HIT2jointNLPlab	−.0869	−.0929	−.0899	**+.0329**	**+.0327**	**+.0328**
	22→22	13↓19	13↓22	**22↑16**	**13↑9**	**13→13**
NTU	−.0692	−.0712	−.0702	**+.0058**	**+.0360**	**+.0209**
	14↓22	15↓21	14↓23	**14↑13**	**15↑14**	**14↑13**
SJTUBCMI	−.1423	−.1206	−.1315	−.0011	**+.0359**	**+.0174**
	17↓20	16↓20	15↓20	**17↓17**	**16↑14**	**15↑13**
III_CYUT_NTHU	−.1618	−.1040	−.1329	**+.1205**	**+.0926**	**+.1066**
	24→24	24→24	24→24	**24↑22**	**24↑19**	**24↑20**

process was repeated for all of the seven teams, and also for $pd = 40$. Thus, we constructed 14 different sets of LOO relevance assessments and evaluated all 24 runs with each of them.

Table 5 summarises our LOO experimental results for $l = 10$. Part (a) of this table shows the results for Mean I-rec, D-nDCG and D♯-nDCG, and the takeaway from this is that the INTENT-1 Chinese DR collection is indeed *not* reusable when these raw (as opposed to condensed-list) metrics are used to evaluate a "new" run. For example, from Table 5(a)(I), we can observe that when the best run from THUIR is evaluated using pd20loo-THUIR, its absolute Mean D♯-nDCG@10 value is smaller than the original one by 0.0867, and more importantly, it is ranked at 10 among the 24 runs even though it is in fact the top performer. That is, the team that has been left out is *heavily underestimated*[8].

---

[8] Note that the rank of III_CYUT_NTHU does not change when its run is evaluated using this team's LOO assessments as the run was ranked at 24 even before leaving out the team.

While Table 5(a)(II) shows that the absolute errors are a little smaller when $pd = 40$ (e.g. the LOO performance in terms of Mean D♯-nDCG@10 for THUIR is smaller than the original performance by 0.0615), the dramatic rank changes do not seem to be alleviated even when $pd = 40$.

Part (b) of Table 5, on the other hand, gives us some hope. It shows the results for the three condensed-list metrics, I-rec′, D-nDCG′ and D♯-nDCG′. For example, when we evaluate the best run from THUIR using pd20loo-THUIR, the run's absolute Mean D♯-nDCG′@10 value is *higher* than the original one by 0.0231, and the run is ranked at 2 when it is in fact the top performer[9]. We say that a condensed-list metric based on a LOO set is *effective in absolute terms* if the absolute difference between the LOO performance and the original performance for the team that has been left out is smaller than the case with the raw metric. For example, in the aforementioned case with THUIR, the absolute error of Mean D♯-nDCG′@10 is 0.0231 while that of Mean D♯-nDCG@10 is 0.0867, so D♯-nDCG′@10 with pd20loo-THUIR is effective in absolute terms. Similarly, we say that a condensed-list metric based on a LOO set is *effective in relative terms* if the absolute rank change of the team that has been left out is smaller than the case with the raw metric. For example, in the aforementioned case with THUIR, the absolute rank change by Mean D♯-nDCG′@10 is $2 - 1 = 1$ while that by Mean D♯-nDCG@10 is $10 - 1 = 9$, so D♯-nDCG′@10 with pd20loo-THUIR is effective in relative terms as well. In Table 5(b), the effective cases are indicated in bold. The results suggest that *condensed-list diversity metrics may be more useful than raw diversity metrics for the purpose of evaluating new systems with an existing diversity test collection.*

The above finding may be true, however, only if the new system to be evaluated is competitive compared to the systems that contributed to the pools. Note that, in Table 5, the condensed-list results for Team III_CYUT_NTHU are not impressive: for example, in Part (II), it can be observed that the absolute error for III_CYUT_NTHU by Mean D♯-nDCG′@10 is as high as 0.1066 (whereas the corresponding error by Mean D♯-nDCG@10 is 0.1329), and that the metric overestimates III_CYUT_NTHU by ranking it at 20 rather than 24. Similar trends can be observed even for the low performers submitted by the top performing team THUIR: Figure 2 visualises the effect of leaving out THUIR with $pd = 20$ for the entire set of 24 runs, which shows that while D♯-nDCG′@10 is more accurate than D♯-nDCG@10 for evaluating the *high performing* runs from THUIR using pd20loo-THUIR, it is no more accurate than D♯-nDCG@10 for evaluating the *low performing* runs from the same team. (Compare the absolute errors of THUIR-D-C-5 and THUIR-D-C-1 with those of THUIR-D-C-3 and THUIR-D-C-4.) Thus, it appears that *condensed-list diversity metrics may overestimate new systems as much as raw diversity metrics underestimate them if the new systems are low performers.* This is probably because low performers contain relevant

---

[9] Recall that, for example, D♯-nDCG′@10 is the same as the raw D♯-nDCG@10 when the original relevance assessments are used, since there will be no unjudged documents involved.

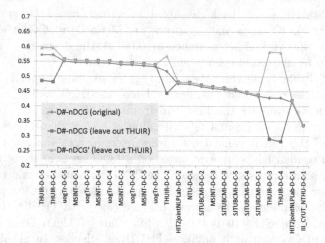

Fig. 2. Original Mean D♯-nDCG ranking vs. Mean D♯-nDCG ranking based on the "leave out THUIR" relevance data ($pd = 20$)

documents at low ranks, and therefore the effect of condensing the list tends to be greater: the relevant documents are "promoted" more dramatically.

Based on the above observation, we have also considered an evaluation method that serves as a compromise between the traditional and the condensed-list methods: the new method computes evaluation metrics based on the condensed-list, but discounts the gain value of each promoted document based on the number of promoted ranks when compared to the original list. The larger the promotion is, the more uncertain we are about the evaluation outcome. However, we have not obtained a promising method that achieves the desired effect and leave it for future work.

5   Conclusions and Future Work

To our knowledge, the present study is the first to have addressed the issue of reusability for diversified search test collections. Although we do not claim that our findings apply to every existing diversity test collection as our study is limited to the case of the NTCIR-9 INTENT-1 Chinese DR test collection, it should be noted that the TREC web diversity test collections were constructed in a similar manner, using a shallow pool depth of either 20 or 25 with a comparable number of participating teams. By conducting additional relevance assessments to achieve a pool depth of 40 for the INTENT-1 collection, we showed that run rankings at the measurement depth of 30 are too unreliable given the pool depth of 20. Thus we recommend that the future INTENT task use the measurement depth of 10 only. Moreover, through leave-one-out experiments for every participating team of the INTENT-1 Chinese task, we showed that condensed-list versions of existing diversity evaluation metrics may be more reliable than the raw metrics for comparing new systems with the contributors of the test collection, However, it appears that condensed-list metrics can be successful only if the new systems to be evaluated are competitive relative to the contributors. These findings should be useful for diversity task organisers as well as researchers who want to reuse existing diversity test collections.

We plan to generalise our findings by examining other diversity test collections from TREC and NTCIR.

References

1. Buckley, C., Dimmick, D., Soboroff, I., Voorhees, E.M.: Bias and the limits of pooling for large collections. Information Retrieval 10(6), 491–508 (2007)
2. Buckley, C., Voorhees, E.M.: Retrieval evaluation with incomplete information. In: Proceedings of ACM SIGIR 2004, pp. 25–32 (2004)
3. Büttcher, S., Clarke, C.L., Yeung, P.C., Soboroff, I.: Reliable information retrieval evaluation with incomplete and biased judgements. In: ACM SIGIR 2007 Proceedings, pp. 63–70 (2007)
4. Carterette, B.: Multiple testing in statistical analysis of systems-based information retrieval experiments. ACM TOIS 30(1) (2012)
5. Carterette, B., Gabrilovich, E., Josifovski, V., Metzler, D.: Measuring the reusability of test collections. In: Proceedings of WSDM 2012, pp. 231–240 (2010)
6. Carterette, B., Kanoulas, E., Pavlu, V., Fang, H.: Reusable test collections through experimental design. In: Proceedings of ACM SIGIR 2010, pp. 547–554 (2010)
7. Carterette, B., Pavlu, V., Kanoulas, E., Aslam, J.A., Allan, J.: Evaluation over thousands of queries. In: Proceedings of ACM SIGIR 2008, pp. 651–658 (2008)
8. Clarke, C.L., Craswell, N., Soboroff, I.: Overview of the TREC 2009 web track. In: Proceedings of TREC 2009 (2009)
9. Clarke, C.L., Craswell, N., Soboroff, I., Voorhees, E.: Overview of the TREC 2011 web track. In: Proceedings of TREC 2011 (2012)
10. Sakai, T.: Alternatives to bpref. In: Proceedings of ACM SIGIR 2007, pp. 71–78 (2007)
11. Sakai, T.: Comparing metrics across TREC and NTCIR: The robustness to pool depth bias. In: Proceedings of ACM SIGIR 2008, pp. 691–692 (2008)
12. Sakai, T.: Comparing metrics across TREC and NTCIR: The robustness to system bias. In: Proceedings of ACM CIKM 2008, pp. 581–590 (2008)
13. Sakai, T., Kando, N.: Are popular documents more likely to be relevant? a dive into the ACLIA IR4QA pools. In: Proceedings of EVIA 2008, pp. 8–9 (2008)
14. Sakai, T., Kando, N.: On information retrieval metrics designed for evaluation with incomplete relevance assessments. Information Retrieval 11, 447–470 (2008)
15. Sakai, T., Song, R.: Diversified search evaluation: Lessons from the NTCIR-9 INTENT task. Information Retrieval (to appear)
16. Sakai, T., Song, R.: Evaluating diversified search results using per-intent graded relevance. In: Proceedings of ACM SIGIR 2011 (2011)
17. Song, R., Zhang, M., Sakai, T., Kato, M.P., Liu, Y., Sugimoto, M., Wang, Q., Orii, N.: Overview of the NTCIR-9 INTENT task. In: Proceedings of NTCIR-9, pp. 82–105 (2011)

18. Voorhees, E.M.: The Philosophy of Information Retrieval Evaluation. In: Peters, C., Braschler, M., Gonzalo, J., Kluck, M. (eds.) CLEF 2001. LNCS, vol. 2406, pp. 355–370. Springer, Heidelberg (2002)
19. Webber, W., Park, L.A.: Score adjustment for correction of pooling bias. In: Proceedings of ACM SIGIR 2009, pp. 444–451 (2009)
20. Yilmaz, E., Aslam, J., Robertson, S.: A new rank correlation coefficient for information retrieval. In: Proceedings of ACM SIGIR 2008, pp. 587–594 (2008)
21. Yilmaz, E., Aslam, J.A.: Estimating average precision with incomplete and imperfect judgments. In: ACM CIKM 2006 Proceedings, pp. 102–111 (2006)
22. Zobel, J.: How reliable are the results of large-scale information retrieval experiments? In: Proceedings of ACM SIGIR 1998, pp. 307–314 (1998)

One Click One Revisited: Enhancing Evaluation Based on Information Units

Tetsuya Sakai[1] and Makoto P. Kato[2]

[1] Microsoft Research Asia, P.R. China
tetsuyasakai@acm.org
[2] Kyoto University, Japan
kato@dl.kuis.kyoto-u.ac.jp

Abstract. This paper extends the evaluation framework of the NTCIR-9 One Click Access Task (1CLICK-1), which required systems to return a single, concise textual output in response to a query in order to satisfy the user immediately after a click on the SEARCH button. Unlike traditional nugget-based summarisation and question answering evaluation methods, S-measure, the official evaluation metric of 1CLICK-1, discounts the value of each information unit based on its position within the textual output. We first show that the discount parameter L of S-measure affects system ranking and discriminative power, and that using multiple values, e.g. $L = 250$ (user has only 30 seconds to view the text) and $L = 500$ (user has one minute), is beneficial. We then complement the recall-like S-measure with a simple, precision-like metric called T-measure as well as a combination of S-measure and T-measure, called $S\sharp$. We show that $S\sharp$ with a heavy emphasis on S-measure imposes an appropriate length penalty to 1CLICK-1 system outputs and yet achieves discriminative power that is comparable to S-measure. These new metrics will be used at NTCIR-10 1CLICK-2.

1 Introduction

The NTCIR-9 One Click Access Task ("1CLICK-1," pronounced *One Click One*) was concluded in December 2011 [17]. In contrast to traditional information retrieval (IR) and web search where systems output a ranked list of items in response to a query, 1CLICK-1 required systems to output one piece of concise text, typically a multi-document summary of several relevant web pages, that fits (say) a mobile phone screen. Participating systems were expected to output important pieces of information first, and to minimise the amount of text the user has to read in order to obtain the desired information. The task was named One Click Access because systems were required to satisfy the user immediately after the user issues a simple query and clicks on the SEARCH button. This task setting fits particularly well to a mobile scenario in which the user has very little time to interact with the system [16].

To go beyond *document* retrieval and design advanced *information* retrieval systems such as 1CLICK systems, the IR community needs to explore evaluation

Y. Hou et al. (Eds.): AIRS 2012, LNCS 7675, pp. 39–51, 2012.

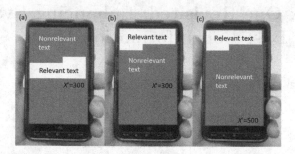

Fig. 1. X-strings: the output of 1CLICK systems

based on *information units* ("iUnits") rather than *document* relevance [1][1]. An iUnit should be an atomic piece of information that stands alone and is useful to the user. At 1CLICK-1, *S-measure* was used to evaluate participating systems based on iUnits: this is a generalisation of the *weighted recall* of iUnits ("W-recall"), but unlike W-recall it takes the *positions* of retrieved iUnits into account.

Figure 1 shows a few conceptual images of texts output by 1CLICK systems, called *X-strings* as 1CLICK-1 systems were required to return a text whose target length is no more than X characters. The X-strings in Figure 1(a) and (b) are both 300-character long ($X' = 300$, where X' is the *actual* length), and they contain exactly the same pieces of information that are relevant to a particular query. However, while the X-string in (a) makes the user read some nonrelevant text before he can get to the relevant text, that in (b) shows the same relevant text first. In this sense, the user can reach the desired information more efficiently with (b) than with (a). While W-recall and traditional "nugget-based" evaluation metrics in summarisation and question answering regard (a) and (b) as equally effective, S-measure rewards (b) more heavily than (a). This *position-sensitive* evaluation can help researchers design effective 1CLICK systems.

S-measure has a parameter called L which represents the user's patience: at 1CLICK-1, where a *Japanese* task was evaluated, L was set to 500 based on the statistic that the average reading speed of a Japanese person is 400-600 characters per minute. Thus $L = 500$ implies that the user has only *one minute* to gather the desired pieces of information. The first objective of the present study is to examine the effect of L on the evaluation outcome of the participating systems at 1CLICK-1. For example, suppose the user only has *thirty seconds* to read the X-string: would the official system rankings change?

The second objective is to complement the official evaluation reported at 1CLICK-1, by proposing a simple extension to the iUnit-based evaluation. Compare Figure 1(b) and (c): the two X-strings contain the same relevant information in the same positions, but the one in (c) contains more nonrelevant text: it may

---

[1] We distinguish the iUnits in the 1CLICK evaluation framework from nuggets used in summarisation and question answering evaluation. As discussed later, the key difference between an iUnit and a traditional nugget is that the former contains *vital string* information [16,17], as we shall explain later.

make the user waste more time. However, as S-measure is a recall-based metric, it cannot differentiate between (b) and (c). Hence we introduce a precision-like metric called *T-measure* and a combination of S-measure and T-measure (or "S" and "T" for short) called $S\sharp$, and demonstrate that they provide new insight into the 1CLICK-1 systems.

2   Related Work

2.1   Evaluating Search

The present study builds on the S-measure framework for evaluating 1CLICK-1 systems [16,17]. There is an analogy between the well-known *normalised Discounted Cumulative Gain* (nDCG) [6] and S: while nDCG evaluates a *ranked list of items* (e.g. URLs) while discounting the value of each item based on their *rank positions*, S evaluates a *textual output* (i.e. the X-string) while discounting the value of each iUnit based on their *offset positions* within the output, to reward systems that satisfy the user quickly. As S assumes that the user's reading speed is constant, its position-based discounting is equivalent to *time-based* discounting, as was recently proposed by Smucker and Clarke [18]. However, the latter concerns web search results, i.e. a ranked list of documents with snippets.

The INEX Snippet Retrieval track[2] evaluates the quality of snippets as a means to judge the relevance of the original documents within the traditional ranked list evaluation framework. $\alpha\text{-}nDCG$, designed primarily for diversified IR evaluation, views both documents and search intents as sets of nuggets [5]. More recently, Pavlu *et al.* [12] have proposed a nugget-based evaluation framework for IR that involves automatic matching between documents and gold-standard nuggets. They are jointly running the NTCIR-10 1CLICK-2 task with Sakai, Kato and Song [16,17] to explore evaluation approaches based on iUnits[3].

2.2   Evaluating Summarisation

ROUGE is a family of metrics for evaluating summaries *automatically* [8]. The key idea is to compare a system output with a set of gold-standard summaries in terms of *recall* (or alternatively *F-measure*), where recall is defined based on automatically extracted textual fragments such as N-grams and longest common subsequences. New automatic summarisation metrics were also explored at the TAC (Text Analysis Conference) AESOP (Automatically Evaluating Summaries of Peers) task[4].

While automatic evaluation methods such as ROUGE are useful for efficient evaluation of summarisers, the S-measure framework builds on the view that automatic string matching between the system output and gold standards is not sufficient for building effective *abstractive* summarisers [16]. Thus, in the

---

[2] https://inex.mmci.uni-saarland.de/tracks/snippet/
[3] http://research.microsoft.com/en-us/people/tesakai/1click2.aspx
[4] http://www.nist.gov/tac/2011/Summarization/

S-measure framework, the identification of iUnits within an X-string is done manually. More importantly, the assessor records the *position* of each iUnit. As we discussed earlier, this enables the S-measure framework to distinguish between systems like Figure 1(a) and (b).

The S-measure framework is similar to the *pyramid method* for summarisation evaluation [11] in that it relies on manual matching. In the pyramid method, *Semantic Content Units* (SCUs) are extracted from multiple gold-standard summaries, and each SCU is weighted according to the number of gold standards it matches with. Finally, SCU-based weighted precision or recall is computed. Just like the automatic methods, however, these methods are *position insensitive*.

2.3   Evaluating Distillation

The DARPA GALE *distillation* program evaluated ranked lists of passages output in response to a query (or a set of queries representing a long-standing information need). Within this framework, Babko-Malaya [3] describes a systematic way to define nuggets in a bottom-up manner from a pool of system output texts. In contrast, the iUnits were defined prior to run submissions at 1CLICK-1 [17].

White, Hunter and Goldstein [20] defined several nugget-based, set retrieval metrics for the distillation task; Allan, Carterette and Lewis proposed a character-based version of *bpref* to evaluate a ranked list of passages [2]. Yang and Lad [21] have also discussed nugget-based evaluation metrics that are similar in spirit to α-nDCG, for multiple queries issued over a period of time and multiple ranked lists of retrieved passages. In Yang and Lad's model, *utility* is defined as *benefit* subtracted by *cost of reading*. Whereas, in the S-measure framework, the cost of reading is used for directly discounting the value of iUnits.

2.4   Evaluating Question Answering

In Question Answering (QA), evaluation approaches similar to those for summarisation exist. *POURPRE*, an automatic evaluation metric for complex QA, is essentially F-measure computed based on unigram matches between the system output and gold-standard nuggets [9]. As in summarisation, the matching between system outputs and gold-standard nuggets can also be done manually. Either way, the main problem with this approach is that *precision* is difficult to define: while we can count the number of gold-standard nuggets present in a system output, we cannot count the number of "incorrect nuggets" in the same output. To overcome this, an *allowance* of 100 characters per nugget match was introduced at the TREC QA track; the NTCIR ACLIA task determined the allowance parameters based on average nugget lengths [10].

S-measure, in contrast, does not require the allowance parameter. While the allowance parameter implies that every nugget requires a fixed amount of space within the system output, the S-measure framework requires a *vital string* for each iUnit, based on the view that different pieces of information require different lengths of text to convey the information to the user (See Section 3.1).

One limitation of S is that it can only evaluate the *content* of the system output, just like all other nugget-based approaches. At 1CLICK-1, *readability* and *trustworthiness* ratings were obtained in parallel with the manual iUnit matches [17], which we will not discuss further in this paper.

3 NTCIR-9 1CLICK-1 Task

3.1 Task and Data

1CLICK-1, the first round of the One Click Access task, was run between March and December 2011. The task used 60 Japanese search queries, 15 for each *question category*: CELEBRITY, LOCAL, DEFINITION and QA. The CELEBRITY and LOCAL queries were selected from a mobile query log; the DEFINITION and QA queries were selected from Yahoo! Chiebukuro (Japanese Yahoo! Answers). The four query types were selected based on a query log study [7]. Two types of runs were allowed: DESKTOP runs ("D-runs") and MOBILE runs ("M-runs"), whose target lengths were $X = 500, 140$, respectively.

For a CELEBRITY query, for example, participating systems were expected to return important biography information. They were expected to return important iUnits first, and to minimise the amount of text the user has to read. For example, the iUnits for Query "Osamu Tezuka" (a famous Japanese cartoonist who died in 1989) represented his date of birth, place of birth, his occupation, the comic books he published and so on. The iUnit that represented his date of birth contained a *vital string* "1928.11.03" because this string (or something equivalent) is probably *required* in order to convey to the user that "Osamu Tezuka was born in November 3, 1928." The length of the vital string is used for defining an "optimal" output and for computing S. Moreover, at 1CLICK-1, each iUnit was weighted based on votes from five assessors.

Only three teams participated in the task, but ten runs based on diverse approaches were submitted to it: Teams KUIDL, MSRA and TTOKU took information extraction, passage retrieval and multi-document summarisation approaches, respectively[5]. Both organisers and participants took part in manual iUnit matching, using a dedicated interface which can record match positions. Every X-string was evaluated by two assessors: in this study, we evaluate runs based on the *Intersection* data (**I**) and the *Union* data (**U**) of the iUnit matches [17]. The 60 queries and the official evaluation results are publicly available[6], and the iUnit data can be obtained from National Institute of Informatics, Japan[7].

For more details on 1CLICK-1, the reader is referred to the Overview paper [17].

---

[5] Note that a limited number of participants is not necessarily a weakness, as the iUnits were extracted *a priori*: the 1CLICK-1 test collection construction did not rely on pooling submitted runs [17].

[6] http://research.nii.ac.jp/ntcir/workshop/OnlineProceedings9/NTCIR/
Evaluations/INTENT/ntc9-1CLICK-eval.htm.

[7] http://research.nii.ac.jp/ntcir/data/data-en.html

3.2   S-Measure and $S\flat$

S-measure was the primary evaluation metric used at 1CLICK-1. Let N be the set of gold-standard iUnits constructed for a particular query, and let $v(n)$ be the vital string and let $w(n)$ be the weight for iUnit $n \in N$. The *Pseudo Minimal Output* (PMO) for this query is defined by sorting all vital strings by $w(n)$ (first key) and $|v(n)|$ (second key) [16]. Thus, the basic assumptions are that (a) important iUnits should be presented first; and (b) if two iUnits are equally important, then the one that can "save more space" should be presented first. The crude assumptions obviously may conflict with text readability, but have proven to be useful [16,17]. Let $offset^*(v(n))$ denote the offset position of $v(n)$ within the PMO. Let $M(\subseteq N)$ denote the set of *matched* iUnits obtained by manually comparing the X-string with the gold standard iUnits, and let $offset(m)$ denote the offset position of $m \in M$. Morever, let L be a parameter that represents how the user's patience runs out: the original paper that proposed S used $L = 1,000$, while 1CLICK-1 used $L = 500$. The former means that the user has about two minutes to examine the X-string, while the latter means that he only has one minute. S is defined as:

$$S\text{-}measure = \frac{\sum_{m \in M} w(m) \max(0,\ L - offset(m))}{\sum_{n \in N} w(n) \max(0,\ L - offset^*(v(n)))}. \tag{1}$$

Thus, all iUnits that appear *after* L characters within the X-string are considered worthless. When L is set to a very large value, S reduces to *weighted recall* (W-recall), which is position-insensitive. Also, as there is no theoretical guarantee that S lies below one, *S-flat* given by $S\flat = \min(1, S\text{-}measure)$ may be used instead. In practice, the raw S values were below one for all of the submitted 1CLICK-1 runs and the "flattening" was unnecessary [17].

4   Research Questions and Proposals

4.1   Effect of the Patience Parameter

The official 1CLICK-1 evaluation used $L = 500$ (one minute) with S. In the present study, we vary this parameter as follows and examine the outcome: $L = 1,000$ (two minutes, the original setting from Sakai, Kato and Song [16]), $L = 250$ (30 seconds) and $L = 50$ (6 seconds). Note that if L is set to an extremely small value, most of the contents of the X-strings will be ignored. This is analogous to truncating ranked lists of documents prior to IR evaluation.

4.2   Evaluating Terseness: T-Measure, $T\flat$ and $S\sharp$

As was discussed earlier, S cannot distinguish between Figure 1(b) and (c). We therefore introduce a precision-like "Terseness" metric for evaluating an X-string of size X':

$$T\text{-}measure = \frac{\sum_{m \in M} |v(m)|}{|X'|}. \tag{2}$$

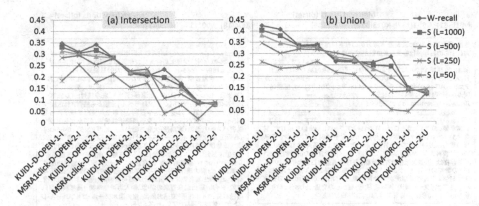

Fig. 2. Effect of the patience parameter L on the system ranking. The x-axis shows runs sorted by Mean S-measure ($L = 500$), i.e. the official ranking.

Note that the numerator is a sum of vital string lengths, and that these lengths vary, unlike traditional nugget precision. As T might exceed one, we also define T-*flat* given by $T\flat = \min(1, T\text{-}measure)$, although in reality T never exceeded one for our data and therefore $T\flat = T$ holds. Finally, following the approach of the well-known F-measure, we can define S-*sharp* as:

$$S\sharp = \frac{(1 + \beta^2)T\flat S\flat}{\beta^2 T\flat + S\flat} \qquad (3)$$

where letting $\beta = 1$ reduces $S\sharp$ to a harmonic mean of $S\flat$ and $T\flat$. However, as we regard S as the primary metric and want T to "enter into the calculation only as a length penalty" [9], we also examined $\beta = 3, 5, 10, 20$. While $\beta = 3, 5$ reflect the practices in QA evaluation [9,10], our experiments suggest that an even higher β may be suitable for 1CLICK, as we shall see later.

To sum up, $S\sharp$ differs from the traditional nugget-based F-measure in the following two aspects: (1) It utilises the positions of iUnits for computing the recall-like S; and (2) Instead of relying on a fixed allowance parameter, it utilises the vital string length of each iUnit for computing the precision-like T-measure.

5 Experiments

5.1 Results on the Patience Parameter

Figure 2(a) and (b) show the effect of L on the overall system ranking with Mean S with **I** and with **U**, respectively. The x-axis shows the runs sorted by Mean S ($L = 500$), i.e. the official ranking. With **I**, Kendall's τ with the official ranking are .87 (Mean W-recall), .96 ($L = 1,000$), .78 ($L = 250$) and .64 ($L = 50$); with **U**, the corresponding values are .82 (Mean W-recall), .96 ($L = 1,000$), .73 ($L = 250$) and .69 ($L = 50$). Thus, $L = 1,000$ (two minutes [16]) produces rankings that are very similar to $L = 500$ (one minute), but $L = 250$

KUIDL-D-OPEN-1 (X'=598) **Matched iUnits (offset) =**
N016(81), N001(232), N013(312), N014(326)

住所:伊賀市霧生2356。交通:大阪方面のお客様は伊賀神戸駅、名古屋方面からのお客様は榊原温泉口駅が便利です。近鉄大阪線・伊賀神戸駅、榊原温泉口駅より、予約制定期送迎バスを運行しています。営業時間:10:00 繰翌18:00。駐車場:あり(10台)、代表電話:0183737642。定休日:火曜。チェックインCheck In/Out:ホテルシャンペール チェックIN 15:00 チェックOUT 10:00 雅楽司 チェックIN 14:00 チェックOUT 11:00。開館時間:9:00繰翌17:00。問い合わせ先TEL:0595-54-1326。メール:aoyama@menard.co.jp。交通:東名阪自動車道から伊勢自動車道を経由し、久居IC下車。国道165号線を西進し、道路沿いにある看板の指示に従いお進みください:。久居ICより約30kmで到着します:。交通:地下鉄南北線、麻生駅4番出口から徒歩1分とアクセスしやすい立地です。営業時間:10:00 繰翌21:00最終受付19:00完全予約制 ※エステ中で留守電の場合はお手数ですが伝言をお願い致します。交通:学園都市線「あいの里教育大駅」から徒歩4分。ご宿泊、もしくはお食事の御予約を頂いたお客様は、近鉄大阪線伊賀神戸駅、及び榊原温泉口駅より当リゾート間の無料の送迎バスがございます。定休日:日曜、祝日 第3土曜日。年中無休。交通:乗車場所:伊賀神戸駅乗車時間:8:20繰翌16:20(事前連絡)。通年ナイター設備:無/無料送迎:無/宿からの交通:徒歩5分。駐車場 Parking:100台(無料)。駐車場:なし。定休日:月曜。

MSRA1click-D-OPEN-2 (X'=632) **Matched iUnits (offset)=**
N004(18), N001(33), N003(48)

〒 518-0295伊賀市霧生2356tel0595-54-1326fax0595-54-1359e-mailaoyama@menard.co.jpurl近鉄上野市駅前tel 0595-24-0270fax 0595-24-0270(社)伊賀上野観光協会伊賀市上野丸之内122-4 だんじり会館内tel 0595-26-7788fax 0595-26-7799最終更新日：2011.5.12客室設備・備品客室設備テレビ、電話、湯沸かしポット、お茶セット、冷蔵庫、ドライヤー、ズボンプレッサー(貸出)、電気スタンド(貸出)、cdプレイヤー(貸出)、加湿器(貸出)、洗浄機付トイレ旅館・ホテルの宿泊予約サイト【ぐるなびトラベル】全国から厳選された旅館・ホテルの宿泊プランをネットや電話で簡単予約貸出車椅子障害者用トイレ住所伊賀市霧生2356交通機関近鉄伊賀神戸駅から30分送迎バスあり(予約・定期便)大阪方面から 松原インターから90分(名阪国道・上野東インターから40分)※収集中[アクセス] ●私鉄近鉄大阪線伊賀神戸駅→タクシー約30分map[住所] 三重県伊賀市霧生2356お気に入りに追加携帯に送る名古屋方面から 名古屋西インターから90分(伊勢自動車道・久居インターから50分)営業時間く休業日など・雅の湯(女湯)入浴可能時間：(通年利用可)眺望：山/浴槽材質：岩・霧生温泉香楽の湯(男湯)入浴可能時間：(通年利用可)浴槽材質：タイル・霧生温泉香楽の湯(女湯)入浴可能時間：(通年利用可)浴槽材質：タイル住所・交通お風呂・室内温水プール・ゴルフコース・レストラン・体験工房などアミューズメント

Fig. 3. X-strings of runs from KUIDL and MSRA1click for the LOCAL query "*Menard Aoyama Resort*" (name of a facility)

(30 seconds) results in substantially different system rankings. In particular, Figure 2(a) shows that while Mean S with $L = 500$ prefers KUIDL-D-OPEN-1 over MSRA1click-D-OPEN-2 and prefers KUIDL-D-OPEN-2 over MSRA1click-D-OPEN-1, Mean S with $L = 250$ has exactly the opposite preferences. This trend is further emphasized by Mean S with $L = 50$.

Recall that S with $L = 250$ *ignores* all iUnit matches between positions 250 and 500 for all of the D-runs. Thus, the above discrepancy between $L = 500$ and $L = 250$ regarding KUIDL and MSRA1click suggests that *while* KUIDL *is good at covering important iUnits,* MSRA1click *is good at presenting the most important units near the beginning of the X-string.* To illustrate this point, Figure 3 shows the actual X-strings of KUIDL and MSRA1click for a LOCAL query "*Menard Aoyama Resort*" (name of a facility). It can be observed that even though KUIDL is superior to MSRA1click in terms of the number of matches with **I** (4 matches vs. 3), MSRA1click is actually very good from the viewpoint of iUnit *positions* as indicated by the underlined texts that correspond to the iUnit matches. With **I**, the S with $L = 500$ for KUIDL is 0.200, and that for MSRA1click is 0.332; whereas, the S with $L = 250$ for KUIDL is 0.120, and that for MSRA1click is 0.528. Thus the difference between two systems is magnified when $L = 250$.

Next, we examine the effect of L on *discriminative power*. Given a test collection with a set of runs, discriminative power is measured by conducting a statistical significance test for every pair of runs [14]. This methodology has been used in a number of evaluation studies [5,13,15,18,19], and is arguably one necessary (but by no means sufficient) condition of a "good" metric. We used a randomised version of two-sided Tukey's Honestly Significant Differences (HSD) test for testing statistical significance, which is known to be more reliable than traditional *pairwise* significance tests [4,15].

Figure 4 shows the *Achieved Siginificance Level (ASL) curves* [14] of S with varying L. Here, the y-axis represents the ASL (i.e. p-value), and the x-axis

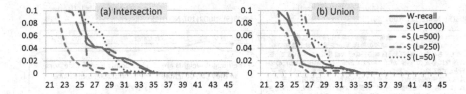

Fig. 4. Effect of the patience parameter L on discriminative power. The y-axis represents the p-value and the x-axis represents run pairs sorted by the p-value.

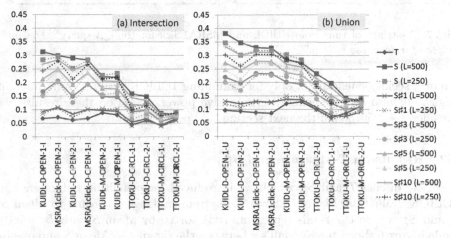

Fig. 5. System ranking by different metrics. The x-axis shows runs sorted by Mean S-measure with $L = 500$, i.e., the official ranking.

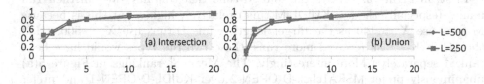

Fig. 6. Effect of β on $S\sharp\beta$: the x-axis represents β and the y-axis represents Kendall's τ with the Mean S-measure ranking

represents the 45 run pairs sorted by the p-value. Metrics that are closer to the origin are the ones that are highly discriminative, i.e. those that provide reliable experimental results. It can be observed that the discriminative power for $L = 250$ is the highest while that for $L = 50$ is low (naturally, as the latter implies looking at only the first 50 characters of every X-string). Moreover, S with $L = 250$ is more discriminative than W-recall. These observations are consistent across **I** and **U**. Thus, at least for the runs submitted to 1CLICK-1, using $L = 250$ (user has 30 seconds) along with the official $L = 500$ (user has one

KUIDL-D-OPEN-1 (X'=441) **Matched iUnits (offset)=**
N003(5), N002(11), N001(30)

「納税の義務・勤労の義務・教育の。納税の義務、勤労の義務、教育の義務』だ。*1政治勤労の義務、納税の義務、教育を受けさせる義務である。そもそも小職が解釈した『日本国民の三大義務』とは、、あくまでも国家の運営に必要な"納税"がゴールとして設定され、納税させるための手段としての"勤労"、・"勤労"の機会を得るための"教育"、といった三段論法で義務が課されて。そもそも小職が解釈した『日本国民の三大義務』とは、、あくまでも国家の運営に必要な"納税"がゴールとして設定され、納税させるための手段としての"勤労"、・"勤労"の機会を得るための"教育"、といった三段論法で義務が課されて。2010年7月15日。2006年2月20日。2008年3月30日。2010年4月23日、2010年3月17日。2011年4月3日。2011年1月30日。|グルメ・旅行の口コミから育児・恋愛等の相談に至るまでの、あらゆる疑問や悩みを質問・相談として投稿し、知識・経験を持った方から回答を得て解決する無料。Q。義務合はやらなければいけない事だが、権利と表裏一体でである。[仕事・キャリア。

MSRA1click-D-OPEN-2 (X'=15) **Matched iUnits (offset)=**
N003(5), N002(10), N001(15)

納税の義務、勤労の義務、教育の義務。

The nugget weights are all 15 (3 points from 5 assessors) so they can be ignored when computing S-measure.
S-measure (L=500) =
((500-5)+(500-10)+(500-15))/((500-2)+(500-4)+(500-6))
=0.988
Vital strings of N003,N002,N001:
納税 (length=2), 勤労 (length=2), 教育(length=2)
T-measure=(2+2+2)/15=0.400

Fig. 7. X-strings of runs from KUIDL and MSRA1click for the QA query "*The three duties of a Japanese citizen.*"

minute) seems beneficial not only for examining 1CLICK systems from different angles but also for enhancing discriminative power. Based on these results, we consider $L = 250, 500$ in the next section.

5.2   Results on T-Measure and $S\sharp$

Next, we discuss T and $S\sharp$, which we introduced for penalising redundancy in 1CLICK evaluation. Figure 5 shows the system rankings according to Mean S, T and $S\sharp$ (where the x-axis represents runs sorted by Mean S with $L = 500$), while Figure 6 shows the Kendall's τ between the ranking by Mean S and one by Mean $S\sharp$ with β (denoted by $S\sharp\beta$). Note that β means "S is β times as important as T" and that $S\sharp 0 = T$ (See Eq. 3).

First, in Figure 5, T rates the four M-runs that contain "-M-" in their run names (especially the two KUIDL-M runs) relatively highly, but this is because M-runs use $X = 140$ as the target length while D-runs use $X = 500$. (Had the 1CLICK-1 task received more runs, these two run types would have been ranked separately.) More interestingly, The Mean $S\sharp$ rankings in Figure 5(a) unanimously prefer MSRA1click-D-OPEN-2 over KUIDL-D-OPEN-1 and prefer MSRA1click-D-OPEN-1 over KUIDL-D-OPEN-2, contrary to the official Mean S ranking. This suggests that MSRA1click was actually better than KUIDL from the viewpoint of terseness. To illustrate this point, Figure 7 shows the X-strings for the QA query "*The three duties of a Japanese citizen*": both KUIDL and MSRA1click managed to capture the three answers and their S values are 0.977 and 0.988, respectively (note that the former underperforms the latter even in terms of S, due to one ill-placed iUnit); whereas, the T values are 0.014 and 0.400, respectively. Thus, T reflects the fact that the X-string of KUIDL is highly redundant while that of MSRA1click is almost perfect. (The figure shows how to compute S and T for the X-string of MSRA1click.) It can be observed that T and $S\sharp$ are useful complements to S for evaluating 1CLICK systems.

Figure 8 shows the ASL curves for our proposed metrics. From the viewpoint of discriminative power, it can be observed that T is very poor, and therefore

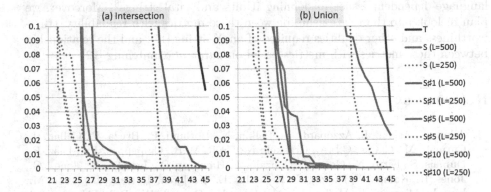

Fig. 8. Discriminative power of S-measure, T-measure and $S\sharp$

that it is safer to set β to a high value when using $S\sharp$. To be more specific, it can be observed that the discriminative power of $S\sharp 10$ is comparable to that of S for both $L = 250$ (shown as dotted lines) and $L = 500$ (shown as solid lines). Since $S\sharp 10$ retains the high discriminative power of S *and* provide new insight to the evaluation as shown in Figures 5 and 6, we recommend $S\sharp 10$ for evaluating 1CLICK systems, along with the original S.

6 Conclusions and Future Work

This paper extended the 1CLICK-1 evaluation framework, where systems were required to return a single, concise textual output in response to a query in order to satisfy the user immediately after a click on the SEARCH button. We first showed that the discount parameter L of S-measure affects system ranking and discriminative power, and that using multiple values, e.g. $L = 250$ (user has only 30 seconds to view the text) and $L = 500$ (user has one minute), is useful: a 1CLICK system which can satisfy the user's information need within one minute may be different from one which can satisfy the need within 30 seconds. Also, S with $L = 250$ appears to be more discriminative than S with $L = 500$ and W-recall, at least for the runs submitted to the 1CLICK-1 task. We then complemented the recall-like S with a simple, precision-like metric called T-measure as well as a combination of S and T, called $S\sharp$. We showed that $S\sharp$ with a heavy emphasis on S (e.g. $S\sharp 10$) imposes an appropriate length penalty to 1CLICK-1 system outputs and yet achieves discriminative power that is comparable to S.

At the NTCIR-10 1CLICK-2 Task, to which over ten teams have registered to participate, T and $S\sharp$ will be used along with S as official metrics. Moreover, at 1CLICK-2, the language scope has been extended to English and Japanese. While the evaluation framework of S, T and $S\sharp$ should apply to any language, it would be interesting to test it in the English subtask as well. There may be language-dependent issues in defining iUnits and vital strings[8]. Moreover, we plan to look into the relationship between these metrics with readability, trustworthiness and other qualities required of an X-string [17], and the relationship between these metrics with metrics based on *automatic* matching [12].

References

1. Allan, J., Aslam, J., Azzopardi, L., Belkin, N., Borlund, P., Bruza, P., Callan, J., Carman, M., Clarke, C.L.A., Craswell, N., Croft, W.B., Culpepper, J.S., Diaz, F., Dumais, S., Ferro, N., Geva, S., Gonzalo, J., Hawking, D., Jarvelin, K., Jones, G., Jones, R., Kamps, J., Kando, N., Kanoulas, E., Karlgren, J., Kelly, D., Lease, M., Lin, J., Mizzaro, S., Moffat, A., Murdock, V., Oard, D.W., Rijke, M.d., Sakai, T., Sanderson, M., Scholer, F., Si, L., Thom, J.A., Thomas, P., Trotman, A., Turpin, A., de Vries, A.P., Webber, W., Zhang, X., Zhang, a.Y.: Frontiers, challenges and opportunities for information retrieval: Report from SWIRL 2012. SIGIR Forum 46(1), 2–32 (2012)
2. Allan, J., Carterette, B., Lewis, J.: When will information retrieval be "good enough"? In: Proceedings of ACM SIGIR 2005, pp. 433–440 (2005)
3. Babko-Malaya, O.: Annotation of nuggets and relevance in gale distillation evaluation. In: Proceedings of LREC 2008, pp. 3578–3584 (2008)
4. Carterette, B.: Multiple testing in statistical analysis of systems-based information retrieval experiments. ACM TOIS 30(1) (2012)
5. Clarke, C.L., Craswell, N., Soboroff, I., Ashkan, A.: A comparative analysis of cascade measures for novelty and diversity. In: Proceedings of ACM WSDM 2011 (2011)
6. Järvelin, K., Kekäläinen, J.: Cumulated gain-based evaluation of IR techniques. ACM Transactions on Information Systems 20(4), 422–446 (2002)
7. Li, J., Huffman, S., Tokuda, A.: Good abandonment in mobile and PC internet search. In: Proceedings of ACM SIGIR 2009, pp. 43–50 (2009)
8. Lin, C.Y.: ROUGE: A package for automatic evaluation of summaries. In: Proceedings of the ACL 2004 Workshop on Text Summarization Branches Out (2004)
9. Lin, J., Demner-Fushman, D.: Methods for automatically evaluating answers to complex questions. Information Retrieval 9(5), 565–587 (2006)
10. Mitamura, T., Shima, H., Sakai, T., Kando, N., Mori, T., Takeda, K., Lin, C.Y., Song, R., Lin, C.J., Lee, C.W.: Overview of the NTCIR-8 ACLIA tasks: Advanced cross-lingual information access. In: Proceedings of NTCIR-8, pp. 15–24 (2010)
11. Nenkova, A., Passonneau, R., McKeown, K.: The pyramid method: Incorporating human content selection variation in summarization evaluation. ACM Transactions on Speech and Language Processing 4(2), Article 4 (2007)
12. Pavlu, V., Rajput, S., Golbus, P.B., Aslam, J.A.: IR system evaluation using nugget-based test collections. In: Proceedings of ACM WSDM 2012, pp. 393–402 (2012)
13. Robertson, S.E., Kanoulas, E., Yilmaz, E.: Extending average precision to graded relevance judgments. In: Proceedings of ACM SIGIR 2010, pp. 603–610 (2010)

---

[8] The new definitions of iUnits and vital strings for the Japanese 1CLICK-2 subtask can be found at http://www.dl.kuis.kyoto-u.ac.jp/ kato/1click2/data/ 1C2-J-SAMPLE-README.pdf.

14. Sakai, T.: Evaluating evaluation metrics based on the bootstrap. In: Proceedings of ACM SIGIR 2006, pp. 525–532 (2006)
15. Sakai, T.: Evaluation with informational and navigational intents. In: Proceedings of WWW 2012, pp. 499–508 (2012)
16. Sakai, T., Kato, M.P., Song, Y.I.: Click the search button and be happy: Evaluating direct and immediate information access. In: Proceedings of ACM CIKM 2011, pp. 621–630 (2011)
17. Sakai, T., Kato, M.P., Song, Y.I.: Overview of NTCIR-9 1CLICK. In: Proceedings of NTCIR-9, pp. 180–201 (2011)
18. Smucker, M.D., Clarke, C.L.A.: Time-based calibration of effectiveness measures. In: Proceedings of ACM SIGIR 2012, pp. 95–104 (2012)
19. Webber, W., Moffat, A., Zobel, J.: The effect of pooling and evaluation depth on metric stability. In: Proceedings of EVIA 2010, pp. 7–15 (2010)
20. White, J.V., Hunter, D., Goldstein, J.D.: Statistical evaluation of information distillation systems. In: Proceedings of LREC 2008, pp. 3598–3604 (2008)
21. Yang, Y., Lad, A.: Modeling Expected Utility of Multi-session Information Distillation. In: Azzopardi, L., Kazai, G., Robertson, S., Rüger, S., Shokouhi, M., Song, D., Yilmaz, E. (eds.) ICTIR 2009. LNCS, vol. 5766, pp. 164–175. Springer, Heidelberg (2009)

A Comparison of Action Transitions in Individual and Collaborative Exploratory Web Search

Zhen Yue, Shuguang Han, and Daqing He

School of Information Sciences, University of Pittsburgh
135 North Bellefield Ave. Pittsburgh 15213 United States
{zhy18,shh69,dah44}@pitt.edu

Abstract. Collaboration in Web search can be characterized as implicit or explicit in terms of intent, and synchronous or asynchronous in terms of concurrency. Different collaboration style may greatly affect search actions. This paper presents a user study aiming to compare search processes in three different conditions: pair of users working on the same Web search tasks synchronously with explicit communication, pair of users working on the same Web search tasks asynchronously without explicit communication and single users work separately. Our analysis of search processes focused on the transition of user search actions logged in our exploratory Web search system called Collab-Search. The results show that the participants exhibited different patterns of search actions under different conditions. We also found that explicit communication is one of the possible sources for users to obtain ideas of queries, and the explicit communication between users also promotes their implicit communication. Finally this study provides some guidance on the range of behaviors and activities that a collaborative search system should support.

Keywords: Collaborative web search, search process, action transition, exploratory search, collaborative information behavior.

1 Introduction

Traditionally, information seeking has been studied as an individual activity. It is now recognized that people do sometime act in a group context when trying to solve information seeking problems (Hansen & Järvelin, 2005). Especially when the search task is exploratory, it may be in the searchers' best interests to collaboratively explore the information space and participate in shared learning (White & Roth, 2009). A collaborative search system need not only support the interaction between a user and the system, but also support the interaction among users. A successful collaborative search system relies on good understanding of the group activities involved in the search process. Studies which seek to describe collaborative search process can help developers understand the range of behaviors and activities that systems need to accommodate. Therefore, understanding the various manifestations of collaborative information behavior involved in search process is crucial for designing and evaluating systems supporting collaborative information seeking.

Y. Hou et al. (Eds.): AIRS 2012, LNCS 7675, pp. 52–63, 2012.

Collaborative information seeking has been studied in various environments including both organizational and Web setting. Our study focuses on collaborative exploratory search in the Web search environment. Golovchinsky et al. (2008) classified the collaboration in Web search using three dimensions – location, concurrency and intent. In terms of location, collaborative web search can be co-located or remotely located. We are interested in remotely located collaborative search, for which a collaborative search system is needed in order to support both collaboration and search. Collaboration in web search can also be synchronous or asynchronous in terms of concurrency. In synchronous collaboration, team members can obtain instant feedback from each other, whereas only those who search later in asynchronous collaboration can benefit from the work of earlier team members. We think that in synchronous collaboration, it is easier for collaborators to explicit communicate with each other such as verbal or text chatting. However, in asynchronous collaboration, the formats of communication are more likely to be implicit such as sharing a document or search history. Golovchinsky's classification of implicit or explicit collaboration in terms of the search intent is somewhat different from our classification of explicit and implicit communication. In his definition, implicit collaboration occurs in collaborative recommendation and filtering systems where users have no idea who the collaborators are and where the information comes from. Explicit collaboration occurs on smaller scales such as in groups of several collaborators. In this study, we are interested in collaboration in a small group (two-member team). Particularly, we want to look into the differences between collaboration with explicit communication and collaboration without explicit communication when team users conducting exploratory search tasks. To serve as a baseline to test the effect of collaboration, individual search should also be included for comparison.

2   Related Works

In this section, we discuss literature related to both individual and collaborative information search processes.

In individual information seeking, researchers explored many methods to investigate single user's search process. There are several well-known search process models in individual information seeking, such as Kuhlthau's (1991) and Ellis' (1993) model. The similarity of these two models is that they are both linear models that present holistic views of information seeking from the initiation stage to the ending stage. Marchionini (1995) proposes an information-seeking process model of eight stages with possible transitions between each of them. Similar to Kuhlthau and Ellis, the information seeking begins with the recognition and acceptance of the problem and continues until the problem is solved. However, this model highlights the likelihood of a stage calling another stage in three types: most likely transitions, high-probability transitions and low-transition probabilities.

In the Web environment, there are also studies examining the transition of search actions. Holscher and Strube (2000) compared action sequences between Internet experts and newbies. Xie and Joo (2010) investigate transitions of search tactics in the

Web search process. They present the most common patterns of search tactic transition at the beginning, middle and ending phases in the search process. When Marchionini (2006) propose the notion of exploratory search, he lists a set of search activities associated with exploratory search process. Later, researchers (White and Roth, 2009) model exploratory search process as iterations between exploratory browsing activities and focused searching activities.

Collaborative information behavior is a relatively new research area compared to individual information behavior research. Researchers conducted studies on collaborative information behavior across a wide variety of domains, including academic (Blake & Pratt, 2002; Talja & Hansen, 2006), industry (Hansen & Jarvelin, 2005) and medicine (Reddy and Spence, 2008). These studies showed that the collaborative information seeking behavior is as common and natural as individual information behavior.

Collaborative information seeking has also been studied in the Web environment. Morris (2008) conducted a survey among 204 information workers about when they used Web search tools collaboratively and on what tasks they usually collaborate with others. In terms of collaborative search process, Evans and Chi (2008) conducted a survey among 150 people using Mechanical Turk to investigate collaborative search strategies involved in before search, during search and after search stage. Halvey et al. (2009) investigated frequency and temporal distribution of user interactions by analyzing log data in an asynchronous collaborative search system for online video search. There are several studies attempt to explore Kuhlthau's ISP model in collaborative setting. Hyldegard (2006) explored ISP model in a group educational setting based on a qualitative preliminary case study. She found that collaborative search process cannot be modeled the same way as individual search process. She suggests that the ISP model should be extended to incorporate the impact of social and contextual factors in relation to collaborative information seeking process. In a follow-up study, Hyldegard (2009) further investigated the group based problem solving process in academic setting and concluded that the ISP model does not fully comply with the collaborative information seeking behaviors. Shah and Gonzalez-Ibanez (2010) also attempted to map Kuhlthau's ISP model to collaborative information seeking. Through a laboratory study with 42 pairs of participants, they investigated similarities and disparities between individual and collaborative information seeking process. Similar to Hyldgard, they also declared that social elements are missing when applying the ISP model in a collaborative setting.

Based on the above literature review, we can see that there are plenty of investigations on search processes in individual user setting. Models of individual search process are well-established. Particularly, search processes have been examined in the Web environment and exploratory search in terms of transitions or sequences of search actions. However, current investigations on collaborative search processes are limited to exploring the application of individual search process model in collaborative setting and particularly focused on Kuhlthau's ISP model. In these studies, only group users are included in the user studies. There is lacking of comparisons of search processes between individual and collaborative search conditions.

3   Experiment Design

Our study was designed as a set of control experiments with human participants. All the experiments were conducted using CollabSearch, a collaborative search system developed by the authors.

3.1   CollabSearch: A Collaborative Search System

CollabSearch[1] is a web search system for either a single user or a group of users. As an integrated collaborative search system, CollabSearch has both search and collaboration features, which can support both explicit and implicit communication among team users. In terms of explicit communication, users can use the chat box on the left side (see Figure 1) to send instant messages to each other. For implicit communication, they can use the shared workspace and shared search history.

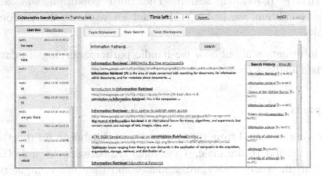

Fig. 1. The web search frame

CollabSearch's interface contains three frames: topic statement, search and team workspace. The topic statement frame shows the task description on which the user is currently working. Team members can also post their comments below the task description. The search frame connects the user's query to Google, and it displays the Google search results. Users can also see their search histories as well as those of their teammates. Users examine search results in the returned list for relevant information, and can save a whole web page or a snippet of the page. All the saved web pages and snippets, collected by the user and the teammates, are stored in the team workspace frame. A notice is displayed at the top when new items are saved to the team workspace. Users can click to view more details of an item in the workspace, or leave comments for an item.

3.2   Experiment Conditions

We identified three experiment conditions. They are:

Condition 1 (CEC): collaborative search with explicit communication. In this condition, two participants P1 and P2 work as a team on the same task at the same time.

---

[1] http://crystal.exp.sis.pitt.edu:8080/CollaborativeSearch/

They can communicate with each other by sending instant text messages, or reading each other's search histories and the collected results shared in team workspace. As we were trying to simulate remotely-located collaboration, P1 and P2 were in the same room but different cubicles, and they couldn't see each other nor talk to each other.

Condition 2 (COC): collaborative search without explicit communication. In this condition, we have a new participant P3 to continue the work of P1. P1 here is the same person as P1 in CEC. When P3 worked on the exploratory search tasks, P1 had already finished the tasks. P3 can see P1's search history and read P1's saved results in the team workspace. Similar design for collaborative search was used in (Paul & Morris, 2009), in which they made the saved results of both P1 and P2 available to P3. However, we removed P2's results because we want to make sure the COC only involved two persons' efforts in order to be fair in the comparison with ECE.

Condition 3 (IND): Individual search. The third condition was devised as a baseline. In this condition, we had a participant P4 work on the exploratory search tasks individually.

More detailed information about how we handle participants' logs in data analysis is introduced in data analysis method.

3.3   Participants

28 participants were recruited from the University of Pittsburgh for this study. Among them, 12 are female and 16 are male. Their ages are between 18 and 60. Despite the age difference, all the participants are students and they use computers on a daily basis. 17 participants are graduate students whereas the other 11 are undergraduates. In terms of operating systems, 7 of them are Mac users and the other 21 are Windows users. According to a question asking them to rate their search experiences from 1-7 with 1 as the least experienced and 7 as the most experienced, the response range from 4-7, thus most of our participants are experienced searchers.

14 of the 28 participants worked under the condition CEC. Since explicit communication will be employed in this condition, we believe that there should be familiarity between the team members. Therefore, we required participants CEC condition to be pairs who know each other. These 7 pairs of participants worked as 7 teams in CEC. They took the role as either P1 or P2. The rest 14 participants were randomly assigned to either COC or IND. For the 7 participants assigned to the COC condition, they took the role of P3. For the other 7 participants assigned to the IND condition, they took the role as P4.

3.4   Search Tasks

Two exploratory web search tasks were used in this study. Both of them had been used in other collaborative web search studies (Paul, 2010; Shah & Marchionini, 2010), so their validity for collaborative search has been examined before. One task

(Shah and Marchionini, 2010) is related to academic work, which asks participants to collecting information for a report on the effect of social networking service and software. The other task (Paul, 2010) is about leisure asking participants to collect information for planning a trip to Helsinki. Morris' (Morris, 2008) identified that travel planning and academic literature search are two common collaborative search tasks. Therefore, both of these two tasks are representative in studying collaborative web search. The task description states the kind of information that the participants need to collect and the goal is to collect as many relevant snippets as possible.

3.5   Experiment Procedure

The experiment procedure was: experiments for CEC conditions are conducted first. 14 participants (7 teams) worked synchronously on the two tasks. During the experiment, after being introduced to the study and the system, and filling out an entry questionnaire to establish their search background, these participants worked on a training task to get familiar with the system for 10 minutes, then worked on task 1 or task 2, depending on the order assigned for each team. They had 30 minutes for each task. At the end of each task, they also worked on a post-search questionnaire for collecting information about their satisfaction levels. Before the end of the experiments, the participants were asked several open-ended questions for their experience with both tasks. The COC and IND experiments were conducted after the CEC experiments. The rest 14 participants were randomly assigned into COC or IND respectively. The experiment procedure in COC and IND is identical to the CEC condition.

4   Data Analysis Methods

4.1   Categorizing User Search Actions

In terms of search process analysis, we are interested in what kind of actions the participants have taken during the whole process of exploratory Web search. Typical types of search actions recognized in this study include Query, View, Collect, Workspace, Topic and Chat, whose details are listed in Table 1. All these search actions were categorized and mapped from the transaction logs recorded in CollabSearch system. Prior to the analysis, we cleaned the transaction log data to remove some meaningless user actions. For instance, the main interface of CollabSearch consists of three tabs: the "Topic Statement", the "Search" and the "Team Workspace". If a participant wants to issue a query after viewing the Team workspace, he or she need to click the tab "Search" and then issuing a query. In this case, the meaningful action transition is from "Workspace" to "Query", rather than considering two steps of action transitions: from "Workspace" to "Search" and then from "Search" to "Query". As a result, we manually removed the action of clicking the "Search" tab.

The purpose for categorizing user search actions is to calculate and compare the frequency of each type of search action in three different conditions.

Table 1. User search actions

Actions	Descriptions
Query (Q)	A user issues a query or clicks a query from search history.
View (V)	A user click a result in the returned result list
Save (S)	A user saves a snippet or bookmarks a webpage
Workspace (W)	A user clicks, edits or comments an item saved in the workspace
Topic (T)	A user clicks the topic statement for view or leaves comments
Chat (C)	A user sends an message to the other user or views the chat history

4.2   Transition Analysis of Search Actions

In order to understand users' interaction during the search process, it is not enough to look at each action in isolation. We need to examine the relationship between different search actions, which we call it the transitions of search actions.

In the transition analysis of search actions, we consider the sequential dependence order of user actions. Each search action has one predecessor action and one successor action. Since all search actions are categorized into six different types, there are total 36 possible action transition pairs, such as from Query to View (Q→V), from View to Save (V→S), and etc. In the first step, to compare the most frequent action transition pairs in three different conditions, we analyzed the percentage of each of the 36 action transition pairs. Then in the second step, we conducted pre-action and post-action analysis for some chosen search actions. Pre-action analysis is defined as, for a given type of search action, analyzing the percentage distribution of its predecessor actions. Similarly, post-action analysis is defined as, for a given type of search action, analyzing the percentage distribution of its successor actions.

5   Results

We conducted three types of transition analysis of search actions and the results are reported, including the frequency distribution of action pairs, the pre-action analysis and the post-action analysis.

5.1   Distribution of Search Action Pairs

Since we have 6 types of search actions, in total there are 36 possible action transition pairs. The frequency distributions of all the action transition pairs in three conditions respectively were plotted in figure 1. The horizontal line denotes each action pair, in which for example Q-V means the action pair transit from Query (Q) to View (V). The abbreviation of each action can be found in Table 1. We can see some high frequent action transition pairs include Q-V, V-V, V-S, S-Q, S-V, S-S, W-W and C-C. They represents typical search behavior pattern. For example, after issuing a query, viewing the results is very likely to happen. And participants may continue viewing several results on the result page. After viewing a result that is relevant, the participant would collect that result. After collecting one result, the participant may issue another query, continue viewing or collecting other results. The reason for the appearance of S-S is that participants could click a link besides each returned Webpage in the result list to directly collect it without opening that page for viewing. In fact participants should have read the abstract of the Webpage before collecting. It is just the viewing cannot be logged by the system. W-W represents continuous actions in "Team workspace" and C-C indicates persistent chatting activities.

For the comparison purpose, we also list the top 6 frequent action (larger than 10%) transition pairs in each of the three conditions respectively in Table 2. Common actions in all three conditions are Q-V, V-V and V-S, which reflects a typical pattern of behavior in both collaborative and individual search. The participant first issues a query, then views the returned results, and then collects the result if it's relevant.

The differences among three conditions are obvious. In CSCW, researchers (Pinelle et al. 2003) have recognized that group activity can be divided up into two categories: taskwork, the actions needed to complete the task, and teamwork, the actions needed to complete the task as a group – "the work of working together." Applying this to the information seeking environment, taskwork are actions that directly related to search and teamwork are actions indirectly related to search. Table 2 shows that there are more "indirect search" than "direct search" actions in CEC. The indirect search tactics such as "C-C", "W-W" and "T-C" are not directly related to user's taskwork but more close to user's explicit and implicit communication for supporting team partners to collaboratively finish tasks more efficiently. Chat represents explicit communication while the workspace represents implicit communication. C-C and W-W are the top two frequent action pairs in CEC while W-W is the third frequent action pairs in COC. It might indicate the explicit communication also promotes implicit communication between team members. W-W appears more often in CEC and COC than in IND because there is only one member in IND. The fact that large percentages of "indirect search" actions (C-C and W-W) in CEC make the percentages of "direct search" action pairs (Q-V, V-V, V-S) in CEC relatively less. This might suggest that given the same amount of time, participants in CEC have less time to devote to the actual search. Differences among three conditions are also existed on Save-related action pairs. In CEC, percentages of V-S and S-S are smaller. It might suggest that participants in CEC needed to reach agreement on what were relevant and they had more stringent criteria on what results to collect.

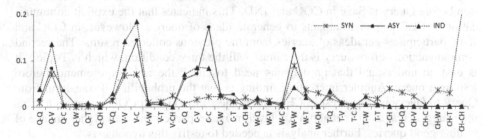

Fig. 2. Percentage distribution of action pairs for both academic task and leisure task in three conditions

Table 2. Most frequent action pairs in three conditions for both academic task and leisure task

Rank	Action (Frequency %)		
	CEC	COC	IND
1	C-C (42.29)	S-S (25.71)	V-S (24.57)
2	W-W (15.57)	V-S (21.29)	V-V (19.14)
3	V-S (14.57)	W-W (17.29)	Q-V (17.86)
4	Q-V (14.36)	Q-V (12.57)	S-V (11.43)
5	V-V (14.21)	V-V (12.14)	S-S (11.00)
6	T-C (7.21)	S-V (10.57)	S-Q (9.57)

5.2 Pre-action analysis

For a given action, the pre-action analysis uncovers the detailed information about what actions precede it. Our analysis focused on pre-actions of Chat in CEC because it is the explicit communication behavior for supporting search process. We also examined pre-actions on Query in all three conditions for better understanding what caused the participants to query in all three conditions.

Pre-chat Analysis in CEC
Figure 3(a) illustrates pre-chat action distribution in CEC. We labeled the transition probability from each type of action to Chat on each link. It is clear that the most common action before chat is chat itself, which suggests the continuousness of chat actions. The second common action before Chat is Topic. The reason might be after viewing topic statement, participants tend to discuss with each other on the task requirements and allocate sub-tasks. The next possible action before Chat is Workspace. This suggest that after viewing, editing or commenting the items saved in the "Team workspace", participants also need to explicitly communicate with each other to inform or discuss the updates and changes in the shared "Team Workspace".

Pre-query Analysis in CEC, COC and IND
Figure 3(b) visualizes the proportion of predecessors of Query action in three different conditions. The visualized result clearly shows the difference in three conditions. The most common action before Query is Chat in CEC while the most common action before Query is Save in COC and IND. This indicates that the explicit communication in CEC helps participants to generate ideas of queries. However, in COC and IND, participants get ideas of queries from the previous collected results. The second common action before query is the same in all the three condition, which is Topic. It is easy to understand that participants need to check the task requirements before issuing a query. Another interesting finding is that the probability of transition from Query to Query is the lowest in CEC and highest in IND with COC in between. This may suggest that as the collaboration level increase, participants had higher chance of issuing good queries. Further analysis is needed to testify this hypothesis.

5.3 Post-Action Analysis

Opposite to pre-action analysis, the post-action analysis reveals the detailed information about what actions succeed a given action. Our analysis focuses on post-actions of Chat in CEC.

Figure 3(c) shows the proportion of each type of action as a successor of Chat action. Unsurprisingly, the most common action after chat is still chat, same as the pre-chat analysis. The second possible action after Chat is Workspace. This suggests that during chat, one participant may inform the other about an updating in the Workspace, and then the participant may go to the "Team Worksapce" to check that update. The third possible action after Chat is Query. This indicates that after chatting with each other, participants may come up with plans or ideas on what query to issue and what information to look for, which is evidence for the benefit of explicit communication in CEC.

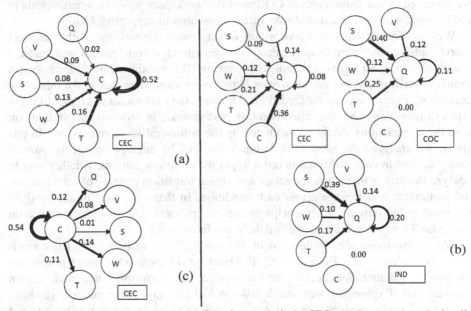

Fig. 3. Action transaction analysis: (a) Pre-chat analysis in CEC, (b) Pre-query analysis all three conditions for both academic task and leisure task, and (c) Post-chat analysis in CEC

6   Discussion and Conclusion

Through the analysis of experiment results, we have observed the following major insights. Having or not having collaboration and having different collaboration styles do affect users' search actions. Issuing queries, viewing and collecting results are actions directly related to search. The rest three actions are indirect search actions. More than half of the actions in CEC condition are related to indirect search activities. Therefore, participants in CEC have fewer actions related to querying and collecting relevant information than participants in COC and IND. However, pre-query analysis demonstrates the possible benefit of explicit communication in CEC condition on helping users to generate queries. The pre-chat analysis revealed that the reasons that trigger the chatting might include needs for discussing task requirements and item collected. Further studies are needed to reveal the effect of explicit and implicit communication on the task performance. Chat represents explicit communication while the Workspace represents implicit communication. Our findings indicate that both types of communication are common in CEC. The fact that participants exhibit higher percentage of Workspace-related actions than participants in COC might indicate that the explicit communication between participants promote implicit communication in the mean time. Although not as many as in CEC, Workspace-related actions are still important proportions of participants' actions in COC. Since users could not explicitly communicate with the other in COC, it is crucial to incorporate in the collaborative search system a shared workspace to support implicit communication. One difference we observed is that participants in COC used the workspace more than participants in IND. This confirms that a shared workspace is essential in supporting COC.

We also found that transitions within direct search actions and within indirect search actions are more frequent than between direct search actions and indirect search actions in all three conditions. However, in CEC condition, transition between indirect search and direct search actions is relatively more than that in the other two conditions. Further studies are needed to understand the reasons caused these different types of transitions and their effects on task performance in order to gain insights on how these transitions should be facilitated by the collaborative search system. In this study, we analyzed the patterns of action transitions by aggregating all the participants' actions in each condition in order to get the common patterns. Another way to analyze the data is to analyze the action and action transition patterns of each individual participant and average across each condition. In that way, we can also analyze the relationship between the participants' actions patterns and their performance on the tasks. These are something we will do in the future work.

We acknowledge some limitations of this study. First, we only considered teams with two members in CEC and COC. It would be different when more people are involved in collaboration. Second, the participants only performed one search session for each task. Exploratory web search tasks in real life might take multiple sessions. In sum, this study provides some guidance for designers on the range of behaviors and activities that a collaborative search system should support. Further studies are needed to fully understand them.

Acknowledgements. This work was supported in parts by National Science Foundation grant IIS-1052773 and III-COR 0704628.

References

1. Blake, C., Pratt, W.: Collaborative information synthesis. In: Proceedings of the 65th ASIST Annual Meeting, pp. 44–56 (2002)
2. Ellis, D.: Modeling the information seeking patterns of academic researchers: A grounded theory approach. Library Quarterly 63(4), 469–486 (1993)

3. Evans, B., Chi, E.: Towards a Model of Understanding Social Search. In: Proceedings of CSCW 2008 (2008)
4. Golovchinsky, G., Pickens, J., Back, M.: A Taxonomy of Collaboration in Online Information Seeking. In: Proceedings of JCDL 2008 Workshop on Collaborative Exploratory Search, Pittsburgh, PA (2008)
5. Hansen, P., Jarvelin, K.: Collaborative Information Retrieval in an information-intensive domain. Information Processing & Management 41(5), 1101–1119 (2005)
6. Halvey, M., Vallet, D., Hannah, D., Feng, Y., Jose, J.M.: An asynchronous collaborative search system for online video search. Information Processing & Management 46(11), 733–748 (2009)
7. Holscher, C., Strube, G.: Web search behavior of Internet experts and newbies 33, 337–346 (2000)
8. Hyldegard, J.: Collaborative Information Behaviour - Exploring Kuhlthau's Information Search Process Model in a Group-based Educational Setting. Information Processing and Management 42(1), 276–298 (2006)
9. Hyldegard, J.: Beyond the Search Process- Exploring Group Members' Information Behavior in Context. Information Processing and Management 45(1), 142–158 (2009)
10. Kuhlthau, C.C.: Inside the search process: Information seeking from the user's perspective. Journal of the American Society for Information Science 42(5), 361–371 (1991)
11. Marchionini, G.: Information seeking in electronic environments, p. 240. Cambridge University Press (1995)
12. Marchionini, G.: From finding to understanding. Communications of the ACM 49(4), 41–46 (2006)
13. Morris, M.R.: A survey of collaborative web search practices. In: Proceeding of the Twenty-Sixth Annual CHI Conference on Human Factors in Computing Systems (2008)
14. Paul, S., Morris, M.R.: Cosense: Enhancing Sensemaking for Collaborative Web Search. In: CHI 2009 (2009)
15. Paul, S.: Understanding together: Sense making in collaborative information seeking. Doctoral Dissertation, Pennsylvania State University (2010)
16. Pinelle, D., et al.: Task Analysis for Groupware Usability Evaluation: Modeling Shared-Workspace Tasks with the Mechanics of Collaboration. Computer-Human Interaction 10(4), 281–311 (2003)
17. Reddy, M., Spence, P.: Collaborative information seeking: A field study of a multidisciplinary patient care team. Information Processing & Management 44(1), 242–255 (2008)
18. Shah, C., Marchionini, G.: Awareness in Collaborative Information Seeking. Journal of American Society of Information Science and Technology 61(10), 1970–1986 (2010)
19. Shah, C., Gonzalez-lbanez, R.: Exploring Information Processes in Collaborative Search Tasks. In: ASIST 2010, Pittsburgh, PA, USA (2010)
20. Spink, A.: A user-centered approach to evaluating human interaction with Web search engines: an exploratory study. Information Processing & Management 38(3), 401–426 (2002)
21. Talja, S., Hansen, P.: Information Sharing. In: Spink, A. (ed.) New Directions in Human Information Behavior. Springer, Netherlands (2006)
22. White, R.W., Roth, R.A.: Exploratory search: Beyond the Query-Response Paradigm. Synthesis Lectures on Information Concepts, Retrieval, and Services 1(1), 1–9 (2009)
23. Xie, I., Joo, S.: Transitions in Search Tactics during the Web-Based Search Process. Journal of the American Society for Information Science 61(11), 2188–2205 (2010)

Query-Oriented Keyphrase Extraction

Minghui Qiu, Yaliang Li, and Jing Jiang

School of Information Systems
Singapore Management University
{minghui.qiu.2010,ylli,jingjiang}@smu.edu.sg

Abstract. People often issue informational queries to search engines to find out more about some entities or events. While a Wikipedia-like summary would be an ideal answer to such queries, not all queries have a corresponding Wikipedia entry. In this work we propose to study query-oriented keyphrase extraction, which can be used to assist search results summarization. We propose a general method for keyphrase extraction for our task, where we consider both phraseness and informativeness. We discuss three criteria for phraseness and four ways to compute informativeness scores. Using a large Wikipedia corpus and 40 queries, our empirical evaluation shows that using a named entity-based phraseness criterion and a language model-based informativeness score gives the best performance on our task. This method also outperforms two state-of-the-art baseline methods.

Keywords: Keyphrase extraction, phraseness, informativeness, language model.

1 Introduction

Online searches generally fall into three categories, namely, informational, navigational and transactional [5]. A recent study has found that the majority of online queries are informational [14], which intend to locate information pertaining to a certain topic. A typical type of informational queries is to simply find out more about a topic such as an entity or an event. For this type of informational queries, instead of showing a ranked list of URLs in the traditional way, a short summary article for the query topic might be a better form to present the search results, from which users can easily digest and further explore. Indeed, recently Google's search results started to include a summary page on the right hand side backed by Google's *Knowledge Graph*, demonstrating the need to automatically summarize information related to a query. But it still remains a challenging task to automatically generate open-domain summaries without supervision.

In this paper, we take a less ambitious step and propose to study the task of finding related keyphrases given a query. This kind of query-oriented keyphrases can be useful for search in a number of ways. For example, to generate an extractive summary of the search results, one may select sentences that maximize the coverage of these related keyphrases. Related keyphrases can also serve as anchor points for further navigation from the original search results in exploratory search.

In Table 1 we show a sample output of our proposed task for the query *Pixar*, where the top-10 keyphrases returned by the best configuration of our proposed method are listed. We can see that these top keyphrases are highly relevant to the query.

Y. Hou et al. (Eds.): AIRS 2012, LNCS 7675, pp. 64–75, 2012.

Table 1. Top-10 keyphrases for the query "Pixar" returned by our method. The descriptions are given by the authors.

Returned Keyphrase	Description
pixar animation studios	full name of the company
john lasseter	CCO of Pixar and Walt Disney animation studios
walt disney pictures	parent company of Pixar
bob iger	CEO of the Walt Disney Company
pixar story	-
walt disney company	owner of Pixar
andrew stanton	director of some Pixar movies
brad bird	director of some Pixar movies
luxo jr.	a Pixar movie
tow mater	the deuteragonist in the Pixar movie "cars"

To find related keyphrases, we transform the task into a query-oriented keyphrase extraction problem where the goal is to extract keyphrases from a set of documents relevant to the given query. Keyphrase extraction has been extensively studied before [12,20,19,22,21,13]. Existing work includes both supervised and unsupervised approaches. Because of the nature of our task, an unsupervised keyphrase extraction method is needed.

Following a framework proposed by Tomokiyo and Hurst [19], we propose a general method for our task which considers both *phraseness* and *informativeness* of a candidate keyphrase. We consider three phraseness criteria based on language models, noun phrase chunking and named entity recognition, respectively. We also consider four informativeness scores using phrase-level Tf-Idf, sum of word-level Tf-Idf, average of word-level Tf-Idf and language models, respectively.

We evaluate the various combinations of the criteria and compare our method with state-of-the-art baselines using 40 queries and a large Wikipedia corpus. We use ground truth both annotated by humans and automatically obtained from Wikipedia articles for evaluation. Experimental results show that the named entity-based phraseness criterion is the best for our task, and the language model-based informativeness score gives the best keyphrase ranking. This configuration of our method outperforms the two state-of-the-art baseline methods we consider.

Our main contributions are twofold. First, we propose to study a new task of query-oriented keyphrase extraction and provide a general solution based on phraseness and informativeness. Second, we empirically compare different phrasesness and informativeness criteria for this task and find a solution better than the baselines that represent the state of the art of keyphrase extraction.

2   Related Work

To the best of our knowledge the task of query-oriented keyphrase extraction has not been well studied. A related line of work is search results clustering, where oftentimes labels for clusters of documents are automatically generated [15,23]. These labels can be seen as phrases related to the query. A major difference between this line of work and our task is that our related keyphrases are not meant as topical labels for a cluster of documents but rather important concepts related to the query. For example, given the query "Pixar," related keyphrases may include key people such as "John Lasseter" (CCO) and "Bob Iger" (CEO of the Walt Disney Company), but these names are not likely to represent different topics for documents relevant to Pixar.

In the information retrieval community, people have studied the task of ranking related entities. Examples include the expert finding problem [6,18,1] and the related entity finding task [2,3], both studied in TREC. The INEX workshop also had an entity ranking track in the past few years [10,7]. Retrieving entities instead of just documents has become an important task for search engines and have been studied in [8,9]. What these tasks share in common is that entities of a specific type (e.g. person) or satisfying a specific description (e.g. airlines using Boeing 747 planes) are being sought after. In contrast, we do not impose such restrictions when extracting keyphrases.

There are generally two approaches to keyphrase extraction. Supervised keyphrase extraction relies on a set of training documents whose keyphrases have already been manually extracted to learn a keyphrase extraction model [12,20]. A major limitation of such methods is clearly the need for a training corpus. A number of unsupervised keyphrase extraction methods have been proposed. In particular, a number of PageRank-based methods such as TextRank [16], SingleRank [22] and ExpandRank [21] have gained much attention. However, in a recent comparative study by Hasan and Ng [13], it shows that a simple Tf-Idf method is more robust than the PageRank-based methods across four different data sets. In our paper we therefore use only the Tf-Idf method as one of our baselines. Another unsupervised keyphrase extraction method which was not compared in [13] was a language model-based method proposed by Tomokiyo and Hurst [19]. The authors stressed the importance of considering both *phraseness* and *informativeness* when extracting keyphrases, and proposed to use the KL-divergence between different language models to measure phraseness and informativeness. We find that for our task indeed we need to consider both phraseness and informativeness, but they can be measured in other alternative ways.

3   Task and Methodology

Given a query q and a large document collection $\mathcal{D}$, we define our task as to return a ranked list of keyphrases that appear in $\mathcal{D}$ and are highly related to q.

Our general approach is to first construct a document set $\mathcal{D}_q$ that is highly relevant to q and then to apply a keyphrase extraction method to identify and rank related phrases occurring in $\mathcal{D}_q$. To construct $\mathcal{D}_q$, a straightforward solution is to retrieve a subset of documents from $\mathcal{D}$ relevant to q using a standard document retrieval method. Our preliminary experiments suggest that using the entire relevant documents for keyphrase extraction may result in many irrelevant phrases which do not co-occur with the query within close proximity and are hence not semantically closely related to the query. We therefore use only paragraphs containing the query from the relevant documents to construct our $\mathcal{D}_q$.

The next step is to extract keyphrases from $\mathcal{D}_q$. There have been many studies on keyphrase extraction, but existing approaches were not designed or evaluated for query-oriented keyphrase extraction, so it is not clear which existing method would work the best for our problem, nor is it clear whether there might be any better method. To answer these questions, we follow a general framework for keyphrase extraction proposed in [19]. In the paper, two criteria are considered, namely, phraseness and informativeness. We study different ways of measuring phraseness and informativeness for our task. In the rest of this section, we first briefly review the notion of phraseness and informativeness, and then present the different phraseness and informativeness measures we consider. We end the section by presenting two baseline methods that represent the state of the art.

3.1 Framework

Tomokiyo and Hurst [19] proposed to use two criteria to judge whether a sequence of words forms a good keyphrase. (1) *Phraseness* measures the degree to which a sequence of words is considered a phrase. For example, "google earth" should have a higher phraseness score than "google owns." (2) *Informativeness* measures how well a phrase illustrates the key ideas in a set of documents. For example, "google earch" is more representative of a collection about Google than "united states." Tomokiyo and Hurst [19] measured phraseness and informativeness by using language models. In this paper, we propose to use different ways to measure phraseness and informativeness. The detailed phrasesness and informativeness measures we used are discussed as follows.

3.2 Phraseness Measures

Language Model-Based: This measure was first proposed in [19]. To measure phraseness, first, a *foreground* corpus $\mathcal{D}_F$ and a *background* corpus $\mathcal{D}_B$ are identified. The extracted keyphrases are supposed to represent $\mathcal{D}_F$. For our task, $\mathcal{D}_q$ is the foreground corpus and the whole document collection $\mathcal{D}$ is the background corpus. Next, a unigram language model and an n-gram language model can be learned from each corpus. If we use θ_F^1, θ_F^n, θ_B^1 and θ_B^n to represent the four language models, respectively, then the *phraseness* of a phrase $\mathbf{w}$ is defined as $\delta_{\mathbf{w}}(\theta_F^n \| \theta_F^1)$, where $\delta_{\mathbf{w}}(p \| q)$ is the *pointwise KL-divergence* between two language models p and q with respect to a sequence of words $\mathbf{w}$, which is defined as follows:

$$\delta_{\mathbf{w}}(p \| q) = p(\mathbf{w}) \log \frac{p(\mathbf{w})}{q(\mathbf{w})}.$$

The general idea here is that if a word sequence is better modeled by an n-gram language model than by a unigram language model, then it is more likely to be a phrase.

We set a threshold τ to filter out n-grams with low phraseness scores. In our experiments, τ is set to 0. This same threshold has been used by Qazvinian et al. [17].

Noun Phrase-Based: The language model-based approach is not the only way to define phraseness. For example, depending on the application, we may be only interested in using noun phrases as keyphrases. In this case, we use noun phrase boundaries as detected by a shallow parser to define candidate keyphrases. In this work, we use a noun phrase chunker[1] to identify noun phrases and filter out those n-grams that are not noun phrases.

---

[1] http://cogcomp.cs.illinois.edu/page/software_view/13

Named Entity-Based: For our task of extracting related keyphrases, we hypothesize that named entities may also be interesting to users. We therefore can define a more stringent phraseness criterion by considering only named entities as candidate keyphrases. In this paper, we use a named entity tagger [11] to identify named entities and filter out those n-grams that are not named entities.

3.3 Informativeness Measures

Language Model-Based: Similar to phraseness, the informativeness of $\mathbf{w}$ can also be defined using language models as $\delta_{\mathbf{w}}(\theta_{\mathrm{F}}^n \parallel \theta_{\mathrm{B}}^n)$ or $\delta_{\mathbf{w}}(\theta_{\mathrm{F}}^1 \parallel \theta_{\mathrm{B}}^1)$. The idea is if a phrase is better modeled by a foreground language model than by a background language model, then it is more representative of the foreground corpus.

Phrase-Level tf-idf-Based: The Tf-Idf scores can also be regarded as an informativeness measure. This is because if a phrase contains words that are frequent in the foreground corpus (i.e. with a high Tf score) but not very frequent in general (i.e. with a high Idf score), then the phrase is more likely to be representative of the foreground corpus. We then define a phrase-level Tf-Idf score as follows to measure the informativeness of a phrase. First, the Tf score of a candidate keyphrase $\mathbf{w}$ is its frequency in $\mathcal{D}_q$. Second, the Idf score of $\mathbf{w}$ is $\log \frac{N}{N_{\mathbf{w}}}$ where N is the size of $\mathcal{D}$ and $N_{\mathbf{w}}$ is the number of documents in $\mathcal{D}$ that contain the phrase $\mathbf{w}$. Note that these definitions are the same as the Tf and Idf definitions for a single word except that here we consider a sequence of words.

Sum of tf-idf-Based: Let us use $(w_1, w_2, \ldots, w_L)$ to denote the sequence of words in phrase $\mathbf{w}$. Using the word-level Tf-Idf scores, we define the informativeness score of $\mathbf{w}$ as $\sum_{i=1}^{L} s(w_i)$, where $s(w_i)$ is the Tf-Idf score of w_i.

Average of tf-idf-Based: In the measure above, a phrase $\mathbf{w}$'s score will be dominated by the most informative words within it. And the criterion will prefer a phrase with more words. To leverage the effect of dominant words and allow phrases with fewer informative words to be ranked high, we define a new informativeness measure by using the average word-level Tf-Idf scores, i.e. $\frac{1}{L} \sum_{i=1}^{L} s(w_i)$.

3.4 Algorithm Outline

With the phraseness and informativeness measures defined above, we now present the outline of our algorithm.

1. **Candidate Phrase Generation:** For a given query q and a document set $\mathcal{D}$, we first extract a relevant document set $\mathcal{D}_q$ by extracting the set of relevant paragraphs containing the query q retrieved from D. From $\mathcal{D}_q$, we then generate all possible n-grams where $n \in \{1, 2, 3, 4\}$.

2. **Keyphrase Filtering:** We filter the n-grams according to one of the following three phraseness measures: language model, noun phrase and named entity based measure. We refer to the resulting candidate keyphrase set as $\mathcal{P}$.

3. **Keyphrase Generation:** Each phrase in candidate keyphrase set $\mathcal{P}$ will be scored by using one of the following four informativeness measures: language model-based informativeness, phrase-level Tf-Idf score, sum of word-level Tf-Idf score, and average of word-level tf-idf score. Then phrases are ranked by their corresponding informativeness scores, resulting in our final keyphrase list.

In summary, in our proposed general method, we consider both phraseness and informativeness, and we allow different ways to define phraseness and informativeness. Note that the language model-based phraseness measure has a phraseness score which can be combined with informativeness scores to rank phrases. But the other two phrassness measures, noun phrase-based and named entity-based, do not have phraseness scores. To make a fair comparison, we use a threshold for language model-based phraseness measure to filter candidate phrases.

3.5   Baselines

We consider the following two baselines for comparison in our experiments.

A Tf-Idf Method: As we have pointed out, Hasan and Ng [13] found that a Tf-Idf-based method is a robust keyphrase extraction method when evaluated on four benchmark datasets. We therefore use their Tf-Idf method as our first baseline.

A Language Model Based Method: Tomokiyo and Hurst [19] proposed to use language models to measure both phraseness and informativeness. The details can be found in Section 3.1. In their method, each phrase is scored by summing up its phraseness and informativeness scores, and then phrases are ranked by their corresponding scores. We use this as our second baseline.

4   Experiments

4.1   Data Set

To evaluate different keyphrase extraction methods for our newly defined task, we need to select a document collection $\mathcal{D}$ and a set of queries $\mathcal{Q}$. We decide to use Wikipedia articles for $\mathcal{D}$ because of the wide coverage of topics in Wikipedia and its rich textual data. We use a version of Wikipedia collection from the ClueWeb09 data set[2]. The data contains 5,945,485 documents, 8,881,880 unique terms and 7,700,294,918 total number of terms. It is indexed by Lemur/Indri, an open-source information retrieval toolkit. We perform Porter stemming and stop word removal during indexing.

We use 40 queries for our evaluation. These queries are the top-ranked pages from two Wikipedia categories based on the numbers of views. The two categories are "companies" and "actors and film makers." For each query, we retrieve the top-100 documents using KL-divergence retrieval model and then extract all the paragraphs containing the query to form $\mathcal{D}_q$. Note that for each query we specifically remove its main Wikipedia article from the set of relevant documents when constructing $\mathcal{D}_q$. This step is important as later we will use the query's corresponding Wikipedia article for evaluation purpose.

---

[2] http://lemurproject.org/clueweb09.php/

4.2 Ground Truth

For each query, a method will return a ranked list of keyphrases. To measure its performance we need to know which phrases are indeed closely related to the query. We consider two ways to obtain the ground truth, one through human annotation and the other through Wikipedia.

First, we created our own manually annotated ground truth. For each query, we got the top-20 keyphrases from each method. We then randomly mixed these keyphrases and asked two judges to perform annotation. Specifically, the judges were asked to score each keyphrase with 0, 1 or 2, where 0 means the keyphrase is not a meaningful phrase or irrelevant to the query, 1 indicates a meaningful but partially relevant phrase, and 2 indicates a meaningful and relevant phrase. We use the Cohen's kappa coefficient κ to measure the agreement between judges. We found that κ ranged from 0.2 to 0.7, which shows fair to good agreement. We used the average scores between the two human judges to form the final ground truth.

We also consider another kind of ground truth automatically obtained from Wikipedia. Specifically, given a query, we obtain its main Wikipedia page. We extract those phrases on this page that are hyperlinks. Because Wikipedia articles are collectively edited by many online users, they represent the general public consensus, and therefore the hyperlinked phrases are presumably highly related to the query. Note that not all these hyperlinked phrases are named entities.

4.3 Evaluation Metrics

With the human annotated ground truth, we adopt the normalized keyphrase quality measure (nKQM) used by Zhao et al. [24]. nKQM is defined in a way similar to the commonly used nDCG measure as follows:

$$n\text{KQM@}K = \frac{1}{|\mathcal{Q}|} \sum_{q \in \mathcal{Q}} \frac{\sum_{j=1}^{K} \frac{1}{\log_2(j+1)} score(\mathcal{M}_{q,j})}{IdealScore(K, q)},$$

where $\mathcal{Q}$ is the set of queries, $\mathcal{M}_{q,j}$ is the j-th keyphrase generated by method $\mathcal{M}$ for query q, $score(\cdot)$ is the average score of human judges, and $IdealScore(K, q)$ is the ideal ranking score of the top K keyphrases of query q.

For the noisy ground truth from Wikipedia, because the relevance score is either 0 or 1, we just compute precision at K for each ranked list of keyphrases for the query. We report $P@5$, $P@10$ and $P@20$ in this paper. Although we could also compute recall values, we do not report them here because we assume that similar to Web search, in our task a user is also usually more interested in the correctness of the top-ranked items rather than the completeness of the retrieved relevant items.

4.4 Methods for Comparison

We have mentioned that we consider two baseline methods. We refer to the Tf-Idf baseline as BL-TI and the language model-based baseline as BL-LM.

As for our own method, we can choose one of the three phraseness criteria to select candidate keyphrases. We refer to the language model-based phraseness criterion as LM, the noun phrase-based criterion as NP and the named entity-based criterion as NE. Each of these phraseness criteria can be coupled with an informativeness score to generate the final ranked keyphrases. We refer to the phrase-level Tf-Idf score for informativeness as P, the sum of the word-level Tf-Idf score as W-S, the average of the word-level Tf-Idf score as W-A, and the language model-based informativeness score as LM. The combinations of these shorthands refer to different configurations of our method. For example, NE-W-A refers to the configuration where we consider candidate keyphrases that are named entities and we rank them using the average word-level Tf-Idf scores.

4.5 Experiment Results

The Effect of Filtering with Phraseness
We first examine whether bringing in a phraseness-based filtering stage in our method can improve the performance. BL-TI is the only method that does not have a phraseness component, and it measures informativeness by the sum of the word-level Tf-Idf scores of a candidate keyphrase. The LM-W-S, NP-W-S and NE-W-S configurations of our method can therefore be regarded as augmenting BL-TI with phraseness-based filtering. We therefore first compare these methods. We report the results based on two ways of evaluation, where automatic evaluation uses Wikipedia hyperlinks as ground truth and manual evaluation is based on human judgment.

Table 2. The effect of filtering with phraseness. † indicates that the result is significantly better than all the results in the other rows at 5% significance level by Wilcoxon signed-rank test.

Method	Automatic Evaluation			Manual Evaluation		
	P@5	P@10	P@20	nKQM@5	nKQM@10	nKQM@20
BL-TI	0.0947	0.0763	0.0618	0.3645	0.3581	0.3309
LM-W-S	0.0368	0.0526	0.0539	0.3116	0.3017	0.2695
NP-W-S	0.1000	0.0974	0.0919	0.4212	0.4033	0.4022
NE-W-S	**0.1892**	**0.1833**†	**0.1721**†	**0.4943**†	**0.4746**†	**0.4842**†

Table 2 shows the performance measures of these different methods. In automatic evaluation, both NP-W-S and NE-W-S improve the baseline BL-TI, while LM-W-S does not outperform the baseline BL-TI. The manual evaluation results in Table 2 are similar to automatic evaluation. Both NE-W-S and NP-W-S outperform the baseline BL-TI, while LM-W-S still has lower performance compared to the baseline. NE-W-S significantly outperforms other competing algorithms for all the metrics except P@5. The results show that both noun phrase and named entity-based phraseness measures are helpful for our task.

Comparison of Informativeness-Based Ranking
Next, we would like to compare the effect of different informativeness scores in ranking the candidate keyphrases. Because we have already found the named entity-based

Table 3. Comparison of different informativeness scores for keyphrase ranking. [†] indicates that the result is significantly better than all the results in the other rows at 5% significance level by Wilcoxon signed-rank test.

Method	Automatic Evaluation			Manual Evaluation		
	P@5	P@10	P@20	nKQM@5	nKQM@10	nKQM@20
NE-W-A	0.2343	0.1853	0.1656	0.2541	0.3119	0.3276
NE-W-S	0.1892	0.1833	0.1721	0.4943	0.4746	0.4842
NE-P	0.3211	0.2694	0.2382	0.5681	0.5796	0.5765
NE-LM	**0.4053**	**0.3472**[†]	**0.2824**[†]	**0.6552**[†]	**0.6468**[†]	**0.6342**[†]

Table 4. Comparisons of our best keyphrase extraction method, two baseline methods and two additional variations of the BL-LM baseline. [†] indicates that the result is significantly better than all the results in the other rows at 5% significance level by Wilcoxon signed-rank test.

Method	Automatic Evaluation			Manual Evaluation		
	P@5	P@10	P@20	nKQM@5	nKQM@10	nKQM@20
BL-TI	0.0947	0.0763	0.0618	0.3645	0.3581	0.3309
BL-LM	0.2947	0.2421	0.1855	0.4725	0.4462	0.4200
BL-LM-P	0.2737	0.2316	0.1816	0.1124	0.1281	0.1523
BL-LM-I	0.1158	0.1158	0.1066	0.4851	0.4589	0.4441
NE-LM	**0.4053**[†]	**0.3472**[†]	**0.2824**[†]	**0.6552**[†]	**0.6468**[†]	**0.6342**[†]

phraseness criterion is the best among the three we consider for our task, here we only compare configurations of our method that use named entities as candidate keyphrases. Table 3 shows the results of these methods. In both automatic evaluation and manual evaluation, NE-LM achieves the best performance, and NE-P's performance is better than NE-W-S and NE-W-A. This shows that for informativeness the language model-based method and the phrase-level Tf-Idf scores outperform other methods. Overall, NE-LM significantly outperforms other competing algorithms on all the metrics except $P@5$.

Comparison with Baselines

Finally, we compare the best configuration of our method with the two baselines. We also show two additional variations of the BL-LM baseline: BL-LM-P uses only the language model-based phraseness scores to rank keyphrases, while BL-LM-I uses only the language model-based informativeness scores to rank keyphrases.

Table 4 shows the comparison, where our method significantly outperforms the two baseline methods, which represent the state of the art. It shows that for our task, using named entities to extract candidate keyphrases and using language model-based informativeness to rank keyphrases achieves the best results. The fact that BL-LM is better than BL-LM-P and BL-LM-I also shows that both phraseness and informativeness are important.

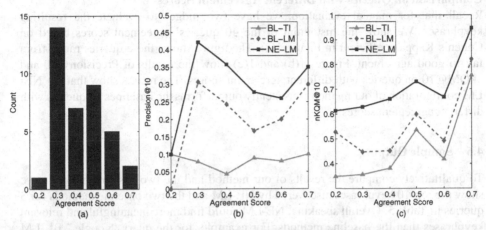

Fig. 1. (a) The histogram of 40 queries' agreement scores, binned into intervals of 0.1. (b) results of Precision@10 on queries with different agreement score. (c) results of nKQM@10 on queries with different agreement score.

Table 5. The top-10 keyphrases of 3 queries

Query	NE-LM	BL-LM	BL-TI
google	tom chavez	search engine	google optimization
	google browser sync	search engine optimization	google maps
	google desktop search	engine optimization	google image
	google inc.	google maps	google desktop
	advanced search	web pages	google mars
	google maps	body copy	google browser
	google earth	deep web	google traffic
	google sky	advanced search web form	google ditu
	google street view	search engine optimization firms	google earth
	google manpower search	spam checking algorithms	google javascript
pixar	pixar animation studios	pixar animation studios	pixar film
	john lasseter	animation studios	pixar animation
	walt disney pictures	pixar animation	pixar films
	bob iger	john lasseter	pixar dvd
	pixar story	pixar film	pixar short
	walt disney company	pixar films	pixar production
	andrew stanton	pixar film references	pixar tradition
	brad bird	toy story	pixar animator
	luxo jr.	walt disney	previous pixar
	tow mater	motor speedway	pixar image
angelina jolie	lara croft	lara croft	tomb raider
	angelina jolie voight	tomb raider	low-budget film
	brad pitt	angelina jolie voight	croft tomb raider
	jonny lee miller	brad pitt	capture film
	ethan hawke	unprecedented media hype	actress angelina jolie
	gia carangi	montreal law enforcement hunt	croft tomb
	robert zemeckis	reported celebrity stories	angelina jolie voight
	mike newell	jonny lee miller	tomb raider videogame
	doug liman	husband jonny lee miller	jolie voight
	john cusack	tv movie true women	actress angelina

Comparison on Queries with Different Agreement Scores

Recall that for manual evaluation, we have two judges to evaluate the resulting keyphrases. We show the histogram of the 40 queries' agreement scores based on Cohen's Kappa coefficient in Figure 1(a). The figure shows these queries range from fair to good agreement. Figures 1(b) and 1(c) show the results of Precision@10 and nkQM@10 on queries with different agreement scores. The results show that the NE-LM configuration of our method consistently outperforms the baselines on queries with different agreement scores.

4.6 Sample Output

To qualitatively compare the results of our method and the two baseline methods, we show the top-10 keyphrases discovered by NE-LM and the two baseline methods for 3 queries in Table 5. Overall speaking, NE-LM could find more meaningful and relevant keyphrases than the baseline methods. For example, for the query "google," BL-LM would find irrelevant phrases like "deep web" and "advanced search web form," and BL-TI would find meaningless phrases like "google traffic."

5 Conclusions

In this paper we studied how to extract a list of keyphrases given a query. Our task was motivated by the observation that for many informational queries in Web search a summary article such as a Wikipedia entry is preferred by online users, and these summary articles usually contain a set of phrases highly related to the query. To address this task of finding related keyphrases, we used unsupervised keyphrase extraction. Inspired by an existing keyphrase extraction method, we proposed a general method that first uses phraseness to select meaningful candidate keyphrases and then ranks them by informativeness. We proposed to measure phraseness using language models, noun phrase boundaries or named entity boundaries, and we proposed to measure informativeness using Tf-Idf scores or language models. We evaluated these different methods on a Wikipedia corpus from ClueWeb09 using 40 queries that represent popular searches on Wikipedia. We found that for our task it is the best to use named entities as candidate keyphrases and the language model-based informativeness scores give the best keyphrase ranking. This method also clearly outperforms two baseline methods that represent the state-of-the-art keyphrase extraction techniques.

Our findings are interesting as they suggest that although keyphrase extraction techniques have been evaluated on a number of benchmark data sets such as scientific literature, when they are applied to novel tasks, new comparison and evaluation needs to be done. Our experiments confirm that both phraseness and informativeness are important criteria to consider.

In the future we plan to look into the problem of characterizing the relations between the related keyphrases and the query. One option is to find support sentences to explain their relations, similar to the work by Blanco and Zaragoza [4]. With such support sentences we can construct more user-friendly summaries for search results.

References

1. Bailey, P., Craswell, N., de Vries, A.P., Soboroff, I.: Overview of the TREC 2007 enterprise track. In: Proceedings of the 16th Text Retrieval Conference (2007)
2. Balog, K., Serdyukov, P., de Vries, A.P.: Overview of the TREC 2010 entity track. In: Proceedings of the 19th Text Retrieval Conference (2010)
3. Balog, K., de Vries, A.P., Serdyukov, P., Thomas, P., Westerveld, T.: Overview of the TREC 2009 entity track. In: Proceedings of the 18th Text Retrieval Conference (2009)
4. Blanco, R., Zaragoza, H.: Finding support sentences for entities. In: SIGIR, pp. 339–346 (2010)
5. Broder, A.: A taxonomy of web search. SIGIR Forum 36(2), 3–10 (2002)
6. Craswell, N., de Vries, A.P., Soboroff, I.: Overview of the TREC-2005 enterprise track. In: Proceedings of the 14th Text Retrieval Conference (2005)
7. Demartini, G., Iofciu, T., de Vries, A.P.: Overview of the INEX 2009 Entity Ranking Track. In: Geva, S., Kamps, J., Trotman, A. (eds.) INEX 2009. LNCS, vol. 6203, pp. 254–264. Springer, Heidelberg (2010)
8. Demartini, G., Missen, M.M.S., Blanco, R., Zaragoza, H.: Entity summarization of news articles. In: SIGIR, pp. 795–796 (2010)
9. Demartini, G., Missen, M.M.S., Blanco, R., Zaragoza, H.: Taer: time-aware entity retrieval-exploiting the past to find relevant entities in news articles. In: CIKM, pp. 1517–1520 (2010)
10. Demartini, G., de Vries, A.P., Iofciu, T., Zhu, J.: Overview of the INEX 2008 Entity Ranking Track. In: Geva, S., Kamps, J., Trotman, A. (eds.) INEX 2008. LNCS, vol. 5631, pp. 243–252. Springer, Heidelberg (2009)
11. Finkel, J.R., Grenager, T., Manning, C.: Incorporating non-local information into information extraction systems by gibbs sampling. In: ACL, pp. 363–370 (2005)
12. Frank, E., Paynter, G.W., Witten, I.H., Gutwin, C., Nevill-Manning, C.G.: Domain-specific keyphrase extraction. In: IJCAI, pp. 668–673 (1999)
13. Hasan, K.S., Ng, V.: Conundrums in unsupervised keyphrase extraction: Making sense of the state-of-the-art. In: COLING, pp. 365–373 (2010)
14. Jansen, B.J., Booth, D.L., Spink, A.: Determining the informational, navigational, and transactional intent of Web queries. IP&M 44(3), 1251–1266 (2008)
15. Leouski, A.V., Croft, W.B.: An evaluation of techniques for clustering search results. Tech. rep., University of Massachusetts at Amherst (1996)
16. Mihalcea, R., Tarau, P.: TextRank: Bringing order into texts. In: EMNLP, Barcelona, Spain (2004)
17. Qazvinian, V., Radev, D.R., Ozgur, A.: Citation summarization through keyphrase extraction. In: COLING, Beijing, China, pp. 895–903 (2010)
18. Soboroff, I., de Vries, A.P., Craswell, N.: Overview of the TREC 2006 enterprise track. In: Proceedings of the 15th Text Retrieval Conference (2006)
19. Tomokiyo, T., Hurst, M.: A language model approach to keyphrase extraction. In: Proceedings of ACL Workshop on Multiword Expressions, pp. 33–40 (2003)
20. Turney, P.D.: Learning algorithms for keyphrase extraction. Information Retrieval 2(4), 303–336 (2000)
21. Wan, X., Xiao, J.: Single document keyphrase extraction using neighborhood knowledge. In: Proceedings of the 23rd National Conference on Artificial Intelligence, pp. 855–860 (2008)
22. Wan, X., Yang, J., Xiao, J.: Towards an iterative reinforcement approach for simultaneous document summarization and keyword extraction. In: ACL, pp. 552–559 (2007)
23. Zeng, H.J., He, Q.C., Chen, Z., Ma, W.Y., Ma, J.: Learning to cluster web search results. In: SIGIR, pp. 210–217 (2004)
24. Zhao, X., Jiang, J., He, J., Song, Y., Achananuparp, P., Lim, E.-P., Li, X.: Topical keyphrase extraction from twitter. In: ACL-HLT, pp. 379–388 (2011)

Using Lexical and Thematic Knowledge
for Name Disambiguation

Jinpeng Wang[1], Wayne Xin Zhao[1], Rui Yan[1], Haitian Wei[2],
Jian-Yun Nie[3], and Xiaoming Li[1]

[1] Department of Computer Science and Technology, Peking University, China
[2] School of International Trade and Economics,
University of International Business and Economics, China
[3] Dpartement d'Informatique et de Recherche Oprationnelle,
Universit de Montral, Montreal, H3C 3J7 Qubec, Canada
{JooPoo,waynexinzhao,r.yan,lxm}@pku.edu.cn,
haataa.wei@gmail.com, nie@iro.umontreal.ca

Abstract. In this paper we present a novel approach to disambiguate names based on two different types of semantic information: lexical and thematic. We propose to use translation-based language models to resolve the synonymy problem in every word match, and to use topic-based ranking function to capture rich thematic contexts for names. We test three ranking functions that combine lexical relatedness and thematic relatedness. The experiments on Wikipedia data set and TAC-KBP 2010 data set show that our proposed method is very effective for name disambiguation.

Keywords: Name Disambiguation, Lexical and Thematic Knowledge.

1 Introduction

Name ambiguity is a common problem when carrying out web searches or retrieving articles from an archive of news articles. For example, the name "Michael Jordan" represents more than ten persons in the Google search results: a basketball player, a professor, a football player and a actor, etc. Along with the rapid growth of the World Wide Web, name ambiguity problem has become more and more serious in areas such as web person search, data integration, link analysis and knowledge extraction.

Based on the task of *entity linking* proposed in the Knowledge Base Population (KBP) track of the Text Analysis Conference (TAC) [1], name disambiguation can be defined as linking mentions of entities within specific contexts to their corresponding entries in an existing knowledge base, e.g., Wikipedia.

Conventionally, name disambiguation methods compute lexical relatedness between mentions (possibly with surrounding contexts) and the document describing one candidate entity, either using vector space model [2,3] or language model [4]. The major challenge of this approach is to resolve name/term ambiguities, which are usually due to *polysemy* and *synonymy*. A synonymous name means that more than one name variations refer to the same entity. To solve the *synonymy* problem, we may consider using query expansion [4] to enrich the query model so that other possible expressions of the

Y. Hou et al. (Eds.): AIRS 2012, LNCS 7675, pp. 76–88, 2012.

same concept or entity can be included. However, the expanded query is still directly matched with words of document, leaving the *polysemy* problem unsolved.

A polysemous name means that it corresponds to more than one entities. The thematic context of the name can be used for its disambiguation. Pilz and Paaß [5], Kozareva and Ravi [6] conducted some preliminary work on name disambiguation using such an idea. Intuitively, different name entities may correspond to different thematic context. For example, "Michael Jordan" (a basketball star) tends to appear in articles related to sports, while "Michael Jordan" (a professor in UC Berkeley) tends to appear in articles related to research. However, Pilz and Paaß [5] used a supervised ranking framework, which requires a set of training data that may not be available for names in general. In a more realistic setting for general ad hoc retrieval, we do not have manually annotated names.

In this paper, we propose to use both lexical knowledge and thematic knowledge for name disambiguation. Generally speaking, lexical knowledge captures more accurate and specific context while thematic knowledge captures more abstract and general context. Specifically, we propose to combine two ranking functions: one is based on translation language models and the other is based on unsupervised topic models. These two functions capture useful information respectively on lexical level and thematic level.

Specifically, the main contributions of this paper are as follows:

1. We leverage the name references on Wikipedia to train a translation model to capture the lexical relationships between terms. Such a translation model can help determine the correspondence between the lexical contexts of a name mention and an entity.

2. We use topic-based ranking function to capture more abstract thematic context. We find that cosine function is more suitable to compute the similarities between topic distributions than symmetric Kullback-Leibler divergence and Kullback-Leibler divergence.

3. In our experiments, we show that the combination of both lexical and thematic information is useful, outperforming the use of a single type of information.

This paper is organized as follows. We first formulate the name disambiguation problem and review the related work in Section 2. Section 3 describes how to leverage lexical and thematic features to enhance the ability to resolve name ambiguities. The experimental results are presented and discussed in Section 4. Finally, we conclude this paper and point out some future work in Section 5.

2   Problem Description and Related Work

2.1   Problem Description

In this paper, we use Wikipedia as a knowledge base. Our goal is to assign the correct Wikipedia article to a name mention found in a text. This task can be used in several places: when a user inputs a query involving a name mention, or when a Web page is created by a user. It is useful to automatically link the name mention to the correct encyclopedia article. In this paper, we use Wikipedia as our encyclopedia data, but our method can be used on other resources.

In general, we consider a name mention as a query. This name mention may appear in some context. Given a name mention (i.e., a person name) q, by matching the surface forms of it we can get a set of candidate entities $\{e_i\}$ from Wikipedia. We denote the context which surrounds q as $D(q)$, and denote the text for entity e in Wikipedia as $D(e)$. Our task is to determine the correct match $\hat{e} \in \{e_i\}$ for name mention q. Table 1 shows an example of one mention together with its context sentence. We adopt a ranking-based approach to this problem, which can be formulated as follows

$$\hat{e} = \begin{cases} e^\star & \text{if } \text{score}(q, e^\star) > \tau \\ nil & \text{otherwise} \end{cases}$$

where $e^\star = \arg\max_{e_i} \text{score}(q, e_i)$, and τ is a threshold which determines whether there is a match or not in $\{e_i\}$. Note that we also consider the case where no correct entity corresponds to the name mention in knowledge base (hereinafter referred to as nil). The key point of this task is to learn an effective score function, i.e., $\text{score}(\cdot, \cdot)$.

Table 1. One mention together with its context sentence

Context sentence ($D(q)$): *Henry is Arsenal's top goal scorer with 226 goals.* **Name mention** (q): *Henry* **Candidate entities** ($\{e_i\}$): *Michel Henry, Thierry Henry, Xavier Henry, Brad Henry* **Correct entity entry in Wikipedia** ($\hat{e}$): *http://en.wikipedia.org/wiki/Thierry_Henry.*	**Wikipedia text of correct entity** ($D(\hat{e})$): *Thierry Daniel Henry (born 17 August 1977) is a French footballer who plays as a striker for New York Red Bulls in Major League Soccer. Henry was born in Les Ulis, Essonne (a suburb of Paris) where he played for an array of local sides as a youngster and showed great promise as a goal-scorer.*

2.2 Related Work

In this section, we will describe the related work in name disambiguation.

Most name disambiguation systems employed methods based on context similarity. Mihalcea and Csomai [7] used cosine similarity to capture the compatibility between name mention and its candidate entities. Bunescu [2] and Cucerzan [3] extended this Bag of Words based method by employing several disambiguation resources, such as Wikipedia entity pages, redirection pages, categories, and hyperlinks. However, their methods rely heavily on word match, and they all suffer from the *synonymy* problem.

There are also some name disambiguation methods that use thematic information of documents. Kozareva and Ravi [6] proposed an approach to disambiguate names using Latent Dirichlet Allocation (LDA) to learn a distribution over topics which correspond to candidate entities. After the topic model of name mention had been trained, they used it to infer the correct entity of ambiguous name mention. Pilz and Paaß [5] used topic model probabilities directly to represent documents, and found the correct entity according to the thematic distance between document and candidate entities. However, Pilz and Paaß [5] adopted a supervised ranking framework, which relied on training data and may not be suitable for evolving topics and new texts.

Recently there are also some name disambiguation methods based on inter-dependency. The idea is that the refered entity of a name mention should be coherent with its unambiguous contextual entities. Medelyan et al's work [8] name mentions with just one matched entity are considered to be unambiguous, and their corresponding correct entity entries are collected and used as context articles to disambiguate the remaining name mentions. This is done by computing the weighted average of lexical relatedness between the candidate entity and its unambiguous contextual entities. Milne and Witten [9] extended this method by adopting typical classifiers to balance the semantic relatedness, the commonness of entries and the context quality.

In this paper, we leverage the name references on Wikipedia to train a translation model to capture the lexical relationships between terms. Such a translation model can help determine the correspondence between the lexical contexts of a name mention and an entity. To capture the more abstract thematic contexts, we use LDA. The two types of contextual information are then combined to produce a final selection of the entity for a mention.

3   The Proposed Method

Our method uses both lexical and thematic relatedness in the score function. In the following subsections, we will first describe the use of lexical relatedness using a translation model. We then describe a thematic model and the combination of lexical and thematic relatedness.

3.1   Using Lexical Relatedness

Previous methods [2][3][4] try to make use of lexical relatedness between name mention and the documents of candidate entities for name disambiguation. However, these methods rely heavily on word match, and they all suffer from the *synonymy* problem. To address this problem, we propose to use translation-based language models. Han and Sun [10] also use translation model in the entity linking task. The major difference is that they utilize translation model only to estimate transition probabilities between an entity name and a mention name, while we apply translation model to all terms. This allows us to relate similar terms in the context around the name mention to those in the candidate entity documents. In addition, we use translation models in a language model based retrieval framework, while they use it to estimate one of the factors in a Bayesian framework.

Before introducing translation-based language models, we first discuss the ranking framework based on language models for name disambiguation.

Language Models Based Ranking Function. Given a name mention q and a candidate entity e, we score e based on the KL-divergence defined as:

$$\text{score}_{lr}(q, e) = -\text{Div}(\theta_q || \theta_e) = -\sum_{w \in \mathcal{V}} p(w|\theta_q) \log \frac{p(w|\theta_q)}{p(w|\theta_e)}, \quad (1)$$

where θ_q and θ_e are the query language model and the entity language model, respectively.

To estimate θ_q, typically we can use the empirical query word distribution:

$$p(w|\theta_q) = \frac{c(w, D(q))}{|D(q)|},$$

where $c(w, D(q))$ is the count of w in $D(q)$ and $|D(q)|$ is the number of words in $D(q)$. Instead of using query expansions like [4], we take a simpler approach by merging name mention with its surrounding context.

To estimate θ_e, we can follow the standard maximum likelihood estimation:

$$p_{ml}(w|\theta_e) = \frac{c(w, D(e))}{|D(e)|},$$

where $c(w, D(e))$ is the count of w in $D(e)$, $|D(e)|$ is the number of words in $D(e)$.

As shown in [11], smoothing is very important for language models in information retrieval. We can extend it by using Dirichlet smoothing:

$$p_{dir}(w|\theta_e) = \frac{c(w, D(e)) + \mu p(w|\theta_C)}{|D(e)| + \mu}, \tag{2}$$

where θ_C is a background language model estimated from the whole collection, and μ is the Dirichlet prior.

The above model relies on a direct word matching. To account for related terms, we use a translation based language model for estimating entity language model θ_e.

Translation-Based Language Models. Translation Model was introduced in information retrieval by [12]. The main idea is to bridge the vocabulary gap between different languages by learning term-to-term probabilities. Similar to that, in our task, we face the problem of vocabulary gap between query document and target entity document.

Assuming a translation model that provides the probability $p(w|w')$, the entity model θ_e can be estimated as follows [13]:

$$p(w|\theta_e) = \beta p_{dir}(w|\theta_e) + (1 - \beta) \sum_{w' \in \mathcal{V}} p(w|w')p_{ml}(w|\theta_e),$$

where β is the self-translation boosting factor and $\mathcal{V}$ is the vocabulary.

In practice, it would be too expensive to enumerate all intermediate terms w' in $\mathcal{V}$, so we take a top-K version of that:

$$p(w|\theta_e) = \beta p_{dir}(w|\theta_e) + (1 - \beta) \sum_{w' \in \mathcal{T}_w} p(w|w')p_{ml}(w'|\theta_e), \tag{3}$$

where $\mathcal{T}_w$ is the set of top K terms ranked by translation probabilities $p(w|\cdot)$. For $w' \in \mathcal{T}_w$, there is a generating probability $p(w|w')p_{ml}(w'|\theta_e)$ computed by following the chain $d \to w' \to w$. Even if term w is not in the entity document of e, $p(w|\theta_e)$ can archive a high value when $D(e)$ includes other terms in $\mathcal{T}_w$.

To learn translation probabilities, we automatically build a parallel training set using Wikipedia, by assuming that a passage including a name and the referred Wikipedia document to be parallel. The Wikipedia data used in our experiments will be described in Section 4.

We adopt a heuristic approximation to estimate the translation model, which is shown efficient by [14]:

$$P(w^S|w^T) = \frac{c(w^S, w^T)}{c(w^T)},$$

where w^S, w^T are terms respectively from "source language" ($D(q)$) and "target language" ($D(e)$). $c(w^S, w^T)$ is the count that w^S and w^T co-occur in the training data, and $c(w^T)$ is the count of term w^T that occurs in the training data.

In practice, it is infeasible to store all the term-to-term translation probabilities, for each term $w \in \mathcal{V}$, we only keep the probabilities of the top K terms ranked by $p(w|\cdot)$, which is consistent with Equation 3.

After learning term translation probabilities, we can use Equation 3 to estimate θ_e, and then rank candidate entities using Equation 1.

3.2 Using Thematic Relatedness

Topic models (e.g., LDA) are unsupervised generative models that learn hidden topics and capture underlying semantic structure of documents. Intuitively, thematic context is useful to help name disambiguation. For example, "Michael Jordan" which appears in a sports article tends to be the baseball star instead of the professor. Pilz and Paaß [5] and Kozareva and Ravi [6] conducted some preliminary work on name disambiguation by leveraging thematic semantics. The main idea is to utilize topic models to learn underlying thematic semantics to resolve disambiguates. However, as we stated earlier, Pilz and Paaß [5] adopted a supervised ranking framework, which relied on training data and may not be suitable for evolving topics and emerging text.

In this section, we first represent entities as topic distribution vectors, and then use an unsupervised ranking function to score those candidate entities.

Topic-Based Entity Representation. In topic models, a document can be represented in a low-dimensional latent space, and the weights of different dimensions are defined as the probabilities of topic distribution given the document. Since we can easily learn the topic-based representation of entity documents, the main difficulty is to represent name mention in topic dimensions. It has been reported that standard topic models (e.g., PLSA and LDA) do not work very well on short text [15], so we take a heuristic method to expand the query following [5]. To assign more importance to the local context, we repeat both name mention and its surrounding context with a window size of 25 for five times, then attach it to the query document [1]. We use the topic distributions of the new generated documents as representations of queries.

We use Latent Dirichlet Allocation (LDA) to infer the probability of an topic in an document. These yield topic distributions of both the query document with name mention q and the candidate entities e as follows:

---

[1] This method is proven to be effective by [5].

- $P(q) = (p_1(q), ..., p_T(q))$: the probability distribution of T topics for the query document.
- $P(e) = (p_1(e), ..., p_T(e))$: the probability distribution of T topics for the entity document.

Topic-Based Ranking Function Given two distributions over topics $P(q)$ and $P(e)$, we define a function to measure the thematic relatedness between them:

$$\text{score}_{tr}(q, e) = \text{TR}(P(q), P(e)) \tag{4}$$

We have three options to define score_{tr} based on the topic-based entity representation. First, we adopt the symmetric Kullback-Leibler divergence [16] which is commonly used in the topic model literature [17] as measuring function:

$$\text{TR}_{skld}(P(q), P(e)) = -\frac{1}{2}\left(\text{Div}(P(q)\|P(e)) + \text{Div}(P(q)\|P(e))\right)$$
$$= -\frac{1}{2}\sum_{i=1}^{T}\left(p_i(q)\log\frac{p_i(q)}{p_i(e)} + p_i(e)\log\frac{p_i(e)}{p_i(q)}\right). \tag{5}$$

The symmetric Kullback-Leibler divergence indicates the similarity degree between two distributions: a small value means they are very similar and vice versa.

The other two alternative functions are Kullback-Leibler divergence and cosine similarity respectively:

$$\text{TR}_{kl}(P(q), P(e)) = -\sum_{i=1}^{T} p_i(q)\log\frac{p_i(q)}{p_i(e)}, \tag{6}$$

$$\text{TR}_{cos}(P(q), P(e)) = \frac{P(q) \cdot P(e)}{\|P(q)\|\|P(e)\|}. \tag{7}$$

3.3 Combining Lexical Relatedness and Thematic Relatedness

The lexical and thematic relatedness defined in the previous section produce two ranking functions for name disambiguation: one is based on translation language models and the other is based on topic models. Generally speaking, lexical feature captures more accurate and specific context while thematic feature captures more abstract and general context. The question now is: can we leverage both features to enhance the ability of resolving disambiguation? The question can be formulated as:

$$\text{score}(q, e) = \text{R}(\text{score}_{lr}, \text{score}_{tr}) \tag{8}$$

where score_{lr} is the lexical relatedness score which defined by Equation 1 and score_{tr} is the the thematic relatedness score which defined by Equation 4.

In this paper, we test three ways to combine both lexical relatedness and thematic relatedness. First, we adopt the linear form:

$$\text{R}_{line}(\text{score}_{lr}, \text{score}_{tr}) = \lambda \cdot \text{score}_{lr} + (1 - \lambda) \cdot \text{score}_{tr}, \tag{9}$$

where λ is a parameter between 0 and 1, which controls the weight of score_{lr} and score_{tr}. Linear form assumes that the total score is the linear combination of the two scores. Each score's contribution to the total score is controlled by λ. The linear form has constant partial derivatives, which means this form has same level of sensitivity towards small and large scores.

The Cobb-douglas form assumes that total score is the product of the two scores. This form is insensitive to small scores and sensitive to large scores:

$$R_{cobb}(\text{score}_{lr}, \text{score}_{tr}) = \text{score}_{lr}^{\lambda} \cdot \text{score}_{tr}^{1-\lambda}, \tag{10}$$

The harmonic form assumes the total score is the weighted harmonic average of the two scores:

$$R_{har}(\text{score}_{lr}, \text{score}_{tr}) = \frac{1}{\frac{\lambda}{\text{score}_{lr}} + \frac{1-\lambda}{\text{score}_{tr}}}. \tag{11}$$

4 Experiments

4.1 Experimental Settings

Data Sets. In Wikipedia, entity pages are connected by hyperlinks: the anchor text is the entity name and the outgoing link points to the referent entity. These entity names are treated as annotated name mentions [9]. In this paper, we treat the pairs of annotated name mention and its corresponding referent entity as parallel data, and use them to train translation models.

We use the July 22, 2011 English version of Wikipedia as knowledge base and extract 16,730 person mentions which contain at least two candidate referent entities (on average 5.23 candidate referent entities for each name) and its corresponding correct referent entity is a person entity. Following [5], we randomly select one fifth of mentions as uncovered mentions (nil) by removing their corresponding correct entities from the collection, and randomly select 60% of the data set for training and the remaining for testing.

We also evaluate our methods on the TAC-KBP 2010 data set [18]. It is the standard test collection used in entity linking task. Its knowledge base was constructed from Wikipedia with 818,741 entries. Our paper is focused on person name disambiguation, so we select all queries of person type in this data set. Then, we get 1,251 queries and 751 of them are used as test data. Some statistics of these data set are shown in Table 2.

Table 2. Statistics of the data sets

data set	queries	%nil	avg. candidate entities
Wikipedia	16,730	20.4%	5.23
TAC-KBP	1,251	56.2%	2.14

Evaluation Metrics. Similar to the standard information retrieval scenario, we adopt precision, recall and F-measure as evaluation metrics. Precision is defined as the ratio between the number of correctly linked queries and the total number of queries which have not been identified as nil. Recall is defined as the ratio between the number of correctly linked queries and the total number of queries which have a corresponding entity in our test collection. In addition, to test the performance on detecting uncovered entities, we also report the accuracy for these mentions separately, denoted as Accuracy$_{nil}$.

Methods to Compare. To examine the effectiveness of our methods, we compare the following methods:

- LR$_{basic}$ and LR$_{trans}$: the methods that use lexical relatedness defined by Equation 2 (basic language model) and Equation 3 (translation-based language model) respectively.
- TR$_{skld}$, TR$_{kld}$ and TR$_{cos}$: the methods that use thematic relatedness defined by Equation 5 (symmetric Kullback-Leibler divergence), Equation 6 (Kullback-Leibler divergence) and Equation 7 (cosine similarity) respectively.
- R$_{line}$, R$_{cobb}$ and R$_{har}$: the methods that combine lexical relatedness and thematic relatedness by Equation 9 (linear form), Equation 10 (Cobb-douglas form) and Equation 11 (harmonic form) respectively.

4.2   Experimental Results

Examining Lexical Relatedness. In Table 3, we compare the basic language model and the model incorporating a translation model. We can see that for all the sizes of context window, the translation-based language model always outperforms the basic language model in terms of F-measure. This result strongly supports our hypothesis that lexical relatedness between terms in the context is highly useful for name disambiguation.

From Table 3, we can also see the impact of context window size. When a context window is used (window > 0), the result is always better than that of no context (window $= 0$) is used. This indicates that context information around the name is highly useful for the task.

If we further zoom into the results, we see LR$_{trans}$ achieves its optimal performance with a relative short context window (i.e, 10) compared with LR$_{basic}$, which needs an

Table 3. Comparison between LR$_{basic}$ and LR$_{trans}$ using different context window size on Wikipedia data set. $K = 700$. Accuracy$_{nil}$ is the accuracy of uncovered queries.

Methods	Metrics	context window size					
		0	10	20	30	40	50
LR$_{basic}$	F-measure	0.417	0.750	0.768	0.772	0.772	0.772
	Accuracy$_{nil}$	0.06	0.056	0.057	0.059	0.058	0.06
LR$_{trans}$	F-measure	0.638	**0.802**	0.787	0.782	0.780	0.779
	Accuracy$_{nil}$	0.119	**0.370**	0.067	0.061	0.062	0.061

optimal window size of 30. LR_{trans} can more effectively leverage lexical information at a relatively small window size. With the increase of context window, more noise is included; therefore the length of context window cannot be too large.

Examining Thematic Relatedness. We now further analyze the impact of topic-based ranking functions TR_{skld}, TR_{kld} and TR_{cos}. In Table 4, we observe that TR_{cos}, which uses cosine similarity, yields the best results. Comparing Table 4 with Table 3, we can see that overall using thematic relatedness is more effective than using lexical relatedness, especially in terms of $Accuracy_{nil}$.

Table 4. Comparison of the performance using different topic-based ranking functions on Wikipedia data set. $T = 150$

Methods	F-measure	Precision	Recall	$Accuracy_{nil}$
TR_{skld}	0.794	0.767	0.823	0.482
TR_{kld}	0.782	0.768	0.796	0.535
TR_{cos}	**0.833**	**0.824**	**0.841**	**0.626**

Examining the Combination of Lexical Relatedness and Thematic Relatedness. Table 5 shows optimal performance using lexical relatedness, thematic relatedness and their combination. We find that by leveraging both lexical information and thematic information, all three combined ranking functions can greatly improve their performances.

In particular, R_{line} achieves the best performance: compared with LR_{best} and TR_{best} (the optimal method of using lexical relatedness and thematic relatedness respectively), it brings 11.5% and 7.4% improvement on F-measure respectively. This indicates that the combination of different types of information is highly useful.

It also shows that lexical information and thematic information can both capture part of evidence between query document and entity document. Performance of TR_{best} is slightly better than that of LR_{best}, but the thematic relatedness is still not enough to capture all evidence for name disambiguation. However, their combination can bring more improvement for name disambiguation.

Table 5. Comparison of the optimal performance using different combined ranking functions on Wikipedia data set. LR_{best} and TR_{best} are the optimal performance of using lexical relatedness and thematic relatedness respectively.

Methods	F-measure	Precision	Recall	$Accuracy_{nil}$
LR_{best}	0.802	0.767	0.841	0.370
TR_{best}	0.833	0.824	0.841	0.626
R_{line} (λ=0.7)	**0.895**	**0.875**	**0.915**	**0.634**
R_{cobb} (λ=0.9)	0.878	0.857	0.900	0.647
R_{har} (λ=0.9)	0.835	0.827	0.843	0.632

Examining Our Method on TAC-KBP Data Set. As our task is similar to entity linking task, we compare our best combined method with state of the art methods [4] on TAC-KBP 2010 data set. The result is shown in Table 6. We can see that our method is better than Gottipati's method, especially on non-nil queries.

Table 6. Experiments result on TAC-KBP data set. Accuracy$_{all}$, Accuracy$_{nor}$ and Accuracy$_{nil}$ are the accuracy on all queries, non-nil queries and nil queries respectively.

Methods	Accuracy$_{all}$	Accuracy$_{nor}$	Accuracy$_{nil}$
Our method	**0.956**	**0.887**	**0.983**
Gottipati's	0.940	0.836	0.981

4.3 Parameter Sensitivity

Recall that we have a few parameters to be tuned for different ranking functions. μ (Dirichlet prior) is empirically set to 2500; β (self-translation boosting factor) is set to 0.5; and τ is the threshold to filter out uncovered queries which automatically set on the training data set. We use the Mallet [19] to implement LDA and we use its default parameter settings. We have three more parameters to be tuned in our methods: K, T and λ. We take a heuristic method to seek the final parameter settings on Wikipedia data set: start with empirical settings, then each time we tune one single parameter and fix the others.

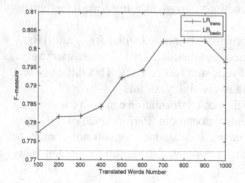 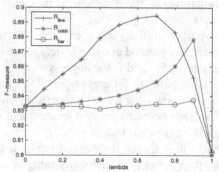

Fig. 1. Sensitivity of number of translated words K in regard to F-measure on LR$_{trans}$

Fig. 2. Comparison of F-measure for three combined ranking functions by varying λ

In Figure 1, we present the results of different K (the total number of translated words) ranging from 100 to 1000 with a step of 100. Recall that K controls the number of top translated words defined in Equation 3. The translated word number K affects the efficiency of the model: a larger K makes the processing more slowly. The optimal K value is 700 for the translation-based language model. We can find that the performance is robust to different settings of K.

In Figure 2, we present the results by varying λ from 0.0 to 1.0 with a step of 0.1. λ controls the weight of these three combined ranking functions. As we can see, the best performance are all achieved when $\lambda > 0.5$. This indicates that lexical relatedness requires a larger weight when considering combination [2].

Besides, we have tried different topic number T in the range from 50 to 500 with a step of 50, and find that the performance has no major change after the number of topics is increased above 150.

5 Conclusions

In this paper we approach the problem of name ambiguity using lexical and thematic information. We compare our approach to two state-of-the-art methods that exploit only lexical information or thematic information. Our approach is able to combine both lexical and semantic information, which can exploit more information than others. We significantly improve the performance of name disambiguation and successfully detect names which are not covered in the knowledge base.

In this paper, linear combination achieves the best performance. However, we believe that there will be more effective methods to combine these two types of information. So we need to explore more combination methods, which is an open question.

Acknowledgments. This work is supported by RenRen Games Grant QXWJ-YX-201206017 and NSFC with Grant No. 60933004, 61073082.

References

1. Dredze, M., McNamee, P., Rao, D., Gerber, A., Finin, T.: Entity disambiguation for knowledge base population. In: Proc. COLING 2010, pp. 277–285 (2010)
2. Bunescu, R.: Using encyclopedic knowledge for named entity disambiguation. In: EACL, pp. 9–16 (2006)
3. Cucerzan, S.: Large-scale named entity disambiguation based on Wikipedia data. In: Proc. EMNLP-CoNLL 2007, pp. 708–716 (June 2007)
4. Gottipati, S., Jiang, J.: Linking entities to a knowledge base with query expansion. In: Proc. EMNLP 2011, pp. 804–813 (2011)
5. Pilz, A., Paaß, G.: From names to entities using thematic context distance. In: Proc. CIKM 2011, pp. 857–866 (2011)
6. Kozareva, Z., Ravi, S.: Unsupervised name ambiguity resolution using a generative model. In: Proc. EMNLP 2011, pp. 105–112 (2011)
7. Mihalcea, R., Csomai, A.: Wikify!: linking documents to encyclopedic knowledge. In: Proc. CIKM 2007, pp. 233–242 (2007)
8. Medelyan, O., Witten, I.H., Milne, D.: Topic indexing with wikipedia. In: Proc. AAAI 2008 (2008)
9. Milne, D., Witten, I.H.: Learning to link with wikipedia. In: Proc. CIKM 2008, pp. 509–518 (2008)
10. Han, X., Sun, L.: A generative entity-mention model for linking entities with knowledge base. In: Proc. HLT 2011, pp. 945–954 (2011)
11. Zhai, C., Lafferty, J.: A study of smoothing methods for language models applied to information retrieval. ACM Trans. Inf. Syst. 22(2), 179–214 (2004)

---

[2] We also tried the combination of normalized scores, and the findings are similar.

12. Berger, A., Lafferty, J.: Information retrieval as statistical translation. In: Proc. SIGIR 1999, pp. 222–229 (1999)
13. Xue, X., Jeon, J., Croft, W.B.: Retrieval models for question and answer archives. In: Proc. SIGIR 2008, pp. 475–482 (2008)
14. Gao, J., He, X., Nie, J.Y.: Clickthrough-based translation models for web search: from word models to phrase models. In: Proc. CIKM 2010, pp. 1139–1148 (2010)
15. Lu, Y., Zhai, C., Sundaresan, N.: Rated aspect summarization of short comments. In: Proc. WWW 2009, pp. 131–140 (2009)
16. Kullback, S., Leibler, R.A.: On information and sufficiency. The Annals of Mathematical Statistics 22(1), 79–86 (1951)
17. Rosen-Zvi, M., Griffiths, T., Steyvers, M., Smyth, P.: The author-topic model for authors and documents. In: Proc. UAI 2004, pp. 487–494 (2004)
18. Heng, J., Ralph, G., Hoa, T.D., Kira, G., Joe, E.: Overview of the tac 2010 knowledge base population track. In: Proc. TAC 2010 (2010)
19. McCallum, A.K.: Mallet: A machine learning for language toolkit (2002), http://mallet.cs.umass.edu

Sells Out or Piles Up? A Sentiment Autoregressive Model for Predicting Sales Performance

Xueni Li, Shaowu Zhang, Liang Yang, and Hongfei Lin

Information Retrieval Laboratory, School of Computer Science and Technology,
Dalian University of Technology, Dalian, China, 116024
lixueni@mail.dlut.edu.cn, zhangshaowu@gmail.com,
yangliang@mail.dlut.edu.cn, hflin@dlut.edu.cn

Abstract. The development of e-commerce has witnessed the explosion of on-line reviews which represent the voices of the public. These reviews are helpful for consumers in making purchasing decisions, and this effect can be observed by some easy-to-measure economic variables, such as sales performance or product prices. In this paper, we study the problem of mining sentiment information from reviews and investigate whether applying sentiment analysis methods can turn out better sales predictions. Based on the nature of various presentations of sentiments, we propose a Latent Sentiment Language (LSL) Model to address this challenge, in which sentiment-language model and senti-ment-LDA are used to capture the explicit and implicit sentiment information respectively. Subsequently, we explore ways to use such information to predict product sales, and to generate an SAR, a sentiment autoregressive model. Extensive experiments indicate the predictive power of sentiment information, as well as the superior performance of the SAR model.

Keywords: Sentiment Mining, Reviews, Sales Prediction, Autoregression.

1 Introduction

Web 2.0 presents an online forum for people to express opinions on products. These discussions not only can influence consumers but also merchants. In both scenarios, opinions suggest early insights into new trends and generate economic values [1]. In addition, previous research offers solid evidence of the decision-making influence of sentiments in reviews.

Some work exists in the area of exploring the influence of sentiment in prediction. Kim et al. [2] discussed that the analysis of network news comments is helpful in predicting the results of U.S. presidential elections. Devitt et al. [3] studied financial market trends by identifying the sentiment polarity. Mishne et al. [4] used posting volumes to predict spikes in sales, while Gruhl et al. [5] and Liu et al. [6] demonstrated that sentiment analysis in online discussions can effectively contribute to better sales prediction than volume only. In this paper, we focus on the mining of sentiments from reviews and investigating ways to use such information to predict sales. The significance of this work lies in pricing, marketing scheme, and so on.

Y. Hou et al. (Eds.): AIRS 2012, LNCS 7675, pp. 89–102, 2012.

Taking book reviews as a case study, we considered ways to predict sales volume through the sentiment information obtained from reviews. We examine book reviews, because special marketing strategies rarely drive sales of books as compared to other categories.

We begin the study by mining sentiments from reviews. Works on this field are plentiful, but conventional text mining methods, such as the classification of reviews, do not apply to this new area. Overall, the goal of our opinion mining is to generate a quantized value that can provide a comprehensive understanding of the sentiments, not the general classification results. Considering the complex nature of language, opinions are presented in two main patterns: explicit expressions and implicit expressions. We apply the sentiment-language model and sentiment-LDA to capture the two types of sentiments, and propose a Latent Sentiment Language (LSL) model. As opposed to the two traditional models, we only focus on sentiment expressions rather than on all words. Therefore, we first construct a sentiment word lexicon in our study. In LSL, the sentiment words compose the feature vectors of reviews, and help extract the subjective nature of the comments.

Past sales performance as another indicator of future sales reflects the influence and trends of the market, and plays an important role in prediction. We capture this effect through the classical autoregressive (AR) model proposed by Box et al. [7] in 1970, which is widely used in many time series prediction problems, including stock prices [8] and box office [6]. Integrating sentiment information and AR, we generate the Sentiment Autoregressive (SAR) model. Abundant experiments conducted on the Chinese book dataset demonstrate the better predictions of the SAR model than the AR model, which also prove the predictive power of sentiments.

In summary, we make the following contributions. First, we propose the probabilistic LSL model to handle the explicit and implicit semantics by sentiment-language model and sentiment-LDA. Second, considering the influence of sentiments and past sales performance together, we present the SAR model for product sales prediction.

Organization of the rest paper is as follows. Section 2 gives a review of related work. Section 3 provides the collection and the statistics of the dataset. Section 4 discusses ways to mine the sentiment using the probabilistic LSL model. In Section 5, we propose the SAR model to predict sales. Section 6 shows the results of the experiments. We conclude the paper in Section 7.

2   Related Work

2.1   Polarity Word Lexicon

Polarity words reflect the basic ideas of the public toward products. Most prior sentiment analysis works start with the extraction of these words. Wiebe [9] tried to collect subjective adjectives using word clustering according to distributional similarity. Hatzivassiloglou et al. [10] gathered sets of semantically oriented adjectives, gradable adjectives, and dynamic adjectives from corpora using a simple log-linear model. Turney [11] measured the semantic orientation of a word by the difference of mutual information between the given word and "poor" or "excellent". Kamps et al. [12]

evaluated the semantic distance between "good" and "bad" through WordNet. Yu et al. [13] judged the polarity of new words through a list of seed adjectives.

These methods are available and practicable. However, employing methods like [9, 10] that based on part of speeches make it easy to omit many sentiment words. Although methods like [11, 12, 13] do not have the above limitation, the results depend on the seed set to some extent. Given these limits, we first adopt the affective lexicon ontology constructed by Xu et al. [14], and then apply the double-channel technique in [15] to extract sentiment words automatically.

2.2   Sentiment Mining

Recently, there is heavy use of review data for polarity analysis and opinion mining. One of the early algorithms employed an unsupervised learning method for sentiment classification [11]. And Pang [16] examined three machine learning algorithms (Naïve Bayes, Maximum Entropy, and Support Vector Machines) for classifying reviews as positive or negative.

Shifting from classification to rating, Pang et al. [17] presented the rating-inference problem, which measures the strength of the sentiment by a rating score (i.e., one to five stars). Zhang [18] also attempted to determine consumers' opinions regarding different rating scales.

Pushing further along the sentiment mining line, Popescu [19] built an unsupervised information extraction system to derive the features of products and the evaluations by reviewers. Liu et al. [20] introduced Opinion Observer, a novel visualized framework to analyze consumer opinions of competing products.

Our work differs from these works in that the goal of our mining is to get representative quantized values of sentiments that can reflect economic worth, rather than discrete classification results or ratings. Here, we present a probability Latent Sentiment Language model to measure the relationship between sentiments and reviews.

2.3   Autoregressive Model

Box et al. [7] proposed the autoregressive (AR) model in 1970. It is a type of random process of time series, which provides a new way to solve the prediction problem, and has garnered great attention. To date, the model has been used in many prediction areas such as stocks [8] and box office [6].

The AR model primarily solves time series problems. Through an analysis of the relationship between historical data in different periods, the model establishes the equation used for the forecast. In our case study, it is reasonable to think that past performance influences the current sales of books. Therefore, we denote the sales of the book at day t by x_t, wherein $t = 1, ..., N$, and use $\{x_t\}$ to denote the time series x_1, $x_2, ..., x_N$. Our goal is to obtain an AR process that can predict sales x_t at day t given the past p days of sales $x_{t-1}, x_{t-2}, ..., x_{t-p}$.

3   Data Collection and Analysis

Here we choose the e-commerce platform Taobao[1] as the data source due to its open policy of both the reviews and sales data. Here, we write a crawler to get reviews and sales data automatically. Relevant to our study, the downloaded data includes 742 days of product sales and 6748 reviews corresponding to the sales dates. Taobao restricts the download to a maximum of 30 days. Therefore, this limit constrains our data collection.

To offer a better understanding of the characteristics of reviews, we carry out a preliminary analysis on the review dataset. Unlike English, there are no separators between two random words; in other words, Chinese word segmentation is necessary. Here we use the ICTCLAS tool[2], a popular Chinese auto-segmentation system. After the participle, Chinese characters are made into different words, and comments just written by punctuations are omitted. Table 1 shows part of the statistics of the other 6381 reviews length (number of words).

As Table 1 shows, the majority of reviews (96%) fall within a length range of 1 to 4, and single-word reviews alone take up about 54% of the total. This suggests that mining and analysis of sentiment words could be very important.

Table 1. Statistics of Taobao customer review data

Len	Number of re-	Len	Number of re-
1	3477	2	1692
3	692	4	305
5	109	6	53

4   LSL: A Probabilistic Approach to Sentiment Mining

4.1   Sentiment Lexicon

Since our work focuses on the mining of sentiment instead of adopting a common "bag of words" (BOW) approach, we mainly pay attention to the sentiment-related words discussed in Section 3. As a result, the first step in our solution is to construct a sentiment lexicon.

Words chosen by people to express sentiments are diverse. In order to guarantee coverage of these terms, we first adopt the affective lexicon ontology constructed by Xu et al. [14], which contains about 13,000 terms consisting of sentiment words, idioms and network terminologies.

Observing that people often express opinions using words located around the product attributes (features), there is a double-channel relationship between sentiment words and features. Thus, we can obtain these words automatically from the corpora by regarding features as indicators. Here, we define a feature list by analysis the

---

[1] http://www.taobao.com/
[2] http://ictclas.org/

review set, words like "service", "delivery" are included. Then by checking if there are any nouns, verbs, adjectives, or adverbs nearby the feature, we extract words that may express opinions. Finally, a list with 1428 terms is generated. Based on the manual judgment, we selected 670 sentiment words and added them into the lexicon, which totaled about 13,670.

4.2 Latent Sentiment Language Model

As a new way for people to express opinions on products, a Nielsen[3] report showed that more people are willing to consult the Internet's word-of-mouth information before they purchase the commodity. Reviews are subjective in nature and opinions usually come in multi-aspects. Therefore, merely classifying the reviews cannot provide a full-scale understanding for customers. When focusing on our goal of sales prediction, mining sentiments from reviews presents unique challenges, and requires a new method of analysis.

To this end, we propose a probabilistic model called Latent Sentiment Language (LSL), with which we can capture the explicit and implicit sentiments, respectively, by the sentiment-language model and sentiment-LDA. A language model representation of a review can "generate" new text by sampling words according to probability distribution. Important words in the review will appear often. In this case, the intuition is that the language model is an approximate representation of what the customer had in mind when he/she wrote the review. When we focus primarily on the sentiment words in a review, the language model also gradually becomes a sentiment carrier of a reviewer. Therefore, the sentiment-language model provides a principled way to quantify the sentiment words associated with natural language. When mining implicit sentiments, sentiment-LDA allows us to adjust the complicated nature of sentiments, wherein reviews reveal a mixture of latent aspects of sentiments. Thus we can model the relationship among these aspects, reviews, and words under a probabilistic framework. Different from the traditional LDA, sentiment-LDA replaces the topics by sentiments. We begin the presentation of LSL by first introducing the sentiment-language model and sentiment-LDA.

Formally, let $R = \{r_1, r_2, ..., r_N\}$ represent a set of reviews for a book, and $W = \{w_1, w_2, ..., w_M\}$ be a set of sentiment words from our sentiment lexicon. The description of the review data can be an $N \times M$ matrix D. We consider a review r_i as generated from a language model by randomly pulling sentiment words out of the words "bucket" W, and representing each row (review) in D as a BOW vector. In other words, the mining of sentiments s expressed in a review r_i can be calculated by the probability of sentiment word generation as follows:

$$p(s \mid r_i) = \sum_{j=1, w_j \in r_i}^{|r_i|} (-1)^{flag} \, p(w_j \mid r_i) \tag{1}$$

Wherein, the sign *flag* indicates the polarity of the sentiment word, the value of 1 implies negative, 0 implies positive, and $p(w_j \mid r_i)$ is defined as:

---

[3] http://www.cr-nielsen.com/marketing/201203/21-1980.html

$$p(w_j \mid r_i) = \frac{f_{w_j,r_i} + u \dfrac{f_{w_j,R_t}}{|c_{R_t}|}}{|r_i| + u} \qquad (2)$$

A widely used method to perform the smoothing, called Dirichlet [21, 22], is applied in (2). It is reflected by the parameter u with value set empirically, and the length of a review $|\ r_i\ |$. R_t denotes the set of reviews posted on day t, $|c_{R_t}|$ represents the total number of words in R_t. $f_{w_j,\,r_i}$ and $f_{w_j,\,R_t}$ denote the number of times word w_j occurs in r_i and R_t respectively.

Once the parameter for the sentiment-language model is selected, we can finish mining sentiments expressed directly in a review. As discussed above, it is not the end of our work. Due to the complex of natural language, semantic analysis techniques are needed in further mining.

To this end, we model the multifaceted sentiment with sentiment-LDA, in which a review is considered as being generated under the influence of a number of latent aspects of sentiments, $Z = \{z_1, z_2, ..., z_K\}$. More formally, the sentiment-LDA process for generating a review is:

1. For each review r_i, pick a multinomial distribution θ_{r_i} from a Dirichlet distribution with parameter α.
2. For each word position in review r_i:

- Pick a sentiment aspect z_k from the multinomial distribution θ_{r_i}.
- Choose a word w_j from $p(w_j \mid z_k, \beta)$, a multinomial probability conditioned on the sentiment aspect z_k with parameter β.

Once we have the distributions, we can get the following probabilities for words in reviews:

$$p_{lda}(w_j \mid r_i) = p(w_j \mid \theta_{r_i}, \beta) = \sum_{k=1}^{K} p(w_j \mid z_k, \beta) p(z_k \mid \theta_{r_i}) \qquad (3)$$

Parameters α and β are corpus-related, assumed to be sampled once in the process of generating a corpus. The variables θ_{r_i} are document-related variables, sampled once per document. And the variables z_k and w_j are word-related sampled once for each word in each document.

Similar to the traditional LDA model, due to the coupling between θ_{r_i} and β, the conditional distribution of latent variables given observed data is intractable to compute. Although the posterior distribution is intractable for exact inference, a wide variety of approximate inference algorithms can be considered for LDA [22]. In this paper, we use variation inference to compute an approximation for the posterior distribution.

By modifying the probability of sentiment words generated from a sentiment-language model using $p_{lda}(w_j \mid r_i)$, we get the following function [22]:

$$p_{lsl}(w_j \mid r_i) = \lambda \frac{f_{w_j, r_i} + u \dfrac{f_{w_j, R_t}}{\mid c_{R_t} \mid}}{\mid r_i \mid + u} + (1 - \lambda) p_{lda}(w_j \mid r_i) \tag{4}$$

$$= \lambda \frac{f_{w_j, r_i} + u \dfrac{f_{w_j, R_t}}{\mid c_{R_t} \mid}}{\mid r_i \mid + u} + (1 - \lambda) \sum_{k=1}^{K} p(w_j \mid z_k, \beta) p(z_k \mid \theta_{r_i})$$

which in turn leads to the results of sentiment analysis:

$$p(s \mid r_i) = \sum_{j=1, w_j \in r_i}^{\mid r_i \mid} (-1)^{flag} \, p_{lsl}(w_j \mid r_i) \tag{5}$$

This summarization reflects the economic value of sentiments. As will be shown in the next section, it can be used to predict future book sales.

5   SAR: A Sentiment Autoregressive Model

5.1   The Autoregressive Model

As a classic model in prediction field, the formula of a basic AR model can be as follows:

$$x_t = \sum_{i=1}^{p} \theta_i x_{t-i} + \varepsilon_t \tag{6}$$

Wherein, θ_i ($i = 1, \ldots, p$) are parameters of the model, and ε_t is an error term (white noise).

As shown in formula (6), the AR model can capture the influence of sales x_{t-1}, x_{t-2}, ..., x_{t-p} on sales x_t at day t. It is necessary to note that the AR model requires that the input of time series must be stationary. In order to guarantee the premise, modeling the time series $\{x_t\}$ properly requires some pre-processing steps. The first step is achieved by a logarithmic method, in which $y_t = \log x_t$. Then, we apply the standard Augmented Dickey-Fuller (ADF) test in Econometrics on $\{y_t\}$ to verify the stationary. After these steps, we get the final time series $\{y_t\}$:

$$y_t = \sum_{i=1}^{p} \theta_i y_{t-i} + \varepsilon_t \tag{7}$$

Wherein, parameters θ_i ($i = 1, \ldots, p$) can be learned from the training data.

5.2   Incorporating Sentiments

The definition of aggregate sentiments on reviews in R_t is

$$s_t = \sum_{i=1, r_i \in R_t}^{\mid R_t \mid} p(s \mid r_i) \tag{8}$$

Where $p(s \mid r_i)$ comes from a LSL model, and R_t denotes the set of reviews posted on day t. Through combining this factor into the modified AR model in (7), we finally obtain our Sentiment Autoregressive model, which considers the two factors and is formulated as follows:

$$z_t = \varphi \sum_{i=1}^{p} \theta_i y_{t-i} + (1-\varphi) \sum_{m=0}^{q-1} s_{t-m} + \varepsilon_t$$

$$= \varphi \sum_{i=1}^{p} \theta_i y_{t-i} + (1-\varphi) \sum_{m=0}^{q-1} \{ \sum_{i=1}^{|R_t|} \sum_{j=1}^{|r_i|} [\lambda \frac{f_{w_j,r_i} + u \dfrac{f_{w_j,R_t}}{|c_{R_t}|}}{|r_i| + u} \quad (9)$$

$$+ (1-\lambda) \sum_{k=1}^{K} p(w_j \mid z_k, \beta) p(z_k \mid \theta_{r_i})] + \varepsilon_t \}$$

Where z_t denotes the predicted sales on day t by the SAR model, parameter q specifies the sentiment information from how many days are considered, others are introduced in Section 4.

5.3   Training the SAR Model

Without known parameters, a model cannot be complete. Training the SAR model aims at learning the set of parameters $\theta_i (i = 1, \dots, p)$ from the training data that consist of the true sales of books. Although many approaches may work for the parameters, in our work, we choose ordinary least squares (OLS) to complete the estimation.

In statistics, OLS is a method for estimating unknown parameters in a linear regression model. This method minimizes the sum of squared vertical distances between the observed responses in the dataset and the responses predicted by the linear approximation. A simple formula can express the resulting estimator especially in the case of a single regressor on the right hand side. After the estimation, the SAR model can adhere to the parameter values.

6   Experiment and Analysis

6.1   Experiment Settings

The data we use in the experiments consist of a set of 6748 reviews of 22 books, and the corresponding 742 days of sales data. The following is the overall process of the experiments.

1. For sales of books, the final 625 daily sales that get through the ADF test are for training, and the other 117 days for testing.
2. Using the training sales, parameters θ_i ($i=1, 2, \dots, p$) for each sale of a day could derive from training a SAR model. Here, we conduct the experiments from $p = 1$ to $p = 6$.

3. Mining the value of sentiments from reviews for each day of one book using the LSL model.
4. Feeding the results of sentiment analysis obtained in step 3, along with the sales of the preceding days, into the SAR model, and evaluating the prediction performance of the SAR model by experimenting with the testing data set.

Without loss of generality, we use the mean absolute percentage error (MAPE) [22] to measure the prediction accuracy, which is standard in regression analysis:

$$MAPE = \frac{1}{N}\sum_{i=1}^{N}\frac{|True_i - Pred_i|}{True_i} \qquad (10)$$

Where N is the total number of predictions made on the testing data, $True_i$ represents the true sale, and $Pred_i$ is the predicted sale.

6.2   Parameter Selection

In the SAR model, several user-chosen parameters provide the flexibility to fine tune the model for optimal performance. They include the number of latent sentiment aspects K, Dirichlet smoothing parameter u, the time orders of the SAR model p, sentiments of q days taken into consideration, and weight regulation parameters φ, λ. We now study the way in which the choice of these parameter values affects the prediction performance.

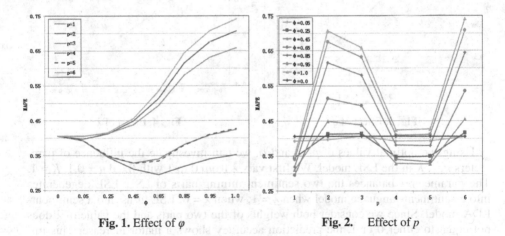

Fig. 1. Effect of φ Fig. 2.   Effect of p

We first vary φ with fixed $u = 0.1$, $\lambda = 0.95$, $q = 3$, and $K = 1$. The parameter φ balances the influences of past sales performance and the sentiments in reviews. When φ equals 1.0, the SAR model considers the sales of preceding days only, which is the basic AR model; when φ equals 0.0, the SAR model only considers the sentiments in reviews, and the value of MAPE is the same and independent of p. As shown in Fig. 1, with the variation of φ from 0.05 to 0.95, the prediction accuracy of SAR can reach better results than AR ($\varphi = 1.0$) even though the parameter p is different, and the SAR achieves its optimal performance at $\varphi = 0.65$ when $p = 1$. It indicates that the integrating of sentiments can indeed lead to more accurate prediction, which helps prove the economic value and predictive power of sentiments. At the same time, the value of MAPE at $\varphi = 0$ also indicates that only sentiment itself can be a good indicator of the book sales, and its performance is much better than AR model when $p = 2, 3, 6$, and almost achieves the same level when $p = 4, 5$.

We then vary the value of p, with $u = 0.1$, $\lambda = 0.95$, $q = 3$, and $K = 1$ to study the influence of the order of the autoregressive model. Fig. 2 shows that the model achieves its best prediction accuracy when $p = 1$. This suggests that prediction closely relates to the sales from recently preceding days. Because p is an intrinsic parameter of the AR model, the trends of MAPE stays stable with varying φ, which reflects the importance of AR in SAR model.

Parameter q represents the number of days to consider sentiments. We vary q to study its effect on the prediction accuracy. As shown in Fig. 3, the best prediction accuracy is at $q = 3$, which implies that the prediction relates to the sentiment information captured from reviews posted on the same day and the immediately preceding two days.

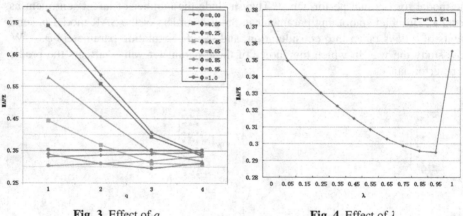

Fig. 3. Effect of q **Fig. 4.** Effect of λ

Using the optimal values of p, q, and φ, we can investigate the influence of parameters u, λ, K in the LSL model. We first vary λ from 0 to 1 with fixed $u = 0.1$, $K = 1$. The parameter λ balances the two sentiment mining parts of LSL. LSL degenerates into a sentiment-language model when $\lambda = 1$; when $\lambda = 0$, LSL turns into a sentiment-LDA model. Since we consider both weights of the two parts and the value of λ does not equal to either 0 or 1, the prediction accuracy shows a major increase. This implies that it is appropriate to mine sentiment from explicit and implicit aspects. As shown in Fig. 4, although the prediction accuracy improves as the λ increases, the difference of MAPE between any two λ is small. This suggests the stationarity of LSL. Fig. 4 also shows that sentiment-language model performs better than sentiment-LDA. It indicates that people are willing to express their opinions and sentiments more directly, and, in future work, more effort should be put into explicit sentiment analysis.

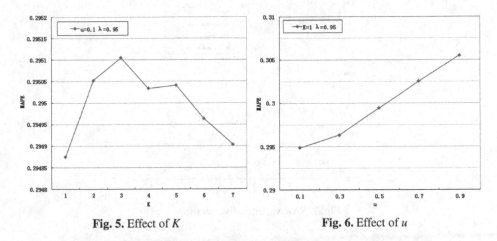

Fig. 5. Effect of K **Fig. 6.** Effect of u

With fixed $u = 0.1$, optimal λ, p, q, and φ, we study how K affects the value of MAPE. As K increases from 1 to 7 in Fig. 5, the prediction accuracy first deteriorates and then improves, and SAR achieves its best performance when $K = 1$. The explanation here is that reviews are short according to Table 1, and the implicit sentiment described with lower dimensional probability vectors is enough. Also, the total number of parameters which must be estimated in the sentiment-LDA grows linearly with respect to the number of latent sentiment aspects K. If K gets too large, it may incur high training costs in terms of space and time.

Parameter u is a factor in Dirichlet smoothing. We vary its value to observe its influence, after the choice of parameter φ, p, q, λ, and K. As seen in Fig. 6, smaller u leads to better results. It suggests that the relative weight of sentiment words plays a more important role than the number of sentiment terms in a review set of one day.

6.3   Comparison with Alternative Sentiment Mining Methods

To verify that the sentiment information captured by the LSL model plays an important role in book sales prediction, we compare SAR with two alternative sentiment mining methods and the basic AR model without the consideration for the sentiment information.

We first conduct experiments to compare SAR against the pure autoregressive model in formula (7). The results are in Fig. 7. We observe the behaviors of the two models as p ranges from 1 to 6. Apparently, SAR consistently outperforms the AR model with increasing p.

We then compare SAR with an autoregressive model that integrates with the language model or the LDA model respective to mining sentiments in prediction. Using the same training and testing data sets as used for SAR, we test the performance of these two models and compare them with SAR. The results are in Fig. 7. Although the two methods yield a moderate performance gain over the pure AR model, the SAR model still dominates performance.

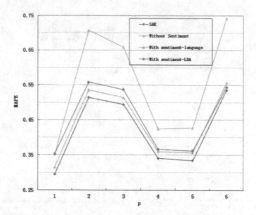

Fig. 7. SAR vs. alterative methods

7   Conclusions and Future Work

The widespread use of reviews as a mean of recording the voice of the public provides a unique opportunity to understand customers' sentiments and to advance business intelligence by analyzing this unsolicited feedback. In this paper, we identify the task in the ongoing research in sentiment analysis, exploring the predictive power of reviews using books as a case study, and studying the problem of predicting sales performance using sentiment information captured from reviews. A center piece of our work is the proposal of LSL, a generative model for sentiment mining, which is different from the traditional classification and extracts sentiments from both the explicit and the implicit aspect. Using LSL as a way of "summarizing" sentiment information from reviews, we develop SAR, a model for predicting sales based on the products' past sales and the sentiments. Our empirical experiments on a book dataset confirm the effectiveness and superiority of the proposed approach.

For future work, there are many interesting research directions to explore. For example, by the analysis of sentiment on a more fine level, such as "happy", "angry", we can explore the differences of their influence degree. In a slightly different twist, there is also much work to consider the reverse effect on how opinions and sentiments will be influenced by the changes in the associated economic variables, such as price, stock, etc.

Acknowledgment. This work is partially supported by grant from the Natural Science Foundation of China (No.60673039, 60973068, 61277370), the National High Tech Research and Development Plan of China (No.2006AA01Z151), Natural Science Foundation of Liaoning Province, China (No.201202031), State Education Ministry and The Research Fund for the Doctoral Program of Higher Education (No.20090041110002).

References

1. Ghose, A., Ipeirotis, P.G., Sundararajan, A.: Opinion Mining Using Econometrics: A Case Study on Reputation Systems. In: 45th Annual Meeting of the Association of Computational Linguistics, pp. 416–423. Association for Computational Linguistics, Stroudsburg (2007)
2. Kim, S.M., Hovy, E.: Crystal: Analyzing Predictive Opinions on the Web. In: 2007 Joint Conference on Empirical Methods in Natural Language Processing and Computational Natural Language Learning, pp. 1056–1064. Association for Computational Linguistics, Stroudsburg (2007)
3. Devitt, A., Ahmad, K.: Sentiment Polarity Identification in Financial News: A Cohesion-based Approach. In: 45th Annual Meeting of the Association of Computational Linguistics, pp. 984–991. Association for Computational Linguistics, Stroudsburg (2007)
4. Mishne, G., Glance, N.: Predicting Movie Sales from Blogger Sentiment. In: 2006 AAAI Symposium on Computational Approaches to Analysing Weblogs, pp. 155–158. American Association for Artificial Intelligence, California (2006)
5. Gruhl, D., Guha, R., Kumar, R., Novak, J., Tomkins, A.: The Predictive Power of Online Chatter. In: 11th ACM SIGKDD International Conference on Knowledge Discovery and Data Mining, pp. 78–87. ACM, New York (2005)
6. Liu, Y., Huang, X.J., An, A.J., Yu, X.H.: ARSA: A Sentiment-Aware Model for Predicting Sales Performance Using Blogs. In: 30th Annual International ACM SIGIR Conference on Research and Development in Information Retrieval, pp. 607–614. ACM, New York (2007)
7. Box, G.E.P., Jenkins, G.M., Reinsel, G.C.: Time Series Analysis. Holden-Day, San Francisco (1976)
8. Enders, W.: Applied Econometric Time Series. John Wiley and Sons, New York (2004)
9. Wiebe, J.: Learning Subjective Adjectives from Corpora. In: 17th National Conference on Artificial Intelligence and 12th Conference on Innovative Applications of Artificial Intelligence, pp. 735–740. AAAI Press/The MIT Press, Cambridge, MA (2000)
10. Hatzivassiloglou, V., Wiebe, J.M.: Effects of Adjective Orientation and Gradability on Sentence Subjectivity. In: 18th Conference on Computational Linguistics, pp. 174–181. Association for Computational Linguistics, Stroudsburg (2000)
11. Turney, P.D.: Thumbs Up or Thumbs Down? Semantic Orientation Applied to Unsupervised Classification of Reviews. In: 40th Annual Meeting of the Association for Computational Linguistics, pp. 417–424. Association for Computational Linguistics, Stroudsburg (2002)
12. Kamps, J., Marx, M., Mokken, R.J., Rijke, M.D.: Using WordNet to Measure Semantic Orientation of Adjectives. In: 4th International Conference on Language Resources and Evaluation, Lisbon, PT, pp. 1115–1118 (2004)
13. Yu, H., Hatzivassiloglou, V.: Towards Answering Opinion Questions: Separating Facts From Opinions and Identifying the Polarity of Opinion Sentences. In: 2003 Conference on Empirical Methods in Natural Language Processing, pp. 129–136. Association for Computational Linguistics, Stroudsburg (2003)
14. Xu, L.H., Lin, H.F., Pan, Y., Ren, H., Chen, J.M.: Constructing the Affective Lexicon Ontology. Journal of the China Society for Scientific and Technical Information 2, 180–185 (2008)
15. Hu, M.Q., Liu, B.: Mining and Summarizing Customer Reviews. In: 10th ACM SIGKDD International Conference on Knowledge Discovery and Data Mining, pp. 168–177. ACM, New York (2004)

16. Pang, B., Lee, L., Vaithyanathan, S.: Thumbs up? Sentiment Classification Using Machine Learning Techniques. In: 2002 Conference on Empirical Methods in Natural Language Processing, pp. 79–86. Association for Computational Linguistics, Stroudsburg (2002)

17. Pang, B., Lee, L.: Seeing Stars: Exploiting Class Relationships for Sentiment Categorization with Respect to Rating Scales. In: 43rd Annual Meeting of the Association for Computational Linguistics, pp. 115–124. Association for Computational Linguistics, Stroudsburg (2005)

18. Zhang, Z., Varadarajan, B.: Utility Scoring of Product Reviews. In: 15th ACM International Conference on Information and Knowledge Management, pp. 51–57. ACM, New York (2006)

19. Popescu, A.M., Etzioni, O.: Extracting Product Features and Opinions from Reviews. In: 2005 Human Language Technology Conference and Conference on Empirical Methods in Natural Language Processing, pp. 339–346. Association for Computational Linguistics, Stroudsburg (2005)

20. Liu, B., Hu, M., Cheng, J.S.: Opinion Observer: Analyzing and Comparing Opinions on the Web. In: 14th International Conference on World Wide Web, pp. 342–351. ACM, New York (2005)

21. Blei, D.M., Ng, A.Y., Jordan, M.I.: Latent Dirichlet Allocation. Journal of Machine Learning Research 3, 993–1022 (2003)

22. Croft, W.B., Metzler, D., Strohman, T.: Search Engines: Information Retrieval in Practice. Addison-Wesley Publishing Company, Boston (2009)

Are Human-Input Seeds Good Enough for Entity Set Expansion? Seeds Rewriting by Leveraging Wikipedia Semantic Knowledge

Zhenyu Qi, Kang Liu*, and Jun Zhao

National Laboratory of Pattern Recognition(NLPR)
Institute of Automation Chinese Academy of Sciences
100190 Beijing, China
{zyqi,kliu,jzhao}@nlpr.ia.ac.cn

Abstract. Entity Set Expansion is an important task for open information extraction, which refers to expanding a given partial seed set to a more complete set that belongs to the same semantic class. Many previous researches have proved that the quality of seeds can influence expansion performance a lot since human-input seeds may be ambiguous, sparse etc. In this paper, we propose a novel method which can generate new, high-quality seeds and replace original, poor-quality ones. In our method, we leverage Wikipedia as a semantic knowledge to measure semantic relatedness and ambiguity of each seed. Moreover, to avoid the sparseness of the seed, we use web resources to measure its population. Then new seeds are generated to replace original, poor-quality seeds. Experimental results show that new seed sets generated by our method can improve entity expansion performance by up to average 9.1% over original seed sets.

Keywords: information extraction, seed rewrite, semantic knowledge.

1   Introduction

Entity Set expansion refers to the problem of expanding a given partial (3~5) set of seed entities to a more complete set which belongs to the same semantic category. For example, a person may give a few elements like "Gold", "Mercury" and "Xenon" as seeds; the entity set expansion system should discover other elements such as "Silver", "Oxygen" etc. based on the given seeds.

These collections of entities are used in many commercial and research applications. For instance, question answering systems can use the expansion tools to handle List questions [1]. And search engines collect large sets of entities to better interpret queries.[2]

Several researches have been proposed for solving this problem, like [3][4][5]. These methods generally include two components: 1) find candidates which may have the same semantics with the given seeds; 2) measure the similarity between each

---

* Corresponding author.

Y. Hou et al. (Eds.): AIRS 2012, LNCS 7675, pp. 103–113, 2012.

candidate and the given seeds. Candidates with higher similarity score will be extracted as results. A typical method starts from several seeds (usually 3-5), then it employ distributional features [6][7] or context patterns [8][9] to find entities of the same category in external data sources such as large corpora of text or query logs.

Table 1. Seeds greatly influences entity set expansion quality

Concept	MAX	MIN	AVG
California Counties	1.000	0.103	0.859
Countries	0.885	0.008	0.667
Elements	0.991	0.026	0.784
F1 Drivers	0.959	0.000	0.456
Roman Emperors	0.804	0.109	0.503
U.S. States	1.000	0.640	0.908

However, since the seeds are provided by human as will, they may be poor-quality and have problems such as being ambiguous or sparse etc.. Taking the three seeds we mentioned at the beginning of this section as instance, seed "Mercury" is ambiguous because it appears as instance of two completely unrelated semantic class *planets* and *elements* with almost equally probability. Seed "Xenon" is so sparse in common corpus that there is a high probability we find very few useful templates by using it.

To study the impaction of seeds, we employ a state-of–art set expansion system [10] to evaluate the performance of different seeds. We use 6 benchmark concepts described in Section 5. For each concept we do 10 trials. In each trial, we randomly select 3 entities as seeds. Table 1 shows the maximum, minimum and average expansion performance of these seeds sets measured by R-precision. We see there is a variation as much as 35% between the max and average performance, confirming that the quality of seeds truly has great influence on the expansion performance. Other studies have come to similar conclusion [2]. Furthermore, previous studies have shown that human editors generally provide very bad seeds [2]. So generating high-quality seeds is very important for entity set expansion.

To avoid seed ambiguity and sparseness, we propose a novel method for generating better seeds in this paper. First, we link original seeds to Wikipedia articles. Second, we measure the quality of seeds by the following three factors and decide which seeds should be replaced: 1) *Semantic Relatedness*. High-quality seeds should have high semantic relatedness among each other; 2) *Ambiguity*. Good seeds should have less ambiguity; 3) *Population*. High-quality seeds shouldn't be sparse. Lastly, we generate new seeds which have high quality by using Wikipedia. In detail, we adopt a three-phase strategy: First, we propose a disambiguation algorithm to identify the articles in Wikipedia which describe the original seeds. Second, by using the semantic knowledge contained in these articles, we measure the quality of original seeds and find out poor-quality seeds. Third, we generate new, high-quality seeds using the category structure and semantic knowledge of Wikipedia.

Specifically, our contributions are:

● We believe the quality of seeds has great influence on entity expansion performance. We identify three factors to measure seed quality. And we present three algorithms to measure these factors respectively.

- We propose a novel method to find out poor-quality seeds and generate high-quality seeds to replace them. The new generated seeds will be used for expanding entities in Web date. Experimental results on data from different domains show that our method can effectively generate high-quality seeds and improve the entity set expansion performance.

The remainder of the paper is organized as follows. Section 2 states the impact of seed set and reviews related work. In Section 3, we introduce Wikipedia as a semantic knowledge base. Section 4 introduces the three factors in measuring seed quality and describes our proposed method in detail. Experimental results are discussed in Section 5. Section 6 concludes this paper and discusses the future work.

2   Problem Statement and Related Work

As mentioned in last section, the problem of seeds rewriting for entity set expansion can be defined as follows:

For a semantic category C, given M entities belong to C, the seed rewriting system should find out which K entities from M given ones have poor quality and generate K new, high-quality seeds to replace them. In this paper we make $M = 3$, which is also used in [10]. Since M is small and there may be high-quality seeds in the original ones, we just replace the most poor-quality seed which means $K = 1$.

For example, suppose we want to find out all elements. And we already know some of them such as "Gold", "Mercury", "Xenon" The seed generation method should be able to find out the one that should be replaced (suppose it is "Mercury") and generate new high-quality seed to replace it (suppose it is "Oxygen").

A similar problem is "better seeds selection". The key point of that problem is to choose K-best seeds from given M ones. Previous studies have proved that methods which solve that problem can also improve the expansion performance. A prominent work about better seeds selection is proposed by Vyas et al [2]. They measure every seed according to the following three factors: 1) Prototypicality, which weighs the degree of a seed's representation of the concept; 2) Ambiguity, which measures the polysemy of a seed; 3) Coverage, which measures the degree of the amount of semantic space which the seeds share in common with the concept. Then they remove the error- prone seeds and return the remaining seeds as results.

Those methods have some limitations: 1) They can only choose relatively better seeds from original ones but cannot generate new, high-quality seeds. If unfortunately original seeds are not high-quality, they can only get a poor performance. 2) In many situations, it is hard to get enough original seeds for selection since seeds are provided by human as will.

To overcome these deficiencies, we propose a novel method to resolve the problem of seeds rewriting. Generally a three-stage strategy is designed to generate new, high-quality seeds: First, we link original seeds to Wikipedia articles which describe them. Second, we present three factors to measure seed quality and propose three algorithms to measure these factors respectively. Then we attempt three ways to decide which seed should be replaced. Lastly, we present a method to generate new high-quality seed and replace the old one. To accomplish this, we use Wikipedia semantic knowledge and web corpus frequency. In the following sections, we will show our method in detail.

3   Wikipedia as a Semantic Knowledge Base

Wikipedia is the largest encyclopedia in the world and surpasses other knowledge bases because of its large amount of concepts, up-to-date information, and rich semantic knowledge. The English version Wikipedia contains more than 4 million articles and new articles are added very quickly.

Because of its large scale and abundant of semantic information, Wikipedia has been widely used in Information Retrieval and Nature Language Processing. In the following subsections, we will introduce some characters of Wikipedia which will be used for our task.

3.1   Wikipedia Articles

In Wikipedia, an article is usually used to describe a single entity. Figure 1 is a snapshot of part of the article "Mercury (element)". The red boxes in the Figure markup links to other articles. Previous study shows that an article in Wikipedia has 34 links out to other articles and receives another 34 links from them on average [11].

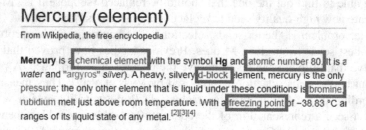

Fig. 1. A Snapshot of A Typical Wikipedia Article

These links can also be used to measure the semantic relatedness between Wikipedia entities. In this paper, we adopt the method described in [11]. Based on the idea that the higher semantic relatedness two entities share, the more common links they have, this method measures semantic relatedness as follows:

$$sr(a,b) = 1 - \frac{\log(\max(|A|,|B|)) - \log(|A \cap B|)}{\log(|W|) - \log(\min(|A|,|B|))} \qquad (1)$$

where a and b are the two entities of interest, A and B are sets of all entities that link to a and b respectively, and W is the entire Wikipedia.

3.2   Wikipedia Anchors

In Wikipedia, anchors refer to the terms or phrases in articles texts to which links are attached. Texts in red boxes in Figure 1 are examples of anchors.

Anchors have a tendency to link to multiple articles in Wikipedia. For example, anchor "Mercury" might refer to a kind of element, a planet, a roman god and so on. Suppose article set D is consisted of the articles that anchor a links to, we can calculate the probability that a links to article d which belong to D as follows:

$$\Pr ob(< a,d >) = \frac{count(< a,d >)}{\sum_{d \in D} count(a \to d^{'})}$$ (2)

where $count(< a,d >)$ is the number of times that anchor a links to article d. In section 4, we will discuss how to use this property to calculate the ambiguity of a seed in detail.

3.3   Wikipedia Category Labels

In Wikipedia, each article has several "Category Labels", which means it belongs to the category. Figure 2 shows the category labels of the article "Mercury (element)".

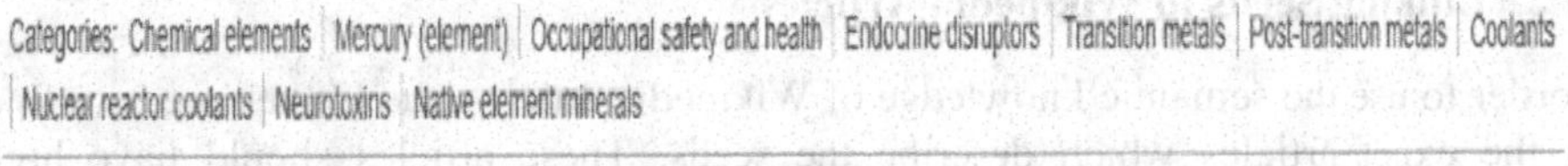

Categories: Chemical elements | Mercury (element) | Occupational safety and health | Endocrine disruptors | Transition metals | Post-transition metals | Coolants | Nuclear reactor coolants | Neurotoxins | Native element minerals

Fig. 2. A Snapshot of Category Labels of A Wikipedia Article

A category label usually indicates a semantic class. So articles that belong to the same category label may belong to a same semantic class with high probability. For instance, "Mercury (element)" has the label "Chemical elements" etc. Oxygen, Gold and many other elements all have category label "Chemical elements". So they have a high probability to belong to the same semantic category. We will show how to use these labels to generate new seeds in section 4.

4   Seeds Rewriting by Leveraging Wikipedia Semantic Knowledge

In this section, we introduce our method in detail and show how to generate new, high-quality seeds by leveraging Wikipedia semantic knowledge. Totally, our system is comprised of three major components: the Linker, the Measure and the Generator. Figure 3 is a schematic diagram of our seeds rewriting system.

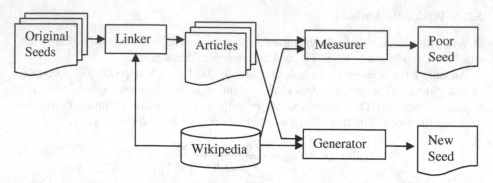

Fig. 3. Flow Chart of the Seeds Rewriting System

The Linker links every seed s in original seed set S to the article d which describes it in Wikipedia. This can be seen as a procedure of disambiguation. The Measurer measures the quality of every seed s from the following three factors: semantic relatedness, population and ambiguity and decide which seed should be replaced based on those three factors. The Generator generates new seeds using Wikipedia category structure and semantic knowledge and returns the most high-quality one as result.

4.1   Linking Seeds to Wikipedia Articles

In order to use the semantic knowledge of Wikipedia articles, the Linker needs to find out the exact articles which describe the seeds. These articles should have high semantic relatedness among themselves because they all describe instances of the same concept. Moreover, the probability of the articles being linked to should also be considered. So we design the following method to solve this problem:

For every seed s_i in S, we use it as an anchor a_i, then we can get an article set Ai includes all articles that a_i links to. So for three input seeds we get three article sets $\{A1, A2, A3\}$. For every possible article group $G:\{A1_i, A2_j, A3_k\}$, we use formula 3 to compute its confidence and choose the group which gets the highest score as result.

$$\text{Conf}(G) = Relatedness(G) + Probability(G) + Category(G) \qquad (3)$$

Here $Relatedness(G)$ is the average relatedness of each two articles in G, which is computed by using formula (1). $Probability(G)$ is the conduct of the probabilities of the articles in G which are computed by formula (2). And for $Category(G)$, if there exists common category label for all the articles, it is set to be 1, if not it is 0. Finally we choose the group that has highest confidence and link seeds $\{s_1, s_2, s_3\}$ to their related articles $\{A1_i, A2_j, A3_k\}$.

When used as an anchor a_i, every seed s_i in S links to about 10 articles in Wikipedia so we can see that the computational complexity is acceptable.

4.2   Seeds Quality Measuring

For every seed s, after linking it to article a in Wikipedia, the Measurer measures its quality from the following three factors: semantic relatedness, population and ambiguity. Then the seed with worst quality is found out and replaced.

4.2.1   Semantic Relatedness

The first factor which affects the quality of expansion is the semantic relatedness between a seed and the target concept. The higher semantic relatedness a seed has with a concept, the better it can represent the concept. So we should replace seeds with low semantic relatedness. Since target concept is unknown, we approximate the semantic relatedness of a seed as the average semantic relatedness of this seed and all other original given seeds:

$$Rel(a) = \frac{\sum_{b \in S, b \neq a} sr(a,b)}{(M-1)} \qquad (4)$$

where S is the given seed set, a is a seed and M is the size of S.

4.2.2   Population

The second factor which determines the quality of a seed is population. Some entities are sparser than other ones. If we use sparse entities as seeds, we may learn fewer templates which may lead to poor expansion performance. So we should replace seeds with low population.

In this paper, we use the following formula to calculate the population of a seed:

$$Pop(s) = \frac{count(s)}{MAX_{s \in S}[count(s')]} \qquad (5)$$

where $count(s)$ of each seed s is the number of web pages returned when we use s as query searching by the Bing API. S refers to the given seeds set.

4.2.3   Ambiguity

The third factor which determines the quality of a seed is ambiguity. As the former example shows, the seed "Mercury" may refers to the element "Mercury (element)", or a planet "Mercury (planet)", or the roman god "Mercury (roman god)". So seed "Mercury" can results in errors during expansion for the concept "Element". In this paper we define ambiguity as the probability that a seed link to the target article.

To calculate the ambiguity of a seed, we use the method described in formula (2):

$$Amb(s) = \Pr ob(<s,a>) \qquad (6)$$

where a is the article which describes s in Wikipedia.

For the three original seeds, we measure their quality from the above three factors. In order to find out influence of different factors, we attempt three ways to decide which seed to be replaced. In each way, we measure one factor. A detailed analysis is shown in Section 5.

4.3   New Seed Generation

The Generator generates new, high-quality seeds. It extracts candidate new seeds using the category structure of Wikipedia and then measures their quality and returns the one with highest quality. A schematic diagram is shown in Figure 4.

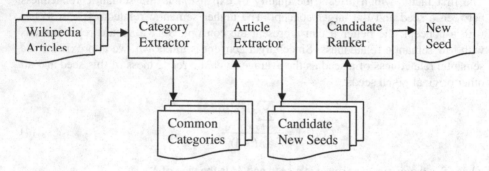

Fig. 4. Flow chart of the new seed generation procedure

Given the three original articles, the Category Extractor gathers their common category labels for further processing. If there is no common category for all three articles, it gathers common category labels for every two articles. The Article Extractor collects all articles belong to the common category labels and using their title as candidate new seeds. The Candidate Ranker uses the following combined formula to measure the quality of every candidate new seed from the three factors discussed in section 4.2. Lastly the Generator returns the one with highest score as the new seed.

$$Qua(s) = Rel(s) + Pop(s) + Amb(s) \qquad (7)$$

5   Experiments

In this section, we analyze the experimental results of our methods. First, we explain our data set and evaluation criteria. Then we discuss the performance of our method.

5.1   Experimental Setup

For evaluating our algorithm, we use 6 lists of named entities chosen from Wikipedia "List of" pages as the gold standard which is the same as [1]. The lists are:

CA counties, Countries, F1 Drivers, Elements, US States and Roman Emperors.

Each list represents a single concept. We use English Wikipedia Ver.20110722[1]. To deal with Wikipedia data, we use the Wikipedia miner[2] toolkit.

To expand the seeds, we employ the expansion algorithm described in [10]. We use R-precision to evaluate the expansion performance. It is also used by [1].

$$R-precision(L) = \frac{RM}{N} \tag{8}$$

Where L is a ranked list of extracted mentions; N is the size of the gold standard set; RM is the num of right mentions in the list.

5.2 Linking Method Evaluation

To evaluate our linking algorithm proposed in section4.1, we make 500 trials for every list. Table 2 shows the linking result. By using our combined disambiguation algorithm, we get 94% linking precision which can meet the need for further processing.

Table 2. Linking precision analysis over six gold standard entity types

Concept	Relatedness	Probability	Category	Combine
California Counties	0.886	0.842	0.916	0.910
Countries	0.270	0.274	0.856	0.946
Elements	0.980	0.960	1.000	1.000
F1 Drivers	0.454	0.402	0.354	0.902
Roman Emperors	0.760	0.752	0.392	0.880
U.S. States	0.136	0.142	0.938	1.000
Average	0.581	0.563	0.726	**0.940**

5.3 Overall Performance

For evaluating the effectiveness of our seed generation method, we do 10 trials for each concept. In each trial, we randomly choose 3 entities as input seeds. Then we use the method described in section 4 to generate a new, high-quality seed to replace the most poor-quality original seed.

Table 3 shows the overall performance. Column3~5 show the expansion performance after using the new seed to replace the most poor-quality seed measured by one single factor. As comparison, Column 2 shows the performance for the original input seeds. We can see that by replacing the original poor-quality seed with the new high-quality seed generated by our method, the average expansion performance can get obviously improved by up to 9.1% in R-precision.

---

[1] http://download.wikipedia.com/enwiki/
[2] http://wikipedia-miner.cms.waikato.ac.nz/

Table 3. Overall R-precision analysis over six gold standard entity types

Concept	Original	Ambiguity	Population	Relatedness
California Counties	0.859	0.976	0.926	0.964
Countries	0.667	0.701	0.652	0.670
Elements	0.784	0.894	0.981	0.889
F1 Drivers	0.456	0.569	0.628	0.549
Roman Emperors	0.503	0.509	0.486	0.554
U.S. States	0.908	0.902	0.858	0.824
Average	0.696	**0.759**	0.755	0.742

5.4   Detailed Analysis

Experimental results in Table 3 show that the expansion performance differs a lot among various concepts. Experimental results published by former study also show this phenomenon [1]. This can be ascribed to the difference in the natures of concepts. Some concepts are more common (such as "Countries" or "U.S. States") or have less ambiguity (such as "Elements"). Entities belonging to them are easier to be found so we get better expansion performance.

We also conclude that the three factors have different impact on the performance of entity set expansion. We get the best performance when replacing the original seed which performs worst measured by ambiguity. This suggests that the ambiguity is the most important factor in measuring the quality of seeds. We see population has almost the same influence. This phenomenon supports our hypothesis that the ambiguity and population are both very important for seeds. As comparison, Semantic Relatedness has less influence.

6   Conclusions and Future Work

In this paper, we propose a novel method for seeds rewriting for entity set expansion. For every input seed, we measure its semantic relatedness, ambiguity and population and decide which one to be replaced. Then we generate new high-quality seed by leveraging Wikipedia semantic knowledge and replace the old one. Experimental results show that our method can improve expansion performance by up to average 9.1% over original input seed sets.

For future work, we plan to use other semantic knowledge provided by Wikipedia like category hierarchy and structural description of entities to help generating better seeds.

Acknowledgements. This work was supported by the National Natural Science Foundation of China (No. 61070106), the National Basic Research Program of China (No. 2012CB316300), the Strategic Priority Research Program of the Chinese Academy of Sciences (Grant No. XDA06030300) and the Tsinghua National Laboratory for Information Science and Technology (TNList) Cross-discipline Foundation. We thank the anonymous reviewers for their insightful comments.

References

1. Richard, W., Nico, S., William, C., Eric, N.: Automatic Set Expansion for List Question Answering. In: Proceedings of EMNLP 2008, pp. 947–954. ACL, USA (2008)
2. Vishnu, V., Patrick, P., Eric, C.: Helping editors choose better seed sets for entity set. In: Proceedings of CIKM 2009, pp. 225–234. ACM, Hong Kong (2009)
3. Marco, P., Patrick, P.: Entity Extraction via Ensemble Semantics. In: Proceedings of EMNLP 2009, pp. 238–247. ACL, Singapore (2009)
4. Luis, S., Valentiin, J.: More Like These: Growing Entity Classes from Seeds. In: Proceedings of CIKM 2007, pp. 959–962. ACM, Portugal (2007)
5. Richard, W., William, C.: Automatic Set Instance Extraction using the Web. In: Proceedings of ACL/AFNLP 2009, pp. 441–449. ACL, Singapore (2009)
6. Patrick, P., Eric, C., Arkady, B., Ana-Maria, P., Vishnu, V.: Web-Scale Distributional Similarity and Entity Set Expansion. In: Proceedings of EMNLP 2009, Singapore, pp. 938–947 (2009)
7. Yeye, H., Dong, X.: SEISA Set Expansion by Iterative Similarity Aggregation. In: Proceedings of WWW 2011, pp. 427–436. ACM, India (2011)
8. Marius, P.: Weakly-supervised discovery of named entities using web search queries. In: Proceedings of CIKM 2007, pp. 683–690. ACM, Portugal (2007)
9. Richard, W., William, C.: Iterative set expansion of named entities using the web. In: Proceedings of ICDM 2008, pp. 1091–1096. IEEE Computer Society, Italy (2008)
10. Richard, W., William, C.: Language-Independent Set Expansion of Named Entities using the Web. In: Proceedings of ICDM 2007, USA, pp. 342–350. IEEE Computer Society (2007)
11. David, M., Ian, H.W.: Learning to link with Wikipedia. In: Proceedings of CIKM 2008, pp. 509–518. ACM, USA (2008)

Combining Modifications to Multinomial Naive Bayes for Text Classification

Antti Puurula

Department of Computer Science, The University of Waikato,
Private Bag 3105, Hamilton 3240, New Zealand

Abstract. Multinomial Naive Bayes (MNB) is a preferred classifier for many text classification tasks, due to simplicity and trivial scaling to large scale tasks. However, in terms of classification accuracy it has a performance gap to modern discriminative classifiers, due to strong data assumptions. This paper explores the optimized combination of popular modifications to generative models in the context of MNB text classification. In order to optimize the introduced classifier metaparameters, we explore direct search optimization using random search algorithms. We evaluate 7 basic modifications and 4 search algorithms across 5 publicly availably available datasets, and give comparisons to similarly optimized Multiclass Support Vector Machine (SVM) classifiers. The use of optimized modifications results in over 20% mean reduction in classification errors compared to baseline MNB models, reducing the gap between SVM and MNB mean performance by over 60%. Some of the individual modifications are shown to have substantial and significant effects, while differences between the random search algorithms are smaller and not statistically significant. The evaluated modifications are potentially applicable to many applications of generative text modeling, where similar performance gains can be achieved.

1 Introduction

The Multinomial Naive Bayes (MNB) model [1, 2, 3, 4] has a number of attractive features for most text classification tasks. It is simple and can be trivially scaled for large numbers of classes, unlike discriminative classifiers. It is generally robust even when its assumptions are violated. Being a probabilistic model, it is very easy to extend for structured modeling tasks, such as multi-field documents and multi-label classes. The main limit to the use of MNB is its strong modeling assumptions. The same simplifying assumptions that make it efficient and reliable also result in a severe performance gap when compared to discriminative classifiers such as Support Vector Machines [5, 6, 7] (SVM).

In practice many applications of MNB use modifications for correcting the strong assumptions, requiring parameters external to the actual model, also called metaparameters. An example of this is the mandatory MNB modification: smoothing of the class-conditional multinomials. Due to the data sparsity problem in natural language, maximum-likelihood estimates for words will result

Y. Hou et al. (Eds.): AIRS 2012, LNCS 7675, pp. 114–125, 2012.

in zero-estimates and correcting this effectively requires a parameterized smoothing technique. Combinations of modifications to MNB has been suggested as a high-performing solution to text classification [2, 4]. This however results in a number of interacting metaparameters, making simple grid searches insufficient to optimize the resulting "black-box" or direct search problem.

With the arrival of both multi-core processors and cloud computing it has become convenient to optimize machine learning systems by using parallel direct search algorithms. This enables optimization of all metaparameters concurrently, with respect to any measure of system performance. Recently random search methods have been advocated for optimizing the metaparameters required by machine learning systems [8]. Random searches offer a paradigm of direct search that makes few assumptions and is ideally suited to the noisy, multi-modal and low-dimensional task encountered here.

In this paper we explore the use of random search methods for optimizing modifications to MNB. We evaluate a total of 7 metaparameters to improve MNB performance with 4 different random search algorithms, including a modified random search proposed in this paper. We compare the results to similarly optimized SVM classifiers across five different datasets. In addition, comparison results are made to unoptimized classifiers and to the individual modifications.

The paper is organized as follows. Section 2 presents MNB, feature transforms and the evaluated modifications. Section 3 presents the random search methods. Section 4 presents results from the optimization experiments on the datasets and Section 5 completes the paper with a discussion.

2 Multinomial Naive Bayes

2.1 Model Definition

MNB is a generative model; a model of the joint probability distribution $p(\boldsymbol{w}, c)$ of word count vectors $\boldsymbol{w} = [w_1, ..., w_N]$ and class variables $c : 1 \leq c \leq M$, where N is the number of possible words and M the number of possible classes. Bayes classifiers use the Bayes theorem to factorize the generative joint distribution into a *class prior* $p(c)$ and a *class conditional* $p(\boldsymbol{w}|c)$ models with separate parameters, so that $p(\boldsymbol{w}, c) = p(c)p(\boldsymbol{w}|c)$. Naive Bayes classifiers use the additional assumption that the class conditional probabilities are independent, so that $p(\boldsymbol{w}|c) \propto \prod_n p(w_n, n|c)$. MNB parameterizes the class conditional probabilities with a Multinomial distribution, so that $p(w_n, n|c) = p(n|c)^{w_n}$.

In summary, MNB takes the form:

$$p(\boldsymbol{w}, c) = p(\boldsymbol{w}|c)p(c) \propto p(c) \prod_{n=1}^{N} p(n|c)^{w_n}, \tag{1}$$

where $p(c)$ is Categorical and $p(\boldsymbol{w}|c)$ Multinomial.

The main strength of MNB is its scalability. Training a MNB is done by summing the counts w_n found for each pair (c, n) in training documents and normalizing these over n to get $p(n|c)$. Omitting the normalization, this has the time complexity $O(D\,e(s(\boldsymbol{w})))$, where D is the number of training documents and $e(s(\boldsymbol{w}))$ is the average number of unique words per document. Classification is done by choosing the class c to maximize $p(\boldsymbol{w}, c)$. This has the time complexity $O(s(\boldsymbol{w})\,M)$, where $s(\boldsymbol{w})$ is the number of unique words in the document.

2.2   Feature Normalization

Feature normalizations such as TF-IDF [9] have been successfully used to correct some of the multinomial data assumptions with MNB [2, 3, 4]. A variety of TF-IDF functions exist in information retrieval literature. The generic version of TF-IDF used here takes the form:

$$w_n = \log[1 + \frac{w_n^u}{s(\boldsymbol{w}^u)}] \log[\frac{D}{D_n})],\tag{2}$$

where w_n^u is the unsmoothed word count, D_n the number of training documents the word n occurs in, and D the number of training documents.

The first factor in the TF-IDF function performs unique length normalization [10] and word frequency log transform. Length normalization corrects for varying document lengths. Other common choices for length normalization are word count and cosine length normalizations, but unique length normalization is more consistent over varying datasets [10]. The word frequency log transform corrects for the "burstiness" effect in word counts, by damping high counts to better fit the Multinomial distribution assumed by MNB [2]. The second factor in the TF-IDF function performs IDF weighting of words. This gives more weight to informative words, causing the weighted counts to better separate the classes.

2.3   Modifications for MNB

The MNB model as defined in Equation 1 is simple, robust and efficient. It however suffers from a number of incorrect assumptions when applied to text classification, and in the following we introduce modifications that have been used in generative modeling. For optimization, the required metaparameters are represented using a vector of scalar values $\boldsymbol{a}$. The original unmodified probabilities are denoted p^u. All modifications listed next can be turned off by setting the associated metaparameter to a non-influencing value.

- a_1 Length scaling (length_scale)
  The length normalization in Equation 2 applies a length normalization before log transformation of counts. Other popular alternatives are applying length normalization after or using a combined function. We can replace the TF factor in Equation 2 with a simple generalization: $\log[1 + \frac{w_n^u}{s(\boldsymbol{w}^u)^{a_1}}]/s(\boldsymbol{w}^u)^{1-a_1}$. With $a_1 = 1$ length normalization is done before log-transforming counts, with $a_1 = 0$ it is done afterwards. Values $0 < a_1 < 1$ produce smooth combinations of these while $a_1 > 1$ and $a_1 < 0$ give more extreme normalizations.

- a_2 IDF lifting (idf_lift)

  A number of IDF variants exist, most popular ones being Robertson-Walker IDF and Croft-Harper IDF [11, 12]. The most suitable IDF function depends on the dataset and word vector preprocessing. Similar to Lee[11], we can replace the IDF factor in Equation 2 with a generalization that covers these two common functions: $\log[\max(1, a_2 + \frac{D}{D_n})]$. $a_2 = 0$ corresponds to Robertson-Walker IDF, $a_2 = -1$ corresponds to unsmoothed Croft-Harper IDF. Values $a_2 > 0$ produce weaker IDF weighting while values $a_2 < -1$ produce steeper IDF weighting.

- a_3 Conditional uniform smoothing (cond_unif)

  Several smoothing methods exist for multinomials, one common choice being interpolation with a uniform distribution. This is included in most smoothing strategies for n-gram language models. Smoothing the conditional with a uniform replaces the Multinomial $p(n|c)$ in Equation 1 by $(1 - a_3)p(n|c) + a_3 U$, where U is the Uniform distribution.

- a_4 Conditional background smoothing (cond_bg)

  The conditional models can be smoothed several times, since different smoothing methods have complementary advantages. This is commonly done in n-gram models, as well as in language models for information retrieval [13]. Interpolation with a class-independent background Multinomial reduces the effect of more common words, similarly to IDF. We can combine this with the uniform smoothing a_3, so that the Multinomial $p(n|c)$ in Equation 1 is replaced by $(1 - a_3 - a_4)p(n|c) + a_4 p(n) + a_3 U$.

- a_5 Prior scaling (prior_scale)

  Scaling of the prior model probabilities is done in generative models for speech recognition, in order to correct differences in the prior and conditional model scores. MNB priors can be easily scaled in the same fashion, so that the prior $p(c)$ in Equation 1 is replaced by $p(c)^{a_5}$.

- a_6 Prior uniform smoothing (prior_unif)

  Like the conditionals, the prior can be smoothed. For instance, this is done in speech recognition systems, where the prior language model is extensively smoothed. Here we attempt interpolation smoothing to a uniform, just like the conditional uniform smoothing. Taking the prior scaling with a_5 into account, this replaces the prior in Equation 1 by $p(c) \propto ((1 - a_6)p(c) + a_6 U)^{a_5}$.

- a_7 Evaluation list pruning (list_prune)

  Information retrieval systems use pruning criteria to reduce the number of considered classes [14]. Using pruning along retrieval is equivalent to using a weak pre-classifier, potentially both improving accuracy and reducing computation. Here we compute a score q_c for each class by summing the TF-IDF weighted word counts having non-zero unsmoothed probability for the class: $q_c = \sum_n w_n \mathbf{1}_{p^u(n|c)>0}$. If this TF-IDF weighted sum of matching words is under a pruning threshold $q_c < a_7$, the class won't be evaluated.

3   Direct Search Optimization with Random Search

In virtually all applications of machine learning some metaparameters are used, most commonly heuristic values used for feature normalization. When only a few metaparameters are involved, simple methods such as grid searches can be used for calibration. But as the systems become more complex, the performance becomes a function of metaparameters with complex interactions. It is then not guaranteed that grid searches or heuristic values will give a realistic measure of calibrated performance. A solution that has been proposed is direct search with random search methods to optimize the performance [8]. In this paper the metaparameter vector a is optimized using four basic random search methods.

Direct search optimization [15, 16] deals with optimization of an unknown function f, when values for the function can be computed with point evaluations $f(a)$. Since the function f is unknown, it can only be approximately optimized. The optimization algorithm of choice for a particular problem depends on many factors, most importantly the amount of points that can be computed and any simplifying assumptions that can be made about the function, such as unimodality and smoothness. As direct search problems are encountered in so many different fields, the literature on the subject is fragmented and consists of hundreds of extensively researched approaches to the problem. Even the problem definition itself goes under different names, including black-box optimization and metaheuristics. Hansen [17] provides a recent evaluation of the most advanced optimization algorithms over a variety of synthetic test functions.

Random search [18, 19] algorithms are an approach to direct search that doesn't rely on strong assumptions about the function. In random searches the function is randomly sampled, guided by information from previous points. This is well suited for low-dimensional optimization, where a small number of randomly generated points can consistently improve the function value. Not relying on strong assumptions makes the algorithms more robust when used in the very multi-modal and noisy functions encountered with real-word optimization problems.

In random search the current best point a is improved by generating a new point $d = a + \Delta$ with step Δ and replacing a with d if the new point is as good or better, if $f(a) \leq f(d)$, $a = d$. The iterations are then continued to a maximum number of iterations I. Many variations of this basic random search exist. An important special case is Pure Random Search, that makes no assumptions and replaces the step generation by uniform sampling of the value ranges. Adaptive Random Searches [19] use stepsizes g_t adapted heuristically according to past points. Steepest Ascent Hill-climbing and other parallelized random searches generate J points per iteration and each point $d_j = a + \Delta_j$ is considered for replacement. Evolution Strategies takes this further, by maintaining a set of Z best points a_z instead of a single point a. Further refinement of this with recombination of best points leads to Continous Genetic Algorithms.

In this paper we evaluate optimization with four variants of parallelizable random search methods. The first one is *Pure Random Search* (prs), sampling the value ranges of the parameters uniformly, currently advocated for optimizing machine learning systems [8]. The second one is a parallelized *Random Search* (rs), using multivariate Gaussian distribution for sampling points and a log-curve stepsize decrease, by multiplying stepsizes g_t by 0.9 after each iteration. Random search with Gaussian step generation is also known as Random Optimization, whereas the log-curve stepsize decreases are very a common strategy used in optimization algorithms, including neural network training algorithms and the Luus-Jaakola random search. The third random search we use is a modification proposed in this paper called *Modified Random Search* (rs+), using four heuristics that are useful with random search methods. As the fourth method we evaluate *Covariance Matrix Adapted Evolution Strategies* (cmaes)[1], a state of the art random search method that forms the basis for many of the leading algorithms for hard optimization problems [20, 17]. This extends Evolution Strategies by using successful steps to adapt the directions of step generation.

The modified random search rs+ proposed here combines four heuristics: multiple best points, adaptive stepsizes, Bernoulli-Lognormal sampling, and mirrored step directions:

- Multiple best points
  Instead of using a single point, each non-degrading iteration replaces the current vector of best points with the Z points a_z sharing the best value. The parallel points d_j for the next iteration are then generated evenly from the vector of best points $a_{1+j\%Z}$.

- Adaptive stepsizes
  Stepsizes can be adapted according to success of previous points. Weak adaptation strategies are preferred with hard functions, as strong adaptation can lead to overadaptation and premature convergence to a local optimum. A common step adaptation strategy is used here to replace the log-curve stepsize decrease used in the baseline random search. Stepsizes are started at half range $g_t = 0.5 * (max_t - min_t)$ and multiplied by 1.2 after an improving iteration and by 0.8 after a non-improving iteration.

- Bernoulli-Lognormal sampling
  Sampling with a Gaussian distribution generates steps that are often too close to current values to make a significant difference. In contrast, some optimization algorithms use Bernoulli-generated step directions with decreasing stepsizes. This resuls in a hypercube sampling pattern around the current maxima, with more meaningful stepsizes, but with the assumption that succesful steps lie in the corners of the sampling hypercube. The Bernoulli-Lognormal proposed here samples the direction for each parameter with a uniform Bernoulli $b_{jt} \in (-1, 1)$, followed by Lognormal sampling of a stepsize multiplier $e^{\mathcal{N}(0,2)}$. Combined with the stepsizes g_t, generating the steps Δ_{jt} takes the form: $\Delta_{jt} = b_{jt} \, g_t \, e^{\mathcal{N}(0,2)}$. Ideally this creates meaningful steps, but does not assume that in successfull steps the parameters are perfectly correlated.

---

[1] http://www.lri.fr/~hansen/pythoncma.html

- Mirrored step directions
  Generating parallel points for each iteration independently is not an optimal exploration strategy for local sampling. If the steps are positively correlated, they will explore a related direction in the function, leaving other directions of the function unexplored. In practice a simple exploration strategy that is commonly used is mirroring or double-shot strategy[21] of steps, so that if $j\%2 = 0$, $b_j = -b_{(j-1)}$. This samples every even step in an opposite direction from the previous step, resulting in better coverage of the search space.

The Bernoulli-Lognormal sampling is novel in this paper, whereas the other heuristics are commonly found across different optimization algorithms. For all optimization algorithms heuristic initial values a_1 are chosen and each metaparameter is constrained to a permissible range $d_{jt} = min(max(a_{1+j\%Z} + \Delta_{jt}, min_t), max_t)$.

4   Experiments

4.1   Experiment Setup

Five standardized and publicly available datasets[2] [22] were used for the experiments. These are R8, R52, WebKb, 20Ng and Cade. To cope with the limited training data for these datasets, random 5-fold segmentation was used to split the training data, reserving 200 documents per fold for a development set. This sums to 1000 documents per dataset for optimizing Accuracy. For all experiments 50x40 (50 iterations with 40 points) random searches were used for optimization. Paired one-tailed t-tests with $p < 0.05$ were used to test for significance of evaluation set results.

TF-IDF was used in all experiments to normalize the features. The Multiclass SVM provided by Liblinear toolkit[3] [7] was used for the SVM comparison. This is based on multiclass formulation of the SVM classifier [5, 6]. The SVM C-parameter was optimized together with the TF-IDF transform for the SVM results.

4.2   Experiment Results

We first evaluated the optimization algorithms, optimizing the 7 metaparameters for MNB and 3 for SVM. Figure 1 shows the mean Accuracies across the training dataset folds for each iteration of optimization. Table 1 shows the corresponding evaluation set accuracies and a baseline "none" result from using unoptimized classifiers. For smoothing the unoptimized classifiers the SVM C-parameter was set to 1 and the MNB cond_bg-parameter was set to 0.1.

---

[2] http://web.ist.utl.pt/~acardoso/datasets/
[3] http://www.csie.ntu.edu.tw/~cjlin/liblinear/

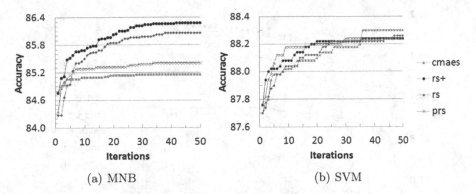

(a) MNB

(b) SVM

Fig. 1. Training set mean Accuracies

The mean differences between the optimization algorithms would suggest that the local random searches rs and rs+ work better in the higher-dimensional MNB optimization task. However, these differences between the algorithms are not statistically significant. Differences between the classifier configurations are significant and substantial. On average the use of optimized modifications improves MNB performance by 4.18 percent absolute, or over 20% relative ($p = 0.010$). For SVM the corresponding improvement is smaller 1.08 and nonsignificant. SVM is significantly more accurate for both unoptimized ($p = 0.009$) and optimized classifiers ($p = 0.036$). The use of optimization reduces the mean difference between SVM and MNB performance by over 60% relative.

Table 1. Evaluation set Accuracies

Dataset	MNB					SVM				
	cmaes	rs+	rs	prs	none	cmaes	rs+	rs	prs	none
WebKb	86.17	**88.11**	86.96	87.54	81.45	**92.69**	92.41	92.41	**92.69**	88.54
R8	95.84	95.89	**96.25**	96.12	91.18	97.76	**97.90**	97.76	97.67	97.62
Cade	59.37	**59.76**	59.42	58.36	58.77	60.60	60.46	60.35	**60.69**	60.08
20Ng	83.79	84.07	83.42	**84.17**	81.83	83.90	83.38	83.42	83.18	**84.39**
R52	92.02	**92.60**	92.33	92.52	86.29	**95.37**	95.21	95.21	95.21	94.28
mean	83.44	**84.09**	83.68	83.74	79.90	**86.06**	85.87	85.83	85.89	84.98

Given the large improvement in MNB performance from the modifications, it would be useful to know which modifications were causing this. As a second experiment, we altered the MNB rs+ configuration by fixing each of the meta-parameters to a default non-influencing value. The model optimizations were otherwise done the same way as in the first experiment. This measures the influence of each modification separately, while taking modification interactions into account. Figure 2 shows the shows the training set mean Accuracies and table 2 shows the resulting evaluation set Accuracies.

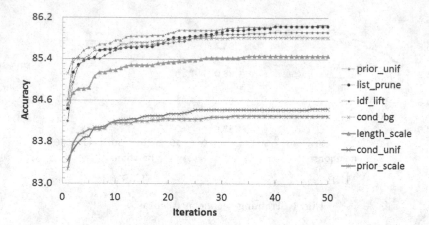

Fig. 2. Training set mean Accuracies for MNB rs+, removing a modification

Comparing the results in table 2 shows that two of the modifications have substantial and significant effects on MNB performance: cond_unif ($p = 0.003$) and prior_scale ($p = 0.016$). Three other modifications have small but non-significant effects: idf_lift, cond_bg and length_scale.

Table 2. Evaluation set Accuracies for MNB rs+, removing a modification

Dataset	cond_unif	prior_scale	length_scale	cond_bg	idf_lift	list_prune	prior_unif
WebKb	86.32	86.17	**85.10**	88.25	88.11	88.47	88.04
R8	**93.79**	94.97	96.35	96.35	96.07	96.21	96.25
Cade	59.76	57.93	59.71	**57.81**	59.24	59.75	59.67
20Ng	**82.80**	83.45	83.49	83.58	83.37	83.41	83.57
R52	**89.49**	91.36	93.11	92.56	92.56	92.60	92.99
mean	**82.43**	82.78	83.55	83.71	83.87	84.09	84.10

5   Discussion

Earlier work in MNB text classification has suggested combination of modifications [4] and used combination with heuristic values [2]. This paper evaluated the combination of 7 basic modifications and proposed the use of direct search algorithms for this task. Averaged over the evaluation sets of 5 standard text classification datasets, a relative error reduction of over 20% was achieved compared to baseline MNBs. The performance gap to similarly optimized SVM classifiers was reduced by over 60%. This raises the question whether generative modelling for text classification could be improved to the same level of performance as discriminative classifiers.

Two of the modifications had large and significant effects on MNB performance. Both scaling of prior probabilities and uniform smoothing of multinomials increased accuracy by over a percent absolute. It should be noted that neither modification is generally used for MNB text classification. Three of the other modifications had small effects as well, but it will take further experiments to see how significant these are. Taking into account earlier work in two-stage smoothing of IR language models [13] and in optimized TF-IDF transforms [11], it is likely these modifications are beneficial. Future research could include modifications such as smoothing by class clusters, as well as the common Dirichlet prior and absolute discounting methods.

Four different random search algorithms were explored to optimize the combination. While the differences between the algorithms were nonsignificant, each of them performed succesfully at this task. Taking the decades of extensive work in the field of optimization into account, it can be said that the prs strategy is preferred when no assumptions can be made about the function. Once assumptions can be made about the optimization problem, more elaborate algorithms such as cmaes and the proposed rs+ are more likely to perform better. It is also likely that complex optimizers such as cmaes will excel once the number of optimized parameters is sufficiently large. However, in practice the best way to improve optimization is simply removing unnecessary parameters and selecting the permissible value ranges correctly.

Optimization with direct search has some similarities to discriminative model training, although not many. Both optimize a measure related to classification performance instead of likelihood. Direct search works on the small set of model metaparameters, whereas discriminative training optimizes the actual model parameters. Optimized metaparameters such as those for smoothing and feature transforms can potentially transfer to similar datasets, whereas the same does not apply to the outcome of discriminative training. An important question left for further research is comparison of discriminative training to direct search optimization.

The use of optimized modifications and direct search algorithms can be highly useful in constructing generative models of text for uses other than text classification. In text modeling many potential methods such as decision trees and mixture models require additional metaparameters to work optimally. In future research we will attempt direct search algorithms for integrating more complex statistical models into text modeling.

References

[1] Lewis, D.D.: Naive (Bayes) at Forty: The Independence Assumption in Information Retrieval. In: Nédellec, C., Rouveirol, C. (eds.) ECML 1998. LNCS, vol. 1398, pp. 4–15. Springer, Heidelberg (1998)
[2] Rennie, J.D., Shih, L., Teevan, J., Karger, D.R.: Tackling the poor assumptions of naive bayes text classifiers. In: ICML 2003, pp. 616–623 (2003)
[3] Kibriya, A.M., Frank, E., Pfahringer, B., Holmes, G.: Multinomial Naive Bayes for Text Categorization Revisited. In: Webb, G.I., Yu, X. (eds.) AI 2004. LNCS (LNAI), vol. 3339, pp. 488–499. Springer, Heidelberg (2004)

[4] Schneider, K.-M.: Techniques for Improving the Performance of Naive Bayes for Text Classification. In: Gelbukh, A. (ed.) CICLing 2005. LNCS, vol. 3406, pp. 682–693. Springer, Heidelberg (2005)

[5] Crammer, K., Singer, Y.: On the learnability and design of output codes for multiclass problems. In: Proceedings of the Thirteenth Annual Conference on Computational Learning Theory, COLT 2000, pp. 35–46. Morgan Kaufmann Publishers Inc., San Francisco (2000)

[6] Keerthi, S.S., Sundararajan, S., Chang, K.W., Hsieh, C.J., Lin, C.J.: A sequential dual method for large scale multi-class linear SVMs. In: Proceedings of the 14th ACM SIGKDD International Conference on Knowledge Discovery and Data Mining, KDD 2008, pp. 408–416. ACM, New York (2008)

[7] Fan, R.E., Chang, K.W., Hsieh, C.J., Wang, X.R., Lin, C.J.: LIBLINEAR: A Library for Large Linear Classification. J. Mach. Learn. Res. 9, 1871–1874 (2008)

[8] Bergstra, J., Bengio, Y.: Random Search for Hyper-Parameter Optimization. Journal of Machine Learning Research 13, 281–305 (2012)

[9] Jones, K.S.: A Statistical Interpretation of Term Specificity and its Application in Retrieval. Journal of Documentation 28(1), 11–21 (1972)

[10] Singhal, A., Buckley, C., Mitra, M.: Pivoted document length normalization. In: Proceedings of the 19th Annual International ACM SIGIR Conference on Research and Development in Information Retrieval, SIGIR 1996, pp. 21–29. ACM, New York (1996)

[11] Lee, L.: IDF revisited: a simple new derivation within the Robertson-Spärck Jones probabilistic model. In: Proceedings of the 30th Annual International ACM SIGIR Conference on Research and Development in Information Retrieval, SIGIR 2007, pp. 751–752. ACM, New York (2007)

[12] Robertson, S., Zaragoza, H.: The probabilistic relevance framework: Bm25 and beyond. Found. Trends Inf. Retr. 3, 333–389 (2009)

[13] Zhai, C., Lafferty, J.: Two-stage language models for information retrieval. In: Proceedings of the 25th Annual International ACM SIGIR Conference on Research and Development in Information Retrieval, SIGIR 2002, pp. 49–56. ACM, New York (2002)

[14] Wang, L., Lin, J., Metzler, D.: A cascade ranking model for efficient ranked retrieval. In: Proceedings of the 34th International ACM SIGIR Conference on Research and Development in Information Retrieval, SIGIR 2011, pp. 105–114. ACM, New York (2011)

[15] Powell, M.J.D.: Direct search algorithms for optimization calculations. Acta Numerica 7, 287–336 (1998)

[16] Luke, S.: Essentials of Metaheuristics. Version 1.2 edn. Lulu (2009), http://cs.gmu.edu/~sean/book/metaheuristics/

[17] Hansen, N., Auger, A., Ros, R., Finck, S., Pošík, P.: Comparing results of 31 algorithms from the black-box optimization benchmarking bbob-2009. In: Proceedings of the 12th Annual Conference Companion on Genetic and Evolutionary Computation, GECCO 2010, pp. 1689–1696. ACM, New York (2010)

[18] Favreau, R.R., Franks, R.G.: Statistical optimization. In: Proceedings Second International Analog Computer Conference (1958)

[19] White, R.C.: A survey of random methods for parameter optimization. Simulation 17, 197–205 (1971)

[20] Hansen, N., Müller, S.D., Koumoutsakos, P.: Reducing the time complexity of the derandomized evolution strategy with covariance matrix adaptation (CMA-ES). Evol. Comput. 11(1), 1–18 (2003)

[21] Brunato, M., Battiti, R.: Rash: A Self-Adaptive Random Search Method. In: Cotta, C., Sevaux, M., Sörensen, K. (eds.) Adaptive and Multilevel Metaheuristics. SCI, vol. 136, pp. 95–117. Springer, Heidelberg (2008)

[22] Cardoso-Cachopo, A.: Improving Methods for Single-label Text Categorization. PhD thesis, Instituto Superior Técnico - Universidade Técnica de Lisboa (October 2007)

Organizing Information on the Web through Agreement-Conflict Relation Classification

Junta Mizuno, Eric Nichols, Yotaro Watanabe, and Kentaro Inui

Tohoku University, 6-3-09 Aramaki Aza Aoba, Aobaku, Sendai 980-8579, Japan
{junta-m,eric,yotaro-w,inui}@ecei.tohoku.ac.jp

Abstract. The vast amount of information on the Web makes it difficult for users to comprehensively survey the various viewpoints on topics
of interest. To help users cope with this information overload, we have
developed an Information Organization System that applies state-of-the-
art technology from Recognizing Textual Entailment to automatically
detect Web texts that are relevant to natural language queries and organize them into agreeing and conflicting groups. Users are presented
with a bird's-eye-view visualization of the viewpoints on their queries
that makes it easier to gain a deeper understanding of an issue. In this
paper, we describe the implementation of our Information Organization
System and evaluate our system through empirical analysis of the semantic relation recognition system that classifies texts and through a
large-scale usability study. The empirical evaluation and usability study
both demonstrate the usefulness of our system. User feedback further
shows that by exposing our users to differing viewpoints promotes objective thinking and helps to reduce confirmation bias.

Keywords: natural language processing, information organization,
recognizing textual entailment, agreement and conflict relations.

1 Introduction

The Web contains vast amounts of potentially useful information on a variety of
topics. Web search engines such as Google help users to locate documents that
are relevant to their topics of interest, however, they often generate millions of
hits, and it is infeasible for users to check them all. Faced with an overflow of
data, users rely on search engine ranking to act as a filter.

Although the relevance between a user's query and the documents retrieved
is reflected in search engine ranking, some users may misinterpret this ranking
as an indication of credibility or trustworthiness, Fallows [8] found that over
1/4 of Web users surveyed selected a search engine's first results because they
thought it was most trustworthy. This misunderstanding gives users who only
read the top search results a skewed impression of the information available
on a given topic, as they may miss important information, such as conflicting
opinions, buried in the search results. In order to find reliable information and
come to a deep understanding about the query, users must survey and evaluate
the documents. However, there is often too much information to be manually

Y. Hou et al. (Eds.): AIRS 2012, LNCS 7675, pp. 126–137, 2012.

feasible. A technological solution is needed to help users organize information on the Web.

To support users in organizing information, a typical approach is information classification and the most popular criteria is sentiment polarity. There has been a lot of research on classifying documents based on sentiment polarity [30,26,24]. A commonplace example of this is the classification of product reviews into *positive* and *negative* categories. Such classifications make it easy for users to evaluate a product's reputation. The study we report in this paper aims at going a step further by generalizing the previous sentiment-oriented classification approaches. We consider a natural language proposition as a user's query, and classify and organize information based on the semantic relation between the query and each relevant document.

Fig. 1 shows an example screenshot from our Japanese Information Organization System for the query, "Hybrid cars are good for the environment." Our system retrieves relevant sentences and classifies them as either AGREEMENT or CONFLICT based on their relation to the query. For example, "Hybrid cars limit emissions and gas usage" agrees with the query and "The more you drive a hybrid car, the more damage it does to the environment" conflicts with it. Finally, a bird's-eye-view of the agreeing and conflicting sentences is shown to the user. In the bird's-eye-view, sentences which agree with the query are classified into AGREEMENT and shown in a red column on the left, and sentences which are conflicting to the query are classified into CONFLICT and are shown in a blue column on the right. To give users an idea of the factual support for each position, EVIDENCE relations for AGREEMENT and CONFLICT are also displayed. For example, "Hybrid cars use an electric motor to reduce gas usage," is evidence for "so they are good for the environment" and is highlighted. Below each relevant sentence, a hyperlink to its source page is given, and summaries of the total number of sentences for each viewpoint are given, making the amount of support clear at a glance. Sentences are ordered by a confidence score, ensuring that the most relevant and reliable information is given the most visible position.

Organizing information based on AGREEMENT-vs-CONFLICT relations to a given natural language propositional query can be considered as a generalization of previous sentiment-oriented classification approaches. For example, finding positive and negative reviews for a given product can be achieved by querying "(product name) is good." Organizing such sentiment information in terms of product features (attributes) can be achieved by issuing more specific queries such as "(product name) is good for the environment." as is the example in Fig. 1. More importantly, information organization based on AGREEMENT-vs-CONFLICT relations can also handle issues which cannot be captured by sentiment orientation. For example, the query "Polar ice caps are melting" is not a matter of good or bad but rather a matter of true or false. AGREEMENT-vs-CONFLICT-based information organization with propositional queries can deal with such a broad range of cases in a uniform manner.

In order to detect the AGREEMENT and CONFLICT relations necessary for information organization, we frame our task as an extension of Recognizing

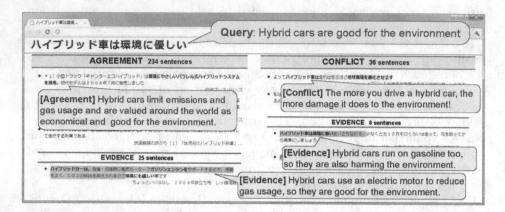

Fig. 1. An example screenshot for the query *Hybrid cars are good for the environment*

Textual Entailment (hereafter "RTE", [5]), whose goal is detecting *entailment* or *contradiction* relations between texts. We treat the former relation as AGREEMENT and the latter as CONFLICT and construct machine learning models to automatically detect these relations between user queries and Web texts.

In this paper, we implement an Information Organization System that uses RTE technology to automatically detect and organize Web texts relevant to a user query into agreeing and conflicting groups. To evaluate our system, we perform two kinds of evaluation. First, we empirically evaluate relation classification performance on a large, open dataset of real world Web texts consisting of 1,467 query-sentence pairs, taking care to use different queries for development and evaluation. Then, we conduct a large-scale usability study with 112 participants to determine the effectiveness of our system for information organization. The study participants were selected in a manner such that they have no knowledge of our research goals and do not exhibit any bias toward our system.

Our major findings are two-fold. First, classification performance evaluation showed that our system's RTE-based semantic relation recognition, when coupled with ranking by classification confidence scores, is sufficient to construct a usable information organization system. In addition, study participants reported that looking at visualizations of the viewpoints on a query lead to a deeper understanding of the query, indicating that users were receptive to our approach. These results support the view that exposing lay users to different opinions reduces *confirmation bias* [15,4].

Our primary contributions are: the development of an information organization system that automatically generates viewpoint visualizations from user queries, an investigation of the level of performance of state-of-the-art RTE technology on Web data, and a large-scale usability study of the effectiveness of our system for information visualization.

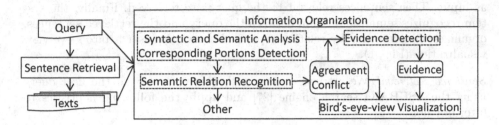

Fig. 2. The system architecture

2   Information Organization System

In this section, we describe our approach to organizing information on the Web. When a user enters a query into our system, relevant documents can be retrieved using an existing web search engine. Then if the semantic relation between the query and the relevant sentence can be recognized, the organized viewpoint visualization (shown in Fig. 1) is easy to generate. Therefore, the most important factor to be discussed is how to recognize the semantic relations between a query and a relevant sentence.

We employed the technology of RTE, a widely-researched topic in the field of natural language processing. The RTE Challenge [5] is a major task where two sentences are classified into ⟨ENTAILMENT⟩, ⟨CONTRADICTION⟩, or ⟨OTHER⟩.

Before describing how to apply the RTE technology to organize information, we need to clarify what sentences are the target of semantic relation classification. While in RTE, entailment is judged between *hypothesis* (H) and *text* (T), we use a natural language query (Q) and a retrieved sentence (T). Q and T are classified into one of four relations: AGREEMENT, CONFLICT, EVIDENCE or OTHER.

The RTE Challenge has successfully employed a variety of techniques in order to recognize instances of textual entailment, including methods based on: measuring the degree of lexical overlap between bag of words [10,31], the alignment of graphs created from syntactic or semantic dependencies [19,18], statistical classifiers which leverage a wide range of features [12], or reference rule generation [29]. In the RTE-5 [2], Iftene *et al.* [14] achieved 68.33% accuracy. In spite of the high accuracy on the RTE Challenge, as de Marneffe *et al.* [6] found in their experiments that RTE on Web texts that real world settings were much more challenging, reporting that "*In a real world setting, it is likely that the contradiction rate is extremely low*".

We tackle this problem by limiting queries to simple clauses. This means that Q consists of a single predicate and its arguments, as in *Mineral water is safer than tap water*. In RTE it is not enough to consider whether the information contained in H is also contained in T, it is also necessary to determine what information could be ignored to recognize entailment. In our case, if Q is a simple clause, the contained information is limited so there is no need to consider what information to ignore in order to detect our target semantic relations.

An overview of our Information Organization System is given in Fig. 2. A short description is as follows. A simple clause natural language sentence is taken as input. Then sentences relevant to the query are retrieved. Finally, the system recognizes semantic relation between the query and the retrieved sentences, organizes the sentences by relation type, and shows the user a bird's-eye-view visualization (Fig. 1).

Sentence Retrieval We retrieve sentences that are relevant to the user query using the TSUBAKI search engine [27] and apply the following heuristics to filter out noise.

1. the sentence contains context all nouns in query
2. the sentence length is greater than 20 words and less than 150
3. the sentence must be a well-formed sentence
4. the part of speech of the last word must be a verb, auxiliary verb, or adjective
5. the sentence contains less than three post-positional particles

Because Q consists of a single predicate and its arguments, the purpose of the filter 1 is to obtain a variety of relevant predicates. The purpose of the other filters is to exclude malformed sentences.

Linguistic Analysis In order to identify semantic relations between Q and T, we first conduct syntactic and semantic linguistic analysis to provide a basis for alignment and relation classification. For syntactic analysis, we use the Japanese dependency parser CaboCha [17] and the predicate-argument structure analyzer ChaPAS [32]. CaboCha splits the Japanese text into phrase-like units called *chunks* and represents syntactic dependencies between the *chunks* as edges in a graph. ChaPAS identifies predicate-argument structures in the dependency graph produced by CaboCha. We also conduct extended modality analysis using the resources provided by Matsuyoshi *et al.* [20], focusing on source, time, modality and polarity because such information provides important clues for the recognition of semantic relations. We conduct sentiment polarity analysis of each word in Q and T using the resources [16,13].

Corresponding Portion Detection Corresponding portions between Q and T are detected based on word similarity. When the content words in corresponding *chunks* are identical or semantically similar then they are aligned. We use the following resources to determine semantic similarity. When a pair of *chunks*, one from Q and the other from T, is found in one of the resources, the *chunks* are aligned. The *chunks* are matched against the resources using a character bi-gram cosine-based similarity measure [23].

Ontologies. We use the Japanese WordNet [3] and Sumida *et al.*'s [28] to check for hypernymy and synonymy between words. E.g. ⟨⟨効果 *kouka* "good effect"⟩ - ⟨作用 *sayou* "effect"⟩ and ⟨イソフラボン *isofurabon* "Isoflavone"⟩ - ⟨健康食品 *kenkou-shouhin* "health food"⟩.

Predicate databases. To determine if two predicates are semantically related, we consult a database of predicate relations [21] and a database of predicate entailments [11] using the predicates' default case frames. E.g. ⟨維持する *iji-suru* "to preserve"⟩ - ⟨守る *mamoru* "to maintain"⟩ and ⟨予防する *yobou-suru* "to prevent"⟩ - ⟨気をつける *ki-wo-tsukeru* "to be careful"⟩.

Semantic Relation Recognition. Once the alignments are successfully identified, the task of semantic relation classification is straightforward. We solve this problem with machine learning by training linear classifier [9]. We used an L2-regularized logistic regression model. As features, we draw on a combination of lexical, syntactic, and semantic information including the alignments from the previous section. The feature set is as follows:

alignments. We define binary function, $ALIGN_{word}(q_i, t_m)$ to be true if and only if the node $q_i \in Q$ has been aligned to the node $t_m \in T$. We also have a feature for the alignment likelihood score.

modality. This feature encodes the possible modalities of predicate nodes. Moods that do not represent opinions (i.e. *imperative, permissive* and *interrogative*) often indicate ⟨OTHER⟩ relations.

antonym. The feature indicates if a given pair is an antonym. This information helps identify ⟨CONFLICT⟩.

negation. To identify negations, we primarily rely on a predicate's *Actuality* value, which represents epistemic modality and existential negation. Aligned predicate pairs with mismatching actuality labels are likely ⟨CONFLICT⟩ or ⟨OTHER⟩.

Evidence Detection. EVIDENCE is identified in Web text after it has been classified into one of the query's viewpoints. Identification is treated as a discourse processing task where presence of explicit cues are used to indicate the presence of EVIDENCE. For example, in the following example, the cue *because* indicates that *Xylitol is effective at preventing cavities* is supporting evidence of *the cavity-causing bacteria streptococcus mutans cannot metabolize it.*

– Xylitol is effective at preventing cavities because the cavity-causing bacteria streptococcus mutans cannot metabolize it.

Visualization. Visualization is carried out as described in Section 1.

3   System Evaluation

Our goal in evaluating the semantic relation recognition performance is to determine the feasibility of applying RTE technology to the information organization task. We also conducted evaluation with a subset of the evaluation data manually annotated with correct alignments (called "Gold" in Table 1) to determine the upper bound to semantic relation recognition performance.

3.1   Evaluation Data

Murakami *et al.* [22] constructed 370 *Q-T* pairs, however their data is small and consisted of only 5 queries. This is insufficient to represent the variety of possible topics in real Web data. Furthermore, their dataset has not been publicly released.

Therefore, we constructed a new, open data set consisting of 20 queries. The new evaluation data set consists of 1,467 Q-T pairs including 1,467 instances consisting of 524 AGREEMENT instances (i.e. 524 pairs of Q-T), 205 CONFLICT, 131 EVIDENCE, and 607 OTHER instances. Annotation was carried out by two native speakers of Japanese with an inter-annotator agreement kappa score of 0.72. The data set is publicly available for download at http://www.cl.ecei.tohoku.ac.jp/stmap/.

The queries are created by an annotator who does not concern the development of our system. A criterion of constructing query is described as following.

- The query is a simple clause as described in Section 2
- When considering query as a question, it must be answerable as a yes-or-no question.
- There must be both agreeing and conflicting opinions for the query on the Web.
- There must be scientific evidence for both agreeing and conflicting opinions. In other words, queries which are not supported scientifically such as superstition are excluded.
- The actual truth of the query does not matter. It can be either true or false.

3.2   Evaluation Results

We conduct 10-fold cross validation over the 20 query open data set described above. In each fold, query-text pairs from 18 queries were used for training, and the other two queries were used for evaluation.

Table 1. System Evaluation Results

setting	#	AGREEMENT		CONFLICT		EVIDENCE	
		Precision	Recall	Precision	Recall	Precision	Recall
System	1,467	0.706	0.769	0.523	0.220	0.586	0.313
Gold	136	0.938	0.918	0.938	0.750		

The results are given in Table 1. *setting* indicates alignment setting (*Gold* is the result of using correct alignments) and # indicates the number of Q-T pairs. In the setting *System*, while ⟨AGREEMENT⟩ is identified with high precision and recall, the performance of ⟨CONFLICT⟩ and ⟨EVIDENCE⟩ are limited. In spite of the limited performance, our goal to organize information is achieved as described in Section 4

To investigate the source of errors, we randomly sampled 10% of instances from the data set and manually annotated them with alignment information. On this data set with correct alignment information, we conducted a 10-fold cross validation, where the correct alignment information was used in both training and testing. The results are shown in the rows indicated by *Gold* in Table 1. These results show a substantial gain was obtained particularly for ⟨CONFLICT⟩

when the system was given correct alignments[1]. These results suggest that it is crucial to further address the issue of alignment in order to improve the overall performance for relation classification. A breakthrough in alignment performance could help make our system practical for arbitrary, user-generated queries.

4   User Evaluation

We conducted a usability study to gain an understanding of the issues that need consideration when deploying an information organization system to real Web users and to provide further evaluation of semantic relation classification in a real world application.

In the usability study, participants were asked to compare our system to existing web search engine[2] in investigating topics with diverse viewpoints on the Web. We prepared 54 user queries in the same way as described in Section 3.1 that can be categorized into three topics: *Society, Health,* and *Environment.* Example queries are shown as follows.

– Society
  - 裁判員になるのを拒否できる *Citizen judge duty can be refused*
  - 血液型で性格が分かる *Blood type predicts personality type*
– Health
  - ミネラルウォーターは水道水より安全だ *Mineral water is safer than tap water*
  - アガリクスは健康に良い *Agaricus is healthy*
– Environment
  - 南極の氷は減っている *Polar ice caps are melting*
  - 地球温暖化によって海面が上昇する *Global warming causes rising sea levels*

We recruited 112 Japanese adults (62 males and 50 females) ranging from 20 to 70 years in age. Almost all participants identified themselves as daily Internet users. In order to avoid bias toward our system, we employed participants indirectly through a recruiting agency, having the agency conduct the evaluation and distribute a survey on completion.

Before starting evaluation, the study participants freely selected 4 queries. Then, in order to avoid bias from system ordering, they alternated the order of system evaluation between each query. Upon completion of the evaluation task, participants answered a two-question survey, rating each system on the Likert scale of 1 (strongly disagree) to 5 (strongly agree).

(1) *Could you find texts on the Web that agrees and disagrees with the query?*
(2) *Could you find texts that contain evidence that supports or opposes the agree the query?*

The results are given in Table 2. They show that satisfaction with our system was greater for both questions. Responses to question (1) showed that over 84.8%

---

[1] For EVIDENCE, we could not conduct the same experiment because the sampled gold data set included only very few instances of EVIDENCE.
[2] We use Google as the web search engine

Table 2. Usability study survey results

System \ Question No.	(1)	(2)
Web search engine	3.54	3.24
Information Organization System	4.06	3.85

of users found the AGREEMENT and CONFLICT viewpoints useful, and question (2) showed that 55.4% of users found EVIDENCE detection useful.

Despite the limited performance of our system in Table 1, it was still found more useful than the web search engine for identifying different viewpoints and their support. We offer two theories for this. First, the classified opinions shown to users are sorted by the system's classification confidence level, so many incorrectly classified results were likely not appeared in the top of the output. Second, users may have been able to find enough results which they knew were correct to give our system the edge in evaluation.

Study participants were also given an opportunity to give feedback during the survey. We received positive responses such as *"The system is useful for finding various viewpoints on the topic."* and *"The system helps me organize and understand information from various sources."*, showing that participants found the goal of Web information organization to be meaningful. For the question *"Do you want to continue to use this system?"*, 70% of the participants responded in the affirmative. This percentage includes people who want to use the system together with the web search engine.

Another response showed that classifying the viewpoints instead of ranking search results is better suited to analyzing information credibility: *"While in Google, I only look some high-ranked results, prevent me from noticing minor viewpoints, but the [Information Organization] system is useful for objective thinking because it gives each viewpoint an equal rank."*

We also received some negative responses concerning user interface problems. The majority referred to trivial issues such colors and font size, however, some raised issues about how the presentation could influence how it is perceived by users. In particular, this important response indicated that displaying information about viewpoint size could create bias toward popular ones: *"The number of opinions seems to indicate that the majority viewpoint is correct."* These sort of comments indicate that the study participants naturally understood what our system was displaying without explanation. While further evaluation is needed to determine the best way of visualizing results, these comments indicate that our approach to information is promising.

Several users expressed the desire to use our system to find viewpoints on arbitrary topics. This raises the question of how search queries should be generated from user input and is an area that requires investigation.

Other responses indicated the importance of identifying the author of texts: *"I want only trustworthy sources."* and *"There are many untrustworthy sources such as weblogs."* Detecting authorship is also important for recognizing EVIDENCE and needs more focus in information organization.

5   Related Work

There have been several studies on classifying search results into a fixed set of categories. For example, [30,26,24] classify documents by sentiment polarity. The WISDOM project [1] represents a more sophisticated approach as it classifies documents both by sentiment polarity and by source category. In both classification approaches, the system classifies a set of documents or sentences relevant to the users' queries into different categories. However, because these approaches do not explicitly consider the relation between document and query, their scope is limited. In order to handle a broader range of queries, for example, why question or yes-or-no questions, we need deeper semantic analysis.

To the best of our knowledge, the only other system that classifies and organizes search results by agreement-conflict relations with the query is Murakami et al. [22]. They addressed the problem of information credibility analysis. For that purpose, they constructed a system that searches for agreeing and conflicting sentences for a given query. In this respect, our system can be seen as similar in nature to their relation classification system. However, they do not address the task of visualizing and presenting to users the results of information organization, and their classification evaluation was limited by their small, closed dataset. In contrast, in this study, we focus on investigating how the state-of-the-art RTE technology performs on large-scale Web data, and we explore its feasibility as a backbone for information organization by implementing a end-to-end system including a user interface and visualization of information organization results and evaluating it in a large-scale usability study with over 100 participants.

Paul et al. [25] also address the task of classifying and summarizing documents by agreement-conflict relations, and they develop a method of identifying representative viewpoints from each group. However, their works assumes a collection of documents for a given controversial topic has already been gathered; they do not have any Web search integration or user evaluation.

Another related project addressing the detection of conflicting relation on the Web is Dispute Finder [7]. This project shows users who are browsing the Web a list of known conflicts whenever a disputed statement is encountered. Dispute Finder builds a database of disputed claims by allowing its users to identify disputed claims on the Web and link them to a trusted source of rebuttal. The goals of Dispute Finder and our system are similar, but Dispute Finder relies on crowd-sourcing to build its dispute database which limits its automation potential, while we use deep natural language processing technology to automatically identify conflicts.

6   Conclusion

Helping users make sense of the many opinions on the Web is an important and challenging task. In this paper, we presented an information organization system that uses state-of-the-art RTE technology to automatically recognize and classify Web texts that are relevant to user queries into agreeing and conflicting groups. In evaluating our system on a new, open data set that is representative of real world Web data, we found that current RTE technology, when coupled with classification confidence level-based scoring, provides sufficient performance to construct a usable system. In conducting a large-scale usability study, we found that our system stood up to the challenge of practical application. Study participants reported that looking at visualizations of the viewpoints on a query lead to a deeper understanding of the query, indicating that users were receptive to our approach. These results support the view that exposing lay users to different opinions reduces *confirmation bias*. In addition, user feedback also showed that there are still practical issues in displaying viewpoint information and automatic query generation that need to be addressed to build a practical tool for information organization.

Acknowledgements. This work was supported by the National Institute of Information and Communications Technology Japan and by JSPS Grant-in-Aid for Scientific Research (23240018 and 23700157).

References

1. Akamine, S., Kawahara, D., Kato, Y., Nakagawa, T., Inui, K., Kurohashi, S., Kidawara, Y.: WISDOM: A Web Information Credibility Analysis Systematic. In: Proc. of the ACL-IJCNLP 2009 Software Demonstrations, pp. 1–4 (2009)
2. Bentivogli, L., Dagan, I., Dang, H.T., Giampiccolo, D., Magini, B.: The Fifth PASCAL Recognizing Textual Entailment Challenge. In: Proc. of Text Analysis Conference, TAC 2009 (2009)
3. Bond, F., Isahara, H., Fujita, S., Uchimoto, K., Kuribayashi, T., Kanzaki, K.: Enhancing the Japanese WordNet. In: Proc. of ALR 2009, pp. 1–8 (2009)
4. Cox, L.A.T., Popken, D.A.: Overcoming Confirmation Bias in Causal Attribution: A Case Study of Antibiotic Resistance Risks. Risk Analysis an Official Publication of the Society for Risk Analysis 28(5), 1155–1172 (2008)
5. Dagan, I., Glickman, O., Magnini, B.: The PASCAL Recognising Textual Entailment Challenge. In: Quiñonero-Candela, J., Dagan, I., Magnini, B., d'Alché-Buc, F. (eds.) MLCW 2005. LNCS (LNAI), vol. 3944, pp. 177–190. Springer, Heidelberg (2006)
6. de Marneffe, M.-C., Rafferty, A.N., Manning, C.D.: Finding Contradictions in Text. In: Proc. of ACL-HLT 2008, pp. 1039–1047 (2008)
7. Ennals, R., Trushkowsky, B., Agosta, J.M.: Highlighting Disputed Claims on the Web. In: Proc. of WWW 2010, pp. 341–350 (2010)
8. Fallows, D.: Search engine users: Internet searchers are confident, satisfied and trusting – but they are also unaware and naïve. Pew Internet & American Life Project (2005)
9. Fan, R.-E., Chang, K.-W., Hsieh, C.-J., Wang, X.-R., Lin, C.-J.: LIBLINEAR: A Library for Large Linear Classification. Journal of Machine Learning Research 9, 1871–1874 (2008)
10. Glickman, O., Dagan, I., Koppel, M.: Web Based Textual Entailment. In: Proc. of the First PASCAL Recognizing Textual Entailment Workshop, pp. 33–36 (2005)
11. Hashimoto, C., Torisawa, K., Kuroda, K., Murata, M., Kazama, J.: Large-Scale Verb Entailment Acquisition from the Web. In: Proc. of EMNLP 2009, pp. 1172–1181 (2009)

12. Hickl, A., Williams, J., Bensley, J., Rink, K.R.B., Shi, Y.: Recognizing Textual Entailment with LCC's Groundhog System. In: Proc. of the Second PASCAL Challenges Workshop, pp. 80–85 (2005)
13. Higashiyama, M., Inui, K., Matsumoto, Y.: Acquiring noun polarity knowledge using selectional preferences. In: Proc. of NLP, pp. 584–587 (2008) (in Japanese)
14. Iftene, A., Moruz, M.-A.: UAIC Participation at RTE5. In: TAC 2009 (2009)
15. Keselman, A., Browne, A.C., Kaufman, D.R.: Consumer Health Information Seeking as Hypothesis Testing. Journal of the American Medical Informatics Association: JAMIA 15(4), 484–495 (2008)
16. Kobayashi, N., Inui, K., Matsumoto, Y., Tateishi, K., Fukushima, T.: Collecting Evaluative Expressions for Opinion Extraction. Journal of NLP 12(3), 203–222 (2005) (in Japanese)
17. Kudo, T., Matsumoto, Y.: Japanese Dependency Analysis using Cascaded Chunking. In: Proc of CoNLL 2002, pp. 63–69 (2002)
18. MacCartney, B., Grenager, T., de Marneffe, M.-C., Cer, D., Manning, C.D.: Learning to recognize features of valid textual entailments. In: Proc. of HLT-NAACL 2006, pp. 41–48 (2006)
19. Marsi, E., Krahmer, E.: Classification of Semantic Relations by Humans and Machines. In: Proc. of ACL 2005 Workshop on Empirical Modeling of Semantic Equivalence and Entailment, pp. 1–6 (2005)
20. Matsuyoshi, S., Eguchi, M., Sao, C., Murakami, K., Inui, K., Matsumoto, Y.: Annotating Event Mentions in Text with Modality, Focus, and Source Information. In: Proc. of LREC 2010, pp. 1456–1463 (2010)
21. Matsuyoshi, S., Murakami, K., Matsumoto, Y., Inui, K.: A Database of Relations between Predicate Argument Structures for Recognizing Textual Entailment and Contradiction. In: Proc. of ISUC 2008, pp. 366–373 (2008)
22. Murakami, K., Nichols, E., Inui, K., Mizuno, J., Goto, H., Ohki, M., Matsuyoshi, S., Matsumoto, Y.: Automatic Classification of Semantic Relations between Facts and Opinions. In: Proc. of NLPIX, pp. 21–30 (2010)
23. Okazaki, N., Ichi Tsujii, J.: Simple and Efficient Algorithm for Approximate Dictionary Matching. In: Proc. of Coling, pp. 851–859 (2010)
24. Pang, B., Lee, L., Vaithyanathan, S.: Thumbs up? Sentiment Classification using Machine Learning Techniques. In: Proc. of EMNLP 2002, pp. 79–86 (2002)
25. Paul, M., Zhai, C., Girju, R.: Summarizing Contrastive Viewpoints in Opinionated Text. In: Proc. of EMNLP 2010, pp. 66–76 (2010)
26. Seki, Y., Evans, D.K., Ku, L.W.: Overview of Multilingual Opinion Analysis Task at NTCIR-7. In: Proc. of the 7th NTCIR Workshop, pp. 185–203 (2008)
27. Shinzato, K., Shibata, T., Kawahara, D., Hashimoto, C., Kurohashi, S.: TSUBAKI: An Open Search Engine Infrastructure for Developing New Information Access Methodology. In: Proc. of IJCNLP 2008, pp. 89–196 (2008)
28. Sumida, A., Yoshinaga, N., Torisawa, K.: Boosting Precision and Recall of Hyponymy Relation Acquisition from Hierarchical Layouts in Wikipedia. In: Proc. of LREC 2008, pp. 2462–2469 (2008)
29. Szpektor, I., Shnarch, E., Dagan, I.: Instance-based Evaluation of Entailment Rule Acquisition. In: Proc. of ACL 2007, pp. 456–463 (2007)
30. Turney, P.: Thumbs Up or Thumbs Down? Semantic Orientation Applied to Unsupervised Classification of Reviews. In: Proc. of ACL 2002, pp. 417–424 (2002)
31. Jijkoun, V., De Rijke, M.: Recognizing Textual Entailment Using Lexical Similarity. In: Proc. of the First PASCAL Challenges Workshop, pp. 73–76 (2005)
32. Watanabe, Y., Asahara, M., Matsumoto, Y.: A Structured Model for Joint Learning of Argument Roles and Predicate Senses. In: Proc. of the ACL 2010 Conference Short Papers, pp. 98–102 (2010)

Exploiting Twitter for Spiking Query Classification

Mitsuo Yoshida[1,*] and Yuki Arase[2]

[1] Graduate School of Systems and Information Engineering, University of Tsukuba,
1-1-1 Tennodai, Tsukuba, Ibaraki, Japan
ceekz@mibel.cs.tsukuba.ac.jp
[2] Microsoft Research Asia,
Building 2, No.5 Dan Ling Street, Haidian District, Beijing, P.R. China
yukiar@microsoft.com

Abstract. We propose a method for classifying queries whose frequency spikes in a search engine into their topical categories such as celebrities and sports. Unlike previous methods using Web search results and query logs that take a certain period of time to follow spiking queries, we exploit Twitter to *timely* classify spiking queries by focusing on its massive amount of super-fresh content. The proposed method leverages unique information in Twitter—not only tweets but also users and hashtags. We integrate such heterogeneous information in a graph and classify queries using a graph-based semi-supervised classification method. We design an experiment to replicate a situation when queries spike. The results indicate that the proposed method functions effectively and also demonstrate that accuracy improves by combining the heterogeneous information in Twitter.

Keywords: Query Classification, Spiking Query, Twitter.

1 Introduction

The frequency of a Web search query naturally reflects the degree of people's interest in it. Therefore, queries that suddenly spike in a search engine can be regarded as gaining more attention from people. We propose a method for *timely* classifying such spiking queries into their topical categories, for example, celebrities and sports. Having categories of spiking queries is useful for search engines in various ways. They may help to immediately improve the relevance of search results and enable to trigger an appropriate vertical search when a query becomes popular. They also benefit search advertisers in presenting relevant advertisements at the time when more people are interested in related topics. Although many query classification methods have been proposed (*e.g.*, [11]), little attention has been paid to spiking queries.

The biggest challenge in spiking query classification is a lack of resources for characterizing them. Although previous methods of query classification have used Web search results and query logs[1], they are not always available in a timely fashion for spiking queries due to the queries' sudden emergence. Broder *et al.* [4] show that about 5% of queries in their experimental dataset are too recent to obtain search results for, and that

---

* This project was conducted while the first author was visiting Microsoft Research Asia.
[1] In this paper, *query logs* include all information associated with a query, even click-through.

Y. Hou et al. (Eds.): AIRS 2012, LNCS 7675, pp. 138–149, 2012.
© Springer-Verlag Berlin Heidelberg 2012

affects the classification accuracy. Apart from this challenge in resource unavailability, the expensive cost of manually labeling queries to train a classifier is a common challenge in query classification tasks, regardless of whether focusing on spiking or general queries.

To tackle these challenges, we propose a method for timely classifying spiking queries by exploiting Twitter[2], which provides a huge amount of super-fresh content on a broad range of topics in real-time. We leverage unique information that Twitter provides: tweets, users, and hashtags. We use tweets that contain spiking queries to evaluate the similarity between queries, assuming that similar queries belong to the same category. In addition, we use information on users and hashtags and evaluate their correlation with queries. We assume that if a Twitter user follows, or in other words, "belongs to" a category, queries that appear in his/her tweets also belong to that category. Likewise, if a query belongs to a category, users who posted tweets containing the query also belong to the same category. The same relationship is also applicable to hashtags. If a hashtag belongs to a category, queries that appear in tweets with the same hashtag also belong to that category, and vice versa. We use a graph-model to integrate such heterogeneous information and adapt a graph-based semi-supervised learning method for classification that requires a smaller amount of training data.

The contribution of our method is twofold. First, we propose a novel method for timely classifying spiking queries by capitalizing on its correlation with Twitter. Second, we leverage unique information that Twitter provides to improve the classification accuracy. We consider not only tweets, but also users and hashtags, to characterize spiking queries and we combine these into a graph-model.

We carefully design an experiment to evaluate the proposed method for classifying spiking queries on the day the queries become popular in a search engine. The results show that the method is effective and that accuracy improves by combining the query similarity and correlation with users/hashtags.

2   Related Work

Many studies have investigated query classification of their topical categories based on Web search results and query logs. They use supervised, semi-supervised, or unsupervised classification methods.

In supervised-learning based approaches, the KDD-Cup 2005 competition featured query classification, where attendees used search result pages and their attributes such as titles and snippets, as well as search engine directories to extract features of queries [11]. Baeza-Yates *et al.* [2] use click-through data to expand an input query and generate a feature based on terms that appear in accessed Web pages. Broder *et al.* [4] focus on rare queries with low frequency that are therefore difficult to classify. The difference between spiking queries and rare queries is the number of relevant Web pages. Broder *et al.* assume that there are still a sufficient number of Web pages to characterize the rare queries, even though their frequency in a search engine is small. Their method classifies search result pages of an input rare query and Web pages linked from them instead of

---

[2] http://twitter.com/

classifying the query itself. Nevertheless, we cannot always assume such resources are available for spiking queries when they spike in a search engine.

These methods use supervised-learning methods and thus have a drawback in that they require a large amount of training data. To relax this requirement, researchers also use semi-supervised or unsupervised learning methods. Shen *et al.* [13] propose building a fine-grained intermediate classifier through which an input query is first classified into intermediate categories based on maximum likelihood estimation, and then the intermediate categories are further classified into target categories with a coarse structure. Beitzel *et al.* [3] generate rules to classify queries based on linguistic knowledge combined with a classifier trained by a supervised-learning method. Xiao *et al.* [10] conduct binary classification that decides whether a query has a predetermined intent, such as job or product intents, using click-through data. Diemert and Vandelle [5] construct a concept graph using their search result pages and query logs, in which they inject target categories for classification. They expand an input query with its search result pages and extract salient categories by matching the expanded query and concepts by random-walk in the graph. Hu *et al.* [7] construct such a concept graph using Wikipedia.

These previous studies depend on relevant Web pages and query logs and they are therefore not applicable for timely classifying spiking queries for which these resources are not always available.

Another stream of related work exploits Twitter for extrinsic tasks such as detecting earthquakes [12], identifying and ranking URLs of trendy Web pages [6], and determining high-quality content from a QA portal [1]. These studies have different goals; however, they show that Twitter is a valuable resource that produces super-fresh content and reflects the trends of the general public. These studies also demonstrate the usefulness of various content in Twitter, *i.e.*, URLs, users, and their social relationships.

3 Problem Statement

We start by formally defining the spiking query classification problem. In this study, we use Japanese queries and tweets, since Twitter has millions of users in Japan and is therefore popular, and we can obtain a sufficient number of tweets to conduct query classification. Although we use Japanese data, our approach is language-independent and is easily applicable to other languages.

3.1 Definition of Spiking Query

We first define a spiking query as a query whose frequency in a search engine spikes; a spike occurs when there is a massive increase followed by a corresponding decrease in the query's frequency [9]. We regard queries spiking once or multiple times as spiking queries and aim to classify them when they become popular. In this study, we choose queries showing spikes based on their frequency history from all queries that we obtain through a toolbar (Bing Bar[3]) installed on Windows Internet Explorer[4]. Our method can handle queries with either single-term or multiple-term structures.

---

[3] http://toolbar.discoverbing.com/
[4] http://windows.microsoft.com/en-US/internet-explorer/products/ie/home

We observe trends in spiking queries on Twitter in comparison with their trends on news pages to investigate whether tweets containing the spiking queries are available in a timely fashion. News pages serve as a good baseline since they are highly responsive to fresh topics, *i.e.*, spiking queries, since they are intended to convey timely information to people. We obtain news data by crawling the main content of Web pages from a news portal (Ceek.jp News[5]) covering local and nationwide news agencies throughout Japan.

A typical example is in Fig. 1. A trend in the spiking query karelog is evident from August 24 to September 15, 2011. karelog is the name of a mobile application released on August 28 in Japan. The frequencies of the query and tweets are shown in a logarithmic scale (for intuitive representation, we add 1 to all frequencies to avoid having a missing value when the raw frequency is 0), while the frequencies of the news pages are raw values. It is clear that the query created a buzz on Twitter with more than 13K tweets simultaneously on the day it spiked in the search engine for the first time. On the contrary, there was only one news page that featured this query on the same day. The spike in news pages did not occur until two weeks later, and on that day, the second spike occurred on Twitter. This shows that Twitter users are amazingly reactive to trendy topics. These observations reveal that Twitter enables us to classify spiking queries in a timely manner.

3.2   Problem Definition of Spiking Query Classification with Twitter

When queries $Q = \{q_1, \ldots, q_n\}$ that are input into a search engine spike, we aim to timely classify a spiking query q_i into a predetermined category in $C = \{c_1, \ldots, c_m\}$ by leveraging unique information that Twitter provides, *i.e.*, tweets, users, and hashtags.

In this study, we assume that one query belongs to one category, because when a query spikes, it is generally triggered by a specific topic such as the release of a new product or new film, which makes the corresponding category dominant. Therefore, we decide the most likely category to be the query's category.

4   Proposed Method

The input used with our method consists of spiking queries, predefined categories, and prior labels of queries $\hat{Y} = \{y'_1, \ldots, y'_n | y'_i \in C \cup \nu\}$ where $y'_j \in C$ if the query q_j is labeled; otherwise, y'_j has a default label ν. Prior labels represent categories of labeled queries and serve as training data in semi-supervised classification. The method finally outputs categories $Y = \{y_1, \ldots, y_n | y_i \in C\}$ assigned to the input queries.

Our first step is to construct a graph. We extract users $U_Q = \{u_1, \ldots, u_l\}$ who posted tweets containing the queries Q, and hashtags $H_Q = \{h_1, \ldots, h_k\}$ that are assigned to tweets containing the queries Q. We match a query and tweet by substring matching. With these queries, users, and hashtags, we construct the query-Twitter graph $G = \{V, E, W\}$. Here, V consists of n query nodes V_q, l user nodes V_u, and k hashtag nodes V_h ($N = n + l + k$), in which we generate the query nodes V_q using the input

---

[5] http://news.ceek.jp/ (This news portal is provided by the first author.)

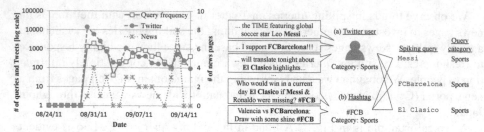

Fig. 1. Trend of spiking query `karelog` in search engine, Twitter, and news pages

Fig. 2. Correlation between query and user (a) and query and hashtag (b); categories of query and user/hashtag affect each other

queries Q. The variable E represents the edges between nodes, and W is an $N \times N$ edge weight matrix. If the $edge(v_i, v_j) \notin E$ between nodes v_i and v_j, then $W_{ij} = 0$. With this query-Twitter graph G, we conduct classification using a graph-based semi-supervised classification method, in which we propagate categories of labeled queries $\hat{Y}$ to the entire graph.

4.1 Graph Construction

Correlation between Query and User. Twitter provides information on who posts tweets, which enables to characterize queries from a novel aspect, *i.e.*, correlation between a query and a user. Fig. 2 (a) illustrates this correlation; we can infer that the category of the user is "sports" if he/she posts many tweets containing queries in a sports category. On the contrary, we can infer that the category of query `Messi` is probably sports if users who belong to the sports category frequently post tweets containing that query.

To employ the correlation, we extract user nodes V_u from Twitter data. For each query q_i, we extract a set of tweets T_i containing the query q_i. Based on T_i, we find a set of users U_i who posted tweets containing the query q_i. Finally, we obtain a unique set of users U_Q from $U_{all} = \{U_1, \ldots, U_n\}$ and use them as user nodes V_u.

We join nodes V_q and V_u by edges E and compute the edge weight matrix W. To introduce the correlation between a query q and user u, we set an edge with weight W_{qu} between them if the user has posted tweets containing the query. Naturally, the more the user posts tweets containing the query, the more likely it is that the user and query belong to the same category. Therefore, we define the edge weight to represent the strength of their correlation. When considering this from the user side, the correlation $\psi(u \to q)$ represents the probability that the user posts tweets containing the query. On the other hand, when considering this from the query side, the correlation $\psi(q \to u)$ represents the probability of the query being tweeted by this specific user. These are computed as

$$\psi(u \to q) = \frac{count(u, q)}{\sum_{q' \in Q} count(u, q')}, \psi(q \to u) = \frac{count(u, q)}{\sum_{u' \in U_Q} count(u', q)},$$

where $count(u, q)$ represents the number of tweets that are posted by the user u and that contain the query q. We then compute the edge weight as $W_{qu} = (\psi(u \to q) + \psi(q \to u))/2$, which represents the strength of the correlation between the user and query.

Correlation between Query and Hashtag. A unique functionality in Twitter is the hash-tag. The hashtag starts with the indicator symbol "#," as in "#FCB," and is occasionally assigned to mark the topic of a tweet. Users share the same hashtag and freely assign it to their tweets, which enables tweets about the same topic to be aggregated. In our data, about 10% of tweets contain hashtags. The hashtag is another useful resource for query classification because it is highly likely that queries contained in tweets with the same hashtag belong to the same category, as Fig. 2(b) shows. This relationship is identical to the correlation between a query and a user.

For each query q_i and tweets T_i containing the query, we find a set of hashtags H_i assigned to T_i. We then obtain a unique set of hashtags H_Q from $H_{all} = \{H_1, \ldots, H_n\}$ and use them as hashtag nodes V_h. Similar to the edge between a query and user, we set an edge with weight W_{qh} between a query q and hashtag h based on their correlation:

$$\psi'(h \to q) = \frac{count'(h, q)}{\sum_{q' \in Q} count'(h, q')}, \psi'(q \to h) = \frac{count'(h, q)}{\sum_{h' \in H_Q} count'(h', q)},$$

where $count'(h, q)$ represents the number of tweets that contain both the hashtag h and the query q. Finally, the edge weight is computed as $W_{qh} = (\psi'(h \to q) + \psi'(q \to h))/2$.

Context Similarity between Queries. In addition to the correlation between a query and user/hashtag, we consider the similarity between queries using tweets in which the queries appear. This common approach has been used in previous studies, as discuss in Sec. 2. The underlying assumption is that queries of the same category are tweeted in a common context; for example, queries belonging to the sports category would be tweeted with the terms "football" and "tournament." We set an edge between query q and q' with weight $W_{qq'}$ based on their similarity $sim(q, q')$.

We generate a vector to represent q using the bag-of-words model with terms extracted from tweets $T_Q = \{T_1, \ldots, T_n\}$ that contain the queries of the classification target. We use tf-idf to compute an element of the vector. Then we compute the similarity $sim(q, q')$ of q and q' using the cosine similarity; this is a conventional method to measure the similarity of two vectors. To prune an edge between dissimilar queries, we introduce a threshold ρ. The edge is set only if $sim(q, q') > \rho$; otherwise, the edge weight is $W_{qq'} = 0$.

When the categories the queries' belonging to are semantically related, the queries may have a similar context even though they belong to different categories. For example, the name of an actress belonging to the celebrities category may appear together with the title of a film starring the actress, even though the title belongs to the movie category. In such a case, queries have similarity with various categories and may confuse the classifier. Therefore, we need to introduce a normalizing factor in the similarity measure. When we focus on q, the similarity is updated as

$$sim'(q \to q') = \frac{sim(q, q')}{\sum_{q^* \in \{q^* | edge(q, q^*) \in E\}} sim(q, q^*)}.$$

This normalization factor makes the similarity directional, and thus, the final edge weight is computed by considering the similarity both from q to q' and from q' to q: $W_{qq'} = (sim'(q \to q') + sim'(q' \to q))/2$.

Balance Edge Weights. We define edge weights between a query and user/hashtag based on their correlation, and an edge weight between queries based on their context similarity. Since these correlations and context similarities are based on different statistical evidence, we may not handle their weights equally. Therefore, we balance their influence by introducing a weighting parameter[6] α.

$$W'_{qu} = \alpha W_{qu}, W'_{qh} = \alpha W_{qh}, W'_{qq'} = (1 - \alpha)W_{qq'}. \tag{1}$$

All of these weights are normalized to range from 0 to 1 as describe previously. The parameter α ranges from $0 \leq \alpha \leq 1$. We evaluate the effect of the parameters in our method, *i.e.*, ρ and α, in Sec. 5.4.

4.2 Graph-Based Semi-supervised Classification

Now that we have the query-Twitter graph, we conduct classification using the graph. We cast the classification problem as a semi-supervised graph labeling problem to achieve a cost-effective classifier. The query-Twitter graph G has N nodes consisting of n query nodes, l user nodes, and k hashtag nodes ($n < k \ll l$). Under the framework of the semi-supervised approach, we assign category labels C to the small number of n_0 query nodes V_{q0}. Other n_1 query nodes are unlabeled ($n = n_0 + n_1$). Our aim is to propagate the labels to unlabeled n_1 query nodes via user/hashtag nodes. In the label propagation, we want nodes connected by a highly weighted edge to have the same label. When a node v_i receives a propagated label y_i and its neighboring node $v_j \in Neighbor(v_i)$ receives a propagated label y_j, our objective function is

$$E(C) = \sum_{i,j} W_{ij}(y_i - y_j)^2 \text{ s.t. } y_k = \tilde{c},$$

where y_k is the label of a labeled node $v_k \in V_{q0}$ that is assigned a category $\tilde{c} \in C$. The final label assignment Y is obtained by minimizing the objective function

$$Y = \underset{Y}{\operatorname{argmin}} E(C).$$

Solutions for this optimization problem have been proposed as graph-based semi-supervised classification algorithms [15,14]. We adapt the modified adsorption algorithm [14] because it is suitable for handling a highly connected graph like ours and prevents densely connected nodes from excessively affecting the lesser connected nodes. In addition, it achieves state-of-the-art performance. We use the implementation distributed by the authors[7].

5 Evaluation

We evaluate the classification accuracy of the proposed method with a realistic setting that replicates the situation when a query spikes.

---

[6] Since the principle of the correlations regarding users and hashtags is the same, we use the same weighting parameter for simplicity.

[7] `https://github.com/parthatalukdar/junto` (Junto v1.2.2)

5.1 Dataset

For this evaluation, we need queries with their frequency history. We use a 1% sample of queries in Japanese that ware input to Bing Bar during three months; from July 1 to September 30, 2011. Our query data consist of a query string, issued date, and frequency on the day. We extract spiking queries and discard those whose frequency is less than 10 since the magnitude of spikes is too small to determine whether the spikes are meaningful or just by chance. The result produced 5,721 unique queries, which we then labeled categories to.

For labeling, we use an on-line dictionary service called Hatena Keyword[8], where only approved users can edit entries and assign categories to them. This is a popular service in Japan and is accessed by more than 5 million people per month. An advantage of using this service is that it has a simple and clear category structure, unlike Wikipedia, which has complex and unstructured categories. As a result, we assign 17 categories to 2,923 queries, as shown in Fig. 3. We sample queries and manually examine the assigned categories, and confirm that their quality is reliable. In this evaluation, we only label queries, not users or hashtags. We plan to label them in a future study.

The Twitter data we use consist of tweets in Japanese posted during the same period. In total, we collect 251M tweets through Twitter's official API[9]. Of these, we use about 45M tweets that contain the spiking queries necessary for the experiments. We carry out preprocessing using MeCab[10] [8], with which we segment Japanese sentences into words in order to compute the similarity (*i.e.*, edge weight) between queries.

5.2 Comparison Method

To evaluate our method in comparison to another method, we need a method that uses a semi-supervised learning based classification approach. In addition, to evaluate accuracy on days when a query spikes, the comparison method should use resources with date information for feature extraction. The latter requirement makes it difficult to compare our method with those used in the previous studies we discuss in Sec. 2 because they depend on resources that do not allow us to replicate their situation in the past, for example, search engine results.

Therefore, we decide to compare our method with a method using a graph constructed with news pages that contain published date information instead of tweets. Since news pages are one of the most responsive resources to fresh topics, this method serves as a baseline to evaluate how timely the proposed method can classify spiking queries. We also compare our method with subgraphs of the query-Twitter graph to analyze the effect of the correlation and context similarity as follows.

1. NewsGraph (Baseline): We use the news archive described in Sec. 3.1. During the three-month experiment period, we collect 402K news pages consisting of 103 sites in total. We construct a graph consisting of query nodes and compute an edge weight based on their context similarity when they appear in news content.

---

[8] http://d.hatena.ne.jp/keyword/
[9] https://dev.twitter.com/docs/streaming-apis/streams/public
[10] http://code.google.com/p/mecab/ (MeCab v0.98 for MS-Windows)

2. QueryGraph: To evaluate the effect of similarity between queries only, we construct a graph consisting of query nodes. This is a sub-graph of the query-Twitter graph, i.e., $G_q = \{V_q, E_q, W_q\}$ where V_q represents query nodes, E_q represents edges among query nodes, and W_q is the edge weight matrix among query nodes.

3. UserGraph: To evaluate the effect of correlation between queries and users/hashtags only, we construct a query-Twitter graph with only edges between query and user/hashtag nodes. This is a sub-graph $G_u = \{V, E_u, W_u\}$, where V includes all nodes, E_u represents edges between query and user/hashtag nodes, and W_u is the edge weight matrix among query and user/hashtag nodes, i.e., $W_{ij} = 0$ if $v_i, v_j \in V_q$.

4. QueryTwitterGraph (Proposed method): This is the proposed query-Twitter graph G that has full features, and is the superimposed graph of QueryGraph and UserGraph.

5.3 Procedure

To replicate the situation when a query spikes in a search engine, we segment queries, tweets, and news pages according to their date information. We set a sliding window of size d-day and use queries spiking in the window to evaluate classification accuracy. As Fig. 4 shows, we use queries spiking in the window to construct a graph with tweets or news pages published in the same window, and then label the queries spiking in the first $d - 1$ days with their categories and then propagate their labels to queries spiking *on* the d-th day (we call the d-th day a test day). For queries that spike multiple times, we regard the day when each query's frequency is maximum as its spike. In this manner, we can evaluate the classification accuracy on the day a query spikes.

We set $d = 28$ ($= 4$ weeks) to avoid any differences in frequency between queries and tweets due to the effect of the day of week. We slide the window from the beginning of the experimental period to the end in one day intervals, i.e., the first test day is July 28. In this setting, we have 65 windows (with test days from July 28 to September 30). Of these, we use the first 28 windows to tune the parameters for graph construction as the development dataset, and the remaining 37 windows to evaluate the accuracy. Overall, each day has about 31.7 queries on average with a standard deviation of 14.3, and each test day has 30.1 queries on average with a standard deviation of 17.7.

We follow the standard evaluation metrics for query classification that were used in the KDD-Cup 2005 competition [11], namely, precision (P) and recall (R), which are defined as:

$$P = \frac{\sum_i \# \text{ of queries are correctly labeled as } c_i}{\sum_i \# \text{ of queries are labeled as } c_i},$$

$$R = \frac{\sum_i \# \text{ of queries are correctly labeled as } c_i}{\sum_i \# \text{ of queries whose category is } c_i},$$

as well as F-score (F1): $F1 = 2PR/(P + R)$.

5.4 Results and Discussion

We first describe the overall classification accuracy, and then show how the different parameters affect the evaluation metrics. Finally, we compare our method with a graph constructed using click-through data.

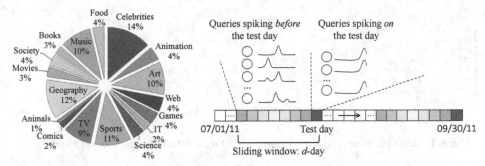

Fig. 3. Distribution of query categories   **Fig. 4.** Experimental procedure

Table 1. Classification accuracy

Method	Precision	Recall	F-score
NewsGraph	45.5	30.3	36.4
QueryGraph	46.1	42.1	44.0
UserGraph	48.9	45.7	47.3
QueryTwitterGraph	**50.9**	**50.1**	**50.5**

Table 2. Comparison with click-through data

Method	Precision	Recall	F-score
ClickGraph	35.4	32.1	33.7
QueryGraph	38.5	25.8	30.3
UserGraph	23.3	17.6	20.1
QueryTwitterGraph	36.8	33.3	35.0

Classification Accuracy. We set the parameters on graph construction (ρ and α) to have the best F-score, as described in the next paragraph. Specifically, $\rho = 0.4$ for News-Graph, $\rho = 0.2$ for QueryGraph, and $\rho = 0.2$ and $\alpha = 0.8$ for QueryTwitterGraph.

Table 1 lists the precision, recall, and F-score of the proposed and comparison methods. The proposed method achieves the best accuracy, with a precision value of 50.9%, recall of 50.1%, and F-score of 50.5%. We conduct a sign test and confirm that Query-TwitterGraph has significantly better classification power than NewsGraph, QueryGraph, and UserGraph ($p \ll 0.01$). In fact, it has about a 20% higher recall value than News-Graph. This result shows that Twitter is useful not only for timely classification but also for widening the coverage of the classifier. Another surprising result is that our method achieves about 5% higher precision than NewsGraph, even though tweets are noisier than professionally edited news text. The results in Table 1 indicate that this is due to the effect of user/hashtag nodes; the fact that UserGraph achieves 3.4% higher precision than NewsGraph demonstrates it.

When comparing QueryGraph and UserGraph, it is impressive that UserGraph achieves about 3% higher precision and about 4% higher recall than QueryGraph ($p = 0.018$ by a sign test). Recall that UserGraph does not depend on any textual resources; it is based purely on the correlation between a query and user/hashtag in terms of their belonging category. This shows that the correlation between a query and user/hashtag is effective evidence of classification. The even better accuracy of QueryTwitterGraph shows that these two graphs complement each other to improve the accuracy.

Effect of Parameters. Next, we evaluate the effect of the parameters in our method. We start with parameter ρ, which controls the number of edges between query nodes. Fig. 5 plots the precision, recall, and F-score on QueryGraph when the value of the parameter ρ is changed in the development dataset. The best F-score (41.6%) is achieved

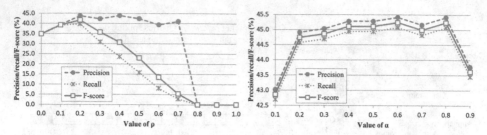

Fig. 5. Effect of parameter ρ (QueryGraph)   **Fig. 6.** Effect of parameter α ($\rho = 0.2$)

when $\rho = 0.2$, along with the best recall (39.7%) and the second-best precision (43.7%). When we increase the value of ρ, the number of edges decreases. Larger ρ disrupts the propagation of labels, and thus both precision and recall drop.

Parameter α balances the effect of the correlation between a query and user/hashtag and that of context similarity between queries in Eq. (1). When α is larger than 0.5, our method places more edge weight on correlation. Fig. 6 plots the precision, recall, and F-score on QueryTwitterGraph when the value of the parameter α (we use $\rho = 0.2$) is changed in the development dataset. The accuracy improves when the value of α is increased, and the best precision (45.4%), recall (45.1%), and F-score (45.2%) are achieved when $\alpha = 0.8$. This result shows the effectiveness of introducing the correlation between a query and user/hashtag, and its contribution is significant.

Comparison with Click-Through Data. Click-through data are a useful resource for characterizing associated queries, since they provide relevant Web pages accessed by people. We obtain click-through data of the experimental queries (four weeks from September 3 to September 30) and use the data to construct a graph, which we then compare with our method. In total, we have 809 queries with click-through data.

We construct a graph (ClickGraph) consisting of query nodes using the click-through data and conduct classification as described in Sec. 4.2. We extract the top-10 most frequently accessed Web pages for each query and compute an edge weight between query nodes based on their context similarity. Due to the limitation in available data, we set the size of the sliding window at two weeks ($d = 14$) to get 14 windows. We use the first 7 windows to tune the parameters and the remaining 7 windows for evaluation. For a fair comparison, we apply the same setting to QueryGraph, UserGraph, and QueryTwitterGraph.

Table 2 lists the precision, recall, and F-score of ClickGraph ($\rho = 0.4$), QueryGraph ($\rho = 0.3$), UserGraph, and QueryTwitterGraph ($\rho = 0.3$, $\alpha = 0.5$). The results indicate that QueryTwitterGraph achieves accuracy comparable to ClickGraph (no significant difference is detected). This result shows that our approach is promising, as it is comparable to a method that uses a highly useful resource such as click-through.

6   Conclusion and Future Work

We propose a method for timely classifying spiking queries by exploiting Twitter. The proposed method achieves high accuracy by leveraging unique information that Twitter provides, *i.e.*, tweets, users, and hashtags.

In the future, we first plan to extend our method and include social network information in Twitter such as follower-followee relationships and behaviors of retweeting and mentioning. Then, we apply multi-label classification. We also plan to collect a large-scale human annotation to obtain categories for spiking queries and conduct a more detailed evaluation. Moreover, we combine our method with a previous approach that uses search result pages and query logs to further improve the classification accuracy.

Acknowledgments. We sincerely thank Mikio Yamamoto, Takashi Inui, Takaaki Tsunoda, Ming Zhou, and Qing Ma for their valuable comments and feedback in this project. This project was supported by JSPS KAKENHI Grant Number 11J01016.

References

1. Agichtein, E., Castillo, C., Donato, D., Gionis, A., Mishne, G.: Finding high-quality content in social media. In: WSDM 2008, pp. 183–194 (2008)
2. Baeza-Yates, R., Calderón-Benavides, L., González-Caro, C.N.: The Intention Behind Web Queries. In: Crestani, F., Ferragina, P., Sanderson, M. (eds.) SPIRE 2006. LNCS, vol. 4209, pp. 98–109. Springer, Heidelberg (2006)
3. Beitzel, S.M., Jensen, E.C., Frieder, O., Lewis, D.D., Chowdhury, A., Kolcz, A.: Improving automatic query classification via semi-supervised learning. In: ICDM 2005, pp. 42–49 (2005)
4. Broder, A.Z., Fontoura, M., Gabrilovich, E., Joshi, A., Josifovski, V., Zhang, T.: Robust classification of rare queries using web knowledge. In: SIGIR 2007, pp. 231–238 (2007)
5. Diemert, E., Vandelle, G.: Unsupervised query categorization using automatically-built concept graphs. In: WWW 2009, pp. 461–461 (2009)
6. Dong, A., Zhang, R., Kolari, P., Bai, J., Diaz, F., Chang, Y., Zheng, Z., Zha, H.: Time is of the essence: improving recency ranking using twitter data. In: WWW 2010, pp. 331–340 (2010)
7. Hu, J., Wang, G., Lochovsky, F., Tao Sun, J., Chen, Z.: Understanding user's query intent with Wikipedia. In: WWW 2009, pp. 471–480 (2009)
8. Kudo, T., Yamamoto, K., Matsumoto, Y.: Applying conditional random fields to Japanese morphological analysis. In: EMNLP 2004, pp. 230–237 (2004)
9. Kulkarni, A., Teevan, J., Svore, K.M., Dumais, S.T.: Understanding temporal query dynamics. In: WSDM 2011, pp. 167–176 (2011)
10. Li, X., Wang, Y.-Y., Acero, A.: Learning query intent from regularized click graphs. In: SIGIR 2008, pp. 339–346 (2008)
11. Li, Y., Zheng, Z., Dai, H.K.: KDD CUP-2005 report: facing a great challenge. SIGKDD Explor. Newsl. 7(2), 91–99 (2005)
12. Sakaki, T., Okazaki, M., Matsuo, Y.: Earthquake shakes twitter users: real-time event detection by social sensors. In: WWW 2010, pp. 851–860 (2010)
13. Shen, D., Sun, J.-T., Yang, Q., Chen, Z.: Building bridges for web query classification. In: SIGIR 2006, pp. 131–138 (2006)
14. Talukdar, P., Crammer, K.: New Regularized Algorithms for Transductive Learning. In: Buntine, W., Grobelnik, M., Mladenić, D., Shawe-Taylor, J. (eds.) ECML PKDD 2009, Part II. LNCS, vol. 5782, pp. 442–457. Springer, Heidelberg (2009)
15. Zhu, X., Ghahramani, Z., Lafferty, J.: Semi-supervised learning using gaussian fields and harmonic functions. In: ICML 2003, pp. 912–919 (2003)

Where Are You Settling Down: Geo-locating Twitter Users Based on Tweets and Social Networks

Kejiang Ren, Shaowu Zhang, and Hongfei Lin

Information Retrieval Lab, School of Computer Science and Technology
Dalian University of Technology, Dalian, China
renkj@mail.dlut.edu.cn,
zhangshaowu@gmail.com, hflin@dlut.edu.cn

Abstract. In this paper, we investigate the advantages of taking two dimensions of tweet content and social relationships to construct models for predicting where people settle down as their profiles reveal city- and town-level data. Based on the users who voluntarily reveal their locations in their profiles, we propose two local word filters - Inverse Location Frequency (ILF) and Remote Words (RW) filter - to identify local words in tweets content. We also extract separately the place name mentioned in tweets using the Named Entity Recognition application and then filter them by computing the city distance. We consider users' friends and 2-hop of followings. In our experiment, we finally combine these two dimensions to estimate user location and achieve an Accuracy of 56.6% within 100 miles in city-level and 45.2% within 25 miles in town-level of their actual location which outperforms the single dimension prediction and the baseline.

Keywords: Geo-location, Social Network, Twitter, Location-based Services, Location Prediction, Text Mining.

1   Introduction

Twitter's open and succinct service allows it to gather vast amounts of data and updates by users who come from different places. The user always inadvertently leaks some dialect words and place names of his/her residence in the process of adding updates. Understanding the geographic features of those update statuses enables the system to push better local advertising, highlight points of interest, show local news, create recommendations for friends living in the vicinity, and even help search engines understand users' search intentions better. In this paper we build textual models of local words and place names based on pure tweets to estimate a user's place of residence, even when the user does not explicitly reveal the place name, or his/her geographic coordinates in the profile.

Living in close geographical proximity may enable people to share common characteristics, provide real-time information and eyewitness updates about events of local interest [12], and recommend local friends to their extended friends. Furthermore, people who have reciprocal relationships are more likely to be geographically

Y. Hou et al. (Eds.): AIRS 2012, LNCS 7675, pp. 150–161, 2012.
© Springer-Verlag Berlin Heidelberg 2012

close [2]. In this paper, we build social network models using users' followers and followings to predict their location. We find that people prefer to follow others who live in close geographic proximity.

In this paper, we propose hybrid probabilistic models combining textual models with social network models to estimate a user's location, and propose two local word filters. In the social networks model, we also consider the 2-hop of a use following's followers to estimate the user's location excepting the user's immediate friends. Finally we predict a user's location using a hybrid probabilistic model by combining the two dimensions.

The remainder of this paper is organized as follows: In Section 2, we review related works. Section 3 introduces the dataset of textual and social networks used in the experiments and estimation metrics in the paper. We introduce our models as well as the estimation algorithm, filter algorithm, and smoothing method in Section 4. We present the experimental results in Section 5. Finally, conclusions and future work are discussed in Section 6.

2   Related Work

Twitter has quickly become the premier platform for sharing real-time information since it first arose. Its convenience lies in a user's ability to post what he/she observes and hears in a local place. Lee et al. [18] proposed a geo-social event detection method to monitor the geographical regularities of local crowd behavior. Yardi et al. [12] examined the relationship between online social network structure and physical geographic proximity and verified that local events are of most interest to local citizens. Vieweg et al. [19] and Lee et al. [12] both discussed event broadcasting by local people.

Ye et al. [8], who used two features of places from explicit patterns of individual places and implicit relatedness among similar places for a binary SVM algorithm, developed a semantic annotation technique for Whrrl to automatically annotate all places with category tags. Lin et al. [9] investigated the factors that influence people to refer to a location and applied machine learning to model people's place-naming preferences. Anastasios et al. [20] analyzed the geo-temporal dynamics of collective user activity on Foursquare and showed that checkins provide a means to uncover people's daily and weekly patterns, urban neighborhood conditions, and recurrent transitions between different activities.

Amital et al. [4] employed the gazetteer approach to identify all geographic mentions within Web pages and to assign a geographic location and confidence level to each geo-locate Web content instance. Fink et al. [10] used both place names and organizational entities in blogs to determine an author's location. Serdyukov[13], Crandall[14], Hays[15], and Gallagher[17] et al. all attempted to predict where photographs originated using user tags and image-textual content. Popescu et al. [16] estimated users' home location by analyzing textual metadata associated with Flickr photos.

The aspects of geo-location users using tweet content in Twitter posts have become an active and promising area of research in the past two years. The most relevant works include Cheng et al. [1], Hecht et al. [6] and Kinsella et al. [5]. Cheng et al. proposed a probabilistic framework for estimating a Twitter user's city-level location based purely on the content of the user's tweets. Hecht et al. studied user behavior in typing information into location field of user profiles, and then used simple machine learning techniques to guess users' locations on the country and state levels. Kinsella et al. [5] created language models of locations using coordinates extracted from geo-tagged tweets and model locations on varying levels of granularity from ZIP code to country level to geo-locate user and single tweets.

Backstrom et al. [3] used the social network structure of Facebook to predict location. Scellato et al. [7] described a supervised learning framework which exploits two linked features of friends-of-friends and place-friends to predict new links among friends-of-friends and place-friends. Li et al. [11] and Kwak et al. [2] studied the geographic features in Twitter.

3   Predicting Location

3.1   Textual Model

The textual model is a probabilistic estimator based on a user's tweets to estimate the location where the user settles down. Typically, the tweets posted by the user contain a great quantity of irrelevant information for location prediction. We next describe our local words filter algorithm and smoothing method.

Local Words Filter Algorithm. In Twitter, many words which appear in tweets have the similar probability in all locations and are distributed consistently with the population across different locations. Since the distribution means that most words provide very little power at distinguishing the location of a user, this even provides a lot of complications for prediction. In addition, some locations have a sparse set of words in their tweets because of the small population of the registered Twitter users or the small number of people updated their status in these locations. In order to improve the estimation accuracy, we must identify these local words in tweets and overcome the tweets' scarcity. Afterward, we are committed to the local words' identification and, at the same time, to overcome the data sparseness.

Before using a filter algorithm to identify local words, we preprocessed the content of the tweets. First, we eliminated the repeat tweets by string matching since we observed that most of these tweets are advertising. Then we removed the link in the tweet using the regular expression, eliminated all occurrences of a standard list of 429 stop words, as well as screen names which start with @ and single-letter words. Finally, we excluded punctuation in the tweets using Lucene Tool[1] and stemmed the word using Snowball[2]. By calculating word frequency, we only considered words that occur

---

[1] http://lucene.apache.org/
[2] http://snowball.tartarus.org/

at least 15 times in order to reduce the impact of incidental words (e.g.,yeeeeeees). Through the above processing, about 46,369 original words were generated from a base set of 521,103 distinct words.

Inverse Location Frequency: The first filter algorithm is the Inverse Location Frequency, which we called the ILF filter. It reflects the importance of the word in the collection of locations. The more locations that a word occurs in, the less discriminating the word is between locations, and consequently, the less useful it will be in location estimation. The form of ILF filter is defined as follow:

$$ILF_w = \log \frac{N}{n_w} \tag{1}$$

where ILF_w is the inverse location frequency for word w, N is the number of locations in training set. And n_w is the number of locations in which word w occurs. We set a threshold for filtering the original words which appear in many locations. After the application of the ILF filter, 19,424 local words were left from 46,369 words.

We used the Remote Words filter, which we called the RW filter, to filter out these remote words. In RW, we calculated the average distance of a location with all other locations for a local word, and eliminated the maximum value which exceeded our threshold in a specific iteration. When the average distance of all locations is less than the threshold, the iteration is over. In our experiment, we set the threshold as 200 miles. Before we operated the RW filter, we removed all words which occurred two times in every location. The remedy for these words is dealt with in the next step of NER. The formula of average distance is calculated as follows:

$$RW = \frac{1}{n-1} \sum_{\substack{i=1 \\ j \neq i}}^{n} | loc_j - loc_i | \tag{2}$$

Named Entity Recognition. We processed each tweet, applying NER and location-entity disambiguation to identify the related locations for the focus location. For each tweet, we exploited the named entity recognizer [21] to extract a location entity mentioned in it. Each entity was matched against the Yahoo! Placemaker[3] and got the latitude and longitude; it was then calculated for the distance of location of the entity to disambiguate the extracted location entity. We left these entities which are within 40 miles of the focus location. There were 10229 entities being recognized, and after disambiguation the total locations were 5702. Because the locations which users reveal in their profiles are place names, we put the words in location entities.

Textual Model Estimator. We used the Maximum Likelihood Estimation to geolocate the users where they are settling down. Given the set of words U_w and location entities U_e extracted from a user's tweets U_T, we proposed the probability of the user being located in city l_i as:

---

[3] http://developer.yahoo.com/geo/placemaker/

$$p(l_i \mid U_w) = \lambda \sum_{w \in U_w} \alpha * p(l_i \mid w) * p(w)$$

$$+ \mu \sum_{e \in U_e} \beta * p(l_i \mid e) p(e) \tag{3}$$

where the $p(w)$ is a *priori* probability which means the probability of the word w in the whole dataset and the a *priori* probability $p(e)$ is equal to 1. The parameters α and β are used to denote the significance of the local word w for estimation location l_i, the other group parameters, λ and μ, are the weight for which the portion is more important for estimation. In our experiment, we set $\lambda=1$ and $\mu=5$. The *priori* probability is calculated: $p(w) = \frac{count(w)}{N \times local(w)}$, $p(l_i \mid w) = \frac{count_i(w)}{N \times count(w)}$ and $p(l_i \mid e) = \frac{count(e)}{N}$. Where $count(w)$ is the number of occurrences of the word w in the whole dataset, N is the total number of the word after filtering via the training set, $local(w)$ stands for the location number where the word w occurs, $count_i(w)$ donates the count of word w in location l_i, and $count(e)$ stands for the total number of entity of recognized location.

Circular-Based Neighborhood Smoothing. There is a problem in that some locations have a few tweets in our training set because of having a small population or users unaccustomed to update status in the locations. The word distribution is sparse in these locations. How to overcome the location sparsity of words in tweets? We used the approach of circular-based neighborhood smoothing to improve the quality of user location estimation. Circular-based neighborhood smoothing considers all geographic neighbors from which the distance is 40 miles to the centre of a location. The circular probability of a word w can be formalized as:

$$p(r_i \mid w) = \sum_{l_j \in S} p(l_j \mid w) \tag{4}$$

where the r_i is the radius of the circle which the center is estimation location l_i, and S is the collection of locations in the round, at the same time including the estimation location. Then, the probability of the word w to be located in location l_i, $p(l_i \mid w)$ can be replaced with $p(r_i \mid w)$.

3.2   Social Network Model

The social network model is a probabilistic estimator based on the user's followers and followings (we called them *friends* consistently and whose locations get from training data) to estimate the location where he/she settles down. We can write down the likelihood of a particular location l_i as:

$$p(l_i \mid U_{SN}) = \frac{\sum \{u_i \mid u_i \in (U_{Fa} \mid U_{Fo}) \wedge u_i \in l_i\}}{N_{Fa} + N_{Fo}} \tag{5}$$

where l_i is the estimation of location, U_{SN} is the social network of user u including all followers U_{Fa} and followings U_{Fo}, N_{Fa} is the total number of followers for user u, and N_{Fo} stands for the number of followings of user u. The equation means how many users in location l_i of user u's followers and followings.

Generally, users do not add friendship connections at random with all other users, but, instead, they tend to prefer other users who are "close" to them in social network. For instance, many links do appear between individuals at closer social distance from each other, with the 2-hop neighborhood of single nodes being the largest source of new ties [7][22]. In Twitter, this phenomenon may be weaker, but it still can be considered. In our experiment, we only considered the number of a user's following's followers. We take into account this portion based on an assumption that the fewer of followers of user u_j (which u_j is a following of user u), the more intimate the relationship between them, the greater the contribution to the prediction. The formula of the likelihood is represented as follows:

$$p(l_i \mid U_{SN}) = \frac{\sum\{u_i \mid u_i \in (U_{Fa} \mid U_{Fo}) \wedge u_i \in l_i\} + \sum\{u_i \mid u_i \in U'_{Fa} \wedge u_i \in l_i\}}{N_{Fa} + N_{Fo} + \sum N'_{Fa}} \qquad (6)$$

where U'_{Fa} is the followers of user u_j, N'_{Fa} is the number of followers for user u_j. We also considered the follower's social networks and the following's followings, but the result is lower.

3.3   Hybrid Model

We proposed a hybrid probabilistic model of combining a textual model with a social networks model. Since the dimensions of the two models are not the same, we first normalized the resulting values of the two models into values in the range [0, 1].

$$p(l_i \mid U) = \frac{p(l_i \mid U) - \min\{p(l_i \mid U)\}}{\max\{p(l_i \mid U)\} - \min\{p(l_i \mid U)\}} \qquad (7)$$

where $\max\{p(l_i \mid U)\}$ and $\min\{p(l_i \mid U)\}$ are the maximum and minimum value of $p(l_i \mid U)$ respectively for all the estimated locations. Then the hybrid probabilistic is

$$p(l_i \mid U) = \omega * p(l_i \mid U_w) + (1 - \omega) * p(l_i \mid U_{SN}) \qquad (8)$$

where ω is the balancing coefficient in the range [0, 1].

4   Experimental Results

4.1   Data Collection

Twitter offers an open API that is easy to crawl and collect data. However, we used two existed corpora: [1] provides tweets and the location of user (we call this data

CHENG) and [2] offers the social network of the user (we call this data KWAK) in Twitter. The commonality of the CHENG and KWAK is that they have the same user ID in Twitter. The data of CHENG provided both the training set and test set. The training set contains 115,886 Twitter users and 3,844,612 updates from the users. All the locations of the users are self-labeled in United States. The test set contains 5,136 Twitter users with over 1000 tweets each of user and the total updates are 5,156,047 from these users. All the locations of users are uploaded from their smart phones with the form of "UT: Latitude, Longitude"[1]. The data of KWAK contain social graphs, mapping tables from numeric IDs to screen names, and restricted user profiles (> 10,000 followers) which are collected from July 6th 2009 to July 31th 2009 in Twitter. In our experiments, we only used the social graph portion which includes 1.47 billion social relations (still calling social graphs KWAK).

4.2   Metrics of Evaluation

We used the metrics which were defined in paper [1] and compared the estimated location of a user versus the actual location based on his/her latitude and longitude coordinates. The first metric is Average Error Distance (*AvgErrDist*) of all test users. The other metric is Accuracy which considers the percentage of users with their error distance categorized in the range of 0-x miles. We regard estimation as city level when x is equal to 100 as mentioned in [1] and [10]; when x is equal to 25, it is town level [3]. *Accuracy@K* means the accuracy metric in the top-k with the least error distance to the actual location which the *ErrDist(u)* is lower x.

4.3   Estimation Methods

1. ILF filter (ILF).We estimated the location of users using the ILF filter to filter the local words in the training set.
2. RW combines with ILF (RW+ILF). An approach that combined the remote word filter with the ILF to select local words in the training set to predict user location.
3. Named Entity Recognition (NER). This is a traditional method to identify a user's location in social networks using the content.
4. Named Entity Recognition augments the two filters (NER+Fs). In this method, we identified the locations from the training set as local words and merged the two local word filters aforementioned.
   All of mentioned approaches combined the circular based neighborhood smoothing to estimate user location in Twitter before the local words filter.
5. Social network predict (SN). For each user, locations were ranked according to the probability that most of their friends are settling down.
6. 2-hop social network (2-hop). We estimated user location using his/her followings' followers and his/her friends.
7. Hybrid estimation (Hybrid_SN, Hybrid_2-hop). Predicting the user's locations used hybrid models combining the two dimensions of textual and social networks.

4.4   Geo-locating on the City Level

For estimating a Twitter user's city-level location, we take into account Cheng et al. [1], whose system, solely based on tweet content, could geo-locate 51% of the 5119 users in the test set within 100 miles of their actual locations and that the *AvgErrDist* across all users was 535 miles as our baseline location estimator. Meanwhile, the traditional location prediction method of NER also is treated as baseline.

First, when we did not use the local word filters, our system only estimated about 8% of 5119 users in their actual locations corresponding to 10% of Cheng's baseline. We observed that the strong positive impacts of the local word filter. With the local word filter ILF alone, we reached an Accuracy of 0.437 which is more than five times as high as the Accuracy without using the local word filter. The filter removed noise resulting from non-local words in the tweet content and significantly affected the quality of user-location estimation. As we continued to use the RW filter after using the ILF to filter out the remote words, the Accuracy increased from 0.437 to 0.501, because the local words, filtered through ILF, were removed from the local focus. The result means that 50% of the test users can be placed in their actual location. We noted that the *AvgErrDist* was reduced significantly, from 652 miles to 516 miles, which is lower than the 535-mile baseline. The result also outperforms the baseline of NER method. When we took the location entities into consideration, the Accuracy reached 0.5098 which is almost identical with the baseline of 0.510. Meanwhile, the Average Error Distance was reduced from 535 miles to 473 miles; the overall estimated error was significantly lower, and the ACC@2, which at most having one location was correct in the first two locations, is 0.635, exceeding the baseline of 0.624.

Table 1. Results for user-location prediction at city level

Method	ACC	*AvgErrDist*(Miles)	ACC@2
Baseline	0.510	535.564	0.624
NER	0.419	499.10	0.591
ILF	0.437	652.449	0.563
RW+ILF	0.501	516.039	0.585
NER_Fs	0.5098	473.617	0.635
SN	0.593	503.588	0.700
2-hop	0.522	573.71	0.643
Hybrid_SN	**0.566**	**442.321**	**0.683**
Hybrid_2-hop	0.560	446.267	0.675

Continuing the observation in Table 1, we found that only using the user friends, the Accuracy could achieve 0.593 and, when using the 2-hop nodes, 52.2% of test users can be geo-located within 100 miles of their actual locations (we do not count the users whose first two or more predicted locations of the probability are equal). However, we did not compare the SN model and 2-hop model with other models, for in SN, we only considered the 704 users who have more than ten followers and followings (whose followers is less than 300) simultaneously and, in the 2-hop model, the user whose followers and followings are more than five are 1421. We still found that the power of using one's social network to estimate the actual locations is strong; to some extent, its ability has exceeded the standard model, purely based on content to predict the user actual location.

Now, let us look at the predictive power of the hybrid model in Table 1. In hybrid models, we calculated the probability of users whose followers or followings are not zero (the Accuracy is about 42% of removing the users whose at least first two estimated locations' score are the same) and set $\omega=0.501$ based on measures when combined with the model of NER_Fs. We also observe the positive impact of combining the two dimensions of textual and social networks. We can see that 2-hop and SN merged with a textual model result in better user-location estimations than only using one dimension. By observing Table 1, we found that the best Accuracy achieved 0.566, which means placing 56.6% of users within 100 miles of their actual location, with an *AvgErrDist* of all users of 442 miles. If the first two locations are considered, 68.3% of 5119 users can predict actual locations.

4.5   Geo-locating on the Town Level

Kinsella et al. [5] considered the users' self-reported location which is extracted from their profiles and tweets using the Yahoo! Placemaker estimated user location on the town level. We treated it, which the Accuracy is 0.362, as our baseline on town-level location estimations. From Table 2, we found that all of the models' (except ILF) predicting accuracy outperform the baseline and NER method. The combining of two local word filters purely based on tweet content provided the overall results: which 40.5% of users placed within 25miles of their actual location. We can see that the location entities have a negative impact on town-level location predictions. The result of estimating user actual location, in which the Accuracy reaches 50.1% of 760 users using the user's followers and following of more than five respectively, is encouraging. It also shown that the hybrid model provided attractive results that 45.2% of 5119 users could be estimated town level actual location.

Table 2. Results of user-location prediction on the town level

Method	ACC	ACC@2
Baseline	0.362	--
NER	0.223	0.377
ILF	0.358	0.458
RW+ILF	0.405	0.477
NER+Fs	0.381	0.489
SN	0.501	0.595
2-hop	0.443	0.537
Hybrid_SN	**0.452**	**0.548**
Hybrid_2-hop	0.445	0.540

4.6   Impact of the Number of User Followers and Followings for Location Estimation

In order to understand the impact of followers and followings on estimating the user's actual location, we investigated the Accuracy while the number of followers and followings is in continuous growth. We have the following four groups of experiments to observe the influence of friends on the prediction Accuracy: 1) the number of followers is growing but ignoring the followings (Fer); 2) the number of followings increases without followers (Fing); 3) the followings and the followers change at the same time (Friends); and 4) the number of friends is in synchronous growth and, meanwhile, the following's follower number is less than 300 (Friends_-300) (we carried out this experiment based on the fact that, if the follower number of a user is on a large scale in Twitter, he/she is likely to be a celebrity, thus bringing noise).

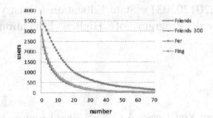
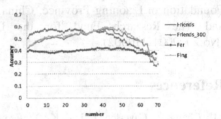

Fig. 1. The number of test users in four conditions

Fig. 2. Accuracy of the four conditions

From Fig.1, we can see that the test users with other three conditions are almost the same variation in different friend number except Fer. However, as the Fig. 2 shows, the estimation of Accuracy varies, and barely around about 40% when the number of followers is growing but ignoring the followings. In conditions 2) and 3), the estimation effect is the same and achieves the maximum when the number is equal, about 42. For the most users, when predicting their actual location using their friends, it is wise to not consider their followings whose own followers' number is more than 300.

5   Conclusion

In this paper, we investigated the problem of geo-locating users, which aimed to estimate user's actual locations via their tweet content and friends in microblogging service of Twitter. Based on the update statuses of a user, we proposed two local word filter methods to filter out noises of tweets content: Inverse Location Frequency (ILF) and Remote Words (RW). ILF filters the words which have a very wide geographic distribution and non-local features. RW is used to eliminate some local words, which are occasionally mentioned in locations far from their local focus. We also separately considered the place name mentioned in tweets to improve the estimation accuracy.

Simultaneously, we attempted to estimate the user's location via where the friends of the user settled most, and also considered the 2-hop friends to refine prediction accuracy. Finally, we combined the two dimensions of textual and social networks to get a better result. The results demonstrated the suitability of our approach and showed that our hybrid estimator can place 56.6% of 5119 users in Twitter within 100 miles of their actual location, 45.2% users geo-located within 25 miles of their actual location.

Next, we plan to investigate users' interactions with each other to refine the accuracy of the estimator, and we are also interested in mining local information from user's self-label tags in futures.

Acknowledgment. This work is partially supported by grant from the Natural Science Foundation of China (No.60673039, 60973068, 61277370), the National High Tech Research and Development Plan of China (No.2006AA01Z151), Natural Science Foundation of Liaoning Province, China (No.201202031), State Education Ministry and The Research Fund for the Doctoral Program of Higher Education (No.20090041110002).

References

1. Cheng, Z., Caverlee, J., Lee, K.: You Are Where You Tweet: A Content-Based Approach to Geo-locating Twitter Users. In: 19th ACM Conference on Information and Knowledge Management, pp. 759–768. ACM, New York (2010)
2. Kwak, H., Lee, C., Park, H., Moon, S.: What is Twitter, a Social Network or a News Media? In: 19th International Cnference on World Wide Web, pp. 591–600. ACM, New York (2010)
3. Backstrom, L., Sun, E., Marlow, C.: Find me if you can: improving geographical prediction with social and spatial proximity. In: 19th International Conference on World Wide Web, pp. 61–70. ACM, New York (2010)
4. Amitay, E., Har'El, N., Sivan, R., Soffer, A.: Web-a-Where: Geotagging Web Content. In: 27th annual International ACM SIGIR Conferenceon Research and Development in Information Retrieval, pp. 273–280. ACM, New York (2004)
5. Kinsella, S., Murdock, V., O'Hare, N.: I'm Eating a Sandwich in Glasgow: Modeling Locations with Tweets. In: 3rd International Workshop on Search and Mining User-Generated Contents, pp. 61–68. ACM, New York (2011)
6. Hecht, B., Hong, L., Suh, B., Chi, B.E.: Tweets from Justin Bieber's Heart: The Dynamics of the "Location" Field in User Profiles. In: 2011 Annual Conference on Human Factors in Computing Systems, pp. 237–246. ACM, New York (2011)
7. Scellato, S., Noulas, A., Mascolo, C.: Exploiting Place Features in Link Prediction on Location-based Social Networks. In: 17th ACM SIGKDD International Conference on Knowledge Discovery and Data Mining, pp. 1046–1054. ACM, New York (2011)
8. Ye, M., Shou, D., Lee, W., Yin, P., Janowicz, K.: On the Semantic Annotation of Places in Location-Based Social Networks. In: 17th ACM SIGKDD International Conference on Knowledge Discovery and Data Mining, pp. 520–528. ACM, New York (2011)
9. Lin, J., Xiang, G., Hong, J.I., Sadeh, N.: Modeling People's Place Naming Preferences in Location Sharing. In: 12th ACM International Conference on Ubiquitous Computing, pp. 75–84. ACM, New York (2010)

10. Fink, C., Piatko, C., Mayfield, J., Chou, D., Finin, T., Martineau, J.: The Geolocation of WebLogs from Textual Clues. In: IEEE International Conference on Computational Science and Engineering, pp. 1088–1092. IEEE Press (2009)

11. Li, W., Serdyukov, P., Vries, A.P., Eickhoff, C., Larson, M.: The Where in the Tweet. In: 20th ACM International Conference on Information and Knowledge Management, pp. 2473–2476. ACM, New York (2011)

12. Yardi, S., Boyd, D.: Tweeting from the Town Square: Measuring Geographic Local Networks. In: 4th International AAAI Conference on Weblogs and Social Media, pp. 194–201. AAAI, California (2010)

13. Serdyukov, P., Murdock, V., Zwol, R.: Placing Flickr Photos on a Map. In: 32nd International ACM SIGIR Conference on Research and Development in Information Retrieval, pp. 484–491. ACM, New York (2009)

14. Crandall, D., Backstrom, L., Huttenlocher, D., Kleinberg, J.: Mapping the World's Photos. In: 18th International Conference on World Wide Web, pp. 761–770. ACM, New York (2009)

15. Hays, J., Efros, A.: IM2GPS: estimating geographic information from a single image. In: IEEE Conference on Computer Vision and Pattern Recognition, pp. 1–8. IEEE Press (2008)

16. Popescu, A., Grefenstette, G.: Mining User Home Location and Gender from Flickr Tags. In: 4th International Conference on Weblogs and Social Media, pp. 307–310. AAAI, California (2010)

17. Gallagher, A., Joshi, D., Yu, J., Luo, J.: Geo-location Inference from Image Content and User Tags. In: IEEE Conference on Computer Vision and Pattern Recognition Workshops, pp. 55–62. IEEE Press (2009)

18. Lee, R., Sumiya, K.: Measuring geographical regularities of crowd behaviors for Twitter-based geo-social event detection. In: 2nd ACM SIGSPATIAL International Workshop on Location Based Social Networks, pp. 1–10. ACM, New York (2010)

19. Vieweg, S., Hughes, A.L., Starbird, K., Palen, L.: Microblogging during two natural hazards events: what twitter contribute to situational awareness. In: 28th International Conference on Human Factors in Computing Systems, may 2010, pp. 1079–1088. ACM, New York (2010)

20. Anastasios, N., Salvatore, S., Cecilia, M., Massimiliano, P.: An Empirical Study of Geographic User Activity Patterns in Foursquare. In: ICWSM 2011: 4th International Conference on Weblogs and Social Media, pp. 570–573. AAAI, California ((2011)

21. Finkel, J.R., Grenager, T., Manning, C.: Incorporating Non-local Information into Information Extraction Systems by Gibbs Sampling. In: 43nd Annual Meeting of the Association for Computational Linguistics, pp. 363–370. ACL, New York (2005)

22. Mok, D., Wellman, B., Basu, R.: Did distance matter before the Internet? Interpersonal contact and support in the 1970s. Social Networks 29(3), 430–461 (2007)

Detecting Informative Messages
Based on User History in Twitter

Chang-Woo Chun[1], Jung-Tae Lee[2] Seung-Wook Lee[1], and Hae-Chang Rim[1]

[1] Dept. of Computer & Radio Communications Engineering,
Korea University, Seoul, Korea
[2] Research Institute of Computer Information & Communication,
Korea University, Seoul, Korea
{cwchun,jtlee,swlee,rim}@nlp.korea.ac.kr

Abstract. Since more and more users participate in various social networking services, the volume of streaming data is considerably increasing. It is necessary to find out valuable messages from huge data archived every moment. This paper investigates the problem of detecting informative messages in Twitter, and proposes effective methods to solve the problem based on User History. Most of the sheer information in tweets has a common defect which is the fact that it is affected by influence of User level within the Twitter network. Our key idea is to leverage each user's history observed from a large scale dataset as features to determine whether a new message is informative or not, compared to their previous messages. This allows us to normalize influence of individual user on tweets and to estimate the probability of informativeness. Experimental results on a real Twitter data show that our method can effectively improve the performance on identifying informative tweets.

Keywords: Social Media, Twitter, Information Filtering, Recommendation System.

1 Introduction

Regarded as one of the most popular social networking services, Twitter, is becoming a new media where people communicate with each other, and disseminate news [1]. Recently, Twitter has become an alternative source of information, and has also inspired many researchers to develop methods for discovering useful posts, known as *tweets*, from huge amount of data that it serves. However, most of tweets produced everyday are generally not worthwhile to read, because many people use the service mainly for daily chattering and posting their status updates [2]. Some previous studies propose to use *retweeted number* of the tweet as a measure of the tweet's importance [5]. However, Suh et al. [4] note that the users number of *followers* strongly affects the retweetability of tweets. And it means that important tweets posted by unpopular users have less chance of getting retweeted. For these reasons, detecting generally informative tweets from the massive Twitter stream becomes a crucial task.

Y. Hou et al. (Eds.): AIRS 2012, LNCS 7675, pp. 162–173, 2012.
© Springer-Verlag Berlin Heidelberg 2012

In this paper, we formulate the task into a binary classification problem, and investigate a wide spectrum of features to study which of them can effectively discriminate informative tweets from mundane tweets. The features used conventionally in related studies can be categorized mainly into three following types: 1) features from tweeting and retweeting practices of users, 2) features from tweet texts, and 3) features from metadata of users. Additionally, we propose a novel set of features that we refer to as User History features. Most of conventional features have a common defect which is the fact that they are affected by influence level of users within the Twitter network. Our key idea behind leveraging User History is to normalize influence of individual user using tweets that they posted in the past. By analyzing how someones new tweet differs from his/her previous tweets, we can make a probabilistic prediction of its informativeness in a fair, quantitative way regardless of how influential the author is in the Twitter network. We demonstrate the effectiveness of the proposed feature set, User History, on a Twitter dataset, by comparing them with conventional feature sets.

2   Related Works

There are several works in literature that aim at finding important messages in Twitter. Naveed et al. [3] apply document length normalization technique, which is important to measure the quality of existing web document, for assessing a short text quality in Twitter. However, it is not appropriate for tweets, since too short text do not have enough clues to fulfill information needs.

On the other hand, Sriram et al. [7] try to classify short texts with domain-specific features to overcome the data sparseness problem. Their filtering-based method classifies incoming tweets by considering information of the author and features within the tweets. They show that authorship plays a crucial role in tweet classification.

In recent years, numerous studies have attempted to use message propagation across the social network. These approaches are assuming that widely spreading messages are important. Information propagation on Twitter is mainly performed via retweeting, which is a behavior of a user taking a tweet someone else has posted and re-posting the same tweet to the user's own followers. *Retweeted number* is the most simple and straightforward indicator for interest of users on a particular tweet [9], so some studies concentrate on predicting retweetability. Suh et al. [4] investigate the relationship of retweeted number against several features like the number of URLs and hashtags in a tweet, the number of followers and followees of the author and so on. They report that *following* relationships and the presence of URLs and hashtags are strong indicators for predicting retweetability. Hong et al. [5] observe that the features, such as user's degree distribution and the fact whether a message has already been retweeted at least once before, contribute significantly to the classification performance.

Uysal and Croft [8] propose a new task which involves personalized tweet ranking. By considering *retweeting preference* of individual user, their method is adequate for specific users rather than general users. Even though such approach

Table 1. Previous feature classes

Class	Features
Propagation	- # of Retweeted, - # of Replies
Message	- # of words, - # of URLs, - # of of hash tags, - Length of text (in byte), - Length text exclude original text (in byte), - Is Mention : e.g. @UserName, - Is Retweeting : e.g. RT @, - Using Qustion Mark : e.g. ?
User	- # of followers, - # of followings, - # of total tweets, - Time passed since account creation (in day)

is merely beyond the scope of this study which is identifying generally informative tweets, we use their feature categories to group our features. There are four feature classes: author-based, tweet-based, content-based, and user-based. We exclude the last class, user-based, since it is introduced for personalized tweet ranking purpose. To summarize, the conventional features used in previous works can be divided into three main categories which can be seen in Table 1.

While the studies mentioned above focus on investigating computationally measurable features, Andre et al. survey what people really want to read in Twitter using questionnaires [2]. Their surveys reveal that informative tweet is among the ones that users feel worth reading. This result strongly motivated us to develop a general framework for identifying informative tweets. The next section presents a clear definition of informativeness and our method to identify informative tweets in details.

3   Detecting Informative Tweets

Our aim is to automatically detect informative tweets as valuable messages. This task is simply formulated into a binary classification problem, determining whether a tweet is informative or not. To define informative tweets, we follow the work of Ni et al. [6] that investigates informative articles in blogs. Although the approach they propose is not suitable for directly processing tweets with extremely short lengths, their concept of informative articles can be applied to tweets. We adapt their definition of informative articles for our classification. To identify tweets effectively, we propose a new feature class, User History, and we show how it works on detecting informative tweets.

3.1 Definition of Informative Tweets

To define informative articles, Ni et al. [6] process surveys which topics and what kind of contents people prefer to read. According to their survey, informative articles are those which have contents of certain specific genres. Since the concept of informativeness is invariant, we define informative tweets based on their work. The contents of informative tweets include following genres:

- News
- Technical descriptions
- Commonsense knowledge
- Objective comments on the events
- Opinions

We bring their definition almost as it is, with slight modification in details to fit characteristics of social networking services. All genres presented in their work are also used in this study. As recency is the most important characteristic in News' genre, not only fresh news tweets but also urgent news tweets are considered as informative tweets. Twitter allows users to post only 140-character-long messages, which are called tweets. Because of the length limit, a tweet can hardly contain enough information, especially in case of two genres, Technical descriptions and Commonsense knowledge. Thus, we additionally take web pages linked by URLs as well as text in tweets into account to determine whether the tweet is informative or not. We add Opinions genre because many recent works on tweets consider opinions as another type of information [7]. In our work, in order to detect generally informative messages, tweets of Objective comments on the events or Opinions genres are treated as informative tweets only when they refer to serious social problems or hot issues.

While building data for our experiments, annotators were asked to read each tweet and first, decide whether it belongs to one of these five predefined genres. Even if, it belongs to one of the genres, annotators could have marked it as not informative in case it has too short text to contain sufficient information or to understand without any specific background. For example, in spite of hot issues, a tweet, "일본지진!!"(earthquake in Japan!!), is regarded not informative because its text is too short text and does not have any further information, such as links to news articles or pictures. In contrast, the tweet, "일본 지진으로 원자력 발전소 붕괴! 링크는 방사선이 인체에 미치 는 영향 http://yfrog.com/gyhobshj(nuclear power plants collapsed by Japan earthquake! This link shows you the impact of radiation on the human body. http://yfrog.com/gyhobshj), was marked as informative.

3.2 User History Features

To estimate informativeness of a tweet, we devise a new feature class, namely User History, with novel characteristics derived from the tweeting behavior of

Table 2. User History class

Category	Features

Distinctiveness of tweets	- **Means** of users' Retweets/Replies/Repliers/Length
	- **Variances** of users' Retweets/Replies/Repliers/Length
	- **Deviations** of Retweets/Replies/Repliers/Length
	   between the tweet and Mean (x - mean)
	- **Standardized** Retweets/Replies/Repliers/Length of the tweet
	   (Z-value of x)
	- **Probabilities** of tweet's of Retweets/Replies/Repliers/Length
	- **Normalized** Retweets/Replies/Repliers/Length of the tweet
	   (x / Max)
	- **Proportions** of tweet's of Retweets/Replies/Repliers/Length
	   (x / Sum)

User Tendency	- Number of Normal/Replying/Retweeting tweets - Sums of Retweets on Normal/Replying/Retweeting tweets - Sums of Replies on Normal/Replying/Retweeting tweets - **Proportions** of Normal/Replying/Retweeting tweets - **Proportions** of Retweets on Normal/Replying/Retweeting tweets - **Proportions** of Replies on Normal/Replying/Retweeting tweets

users. It consists of two categories of features: Distinctiveness of tweets and User tendency. We descript these categories with motivations and roles in the next subsections. The whole features of User History class are shown in Table 2.

Distinctiveness of Tweets. Since informative tweets are considered as the most worthwhile tweets to read, they will receive far more attention from other users than the users mundane tweets. Other users interest in a tweet is shown through Retweets[1], Replies[2], and Repliers[3] of that tweet. Therefore, between informative tweets and not-informative tweets would be distinct differences of Retweets, Replies, and Repliers. We utilize these differences as features to classify tweets.

The level of attention which each user attract from others is various. From crawled Twitter corpora, we can obtain individual user's tweeting records, such as Retweets, Replies, and Repliers. Our premise is that these records reflect how the community of Twitter users perceives the informativeness of individual tweets. However, to estimate influence of individual user, it may be not fair to directly use sheer statistics of Retweets, Replies and Repliers without any processing, since less influential users tend to receive fewer responses on their tweets. It is why we model each users influence by using tweets that he/she posted in the past.

---

[1] Retweets means how many times a tweet is retweeted.

[2] Replies is the number of replies which the tweet receives.

[3] Repliers is the number of distinct users who retweet or reply on the tweet avoiding duplication.

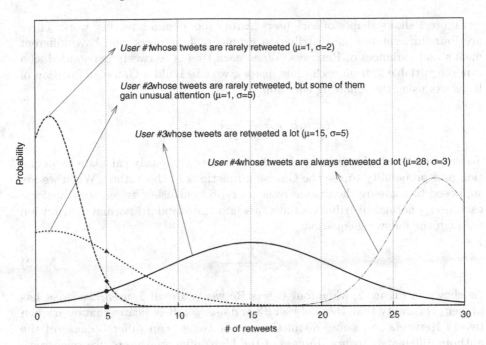

Fig. 1. An example of User History models of different users

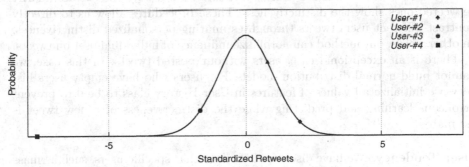

Fig. 2. Comparement of tweets which are retweeted 5 times

To model each users influence, we define a users history as a set of the entire tweets which belong to that user. Since each users history contain the statistics of Retweets, Replies, Repliers and Length[4], it is possible to calculate each users means and variances of Retweets, Replies, Repliers and Length. Assuming that the statistics of each element (Retweets/Replies/Repliers/Length) follows a normal distribution, we can construct multiple Gaussian functions as models of users influence. These functions provide important features for measuring distinctiveness of an incoming tweet, such as the deviation of Retweets and the probability of Length.

---

[4] Length is character-length of a tweet.

Figure 1 shows shapes of each users history model and how they work. There are four different normal distribution graphs of four users who have different means and variances of Retweets. When each tweets Retweets is regarded as a random variable x, from each users history, we can build a Gaussian function of Retweets using the following equation:

$$f(x; \mu, \sigma^2) = \frac{1}{\sigma\sqrt{2\pi}} e^{-\frac{1}{2}(\frac{x-\mu}{\sigma})^2} \tag{1}$$

To measure the distinctiveness of a new tweet, we can easily calculate its deviation and probability to use the Gaussian functions of the author. When we are supposed to compare the distinctiveness of different users' tweet, we transform each user's normal distribution functions into the standard normal distribution through the following equation:

$$Z = \frac{X - \mu}{\sigma} \tag{2}$$

As shown in Figure 2, when four tweets Retweets are all 5, user#2's tweet has higher probability than those of user#3 and user#4. It indicates that even when tweets Retweets are same, distinctiveness of tweets can differ because of the authors different influence. In case of the high influential users like celebrities, among their tweets which are usually retweeted a lot, only the extraordinarily retweeted ones show the distinctiveness. These procedures allow us to directly contrast different users tweets through estimating personalized distinctiveness. In other words, our method can normalize influence of individual user on tweets.

There is an exceptional case, users without posted tweets. In this case, we cannot build normal distribution models. For users who have empty users history, we initialize all values of features in User History class to be 0 to prevent erroneous learning and predicting while the distinctiveness of a new tweet is estimated.

User Tendency. We have discovered the fact that specific users, such as mass media, frequently post informative tweets. Also there are certain users who always retweet informative tweets. There are fairly distinguished from general users. Above facts can be helpful to identify informative tweets. So, to capture tendencies of users tweeting, we additionally leverage several features which are associated with individual users tweeting behaviors.

First, we assort tweets into three types: Normal, Retweeting, and Replying tweets. Normal tweet means a tweet written by user him/herself in an open space. Retweeting tweet means a tweet which user has retweeted someone elses message, and Replying tweet is a tweet written to reply someones tweet. Each user has various tweeting behavior. Generally, Twitter accounts of the mass media post news and issue tweets deemed as informativeness. But they do not post Retweeing or Replying tweets. Some users called social hubs usually retweet informative tweets, nonetheless hardly write their tweets. Although chatty users post a lot of Normal and Replying tweets, almost the whole their tweets are not retweeted. So we propose some proportional features representing which types of tweets are mainly posted and which types of tweets are mostly retweeted and replied. All of the features in this category are shown in Table 2.

4 Experiment

4.1 Dataset and Training Instances

We conduct our experiments on Korean Twitter dataset collected from November 2010 to March 2011. The whole dataset contains 337,028,356 tweets and 3,662,778 users.

Though we first had randomly picked samples from the entire dataset to build training and testing instances, we found that the portion of informative tweets was surprisingly low; only 0.5% of randomly sampled tweets were judged to be informative by human annotators. As it is very challenging to learn a useful classifier when training data is highly skewed, we need to use alternative sampling method in order to increase the proportion of informative tweets. By our intuition, we devise a simple sampling algorithm like the following formula:

$$Priority(x) = \#Retweet(x) + \#Reply(x) + \#Replier(x) + \#URL(x) +$$
$$\#Hashtag(x) + Length(x) \tag{3}$$

where x denotes a *tweet*.

First, all the tweets were ranked in descending order of the priority. Then three annotators were asked to determine the informativeness of the top 1,000 tweets. Tweets are labeled in a way that more than two annotators agree. The average of each annotator agreement is 0.86 and over.

With this sampling method, we could get a result with much higher portion (21.7%) of informative tweets. Among the 1,000 labeled tweet instances, the 200 most recently created tweets are chosen as a testing set whereas the remaining tweets are used as a training set. The portions of informative tweets in the training set and testing set are 20% and 29%, respectively. We build User History models to gather four-month's data of the users. For experiments, we use the maximum entropy model as the classifier.

4.2 Informativeness Classification and Ranking

We investigate the usefulness of each conventional feature set separately, and then integrate them as a unified baseline. Our proposed method is the combination of the baseline features and the selection of features in User History class. To select useful features from User history, we arrange them in descending order of an absolute correlation coefficient between feature values and classes in the training set, and chose features over the threshold (0.3). Our selection shows the highest classification performance on 10-fold cross validation in the training set. In User History class, the deviation of Retweets and standardized

Table 3. Performance of classifying informative tweets

Feature Class	Accuracy	Precision	Recall	F1
Propagation	0.690	0.421	0.276	0.333
Message	0.835	**0.821**	0.552	0.659
User	0.685	0.432	0.276	0.337
Baseline(P+M+U)	0.800	0.725	0.500	0.591
Proposed(B+History)	**0.860**	0.778	**0.724**	**0.750**

Table 4. Performance of ranking informative tweets

Feature Class	P@1	P@10	P@30	R-Precision	AveragePrecision
Propagation	0.000	0.500	0.400	0.362	0.262
Message	1.000	0.700	0.767	0.621	0.592
User	0.000	0.700	0.433	0.466	0.305
Baseline	1.000	0.800	0.767	0.707	0.647
Proposed	**1.000**	**0.900**	**0.800**	**0.759**	**0.764**

Retweets especially play a key role in classification perfomance, with more than 0.4 correlations.

We have experimented on the testing set with each conventional feature classes, the unified baseline, and the proposed method in order to know how much each method is helpful for classifying informative tweets. Table 3 shows the classification performance. We use Accuracy, Precision, Recall and F1 as evaluation metrics.

On the classification task, proposed method outperforms almost all of measures. As we can see, Message class is considerably helpful for precision, and shows strangely better performance than a baseline that includes it. We expect that this is because user influence interferes while retrieving some tweets written by unpopular users. Even though our method shows slightly low performance than the message class in terms of precision, our method significantly improves recall. Compared to the baseline, proposed method improves Accuracy, Precision, Recall, and F1 by 8%, 7%, 45% and 27%, respectively. These results mean that User History features, such as deviations of retweets and replies, are particularly helpful to find candidates of informative tweets.

We also have conducted more experiments in terms of ranking to evaluate the effectiveness of method for detecting informative messages. For ranking, we sort tweets by descending order of the probability given by the maximum entropy classifier. Precision@k, R-Precision, and Average Precision are used as evaluation measures. Since there are 58 informative tweets in testing data, R-precision is identical with Precision@58. The results are shown in Table 4 and Figure 3.

As well as on the classification task, our proposed method is still the best on the ranking task. Although the Message class shows high precision on the classification task, it frequently fails to locate informative tweets in the high

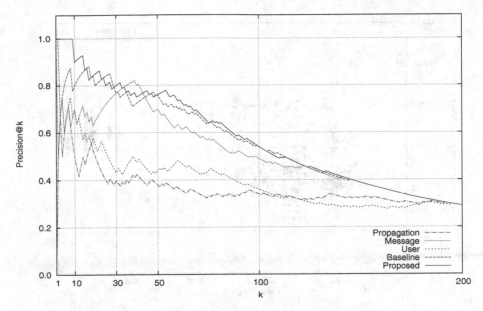

Fig. 3. Precision@k graphs

rank position compared to the baseline. This means that non-informative tweets hold the high ranks when message-based features are only used. In contrast, proposed method using User History consistently shows outstanding performance especially in terms of average precision. The proposed method achieves very high average precision which improves 18% compared to the baseline. This indicates that User History, which normalize each users influence and estimate importance of a tweet, makes informative messages hold a high rank. So our method is available to apply to information retrieval system and we can also expect promising performance on searching informative messages in social networking services.

Additionally, the improvement of our method is statistically significant (at the level of p-value less than 0.05). This demonstrates that our method based on User History is effective for detecting informative tweets which are worthwhile to read.

4.3   Normalizing Effects of User History

Figure 4 shows the retweets statistics of tweets with regard to the influence of the author, measured by the number of followers. We can observe that most of the tweets written by highly influential users are more frequently retweeted regardless of the tweets informativeness. If we use a sampling methods utilizing retweet frequency to gather instances from Twitter corpus, most of tweets necessarily belong to a small number of specific users, such as celebrities and sports stars. This is the inherent defect of conventional features like Retweets

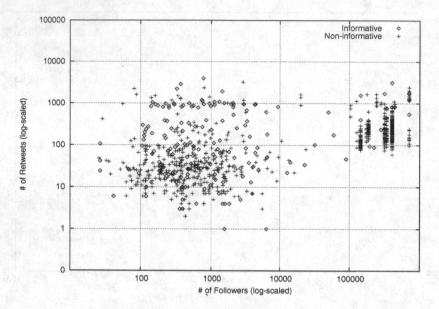

Fig. 4. Retweets statistics of users

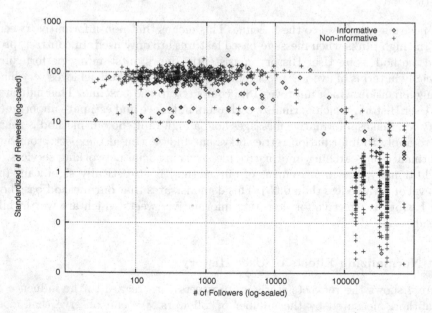

Fig. 5. Standardized Retweets statistics of users

and Replies. Figure 5 shows the *standardized* retweet statistics of tweets. Note that the retweet values for authors with relatively less followers are promoted, and the values of influential users are degraded. These two figures demonstrate how our method affects history statistics of users.

5   Conclusion

In this paper, we define a new task involving detection of informative messages which can extend the benefits to the general users in the Twitter stream. We propose a novel feature set, named User History, focused on tweets' distinctiveness and users' tendencies. Experimental results on Twitter dataset show that the proposed method improved the performance of both classification and ranking of informative tweets, especially in terms of Recall and Average Precision. This approach can contribute to the improvement of information retrieval system on the social media, and also can be utilized by contents curators [10] who select and organize valuable posts to diffuse them over the social networking services.

Though our method is appropriate to detect informative tweets, our model has some limitations, such as too simple parameter estimation for unseen users. For future work, we plan on improving the model by adapting and investigating the other probability distributions, such as Poisson and Beta distributions, for more reliable probability estimation.

Acknowledgment. This work was supported by the National Research Foundation of Korea(NRF) grant funded by the Korea government(MEST) (No. 2012033342).

References

1. Kwak, H., Lee, C., Park, H., Moon, S.: What is Twitter, a social network or a news media? In: WWW 2010, pp. 591–600 (2010)
2. André, P., Bernstein, M., Luther, K.: Who gives a tweet?: evaluating microblog content value. In: CSCW 2012, pp. 471–474 (2012)
3. Naveed, N., Gottron, T., Kunegis, J., Alhadi, A.C.: Searching microblogs: coping with sparsity and document quality. In: CIKM 2011, pp. 183–188 (2011)
4. Suh, B., Hong, L., Pirolli, P., Chi, E.H.: Want to be Retweeted? Large Scale Analytics on Factors Impacting Retweet in Twitter Network. In: SOCIALCOM 2010, pp. 177–184 (2010)
5. Hong, L., Dan, O., Davison, B.D.: Predicting popular messages in Twitter. In: WWW 2011, pp. 57–58 (2011)
6. Ni, X., Xue, G.-R., Ling, X., Yu, Y., Yang, Q.: Exploring in the weblog space by detecting informative and affective articles. WWW 2007, pp. 281–290 (2007)
7. Sriram, B., Fuhry, D., Demir, E., Ferhatosmanoglu, H., Demirbas, M.: Short text classification in twitter to improve information filtering. In: SIGIR 2010, pp. 841–842 (2010)
8. Uysal, I., Croft, W.B.: User oriented tweet ranking: a filtering approach to microblogs. In: CIKM 2011, pp. 2261–2264 (2011)
9. Yang, Z., Guo, J., Cai, K., Tang, J., Li, J., Zhang, L., Su, Z.: Understanding retweeting behaviors in social networks. In: CIKM 2010, pp. 1633–1636 (2010)
10. Bhargava, R.: Manifesto for the content curator: The next big social media job of the future? (2009), http://rohitbhargava.typepad.com/

A Study on Potential Head Advertisers in Sponsored Search

Changhao Jiang[1,*], Min Zhang[1], Bin Gao[2], and Tie-Yan Liu[2]

[1] State Key Laboratory of Intelligent Technology and Systems,
Tsinghua National Laboratory for Information Science and Technology,
Department of Computer Science and Technology, Tsinghua University,
Beijing, 100084, P.R. China
jch.cst@gmail.com, z-m@mail.tsinghua.edu.cn
[2] Microsoft Research Asia, 13F, Bldg 2, No. 5, Danling St,
Beijing, 100080, P.R. China
bingao@microsoft.com, tyliu@microsoft.com

Abstract. This paper studies the advertisers from whom the search engine may increase the revenue by offering an advanced sponsored search service. We divide them into head and tail advertisers according to their contributions to the search engine revenue. Data analysis shows that some tail advertisers have large amount of budgets and low budget usage ratios, who aimed to achieve the planned campaign goals (e.g., a large number of clicks), but they finally failed in doing so due to wrongly-selected bid keywords, inappropriate bid prices, and/or low-quality ad creatives. In this paper, we conduct a deep analysis on these advertisers. Specially, we define the measures to distinguish potential head advertisers from tail advertisers, and then run simulation experiments on the potential head advertisers by applying different improvements. Encouraging results have been achieved by our diagnosing approaches. We also show that a decision tree model can be implemented for a better improvement to those advertisers.

1 Introduction

Sponsored search has become one of the most profitable business models on the Internet. It helps the advertisers achieve a considerable amount of revenue by bringing search users, i.e. potential customers, to the advertisers' websites.

In sponsored search, when a query is submitted, the search engine will show some selected ads along with the organic search results. If an ad is shown on the search result page, we say that the ad has an *impression*. The selection of such ads is based on several factors such as the bid keywords, the bid prices, and the ad quality (including ad relevance). If an ad is clicked by a user, the search engine will charge the advertiser a certain amount of money (i.e., the cost for that click), according to the pricing model in the auction mechanism. This is the

---

* This work was performed when the first author was an intern at Microsoft Research Asia.

major revenue source for the search engine. Thus it can be seen that the revenue of the search engine is related to the number of *impressions*, *click-through rate* (CTR), and the *cost per click* (CPC) of the ads. Very limited study has been done on increasing search engine revenue by identifying the advertisers with high potential in revenue contribution and helping them improve their performance.

In a sponsored search system, it is difficult for all the advertisers to achieve their desired campaign goals. As a result, their contributions to the search engine revenue vary largely. Furthermore, the bad performances (and small revenue contributions) of the advertisers may lie in different situations. For example, some advertisers are satisfied though not so much traffic is achieved; however, some advertisers are ambitious and they desire more attentions. In order to level up the effectiveness of the entire sponsored search system, we argue that the first step to help low-performance advertisers is to identify the unsatisfied advertisers who have potential to be improved.

In this paper, we make investigations on advertisers' performances and conduct potential head advertiser identification and diagnosis in sponsored search. In particular, we would like to answer the following four questions by our study.

- What are the most significant differences between advertisers with large revenue contributions to the search engine and those with small revenue contributions?
- Which group of advertisers has the great potential to improve their performances?
- What are the reasons for these advertisers' low performances?
- How can the sponsored search system identify the primary cause for the bad performed advertisers and provide the personalized suggestions to them?

We have conducted an intensive study on a large-scale sponsored search dataset from a commercial search engine. Characteristics of the two groups are investigated (see Section 3). Specifically we find that less than 10% advertisers contribute 90% revenue to the search engine. We call these advertisers *head* advertisers. The remaining advertisers are called *tail* advertisers.

Tail advertisers with budgets no less than those of the head advertisers can be regarded as *potential head* advertisers[1]. They are willing to perform like the head advertisers but finally failed in doing so. An interesting observation by our diagnosis on potential head advertiser is that the biggest gap between head and tail advertisers lies in the number of impressions, while the differences between other aspects (i.e., CTR and CPC) are not so significant.

We design different improvement strategies and conduct simulations on the real sponsored search data to verify the effectiveness of the improvement strategies. The experimental results show that the performances of the potential head advertisers are greatly boosted in all scenarios. In the end, we implement a decision tree model to identify the primary failure reason for these potential head

---

[1] There are some accounts managed by advertising agencies. To avoid the influence from these accounts, we identified the agency-associated advertisers and removed them from our corpus.

advertisers. With such a methodology, the search engine can provide personalized suggestions (e.g., keyword suggestions) for the potential head advertisers and help them achieve the campaign goals as much as possible.

To sum up, the main contributions of our work are listed as follows. (i) We conduct an intensive comparison between head and tail advertisers according to their contributions to the search engine revenue. To the best of our knowledge, it is the first piece of work on this kind of study in the literature. Our study shows that the biggest difference between head and tail advertisers is the number of impressions but not CTR or CPC. (ii) It is the first reported study, as far as we know, that cares about identifying and helping the advertisers with high potential to contribute more revenue to the search engine.

2   Related Work

It is always an essential goal for search engines to improve their revenue in sponsored search. To achieve the goal, there are two branches of approaches: one is to optimize the search engine ads delivery system, and the other is to help the advertisers improve their performance.

In the first branch, a lot of work has been done on optimizing auction theory, ranking strategy, ads relevance calculation, click-through rate prediction, and keyword matching algorithm. Some work focuses on auction and ranking mechanisms [4,11,7,1,14,6]. For example, Feng et al [7] compared several mechanisms for allocating sponsored slots and proposed a rank-revision strategy that weighted clicks on lower ranked items more than those on higher ranked ones. Some other work [1,2,16,5] focuses on optimizing the search engine's ad recommendation in order to satisfy the users and get optimized revenue at the same time. Besides, a lot of work focuses on the prediction of relevance or CTR and the construction of the click model [15,8,12,17,9]. Hillard et al [9] presented a relevance prediction method using translation models to learn user click propensity from sparse click logs.

In the second branch, only a little work has been done on improving advertiser performance. In [3], Brogs et al studied a natural bidding heuristic in which advertisers attempt to optimize their utility by equalizing their return-on-investment (ROI) across all keywords. They come up with good results based on an assumption that advertisers are well-informed and familiar with sponsored search. However, this assumption is not as sound in many cases as expected. Our preliminary study on a commercial search engine log shows that a large fraction of advertisers aim to achieve good performance by committing a lot of budgets but eventually fail in doing so due to wrongly-selected bid strategy and/or low-quality ads.

To sum up, (i) there is little work in the literature analyzing the impact of different factors to sponsored search advertisers in terms of their contribution on search engine revenue; quite a lot of work has been done on CTR prediction and relevance prediction, while little is emphasized on the significant impact of impressions. (ii) We have not found previous work that scientifically makes intensive study on helping advertisers with small revenue. (iii) We proposed a system to diagnose the advertiser performance on sponsored search, separate them into different categories, and find the most primary issues for them to improve. To the best of our knowledge, such kind of diagnosis has not been reported before in sponsored search.

3   Data Analysis on Performances of Advertisers

We have performed an intensive study on the performances of the advertisers based on the sponsored search data obtained from a commercial search engine. We use two kinds of data in our study: one is the auction log that records the submitted queries and the corresponding ad impressions and clicks, and the other is the advertiser database that records bid keywords, bid prices, and the budget for each advertiser. The data was collected in four successive weeks in Oct 2010, containing more than two hundred thousand active advertisers[2].

3.1   Definitions of Head and Tail Advertisers

As discussed in the introduction, we divide the advertisers into two groups in our study, according to their contributions to the search engine revenue. The criterion is as follows.

- HEAD ADVERTISER: if an advertiser contributes more than R revenue to the search engine in a certain period, the advertiser is called a head advertiser.
- TAIL ADVERTISER: if an advertiser contributes no more than R revenue to the search engine in a certain period, the advertiser is called a tail advertiser.

Thus, the selection of R is crucial to the analysis of head and tail advertisers. Different settings of R will significantly influence the properties of the two groups. For example, it will affect the portion of search engine revenue contributed by head advertisers (denoted as $RevCoverage$ for ease of reference), the percentage of head advertisers among all the advertisers (denoted as $AmtCoverage$), and the stability of the definitions on head and tail advertisers along with time (denoted as $Stability$).

Mathematically, the above properties can be defined as follows. Suppose we have a series of successive periods (e.g., weeks) of data on the budget and spending of the advertisers. Let $H_i(R)$ be the set of head advertisers in the i-th week given R, and let A_i denote the set of all advertisers in the same week. Let a denote a single advertiser and $Rev_i(a)$ denote his/her contribution to the search engine revenue in the i-th week. Then we have

$$RevCoverage_i(R) = \frac{\sum_{a \in H_i(R)} Rev_i(a)}{\sum_{a \in A_i} Rev_i(a)}. \tag{1}$$

---

[2] Each account in sponsored search system is regarded as an individual advertiser.

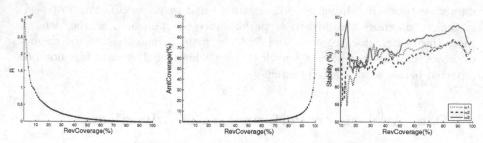

Fig. 1. R on different **Fig. 2.** *AmtCoverage* on **Fig. 3.** *Stability* Distribu-
RevCoverage   *RevCoverage*   tion

$$AmtCoverage_i(R) = \frac{|H_i(R)|}{|A_i|}. \tag{2}$$

$$Stability_i(R) = \frac{|H_i(R) \cap H_{i+1}(R)|}{|H_i(R) \cup H_{i+1}(R)|}. \tag{3}$$

According to Figure 1, R decreases fast as *RevCoverage* increases, indicating that a small R is needed if we want to obtain a large revenue coverage. Based on information in Figure 2, sponsored search seems to be a long-tail market: no more than 10% advertisers contribute more than 90% revenue to the search engine. According to Figure 3, as *RevCoverage* raises, the stability first rises and then drops, and the *RevCoverage* of 90% corresponds to a peak on the curves with different pairs of successive weeks.

From the above observations, we find the *RevCoverage* of 90% seems to be a good threshold to distinguish head and tail advertisers. The corresponding R can be determined by its monotonous relationship with *RevCoverage* in Figure 1. In our study[3], $R = 339.8$. Thus, we get these head advertisers (6.75%) and the tails (93.24%).

3.2   Reason for Bad Performance

After dividing the advertisers into head and tail, we want to identify the main reason for these advertisers' low performances. As we know, the contribution of an advertiser a to the revenue of search engine can be approximately computed as

$$Revenue_a \sim Impression_a \times CTR_a \times CPC_a, \tag{4}$$

where CTR_a denotes the advertiser's average click-through rate, and CPC_a denotes the advertiser's average cost per click.

Therefore, we should study the three factors when analyzing the performance of an advertiser. We compute the average values of these factors from the data used in our study in Table 1.

---

[3] Note that throughout the paper we multiplied all the amount related numbers (bid, revenue, etc.) by a positive value, according to the confidential policy of the commercial search engine.

Table 1. Average properties for head and tail advertisers

TYPE	|Impression|	CTR	CPC
Head	5.26E+5	0.0240	1.4737
Tail	7.92E+3	0.0157	1.1436
Head/Tail Ratio	66.41	1.5317	1.2886

We may find the average impressions of head and tail advertisers differ largely, while the average CTR and CPC do not have significant differences. Similar observations can be drawn from not only average value but also their distribution (see Figure 4). We also calculated the correlation between the distributions by following formula.

$$Cor(X,Y) = \frac{\sum_i (x_i - \bar{x})(y_i - \bar{y})}{\sqrt{\sum_i (x_i - \bar{x})^2 \sum (y_i - \bar{y})^2}} \tag{5}$$

Here $X = \{x_1, x_2, \cdots, x_n\}$ and $Y = \{y_1, y_2, \cdots, y_n\}$ are the vectors of two distributions, and $\bar{x}$ and $\bar{y}$ are the mean of the elements in the two vectors.

Figure 4 illustrates the distribution of Impressions, CTRs, and CPCs. The distributions are calculated from advertiser database for one week. We divide the number of Impression/CTR/CPC into logarithmic intervals. Specifically, the correlation between head and tail is -0.21. CTR and CPC do not have large difference between head and tail, and the corresponding correlations are 0.942 and 0.873 respectively. P-HEAD denotes the group of potential head advertisers, which will be discussed in next subsection.

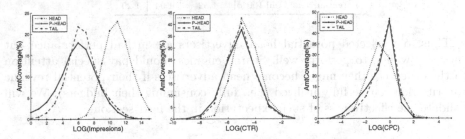

Fig. 4. Distributions of Impression/CTR/CPC

Thus, the impression number is regarded as the major difference. The small revenue contribution of the tail advertisers is mainly caused by limited impressions. In order to increase the revenue contribution, we need to find effective ways to help some of tail advertisers increase their impressions.

3.3 Potential Head Advertisers

After we find the main reason for bad performance, we focus on finding the group of advertisers that has potential to get improved. Though tail advertisers are in bad performances, not all of them have the incentive to increase their impressions. For example, some advertisers would not like to pay much in sponsored search because they have achieved their goals even in the limited traffic or they cannot afford more clicks. Meanwhile, the number of budget is set by each advertiser so that the search engine cannot charge the advertiser more than it. Thus, the budget might reflect the maximum willing cost of an advertiser. The advertisers with high enough budgets are regarded as the candidates with high potential to get improved performance. We call them *potential head* advertisers.

– POTENTIAL HEAD ADVERTISER: if the budget of a tail advertiser is larger than R (the same R in the definition of head advertiser) in a certain period, the advertiser is called a potential head advertiser.

We can see from Figure 4 that the properties of potential head advertisers are similar to those of the tail advertisers.

In our real world data, we find some agency-related advertisers. As some agencies might use other means but not budget (e.g., principled bidding and keyword selection strategies) to control the cost, we filtered these advertisers out of our potential head advertiser corpus. After that, the average impression number, CTR, and CPC of the potential head advertisers do not have significant difference from those of the tail advertisers (as following table shows).

Table 2. Average properties for tail and potential head advertisers

TYPE	Impression	CTR	CPC
Tail	7.92E+4	0.0157	1.1436
Potential Head	1.20E+5	0.0163	1.4752
Potential Head/Tail Ratio	1.52	1.04	1.29

Thus, our selected potential head advertisers are in bad performance but they are willing to perform well. Search engines should pay special attention to them because they might become head advertisers if their potential revenue contribution can be fully utilized (i.e., fully consuming their budgets). We will validate the effectiveness of such expectation in the next section.

4   Impression Improvement

To improve the advertisers' performances, we analyze the reasons for the bad performance and proposed some algorithms to deal with these problems. The experiment results show that potential head advertisers are very likely to be improved towards better performance.

4.1   Analysis for Low Impressions

In sponsored search, low impressions might be caused by several factors. To better understand these factors, let us have a look at the process of ad selection and delivery strategies. In a typical sponsored search system, when a user submits a query, the ad platform in the search engine will first check the ads and the keywords in its ad database. The ads which bid keywords match the query (according to a specific matching algorithm) will be selected as candidates for the auction. With a ranking mechanism [7], a rank score will be computed for each candidate ad based on the ad quality [8] (including the ad relevance to the keywords) and the bid price of the ad. Then the candidate ads are ranked according to the descending order of the scores, and the top-ranked ads will be shown to the users in the search result page.

From the above process, we can see that three factors will affect whether an ad will be shown or not: bid keyword (which affects the matching and filtration), bid price (which affects the ranking score), and the ad quality (which affects the ranking score). Then a question arises for a potential head advertiser, i.e., which of the above three factors is the primary cause of his/her bad performance? In other words, if the advertiser wants to improve his/her campaign performance, which factor(s) should he/she consider with the highest priority? We will try to answer this question in the next subsection.

4.2   Distribution of Different Reasons

We sampled 2,000 advertisers from the set of potential head advertisers, each of which are with plentiful information in their performance data and search log data. We asked three human experts to label the advertisers to the following five categories (with distribution percentage in the bracket). The experts were experienced advertiser campaign analysts in the customer service group of the commercial search engine.

- SUCCESS: the advertisers are judged to be in healthy status and no improvement is needed (7%).
- BAD KEYWORDS: the advertisers should improve their keyword selection (24%).
- LOW PRICES: the advertisers should increase their bid prices (33%).
- LOW QUALITY: the advertisers should improve their ad quality (31%).
- UNCERTAIN: the human experts cannot identify the primary problem for the advertisers (5%).

4.3   Validation of Potential Head Advertisers

Firstly, we propose an example method to simulate the endogenous improvement of the advertisers for each of the reasons in Section 4.1. Then we conduct our simulation experiments in the real one-week auction log. Specifically, We denote historical CTR as *quality score* of the ad and the rank score is set as bid price multiplied by *quality score*. Ads which are ranked at top 8 positions in each auction will have one impression. By implementing the same methods on different advertiser sets (head, tail, and potential head), we validate that potential head advertisers can be better improved.

Keyword Suggestion. We improve keyword selection by adding ten extra key-words into each order[4] according to their similarities with the original keywords. In particular, we compute the keyword similarity using the algorithm proposed by Glen Jeh *et al* in [10]. The ten queries that have top similarities with the original bid keywords in the order are selected as additional bid keywords to participate in the related auctions.[5]

Bid Price Tuning. Bid prices are set by the advertisers themselves, and thus it is unreasonable to select an ad-hoc new price for an advertiser directly. Though it is difficult to guess the advertisers' acceptable maximum bid price, the price changes should obey the rules or the habits of the advertisers. We build a matrix M to record the probabilities of bid price switch[6] along with time. Thus we have,

$$M(i,j) = Probability(P_{t+1} = j | P_t = i, j > i), \tag{6}$$

P_t stands for the bid price in time t. The condition $j > i$ is to make sure this is a bid price *incremental* switch.

Quality Improvement. Quality score is calculated by complex methods like [8]. It might be very difficult to quantitatively calculate these improvement po-tentials. For simplicity, in our experiments, we simply use the best quality score in the ad's history as its improved quality score, which is achievable for the corresponding advertiser.

Result Analysis. To evaluate the effectiveness of the improvement strategies, we define a special metric denoted as $M1$. Suppose $\mathbf{G}, \mathbf{T}$, and $\mathbf{P}$ denote the adver-tiser set of Head, Pure Tail (Potential Head Advertisers excluded from Tail Ad-vertisers), and Potential Head, respectively. For each set $s \in \{\mathbf{G}, \mathbf{T}, \mathbf{P}\}$ and each method $m \in \{$Keyword Suggestion, Bid Price Tuning, Quality Improvement$\}$, $M1$ is defined as,

$$M1(s,m) = \frac{\sum_{a \in s} NewImp_m(a)}{\sum_{a \in s} OriImp(a)}. \tag{7}$$

In the above definition, $OriImp$ and $NewImp_m$ denote the impression number before and after applying the improvement method m. Thus, $M1$ reflects the performance improvement on each advertiser set.

From Table 3, we can see that the performances of potential head advertisers are better improved than the other two groups by all three improvement meth-ods. Among the three methods, we can find that Keyword Selection provides the

---

[4] In the commercial search engine, an order contains a group of keywords and a group of ads. Each ad in the order will be taken into consideration in the auction when any keyword in the order is triggered by a query.

[5] In the new auctions, the bid prices for the new keywords are set as the average bid price of the original bid keywords in the order.

[6] The bid price switch is assumed as a Markov process and obey Markov property. At any time, the next time bid price is only decided by the current price and situation.

Table 3. Improvement Results on M1

Method	Head	Pure Tail	Potential Head
Keyword Selection	250.25	322.22	**392.32**
Bid Price Tuning	1.21	13.79	**53.92**
Quality Improvement	4.60	45.97	**108.33**

best improvement. The most probable reason is that most of the queries in keyword similarity matrix are popular ones. It means that once an advertiser wins auctions on these keywords the impression number will increase a lot. As the long tail theory holds in query submission, it is reasonable that head advertisers are also improved significantly. For the other two methods, as head advertisers have been competitive enough, high performance improvement is hard to achieve. However, potential head advertisers are better improved than both head and pure tail advertisers. Therefore, potential head advertisers are easily found by simple rules and can be improved by much via certain strategies.

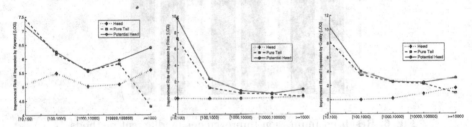

Fig. 5. M1 by improving keyword selection, bid price, and quality score

Figure 5 illustrate the improvement results (using M1 metric) by improving keyword selection, bid price, and quality score respectively, throughout different original impression numbers. Specifically, we put the advertisers into different buckets according to their original impression numbers, and then we calculated the logarithmal M1 values on different sets and plot these figures. We can find that potential head advertisers are better improved compared with the other groups.

Diagnosis System. In order to identify the primary factor for a potential head advertiser to improve, we build an advertiser diagnosis system according to the labeled data mentioned in Section 4.2.

Firstly, we extracted different categories of features for the potential head advertisers. That is, *Key performance indicators* (impression number, click number, CPC, etc.), *Advertiser account attributes* (budget, campaign numbers, etc.), *Advertiser ads attributes* (average quality score, average bid, etc.), and *Advertiser auction properties* (number of auctions the advertiser participated in, etc.).

Then, we trained a C4.5 [13] decision tree model from the label data. The average precision of the model was 92% in five-fold cross validation, showing that the diagnosis classifier works well in identifying the primary factors for the potential head advertisers.

To further investigate the effectiveness of the proposed advertiser diagnosis system, we conducted a set of comparison experiments. We randomly select 1,000 advertisers from the categories BAD KEYWORDS, LOW PRICES, and LOW QUALITY respectively as three *target sets*, and applied the corresponding improving strategies as described above. These improving strategies were also applied in a *comparison set*, which contained another 1,000 advertisers randomly selected from the potential head advertisers. The experimental results are shown in Figure 6. We can see that the target suggestions work very effectively: the improvement obtained on the three target sets are much larger than those obtained on the comparison set.

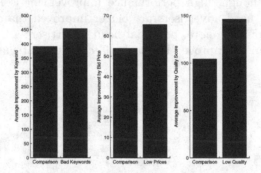

Fig. 6. Result on Impression Improvement by different methods

5   Conclusions and Future Work

This paper is a study on the sponsored search advertisers who perform badly but still have high potential to get improved. We made an intensive analysis on the differences of head and tail advertisers, and proposed a simple way to identify the potential head advertisers. To evaluate the effectiveness of our method, we conducted a group of simulation experiments on a real world dataset. We also proposed a decision tree model to diagnose the primary reason of the bad performing advertisers and thus to provide suggestions for the advertisers to get improvements. The experiment results show that potential head advertisers have better properties and the improvement can be enlarged by our diagnosis system.

For future study, there are two aspects to improve our work. On one hand, we will conduct more study on our improvement methods. On the other hand, we will focus on enhancing the diagnosis system. First, we will consider more features and employing advanced classifiers. Then we will investigate the diagnosis in the ad level instead of in the advertiser level.

References

1. Abrams, Z., Mendelevitch, O., Tomlin, J.: Optimal delivery of sponsored search advertisements subject to budget constraints. In: Proceedings of the 8th ACM Conference on Electronic Commerce, EC 2007, pp. 272–278. ACM, New York (2007)
2. Anastasakos, T., Hillard, D., Kshetramade, S., Raghavan, H.: A collaborative filtering approach to ad recommendation using the query-ad click graph. In: Proceeding of the 18th ACM Conference on Information and Knowledge Management, CIKM 2009, pp. 1927–1930. ACM, New York (2009)
3. Borgs, C., Chayes, J., Immorlica, N., Jain, K., Etesami, O., Mahdian, M.: Dynamics of bid optimization in online advertisement auctions. In: Proceedings of the 16th International Conference on World Wide Web, WWW 2007, pp. 531–540. ACM, New York (2007)
4. Borgs, C., Chayes, J., Immorlica, N., Mahdian, M., Saberi, A.: Multi-unit auctions with budget-constrained bidders. In: Proceedings of the 6th ACM Conference on Electronic Commerce, EC 2005, pp. 44–51. ACM, New York (2005)
5. Chatterjee, P., Hoffman, D., Novak, T.: Modeling the clickstream: Implications for web-based advertising efforts. INFORMS on Marketing Science 22, 520–541 (2003)
6. Dütting, P., Henzinger, M., Weber, I.: An expressive mechanism for auctions on the web. In: Proceedings of the 20th International Conference on World Wide Web, WWW 2011, pp. 127–136. ACM, New York (2011)
7. Feng, J., Bhargava, H.K., Pennock, D.M.: Implementing sponsored search in web search engines: Computational evaluation of alternative mechanisms. INFORMS J. on Computing 19, 137–148 (2007)
8. Graepel, T., Candela, J., Borchert, T., Herbrich, R.: Web-scale bayesian click-through rate prediction for sponsored search advertising in microsoft's bing search engine. In: Proceedings of the 27th International Conference on Machine Learning (2009)
9. Hillard, D., Schroedl, S., Manavoglu, E., Raghavan, H., Leggetter, C.: Improving ad relevance in sponsored search. In: Proceedings of the Third ACM International Conference on Web Search and Data Mining, WSDM 2010, pp. 361–370. ACM, New York (2010)
10. Jeh, G., Widom, J.: Simrank: a measure of structural-context similarity. In: Proceedings of the Eighth ACM SIGKDD International Conference on Knowledge Discovery and Data Mining, KDD 2002, pp. 538–543. ACM, New York (2002)
11. Kominers, S.D.: Dynamic Position Auctions with Consumer Search. In: Goldberg, A.V., Zhou, Y. (eds.) AAIM 2009. LNCS, vol. 5564, pp. 240–250. Springer, Heidelberg (2009)
12. König, A.C., Gamon, M., Wu, Q.: Click-through prediction for news queries. In: Proceedings of the 32nd International ACM SIGIR Conference on Research and Development in Information Retrieval, SIGIR 2009, pp. 347–354. ACM, New York (2009)
13. Quinlan, R.: C4.5: Programs for Machine Learning. Morgan Kaufmann Publishers (1993)
14. Radlinski, F., Broder, A., Ciccolo, P., Gabrilovich, E., Josifovski, V., Riedel, L.: Optimizing relevance and revenue in ad search: a query substitution approach. In: Proceedings of the 31st Annual International ACM SIGIR Conference on Research and Development in Information Retrieval, SIGIR 2008, pp. 403–410. ACM, New York (2008)

186   C. Jiang and T.-Y. Liu

15. Richardson, M., Dominowska, E., Ragno, R.: Predicting clicks: estimating the click-through rate for new ads. In: Proceedings of the 16th International Conference on World Wide Web, WWW 2007, pp. 521–530. ACM, New York (2007)
16. Wang, C., Zhang, P., Choi, R., D'Eredita, M.: Understanding consumers attitude toward advertising. In: Proceedings of the Eighth Americas Conference on Information Systems (2002)
17. Xu, W., Manavoglu, E., Cantu-Paz, E.: Temporal click model for sponsored search. In: Proceeding of the 33rd International ACM SIGIR Conference on Research and Development in Information Retrieval, SIGIR 2010, pp. 106–113. ACM, New York (2010)

Feature Transformation Method Enhanced Vandalism Detection in Wikipedia

Tianshu Chang, Hongfei Lin, and Yuan Lin

Information Retrieval Laboratory, School of Computer Science and Engineering,
Dalian University of Technology, Dalian 116023, China
tianshuchang@mail.dlut.edu.cn, hflin@dlut.edu.cn,
yuanlin@mail.dlut.edu.cn

Abstract. A very example of web 2.0 application is Wikipedia, an online encyclopedia where anyone can edit and share information. However, blatantly unproductive edits greatly undermine the quality of Wikipedia. Their irresponsible acts force editors to waste time undoing vandalisms. For the purpose of improving information quality on Wikipedia and freeing the maintainer from such repetitive tasks, machine learning methods have been proposed to detect vandalism automatically. However, most of them focused on mining new features which seem to be inexhaustible to be discovered. Therefore, the question of how to make the best use of these features needs to be tackled. In this paper, we leverage feature transformation techniques to analyze the features and propose a framework using these methods to enhance detection. Experiment results on the public dataset PAN-WVC-10 show that our method is effective and it provides another useful method to help detect vandalism in Wikipedia.

Keywords: Wikipedia, Vandalism, Classification, PCA.

1 Introduction

Wikipedia[1], the collaboratively edited encyclopedia, is among the most prominent websites on the Internet today. Every visitor of a Wikipedia Web site, even anonymous ones, can participate immediately in the authoring process: articles are created, edited, or deleted without the need for authentication. Through this way, a large scale database is constructed with over 15 million articles in over 270 languages, as measured in June 2012[2]. According to a recent comparison conducted by Nature, the scientific entries in Wikipedia are of quality comparable to those in the established Britannica encyclopedia[3].

However, allowing anonymous edit destructive in its removal of content. Wikipedia administrators ask volunteers into joining the battle to detect vandalism and revert the malicious modified content manually. As Wikipedia continues to grow, it

---

[1] http://www.wikipedia.org
[2] http:// stats.wikimedia.org/EN/
[3] http://www. nature.com/news/2005 /051212/full/438900a.html

Y. Hou et al. (Eds.): AIRS 2012, LNCS 7675, pp. 187–198, 2012.
© Springer-Verlag Berlin Heidelberg 2012

will become increasingly infeasible for users to manually police articles. Prominent anti-vandal bots such as ClueBot and VoABot II are also used to detect vandalism in Wikipedia, which use lists of regular expressions and user blacklists. These regular expression rules are created manually and are difficult to maintain.

Since 2008, machine learning methods have been introduced to solve the problem. Various features are collected and used to train a classifier to detect vandals including text features, reputation features and metadata features. Adler et al. [2] combined these features and train a classifier which outperforms each individual classifier trained on only one group of these features, which means that there is still much room for improvement on vandalism detection task.

The motivation of our research comes from the observation that most previous work tried to increase the number of features used for vandalism detection. However, larger feature set will unavoidably increase the complexity of the method and may bring possible burden for feature extraction. Another possible way to improve detection success is to make better use of the available features. So the question we examine in this paper is: Would it be better to make the best use of features we already have at hand and get almost the same result?

In this paper, we present a novel solution to automatically detect vandalism. Different from the previous work, after obtaining enough effective features, our method focuses on transforming the available features to create a new feature space. We use the techniques of multivariate statistics for feature transformation. A binary classifier is then trained to detect vandalisms. Our experiment results show an improvement of approximately 2%~25% at different evaluation metrics over the baseline. The main contributions of this paper are listed below.

It is a double-edged sword. Often times Wikipedia's freedom of editing has been misused by some editors. Nearly 7% of edits are vandalism [1], i.e. revisions to articles that undermine the quality and veracity of the content. Vandalism is defined as any edit which is non-value adding, offensive, or destructive in its removal of content. We explore the different groups of features used in previous studies, and the most efficient and best performing features are selected to construct our original feature set.

- Feature transformation methods are studied and applied to transform features before training the classifiers for detection task. To ourknowledge, such an approach has not been used in the vandalism detection task.
- Principle Component Analysis is studied and different kinds of utilizations of transformation using PCA are tested with both the entire data and the partial data to find an optimal solution.
- We combine feature selection methods with PCA to expand the feature set though PCA is often used to reduce the dimensions.

The rest of the paper is organized as follows: Section 2 describes some previous work on vandalism detection in Wikipedia. In Section 3 we will give detailed description of the proposed framework. Section 4 presents the experimental setting, the results and some with analyses. Finally, we conclude our work in Section 5.

2   Related Work

Wikipedia vandalism detection is a novel research topic related to social media quality and reliability. Before it is explicitly put forward, people mostly focused on the more general tasks such as information quality assessment and reputation systems. The previous work done on Wikipedia articles' information quality provides valuable references for our current research, such as featured article identification [3] and quality flaws distribution in Wikipedia [4].

The early approach being applied in Wikipedia for automatic vandalism detection use heuristics. These strategies include the amount of text inserted or deleted, the amount of uppercase letters and the frequency of vulgarisms detected via regular expressions [5-6]. Automated bots (e.g., Cluebot), filters (e.g., abusefilter), and editing assistants (e.g., Huggle and Twinkle) all aim to locate acts of vandalism. Due to the need of creating lists of regular expressions and manually adjusting weights and thresholds, these earlier heuristics based systems are difficult to maintain.

The Wikipedia vandalism detection research community begins to concentrate on machine learning approaches since 2008. Potthast et al. [7] are among the first to cast the vandalism detection task as a machine learning classification problem. They inspect a small sample of Wikipedia and create a feature set based on the content and metadata and train a classifier using logistic regression. Smets et al. [8] wrap all the content in diff text into a bag of words, disregarding grammar and word order. They use Naïve Bayes to detect vandalisms.

Many researchers apply natural language processing techniques in their learning model. Their researches on vandalism detection have explored features based on text, such as stylometric analysis based text features [9], text stability-based approach [10]. Wang et al. [11] propose a Web-based shallow syntactic semantic modeling method, which utilizes Web search results as resource and trains topic-specific n-tag and syntactic n-gram language models to detect vandalism. Chin et al. [12] adopt an active learning model to solve the problem of noisy and incomplete labeling of Wikipedia vandalism, and propose a solution with additional features using active learning and statistical language model. Similar to Chin, Mola-Velasco [13] proposes a topic-sensitive and language-independent method. With his method, given an edit, a set of related articles can be retrieved. These articles are used to build a language model.

Reputation systems are initially designed to assess the information quality of the content, but now they are used in vandalism detection task to evaluate the reputations of both the article and the contributor [14]. Since different parts of an article are contributed by different users, the reputation degree computed from their previous behaviors may affect the judgment whether their current edits are offensive or not. Adler et al. [15] analyze the reputation scores generated by a well-known Wikipedia reputation system WikiTrust for both contents, edits and authors, and transform them into feature vectors. The classifier trained by reputation features achieves a recall of 80% but a precision of only 40%.

Revision metadata have been proved to be an effective feature for Wikipedia vandalism detection. West et al. [16] extract features from the revision history of

Wikipedia, including the time-of-day, time-of-week when edits happened, time-since last OE (offensive edit) and so on. A system named STiki is implemented for real-time vandalism detection online.

To our knowledge, among all the previous research mentioned above, none of them has used advanced feature processing techniques. These techniques are well studied in classification and pattern recognition. In this paper, we apply a feature transformation technique to vandalism detection.

3   Feature Transformation Based Vandalism Detection

We treat the problem of Wikipedia vandalism detection as a classification task. For the purpose of recognizing vandalisms from regular edits, features are extracted to train a classifier in which vandalisms are considered as positive training examples and regular edits as negative training examples.

In contrast to previous work, our approach adds a post processing stage before training the classifier. Therefore, feature transformation techniques like PCA, SVD can be used to analyze the properties of the feature distribution. After the transformation, additional features are generated which represent the combined effect or the latent properties of features. Next the feature selection method is used to select the best performing features from the newly generated feature space which combines the additional features and the original features. Finally we use these features to train a classifier.

3.1   Original Feature Set Construction

Wikipedia revision histories provide the Pagediff page[4], which shows the differences between two versions of the article. On the Pagediff page, the numbered version and the subsequent version of the content are presented symmetrically in the same page with different parts highlighted. Features are mostly extracted from this page. The two versions of content are also used for feature extraction like the language model computing and KL distance computing.

Using all these data, 59 features are extracted to create the initial feature set in our system. Text features take the highest percentage. Many of them are used by Potthast and the PAN completion teams. Language features are computed based on the manually constructed dictionary for bad and good words. Revision metadata features are optionally added such as time-of-day, time-since-last edit. We also include language model features because sometimes vandalism comes with some unexpected words so we expect to see sharp changes. Conversely, deleting unexpected words can be an indicator of legitimate edits. Therefore we calculate Kullback–Leibler distance (KLD) between two language models for both unigram and bigram. For a complete feature set, see Table 1.

---

[4] http:// en.wikipedia.org/wiki/Template:Diff

Table 1. List of features used in our system

Class	Features	Descriptions
Text Features	AvgTF	Average frequency of inserted words in the new revision.
	CharacterDiversity	Different characters compared to the length of inserted text
	Digit_ratio	Digit to all characters ratio
	LongestCharSquence	Longest consecutive sequence of the same character
	LongestWord	Length of the longest word in inserted text
	Non-alphanumeric ratio	Non-alphanumeric to all characters ratio
	Size_ratio	Size of the new revision relative to the old revision
	SizeIncrement	Absolute increment of article size
	Upper to all ratio	Ratio of uppercase letters to all letters
	Upper to lower ratio	Ratio of uppercase to lowercase letters ratio
	PronFrequency	First and second person pronouns frequency
	PronImpact	The increased percentage of pronouns
	SexFrequency	Frequency of non-vulgar sex-related words
	SexImpact	The increased percentage of sex-related words
	VulgarismsFrequency	Frequency of vulgar and offensive words
	VulgarismsImpact	The increased percentage in vulgar words
	CompressRate	Compression rate of inserted text using the LZW method
	TransCompressRate	Compression rate of the new version to the old version
Language Model features	Perplex	Perplexity value from language model *diff*
	Entropy	Entropy value from language model *diff*
	OOV_Num	Number of unknown words from *diff*
	OOV_Per	Percentage of unknown words from *diff*
	Bigram_Hit	Number of known bigrams from *diff*
	Bigram_Per	Percentage of known bigrams from *diff*
	Unigram_Hit	Number of known unigrams from *diff*
	Unigram_Per	Percentage of known unigrams from *diff*
	KLdistanceUnigram	KL distance between the versions using unigram LM
	KLdistanceBigram	KL distance between the versions using bigram LM
Metadata features	CommentLength	Length in characters of the edit summary
	Anonymous	Whether the editor is anonymous or not
	Blanking	Whether the article is blanked
	CommentExist	Whether the comment exists
	IPCountryNumber	The IP address of the author
	TimeSincePage	Time since the article last changed
	PrevSameAuthor	Whether the two versions are submitted by the same author
	CommRevert	Whether this edit is a' revert'

3.2   Feature Transformation

Feature transformation is a process through which additional features can be generated from the original feature set. There are two classes of features generated by feature transformation methods. The first kind of features can aggregate the information given by the original set. For example, we have the feature of length and width of a rectangle in our original feature set which can represent the attributes of the object. However, the area of this rectangle is some kind of missing information about the latent relationships between features, and it can be added after a transformation method is applied. Another kind of features in the transformed space maybe computed with mapping functions. Linear or nonlinear projection can be applied to original features. These features are often used to replace the old ones for the purpose of dimensional reduction.

Based on the idea of feature transformation, we propose a different solution for the problem of vandalism detection. Our model of training a detector is divided into three steps including feature construction (a), feature transformation (b) and classifier training (c), which are illustrated in Figure 1.

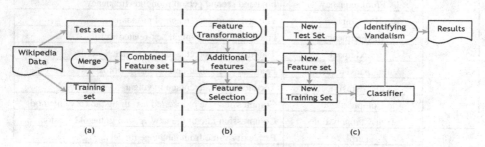

Fig. 1. System workflow for Wikipedia Vandalism Detection.

Notice that most previous work on Wikipedia vandalism detection focused on parts (a) and part (c). We try to introduce some feature transformation methods to enhance the performance of vandalism detector. Our feature transformation methods are based on Principle Component Analysis.

Principle Component Analysis (PCA) is the best, in the mean-square error sense, linear feature transformation [17] that we can achieve. It is an unsupervised projection method that computes linear combinations of the original attributes that are constructed to explain maximally the variance.

Let $X = \{X_1, X_2, ..., X_n\}$ be a feature set, and each dimension X_i is an attribute. The goal of PCA is to find such a group of mappings.

$$\begin{cases} F_1 = a_{11}X_1 + a_{12}X_2 + \cdots + a_{1n}X_n \\ F_2 = a_{21}X_1 + a_{22}X_2 + \cdots + a_{2n}X_n \\ \quad\quad ...\,... \\ F_m = a_{m1}X_1 + a_{m2}X_2 + \cdots + a_{mn}X_n \end{cases} \quad\quad (1)$$

where $F = \{F_1, F_2, ..., F_m\}$ are the principle components of our feature set, that we call the additional features. a_{ij} is the projection coefficient. When looking for the first principal component F_1 , we look for a vector $a_1 = \{a_{11}, a_{12}, ..., a_{1n}\}$ such that $a_1 a_1^T = 1$ and the variance of $a_1 X^T$ is maximal.

$$a_1 = \arg\ max_{||a=1||}\ Var\{a_1 X^T\} \tag{2}$$

Assuming the covariance matrix of X is $Cov(x)$ and X has been normalized, the covariance matrix is shown in Equation (3).

$$Cov(x) = \frac{1}{n}\sum_{i=1}^{n} X_i X_i^T \tag{3}$$

It has been proved that the principal components F_i are equal to eigenvectors of the variance matrix $Cov(x)$ [18]. If $diag\{\lambda_1, \lambda_2, \dots, \lambda_m\}$ is the diagonal matrix of the ordered eigenvalues $\lambda_1 > \lambda_2 > \dots > \lambda_m$, then the additional features are the corresponding eigenvectors of the m largest eigenvalues. Afterwards, the dimensionality of transformed space m could be determined. According to the normal procedure of matrix decomposition, we can get m eigenvalues where $m=n$. However, some of them do not contain any useful information for the covariance matrix. In practice, we usually choose m by the threshold of cumulative contribution rate (85% ~ 95%) or the value of eigenvalue (greater than 1). In our approach, all the eigenvectors are considered and added to our feature set, which is then used in the next classification step

If the additional features are all added to train the classifier, we can foresee the unsatisfactory performance caused by the possible noise. Therefore, a feature selection step is necessary here to filter out the useless ones and keep the features with strong ability of classification. We use a gain-ratio based feature selection process to select the final feature set.

With the transformed features, we are ready to train a classifier. We are now faced with the problem of imbalance labeled data: vandalisms only account for less than 10% and of the data and there are many outliers. For this reason Random Forest and LogitBoost with Decision stump as base learner are preferred because of their capbility of implicit feature selection, generalization properties and a low number of parameters. Both methods have been implemented in Weka[5] which we use in our experiments.

4   Experiment and Results

4.1   Experiment Setup

A large-scale corpus has been created in the PAN 2010 competition. This corpus makes the experiments easily reproducible. The corpus contains a week's Wikipedia edits (PAN-WVC-10)[6]. It is generated by Crowdsourcing task, and every edit has been labeled by at least three annotators.

We download the entire corpus of PAN-WVC-10 and use this public dataset to run the experiment. The corpus contains 32,452 edits on 28,468 Wikipedia articles, among which 2391 vandalism edits have been identified. An edit refers to a revision on a numbered article and reflects the differences between the two versions.

---

[5] http://www.cs.waikato.ac.nz/ml/weka/
[6] www.webis.de/research/corpora/pan-wvc-10

The classifier using only the original feature set listed in section 3.1 is our baseline method, which is a commonly used approach to detect vandalism. Another baseline method uses a classifier trained on the feature set created by Potthast [7]. We apply two different kinds of PCA-based methods for feature transformation in the experiment which are listed below

- **PCA-base method.** Using the entire collection (training set + test set) as the input to PCA, and new training set and the new test set are generated by the projection computed on the entire collection;
- **Partial PCA-based method.** Using only the test set as the input to PCA, so that a group of mappings are generated only for the test set. We then map both training and test set into a feature space and generate new training set and new test set.

The idea of proposing the Partial PCA-based method emanates from the real world online detection. If we can extract the principle components from the real world data (test set in our experiment), and map the training data into the same space, then the task of vandalism detection will be more realistic.

Besides PCA which is introduced in section 3.2, we also attempt to use some other feature transformation methods including the Singular Value Decomposition (SVD), and Non-negative Matrix Factorization (NMF). The experiment with feature transformation method of SVD uses the vectors in matrix U ($A = U\Sigma V^T$) as the additional features. Similarly, the experiment with NMF uses the vectors in matrix W ($A = WH$) as the additional features. Because the NMF method requires the non-negative matrix as input variable, we normalize the data with *atan* normalization function which is shown in Equation (4) and let the training data be in the range between 0 and 1.

$$\mathcal{F}(x) = \arctan(x) \times 2/\pi \tag{4}$$

We conduct 5-flod cross validation for all groups of experiments to get an average results of our classifiers' performance.

4.2   Evaluation Metrics

Vandalism detection can be considered as a binary classification problem, we want to determine whether the input instance is vandalism. The evaluation metrics of information retrieval including precision, recall and F-score are adopted here. Besides, AUC-ROC (The area under the receiver operating characteristic curve) is also reported here to compare the performance of different classifiers. The ROC curve is plotted with FP as the horizontal axis and TP as the vertical axis. In the case of vandalism detectors, TP is the number of edits that are correctly identified as vandalism (true positives), and FP is the number of edits that are untruly identified as vandalism (false positives).

4.3   Results and Discussion

The experimental results by feature transformation method based classifiers are shown in Table 2 and Table 3, in which the number of trees in Random Forest classifier and iteration number in LogitBoost are set to 100. The results of baseline1 are calculated with the features used by Potthast [7], and the results of baseline2 are calculated with the feature set in section 3.2. Both of the baseline models are not aggregated with features transformation methods.

Table 2. The performance of Random Forest classifiers

Approach	Precision	Recall	F-score
Baseline1	0.753	0.422	0.541
Baseline2	0.830	0.427	0.565
SVD-based	0.853	0.428	0.570
NMF-based	0.829	0.436	0.572
PCA-based	0.854	0.432	0.574
Partial PCA-based	**0.862**	**0.434**	**0.581**

Table 3. The performance of LogitBoost classifiers

Approach	Precision	Recall	F-score
NMF-based	0.758	0.409	0.531
Baseline1	0.765	0.411	0.565
Baseline2	0.766	0.477	0.591
SVD-based	0.771	0.479	0.591
PCA-based	0.776	0.513	0.618
Partial PCA-based	**0.787**	**0.516**	**0.623**

As we can see from the results, the precisions of all the methods except baseline1 are higher using Random Forest classifier than LogitBoost. However, the latter outperforms the others on recall. In terms of the overall performances, the average F-score, the average F-score 0.587 of LogitBoost is higher than 0.567 of Random Forest. This shows that the LogitBoost classifiers have better overall performance.

Among the feature transformation methods which are used in our experiments, Partial PCA-based classifiers rank the first for both classification methods. In the first group of results by Random Forest, it improves the baseline1 method by 14% at precision, 2% at recall, and 7% at F-score and improves the baseline2 method by 4% at precision, 2% at recall and 3% at F-score. The numbers in LogitBoost classification group respectively 3%, 25%, 10% for baseline1 and 3%, 8%, 5% for baseline2. The second best performed method is PCA-based which also makes an obviously improvement in comparison with baselines.

Some of the other feature transformation methods improve the performance (SVD, PCA, Partial PCA) of the vandalism detector but some do not (NMF). This is due of the different natures of those methods. For example, vectors extracted by NMF are not orthogonal, because of which there will be much noise in the vectors. On the other hand, all the feature transformation methods we used in this paper are unsupervised, which means the class label is not considered during the analysis for our training samples. As a result, the additional features created by those methods are not selected or computed to be the discriminative features. However, the linear method of PCA has been proven to be effective and efficient and it can be used to transform features in the vandalism detection on Wikipedia. The AUC-ROC results for our best performing classifiers (LogitBoost) are also listed in Table 4 in comparison with the baseline models

Table 4. The AUC-ROC results for our proposed methods and baseline methods

Approach	AUC-ROC
Baseline1	0.926
Baseline2	0.936
NMF-based	0.936
SVD-based	0.939
PCA-based	0.943
Partial PCA-based	**0.947**

The Partial PCA-based method is proved to be the best performed solution through our experiments, and the runtime efficiency of Partial PCA is obviously better than the normal PCA-based method because of the use of a subset for PCA computing (only one fifth of the original training time for 5-fold cross validation). The two PCA-based methods outperform others because the effective latent space can provide more discriminative features through our feature selection process which combine some features from our raw features according to their latent similar meaning. As a result, the new features can enhance the vandalism detectors.

4.4   Detailed Analysis

We ask some questions in order to examine the properties of our proposed detection framework, and the answers to those questions are given in detail to help understand why the PCA based methods achieve the best results.

— *How can the additional PCA features are interpreted?*
PCA-based feature extraction for classification can be treated as means of constructive induction. When generating new features, it can produce new (emerging) concepts which in turn may lead to a new understanding of a problem and can produce additional knowledge, including new concepts and their relationship to the primary concepts. For example, we get the new feature generated by the features which share almost the same information of using bad words (*SexFrequency, Vandalism Frequency etc.*), and this new feature can provide the total degree of users' negative impact which is defined as a linear-weighted-sum of those features. All these newly introduced features like the "total degree of users' negative impact" bring us new and useful information which we cannot find in the raw feature set and finally lead to the improvements in the results.

— *Why the Partial-PCA more suitable for the task of vandalism detection?*
The difference between the Partial-PCA method and the PCA method is the different original feature set for transformation they use. When using test set for PCA, the projections obtained from test set can better reflect the real distribution of featured vandalisms. For example, the styles of vandalisms perhaps have evolved with time. If we use the Partial-PCA to extract features based on the latest three months' data, we can better catch the evolved habits of the vandals.

— *What is the computational complexity of the proposed methods?*
As described in section 3.2, we can see PCA is a linear method, which means that the speed of the algorithm is only relevant to the feature dimension, and the computing complexity only linearly increases with the size of the feature set. Therefore, for a large dataset like Wikipedia with only less than 100 features, it is appropriate to apply PCA. At the same time, the additional training time of PCA-based feature transformation can be neglected relative to the improvement in performance, which the time ratio of the two stages is almost 10:1(training classifier vs. PCA).

5   Conclusion and Future Work

We have proposed a new idea to help detect vandalism in Wikipedia more efficiently. The feature transformation approaches are explored in the vandalism detector in order to generate additional effective features. The linear Principle Component Analysis is proved to be a good choice among the others studied in this paper. With the limitation of our feature set, the performance of our proposed classifiers may not be as good as the best state-of-art approach, which involved a large number of features, but the idea of making good use of the features is still a valid one and deserve further studied in vandalism detection on Wikipedia.

In our future work, a more comprehensive feature set will be created. More feature transformation methods will be tested to extend the feature set such as the non-linear methods with kernel functions. More effective methods for aggregating features will also be studied.

Acknowledgement. This work is partially supported by grant from the Natural Science Foundation of China (No.60673039, 60973068, 61277370), the National High Tech Research and Development Plan of China (No.2006AA01Z151), Natural Science Foundation of Liaoning Province, China (No.201202031), State Education Ministry and The Research Fund for the Doctoral Program of Higher Education (No.20090041110002).

References

1. Potthast, M.: Crowdsourcing a wikipedia vandalism corpus. In: 33rd International ACM SIGIR Conference on Research and Development in Information Retrieval, Geneva, Switzerland, pp. 789–790. ACM, New York (2007)

2. Adler, B.T., de Alfaro, L., Mola-Velasco, S.M., Rosso, P., West, A.G.: Wikipedia Vandalism Detection: Combining Natural Language, Metadata, and Reputation Features. In: Gelbukh, A. (ed.) CICLing 2011, Part II. LNCS, vol. 6609, pp. 277–288. Springer, Heidelberg (2011)
3. Lipka, N., Stein, B.: Identifying Featured Articles in Wikipedia: Writing Style Matters. In: 19th International World Wide Web Conference, Raleigh, USA, pp. 1147–1148. ACM, New York (2010)
4. Anderka, M., Stein, B.: A Breakdown of Quality Flaws in Wikipedia. In: The 2nd Joint WICOW/AIRWeb Workshop on Web Quality, Lyon, France, pp. 11–18. ACM, New York (2012)
5. Carter, J.: ClueBot and Vandalism on Wikipedia (2010),
   http://www.acm.uiuc.edu/~carter11/ClueBot.pdf
6. Rodríguez Posada, E.J.: AVBOT: Detecció y correcció de vandalismos en Wikipedia. NovATIca (203), 51–55 (2010)
7. Potthast, M., Stein, B., Gerling, R.: Automatic Vandalism Detection in Wikipedia. In: Macdonald, C., Ounis, I., Plachouras, V., Ruthven, I., White, R.W. (eds.) ECIR 2008. LNCS, vol. 4956, pp. 663–668. Springer, Heidelberg (2008)
8. Smets, K., Goethals, B., Verdonk, B.: Automatic vandalism detection in Wikipedia: Towards a machine learning approach. In: AAAI Workshop on Wikipedia and Artificial Intelligence: An Evolving Synergy, Chicago, Illinois, USA, pp. 43–48 (2008)
9. Harpalani, M., Hart, M., Singh, S., Johnson, R., Choi, Y.: Language of Vandalism: Improving Wikipedia Vandalism Detection via Stylometric Analysis. In: Proceedings of the 49th Annual Meeting of the Association for Computational Linguistics, Portland, Oregon, pp. 83–88. ACM, New York (2011)
10. Wu, Q., Irani, D., Pu, C., Ramaswamy, L.: Elusive Vandalism Detection in Wikipedia: A Text Stability-based Approach. In: Proceedings of the 19th ACM International Conference on Information and Knowledge Management, Toronto, Ontario, Canada, pp. 1897–1800. ACM, New York (2010)
11. Wang, W.Y., Mckeown, K.R.: "Got you!": automatic vandalism detection in Wikipedia with web-based shallow syntactic-semantic modeling. In: Proceedings of the 23rd International Conference on Computational Linguistics, Beijing, China, pp. 1146–1154. Association for Computational Linguistics Stroudsburg, PA (2010)
12. Chin, S.C., Street, W.N., Srinivasan, P., Eichmann, D.: Detecting Wikipedia vandalism with active learning and statistical language models. In: Proceedings of the 4th Workshop on Information Credibility, Raleigh, North Carolina, USA, pp. 3–10. ACM, New York (2010)
13. Mola-velasco, S.M.: Wikipedia Vandalism Detection. In: Proceedings of the 20th International Conference Companion on World Wide Web Conference, Hyderabad, India, pp. 391–395. ACM, New York (2011)
14. Adler, B., Alfaro, L.: A Content-Driven Reputation System for the Wikipedia. In: Proceedings of the 16th International World Wide Web Conference, Banff, Alberta, Canada, pp. 261–270. ACM, New York (2007)
15. Adler, B.T., Alfaro, L., Pye, I.: Detecting Wikipedia Vandalism using WikiTrust. Lab Report for PAN at CLEF (2010)
16. West, A.G., Lee, I., Kannan, S.: Detecting Wikipedia Vandalism via Spatio-Temporal Analysis of Revision Metadata. In: Proceedings of the Third European Workshop on System Security, Paris, France, pp. 22–28. ACM, New York (2010)
17. Fodor, I.K.: A survey of dimension reduction techniques. Technical report UCRL-ID-148494, LLNL (2002)
18. Wold, S., Esbensen, K.: Principal component analysis. Chemometrics and Intelligent Laboratory Systems 2(1-3), 37–52 (1987)

PLIDMiner: A Quality Based Approach for Researcher's Homepage Discovery

Junting Ye, Yanan Qian, and Qinghua Zheng

SPKLSTN Lab, Department of Computer Science and Technology,
Xi'an Jiaotong University, Xi'an,710049, China
yejuntingben@gmail.com, yananqian@yeah.net, qhzheng@mail.xjtu.edu.cn

Abstract. Researchers' high quality homepages are important resources in academic search because they provide comprehensive and up-to-date information about researchers. Meanwhile, low quality homepages widely exist. A case study shows that 57.8% of all homepages retrieved among top 10 results from Google are low quality and 95% top researchers own out-of-date homepages. Besides, some academic portals generate dynamic homepages introducing researchers. These homepages are not maintained by researchers and may contain incorrect information. The quality of discovered homepages can not be ensured by existing work, which decreases the efficiency of academic search. It is difficult to define a high quality homepage from a quantitative perspective. Instead, on the basis of analyzing labeled high quality homepages, we propose "informative researcher's homepage", at least consisting of identifiable information (introducing a researcher's basic information) and publication list (listing his/her corresponding publications), as an estimation for high quality homepage. Based on the observation that informative researchers' homepages are organized in two ways, integrated and scattered, we propose an effective discovering model, PLIDMiner, with F1 scores over 0.9 on labeled data. Our model can also be applied to verify homepages' quality. We crawl thousands of homepage resources from popular academic portals and assess their overall qualities. It turns out that nearly 25% of homepage resources in these portals are not informative, which strengthens our motivation.

Keywords: Researcher's homepage, Quality based, Machine learning.

1 Introduction

A researcher's homepage is a web page maintained by academic researcher and it includes researcher's identifiable information (such as name, affiliation) and research activities (such as research subjects and publications) [1]. Many researchers build their own homepages on the Web mainly for describing their researches and contributions, or announcing their new papers [2]. These homepages are valuable and reliable information sources for academic search. To access researcher's homepage, the most Web users query a researcher's name or related information in search engines and select useful homepages from the returned results.

However, more and more homepages with low quality emerged. In a case study, we randomly chose 20 researchers' in the top 100 researchers[1] in each discipline from

---

[1] This rank is based on researcher's H index.

Y. Hou et al. (Eds.): AIRS 2012, LNCS 7675, pp. 199–210, 2012.
© Springer-Verlag Berlin Heidelberg 2012

Microsoft Academic Search[2]. We applied "author name + homepage" as queries (i.e. "Scott Shenker homepage") in Google to find homepages. We regard homepages being updated manually after 2011 and contain various aspects of information as high quality homepages, while the other homepages' qualities are viewed as low. To sum up, there are 27 high quality homepages(4 researchers have more than 1 high quality homepages) and 37 low quality homepages. The case study also shows that 95% researchers have at least one low quality homepage and these pages comprise 57.8% of all homepages retrieved. These low quality homepages provide the Web users with incorrect or incomplete information. On one hand, more and more dynamically generated researchers' homepages are emerging because of the development of numerous academic portals, like DBLife[3] [3] and ArnetMiner[4] [4, 5]. The information in these homepages is less reliable because they are automatically generated and unavoidably contain incorrect information. The name ambiguity is a typical example: if different authors share the same name, academic portals often mix their publications together in a single list [6, 7]. Qian et al. pointed out that the name ambiguity problem is unavoidable in dynamically generated homepages [8]. On the other hand, low quality researcher's homepage emerges if a researcher has several different versions of homepages. Generally, only one or two of them are high quality while the others are not. Fig. 1 is an example of Prof. Scott Shenker, a well-known researcher of computer science in UC Berkeley, who has a high quality homepage and a low quality homepage. It is obvious that the high quality homepage is more informative and useful than the low quality one.

With the growing emergence of dynamically generated homepages and accumulation of out-of-date homepages, the increasing number of low quality homepage decrease the

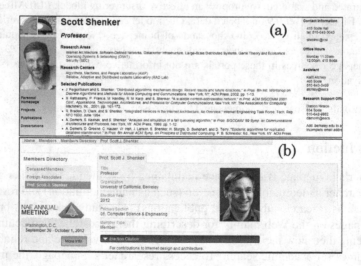

Fig. 1. (a) An example of high quality homepage; (b) An example of low quality homepage.

[2] http://academic.research.microsoft.com/
[3] http://dblife.cs.wisc.edu/
[4] http://arnetminer.org/

efficiency of academic search. If low quality homepage is selected by the Web users after querying in search engines, he/she doesn't get the wanted information and has to continue to try other homepages. In particular, after we verify several popular academic portals, we find nearly 25% homepage resources in them are not informative. In other words, if a user clicked four homepage resources, one of them is not informative and he/she may have to find other ways to access a more satisfying one. Therefore, discovering informative homepages of researchers can improve academic search efficiency, and it is also an important task for academic search. To the best of our knowledge, this is the first research work of researchers' homepages discovery explicitly considering their qualities.

In this paper, we define Informative Researcher's Homepage (IRH) as at least consisting of identifiable information (including email, affiliation, office number, fax, address) and publication list (a list of citation records, consisting of author names, title, conference or journal name, publishing year) because of their importance and universality. Identifiable information (ID) is basic and essential since it is the key information to introduce homepage owner's basic information and to make the owner easily distinguished from other researchers. Publication list (PL) is also important because the Web users learn about state-of-the-art theory and technology from them [2, 9]. Meanwhile, in the case study we mentioned above, we find that ID and PL are the two most universal elements among all information in high quality homepages. 100% contain ID and 92.6% contain PL while other information like "Experience" and "Honors" exist in less than 60% of high quality homepages. It means that the priority order of building homepage is first ID, then PL, and then others. In other words, if a homepage contains various information, it is most probable to include both ID and PL. Low quality homepages have the same order. However, only 18.9% of them contain both ID and PL. Therefore, it is optimal to choose ID and PL as requisite components for IRH, which is used as an estimation for high quality homepage.

In this paper, we propose PLIDMiner to find IRH based on following observations. First, IRHs mainly exist in two ways: ID and PL are in the same web page (Integrated IRH) while ID and PL are in different web pages (Scattered IRH). Second, in terms of discovering directions, we can either first find ID, then PL, or the other way around.

The contributions of this paper are: (1) The quality factor is getting more and more important because of the growing emergence of low quality homepages. We are first to focus on the quality of researchers' homepages. We find universal and essential information in high quality homepages and regard them as requisite components for IRH. (2) We have proposed an effective model based on machine learning for IRH discovery on the basis of IRH's unique features including URL, page structure and page content. Experiment results on labeled data show that our model performs well with F1 scores over 0.9. (3) We have verified homepage resources in popular academic portals to prove the necessity of finding IRH. Plenty of homepage resources (directing to homepages through URL links) in Google Scholar, Microsoft Academic Search and ArnetMiner are verified based on PLIDMiner. It turns out that all academic portals have potentials to improve their high of homepage resources. Meanwhile, Google Scholar and ArnetMiner have better homepage resources than Microsoft Academic Search.

The rest of the paper is organized as follows: In Section 2, we introduce the previous work relevant to researcher's homepage discovery. In Section 3, we elaborate PLID-Miner in detail and in Section 4 we show the experiment results of discovery models and overall assessment of three academic portals. At last, we present our conclusion and future work in Section 5.

2   Related Work

The most relevant task of homepage discovery is entry page finding. Entry page is the central page of an organization's web site, which functions as a portal for its information. It is a fundamental issue in information retrieval research community and has attracted tremendous attention. Xi and Fox [10] used machine learning approaches (like decision tree and logistic regression model) to predict the correct entry page in response to a user's entry page finding query, with considering extensive tagged field and URL. Kraaij et al. [11] enhanced IR systems' performance for entry page search by assigning a prior probability to three non-content features of web pages: page length, number of incoming links and URL form. Upstill et al. [12] compared the usefulness of query-independent evidences like document content, anchor text, and URL-type classification and concluded that a general entry page finding system should combine these evidences.

There is considerable work focused in researcher's homepage discovery. An early effort for homepage discovery can be dated back to 1996 resulting in a popular real-world system called Ahoy, in which Shakes et al. [13] used three external systems and applied simple yet successful heuristics to retrieve homepages. Fang et al. [14] proposed a conditional undirected graphical model to predict the labels of homepages by capturing the dependence of them. Since publication list is important information in researcher's homepage, Yang et al. [9] used citations as queries to find publication list web pages from search results. Yang and Ho [2] parsed publication lists after noticing that citation records are usually represented as nodes at the same level in the DOM tree and citation records in the same page are presented by similar HTML tags.

Researchers' homepages are useful resources in tackling the problem of name ambiguity in academic portals. Kang et al. [7] acquired coauthors information from search engine by querying labeled coauthor pairs. Inspired by the concept of TF-IDF, Tan et al. [15] proposed "Inverse Host Frequency (IHF)", which is based on the observation that pages from rare web sites are stronger source of evidence than pages from common web sites. Pereira et al. [16] used IHF to weigh the documents crawled from the Web and extracted useful evidence for author disambiguation.

Homepages are also very valuable resources in academic social network building. Tang et al. [17] gives a formalization of finding, extracting, and fusing the semantic-based profiling information of a researcher from the Web. Specifically, it finds the "relevant documents" from the Web using Support Vector Machines (SVM). Culotta et al. [17] used filters to eliminate noise data retrieved from search engines by using author names as queries. Matsuo et al. [18] and Mori et al. [19] extracted relations of researchers by counting the co-occurrences of researchers' names from the results retrieved from search engine and used the information to predict the relation type (like co-authors of a paper, co-members in a project) of researchers.

3   Our Model: PLIDMiner

3.1   Description

As defined in Sect. 1, an IRH at least consists of two key information elements: identifiable information (ID) and publication list (PL). Author names and publication titles are useful hints to query homepages through search engines [1, 20]. For author names queries, search engines return researchers' homepages containing the authors' ID and other name related pages, such as the authors' social network pages. When we query publication titles, search engines fetch researchers' homepages containing PL and other publication related pages, such as citation pages dynamically created by academic portals or publications' download pages. Inspired by these two search patterns observed, we propose PLIDMiner containing two directions' discoveries for IRH. For ID to PL direction, web pages are processed by following components: ID Classifier, Sub-PL Classifier, ID to PL URL Finder and PL Classifier. Symmetrically, PL Classifier, Sub-ID Classifier, PL to ID URL Finder and ID Classifier comprise the discovering procedure in PL to ID direction. Fig. 2 illustrates the procedure of IRH discovery in ID to PL direction. IRH discovery in PL to ID direction is similar to this.

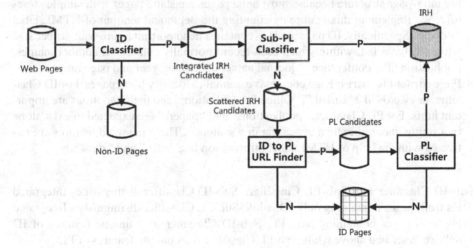

Fig. 2. Procedure of Finding IRH in ID to PL Direction

In Fig. 2, the function of each step is: (1) ID Classifier determines whether the input web page contains ID; (2) Sub-PL Classifier is expected to find Integrated IRH among ID pages using subset features of PL Classifier; (3) ID to PL URL Finder explores potential URLs of PL pages in the input ID pages for the discovery of Scattered IRH; (4) PL Classifier determines whether these PL candidates actually contain PL. Take the low quality homepage in Fig. 1-(b) as an example, it is first predicted true by ID Classifier because of identifiable information like "title"and "organization", then false by Sub-PL Classifier since there is no PL in this page. In the third step, potential URLs to PL are found and the corresponding pages are crawled. However, these pages are predicted as non-PL pages by PL Classifier in the last step. So this page is only an ID page, but not IRH. The functions of steps in PL to ID direction are similar.

3.2 PLIDMiner's Components

There are six components involved in IRH discovery model. We will introduce them in detail in three groups.

ID Classifier and PL Classifier. In both ID Classifier and PL Classifier, features are extracted from three aspects: URL features, structural features and page content features. The detailed description of these three feature subsets are as follows:

- **URL features:** URL features play a significant role in classifying pages. By analyzing homepage URLs, we have found the following features to identify homepages. (1) whether the host name ends with ".edu" since the majority of homepages are in universities' web sites; (2) whether the URL contains the notation "~" since many researchers prefer to use it in their URLs. Specifically, URLs of ID pages tend to contain keywords like "index", while "pub" or "paper" for PL pages.
- **Structural features:** Homepages often show unique visual characteristics than other web pages. The dominant feature for both classifiers is whether the page has simple intra-page structure because most homepages are static pages with simple structure. We implement this feature by counting the depth and amount of HTML label "<div>". Specifically, ID page tends to contain a head portrait figure with proper size while PL has a list, within which each record normally consists of authors' names, publication title, conference or journal name, publishing year and page range.
- **Page content features:** Page content is essential to identify homepages. For ID Classifier, keywords like "Email:", "Office:", "Affiliation:" and their variations are important hints. For PL Classifier, "publications" and "papers" are expected to exist alone in a single line, rather than appearing in a sentence. These keyword features are extracted with the help of HTML label information like "<h>" and "<a>".

Sub-ID Classifier and Sub-PL Classifier. Sub-ID Classifier distinguishes Integrated IRHs from pages containing only ID, while Sub-PL Classifier distinguishes Integrated IRHs from pages containing only PL. Sub-ID Classifier uses unique features of ID which are described above while Sub-PL Classifier uses unique features of PL.

If we want to discover Integrated IRH, why not simply combine ID and PL as one classifier? We ignored this idea because we prefer to distinguish low quality homepages from non-homepages. These two types of pages are significantly different and this will contribute to the verification of homepage resources in academic portals. We use Sub-ID Classifier (or Sub-PL Classifier) instead of ID Classifier (or PL Classifier) in step 2 of PLIDMiner because ID and PL Classifiers share common features. It will decrease the performance if we simply use ID and PL Classifiers instead of Sub-ID and Sub-PL Classifiers.

URL Finders. ID to PL URL Finder finds possible URL links in ID pages, which direct to PL pages. The function of PL to ID URL Finder is similar. URLs from homepages can be generally divided into three types: (1) homepage-relelated URLs; (2) homepage-unrelated URLs; (3) parent and grandparent directory URLs. Homepage-relelated and unrelated URLs are obtained directly from web page content, while parent and grand-parent directory URLs are deduced from original URLs. ID to PL URL Finder finds candidates among homepage relevant and irrelevant URLs, while PL to ID URL Finder uses all these three types of URLs based on the observation that parent or grandparent page of PL always contains ID. We use the URL features in ID Classifier and PL Classifier as heuristic to find candidate URLs. For the balance of performance and expense on the basis of comparison, the pages of top 3 candidate URLs are crawled for further process.

4   Experiment

4.1   Experimental Setting

Data Set. For data preparation, we automatically crawled 750 researchers' homepages through Microsoft Academic Search. These researchers are among Top 160 in 15 disciplines based on their H index. Among these homepages, we manually labeled 391 ID pages, 328 PL pages, 255 Integrated IRHs and 150 Scattered IRHs. For negative instances, we randomly chose 100 researchers out of 750 to get their names and publication titles information. We used authors' names and publication titles as queries for Google, and crawled the top 10 search results. Then 444 name related non-homepages and 426 publication title related non-homepages were labeled. The reason for choosing these related pages as negative instances instead of random pages on the Web is that these negative instances are more likely to be regarded as homepages, though this choice may lower PLIDMiner's performance.

Evaluation Measures. We use standard evaluation measures in classification: precision p, recall r and the combined measure $F1$ of the classifier. The terms true positives TP, true negatives TN, false positives FP, and false negatives FN compare the results of the classifier under test with human judgments (label of pages). The terms "positive" and "negative" refer to the classifier's prediction, and the terms "true" and "false" refer to the external judgments. The definitions of p, r and $F1$ are as following equations:

$$p = \frac{TP}{TP + FP} \tag{1}$$

$$r = \frac{TP}{TP + TN} \tag{2}$$

$$F1 = \frac{2pr}{p + r} \tag{3}$$

4.2   Baselines Description

IHF Model. IHF is an estimation of web pages' uniqueness (refered in Sect. 2). We observed that high quality homepages are usually located in more rare domains. Therefore, IHF can be a good predictor to find true IRH from candidates. Besides, different pages in the same homepage generally share the same host name, so the calculation of a homepage's IHF value can ignore its inter-page structure. For comparison, we use IHF as a simple but practical way to find IRH.

To calculate frequency, we need to form a corpus to establish IHF values. First, we crawl 2,000 author names and 2,000 publication titles from Microsoft Academic Search (MAS). These author names are among the top 160 researchers in 15 disciplines ranked by their H index and the publication titles are randomly acquired from their homepages automatically generated by MAS. Then we use these author names and publication titles as queries in Google. The top 10 returned URLs from Google are truncated to its host name. The calculation of IHF value is based on this corpus. If a hostname h has frequency $f(h)$, then its IHF value is computed as (4).

$$IHF(h) = log_2 \frac{max_h f(h) + 1}{f(h) + 1} + 1 \qquad (4)$$

Single Page Model. Researcher's homepage is a type of entry page. As described in Sect. 2, there are tremendous work focused in entry page discovery in information retrieval research community. In previous work [10–12], various features of one single page are considered. These features are effective to find entry page. We adopt three query-independent features, which are frequently used in finding entry page.

- **Page Length:** Statistics in [11] shows that a medium length (60-1000 words) of page is a good predictor for entry page. We eliminate HTML label information (like <a>, <div>) in web page, which is not presented for the Web users.
- **InLink Number:** Entry pages tend to have a higher number of inlinks than other pages (i.e. they are referenced more often). We use a query tip in Google to retrieve the inlinks number of a particular web page. If we use queries like "links:www.hawking.org.uk", Google will return pages referring to site "www.hawking.org.uk" and their total amount.
- **URL Type:** A URL normally consists of the name of the server, a directory path and a filename. Pages at the top level of a specific server are often entry pages. As we descend deeper into the directory tree, the possibility of entry pages decreases. Four types of URLs are considered:
  - **root:** a host name, optionally followed by 'index.html'.
  - **subroot:** a host name, followed by a single directory, optionally followed by 'index.html' name.
  - **path:** a domain name, followed by an arbitrarily deep path, but not ending in a file name other than 'index.html'.
  - **file:** anything ending in a filename other than 'index.html'.

4.3 Performance and Analysis

PLIDMiner's Components. The overall performance of PLIDMiner depends on its components' effectiveness. Ten folds cross-validation is used to assess these components' performances on labeled data. Some of these data are reused. For example, some

ID pages as positives for ID Classifier are also used as negatives for Sub-ID Classifier. This reusability is reasonable because these classifiers are trained independently. Four machine learning techniques (SVM, Naive Bayes, Decision Tree and KNN) are compared to obtain optimal model. From the comparison results in Table 1, SVM achieves four highest F1 scores out of six. Besides, for the rest two classifiers, SVM achieves only 0.002 lower than the best ones. Based on the performances of the components and the convenience of automation, we apply SVM in all components in PLIDMiner.

Table 1. F1 Scores of PLIDMiner's Components

	SVM	Naive Bayes	Decision Tree	KNN
ID	**0.935**	0.926	0.934	0.915
PL	0.947	0.935	0.913	**0.949**
Sub-ID	**0.864**	0.851	**0.864**	0.847
Sub-PL	**0.863**	0.857	0.829	0.837
ID URL	**0.911**	0.899	0.909	0.906
PL URL	0.883	0.868	**0.885**	0.872

Baselines and PLIDMiner. Quality based IRH discovery differs from traditional ones, in which homepage is generally represented by one single page. To find an IRH, we should at least discover both ID and PL. ID and PL may locate in the same page, or in different pages. This requires relatively more complex model, which considers the inter-page structure. To validate the performance of PLIDMiner, we use two baselines (IHF Model and Single Page Model), involving four features with about 22 feature values. We implement experiment for each baseline and their combined model. We also use the same four machine learning techniques to assess their performances. Table 2 shows that Single Page Model (SPM) has a much better performance than IHF Model. Their combined model is slightly better than SPM. It is interesting that four techniques get the same F1 score for IHF Model. It is probable that there is only one feature (IHF) used, though we use 9 feature values for discretization. PLIDMiner achieves higher F1 scores than any of these (see Fig. 3), with 0.948 in ID to PL direction and 0.981 for the other direction. It may be confusing that PLIDMiner achieves 0.944 F1 score while its ID Classifier only gets 0.935 (see Table 1) and other components perform even lower. The reason is the components' performances in Table 1 are trained independently while their performances get considerable improvement in PLIDMiner where one component's output is its next component's input. In other words, their performances become conditional, contributing to much higher precision and slightly lower recall. Discovery in PL to ID direction performs better than discovery in ID to PL direction. The reason is PL Classifier performs better than ID Classifier (see Table 1) and PL Classifier is the first step of PL to ID direction's discovery while ID Classifier is the first step in the direction of ID to PL. First step in discovery process influences most because of its largest input, so the performance difference appears.

Table 2. F1 Scores of Baselines

	SVM	Naive Bayes	Decision Tree	KNN
IHF	0.744	0.744	0.744	0.744
SPM	0.858	**0.860**	0.858	0.857
IHF&SPM	0.858	**0.869**	0.852	0.853

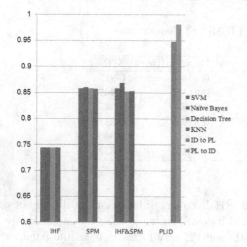

Fig. 3. F1 Scores Comparison among Baselines and PLIDMiner

Fig. 4. Verification of Resources in Academic Portals

4.4 PLIDMiner's Application

With the increasing needs of academic users, more and more academic portals have emerged in recent years. Some of these provide comprehensive introduction for users including affiliation, publications and homepages links. Most of these links are acquired through users' annotation or automatic mining. To assess these resources, we apply PLIDMiner in ID to PL direction to verify these homepage resources in academic portals based on the fact that users expect the homepage links to direct to ID pages. Users have to confirm whether the researcher is whom they want or not.

As shown in Fig. 2, the pages processed by PLIDMiner in ID to PL direction can be naturally divided into three categories: (1) Tier 1: IRH; (2) Tier 2: ID pages; (3) Tier 3: non-ID pages. We crawled 529, 1,816 and 2,116 homepage resources from Google Scholar (GS), Microsoft Academic Search (MAS) and ArnetMiner (AM) respectively. The resources in MAS and AM are directly crawled from their sites while the ones in GS are accessed via Jens Palsberg's Homepage[5]. All these homepages belong to top influential scientists in all fields according to their total citation times or H index. From Fig. 4, we can find that each academic portal has different proportion in each tier. To assess overall quality, we quantify score S as (5).

---

[5] http://www.cs.ucla.edu/~palsberg/h-number.html

$$S = \sum_{i=1}^{3} \omega_i * P_i \tag{5}$$

P_i is the percentage of Tier i homepage. ω_i is the weight to evaluate the importance of Tier i homepage. By experience, we set 1, 0.5 and 0 to ω_1, ω_2 and ω_3 respectively. As a result, GS and AM achieve about 0.05 higher scores than MAS.

5 Conclusion and Future Work

Researchers' homepages are essential resources in academic search. However, the quality of researchers' homepages has not been fully discussed before. Based on users' searching patterns, we proposed IRH discovery model in two directions: ID to PL and PL to ID. Each direction has achieved excellent performance. This provides multi-ways to access IRH with the help of search engines. We applied PLIDMiner to verify resources in popular academic portals. Normally, the homepage resource should direct to the ID page, so we used ID to PL direction's discovery as the basis of verification. We crawled plenty of homepage resources from Google Scholar, Microsoft Academic Search and ArnetMiner. We assess their overall quality of homepage resources and find that Google Scholar and ArnetMiner have better homepage resources than Microsoft Academic Search.

Researchers' homepages are fundamental and significant information for many areas. As future work, we intend to expand our work in the following three aspects. First, automatic homepage resource refinement with the assist of search engines: this paper provides two ways to discover IRH effectively and give practical implementations of homepages resource verification for popular academic portals. These approaches contribute to refining the resources. Second, building academic social network: Useful information, such as E-mail, affiliation and author's full names, is extracted by using regular expression in PLIDMiner. This will contribute to building academic social network with comprehensive information and relations of researchers. Third, author disambiguation in academic portals: The publication list extracted from discovery model provides reliable and strong evidence for author disambiguation. It is obvious that all the publications in PL of IRH belong to a particular researcher.

Acknowledgements. The research was supported in part by National Science Foundation of China under Grant Nos. 60825202, 61173112, 61070072; National Science and Technology Major Project under Grant Nos. 2010ZX01045-001-005, National High Technology Research and Development Program 863 of China under Grant No. 2012AA011003; Cheung Kong Scholar's Program; Ministry of Education of China Humanities and Social Sciences Project; Key Projects in the National Science and Technology Pillar Program under Grant Nos. 2011BAK08B02, 2011BAK08B05, 2009BAH51B02.

References

1. Kang, I.-S., et al.: Construction of a Large-scale Test Set for Author Disambiguation. Information Processing and Management 47, 452–465 (2011)
2. Yang, K.-H., Ho, J.-M.: Parsing Publication Lists on the Web. In: Proceedings of IEEE/WIC/ACM International Conference on Web Intelligence and Intelligent Agent Technology, vol. 1, pp. 444–447 (2010)
3. Doan, A., Ramakrishnan, R., et al.: Community information management. IEEE Data Engineering Bulletin 29, 64–72 (2006)
4. Li, J., Tang, J., et al.: Arnetminer: Expertise Oriented Search Using Social Networks. Frontiers of Computer Science in China, 94–105 (2008)
5. Tang, J., Zhang, J., et al.: ArnetMiner: Extraction and Mining of Academic Social Networks. In: Proceedings of the Fourteenth ACM SIGKDD International Conference on Knowledge Discovery and Data Mining, pp. 990–998 (2008)
6. Torvik, V., Weeber, M., et al.: A probabilistic similarity metric for Medline records: A model for author name disambiguation. Journal of the American Society for Information Science and Technology 56, 140–158 (2005)
7. Kang, I.-S., Na, S.-H., et al.: On co-authorship for author disambiguation. Information Processing and Management 45, 84–97 (2009)
8. Qian, Y., Hu, Y., et al.: Combining machine learning and human judgment in author disambiguation. In: International Conference on Information and Knowledge Management, pp. 1241–1246 (2011)
9. Yang, K.H., Chung, J.M., et al.: PLF: A Publication list Web page finder for researchers. In: IEEE/WIC/ACM International Conference on Web Intelligence, pp. 295–298 (2007)
10. Xi, W., Fox, E.A., Tan, R.P., Shu, J.: Machine Learning Approach for Homepage Finding Task. In: Laender, A.H.F., Oliveira, A.L. (eds.) SPIRE 2002. LNCS, vol. 2476, pp. 145–159. Springer, Heidelberg (2002)
11. Kraaij, W., Westerveld, T., Hiemstra, D.: The importance of prior probabilities for entry page search. In: Proceedings of the 25th Annual International ACM SIGIR Conference on Research and Development in Information Retrieval, pp. 27–34 (2002)
12. Upstill, T., Craswell, N., et al.: Query-independent evidence in home page finding. ACM Transactions on Information Systems 21, 286–313 (2003)
13. Shakes, J., Langheinrich, M., et al.: Dynamic reference sifting: A case study in the homepage domain. Computer Networks and ISDN Systems 29, 1193–1204 (1997)
14. Fang, Y., Si, L., et al.: Discriminative graphical models for researcher's homepage discovery. Information Retrieval 13, 618–635 (2010)
15. Tan, Y.F., Kan, M.Y., et al.: Search engine driven author disambiguation. In: Proceedings of the 6th ACM/IEEE-CS Joint Conference on Digital Libraries, pp. 314–315 (2006)
16. Pereira, D.A., Ribeiro-neto, B.A., et al.: Using web information for author name disambiguation. In: Proceedings of 9th ACM/IEEE Joint Conference on Digital Libraries, pp. 49–58 (2009)
17. Culotta, A., Bekkerman, R., et al.: Extracting social networks and contact information from email and the Web. In: Proceeding of Conference on Email and Anti-Spam (2004)
18. Matsuo, Y., et al.: Mining Social Network of Conference Participants from the Web. In: IEEE/WIC International Conference on Web Intelligence, pp. 190–193 (2003)
19. Mori, J., Tsujishita, T., Matsuo, Y., Ishizuka, M.: Extracting Relations in Social Networks from the Web Using Similarity Between Collective Contexts. In: Cruz, I., Decker, S., Allemang, D., Preist, C., Schwabe, D., Mika, P., Uschold, M., Aroyo, L.M. (eds.) ISWC 2006. LNCS, vol. 4273, pp. 487–500. Springer, Heidelberg (2006)
20. Kang, I., Kim, P., et al.: A largescale testset for authordisambiguation. Journal of the Korea Contents Association, 455–464 (2009)

Exploiting and Exploring Hierarchical Structure in Music Recommendation

Kai Lu, Guanyuan Zhang, Rui Li, Shuai Zhang, and Bin Wang

Institute of Computing Technology, Chinese Academy of Sciences,
Beijing 100190, China
{lukai,zhangguanyuan,lirui,zhangshuai01,wangbin}@ict.ac.cn

Abstract. Collaborative Filtering (CF) approaches have been widely applied in music recommendation as they provide users with personalized song lists. However, CF methods usually suffer from severe sparsity problem which greatly affects their performance. Previous works mainly use music content information and other external resources to relieve it, while they ignore that music entities are multi-typed and items are tied together within a hierarchy. E.g., for a track, we can identify its album, artist and associated genres. Therefore, in this paper, we propose a framework which utilizes the hierarchical structure in two ways. On one side, we exploit the hierarchical links to find more reliable neighbors; On the other side, we explore the effect of hierarchical structure on users' potential preferences. In a further step, we incorporate the two aspects seamlessly into an integrated model which could make use of the advantages of both sides. Experiments conducted on the large-scale Yahoo! Music datasets show: (1) our approach significantly improves the recommendation performance; (2) compared with baselines, our approach is much more powerful on the even sparser training data, demonstrating that our approach could effectively mitigate the sparsity issue.

Keywords: Music Recommendation, Collaborative Filtering, Hierarchical Structure, Neighborhood Model, Latent Factor Model.

1 Introduction

As the fast development of the Internet, users are inundated with choices, e.g., there are usually thousands of different types of products for the users to choose in an E-commerce web site or music service. Thus providing appropriate guidance and recommendation is very important to enhance users' satisfaction and loyalty. Driven by this demand, recommender systems (RecSys) which provide personalized services are becoming more and more popular in the latest two decades. The well-known Internet service providers such as Amazon, Google, Yahoo! and Netflix have adopted such recommendation engines.

The goal of music recommendation is providing a list of songs for listeners which they are more likely to enjoy. Deploying music recommendation engines could greatly satisfy users' personal needs, increase their on-line time, and reserve

Y. Hou et al. (Eds.): AIRS 2012, LNCS 7675, pp. 211–225, 2012.

huge potential commercial profits. Therefore, it is very meaningful to conduct music recommendation and actually there are many successful music Recsys in our lives, such as Yahoo! Music, Pandora, Last.fm, Ping and so on.

In additions, many institutes and researchers have worked out various approaches to advance music recommendation performance. Among these methods, collaborative filtering (CF) based models which use interactions between users and items are more effective and popular. CF based methods make recommendations only employ users' explicit feedback (e.g., ratings, thumbs up or down) and implicit feedback (e.g., purchasing histories, watching or listening records) of the past, then will get accurate recommendation lists without relying on item attributes and domain knowledge too much. Therefore, CF has been successfully applied in the practical systems such as Amazon, Netflix and so on.

However, CF usually faces several issues in practical applications. Among them, one distinct problem is that users' feedback data is severe sparse and it brings about great difficulty to make accurate recommendation on them. Therefore, utilizing appropriate approaches to mitigate sparsity issue will observably improve the recommendation performance.

This paper attempts to explore the characteristics of music itself to mitigate the severe sparsity problem in CF based recommendation. We observe that music possess plenty of useful information. One distinctive feature that music entities are multi-typed, except for the lyrics, rhythms, pitch, timbre and others, there are some hierarchical structure in the music. Usually, music items can be categorized by their types into: tracks, albums, artists and genres. The items in different types are tied together through hierarchical links, e.g., each album belongs to a specific performing artist, and it might have several tracks and several possible genres. Our approach mainly utilize the hierarchical structure from two aspects. On one side, we exploit the hierarchical links to find more reliable neighbors; on the other side, we explore the effect of hierarchical structure on users' potential preferences. In a further step, we incorporate two aspects into an integrated model which could take in the advantages of both sides.

It is noteworthy that our models are not only limited to music recommendation,but also can be effectively used in any situations with hierarchical information. E.g., they can be used in movies recommendation, in which we can find hierarchical links among actors, directors and movie genres.

The rest of this paper is organized as follows. Related work is introduced in Section 2. We present our approach in Section 3, and evaluate related models in Section4. Finally, we conclude in Section 5.

2   Related Work

The approaches for music recommendations can be generally divided into two categories: content based methods and CF based methods. As there is plenty of useful information in the music, such as genre, lyric, timbre, pitch, rhythm and so on, various content based methods have been explored. E.g., signal filtering based methods are often used to make recommendations [1] as a song is relatively short and has regular rhythm. Among these methods, Mel-Frequency Cepstral Coefficients (MFCC) [2] which is designed to capture short-term spectral-based features and establish item-to-item similarity matrix is highly popular. Demographic information and discography (e.g., Last.fm and Pandora) also have been used for music recommendations [3]. Some other researchers also try to use context information, item textual attributes, social tags and other annotations to provide recommendations [4].

Because content based recommendation methods do not consider users' history behaviors, they could not meet users' personal needs enough. Therefore, CF methods which directly build models on users' behaviors are more popular and effective. Users usually express their preferences in two ways: explicit feedback (ratings, purchase histories) and implicit feedback (listening histories, browsing histories). Ratings based CF methods [5,6,7] are most popular in the past 20 years, especially as Netflix Prize contest's successful holding. Implicit feedback [8] methods also attract many attentions recently as most of the time users interact with systems through clicking, browsing or listening, while they are reluctant to give ratings. Generally speaking, CF models include neighborhood models and latent factor models [9].

Neighborhood models mainly concern the similarities between items (users), and then make predictions based on the ratings of the k-nearest neighbors. Neighborhood models can also be categorized into two types: user-based neighborhood (user-knn) and item-based neighborhood (item knn). Compared to user-knn, item-knn is more effective in many situations, such as in the Netflix [10] and Movielens movie ratings [5], because usually users' numbers are far more than that of items. In addition, item-based neighborhood has better scalability [10] and is more capable to explain the recommendation reasons, so item-knn is more popular. The key step in neighbor based models is how to calculate the similarities between users (items), and a better method named Pearson Correlation Coefficient (PCC) [5] is often used to compute similarities. The prediction of user u's possible rating for item i is taken as a weighted average ratings of u's (or i's) neighbors [5]. Some other neighborhood methods such as the interpolated-weight item-based neighborhood model [10] and temporal item-based neighborhood model [6] were shown even better predication accuracy in the Netflix prize contest.

Latent factor models try to disclose the latent features that can explain the observed ratings, and it shows better ability to overcome sparsity than that of neighbor based methods [7]. One of the most common latent factor model is known as Singular Value Decomposition (SVD). In this model, each user u is tied with a user-factor vector p_u, and each item is related with an item-factor vector q_i, the predication for user u on item i is made by taking the inner product of vector p_u and q_i : $\hat{r}_{u,i} = p_u^T q_i$. Since Simon Funk published a detailed implementation of a regularized SVD with separate feature update [11], many other SVD extension algorithms have been created which show much better predication accuracy, such as NSVD [12], RISMF [13], Biased-SVD [7] and so on.

Biased-SVD was proposed by Koren in the Netflix contest through adding user and item bias into the original SVD. The rating predication by user u on item i is given by:

$$\hat{r}_{ui} = \mu + b_i + b_u + p_u^T q_i \tag{1}$$

where μ is the average value of all the known ratings, b_u and b_i are user and item bias respectively.

Various solutions have been explored to mitigate the severe sparsity problem in CF based recommendations. Earlier work [14] try to fill in missing ratings to make the rating matrix denser, while it relies on imputation which will become more costly as the data growth and the prediction results are instable. In [15], the authors incorporate CF and music acoustic contents together to improve similarity measurement accuracy, while it relies too much on the music feature extraction technologies. Some other researchers also consider the affect of music taxonomy information on users bias and items bias [16,17], yet they ignore the internal links of music hierarchical structure and its effect on users' latent interests. The KDD-Cup 2011 contest which aims for predicting users' preferences on music attracts thousands teams to participate in [18], while most of the teams only use existing methods and try to ensemble more models to get more accurate results[19,20]. Therefore, in this paper, we mainly focus on exploiting the hierarchical structure to find more reliable neighbors and exploring its affect for users' potential preference for the latent related items that he has not rated.

3   Our Approach

3.1   Preliminaries

We denote u, v for users and i, j for items, and r_{ui} is the known value that user u rates for item i. To distinguish with the known ratings, we denote $\hat{r}_{ui}$ for the predicated value of r_{ui}. For the pairs (u,i) with known ratings are stored in the set $R = \{(u,i)|r_{ui} is known\}$. V and T indicate the validation ratings and test ratings respectively. $R(u)$ is defined as the item sets that u has rated and the item ratings are known. To reveal music taxonomy information and music hierarchical structure, we denote $type(i)$ is item i's type, which represents as $type(i) \in \{track, album, artist, genre\}$. For each item i whose type is track, we denote $album(i)$, $artist(i)$ and $genre(i)$ as its album, artist and genre set respectively, we denote similarly for other types. For each user u, we denote $album(u)$ are the items that u might have interests in which generated by the album sets he has rated.

3.2   Hierarchical Structure Based Neighborhood Model

As we have introduced in the above sections, music has plenty of useful characteristics. For the music taxonomy information, we can classify it into four categories: track, album, artist and genre. Based on the records that the artists have released, we can deduce several hierarchical links from them. E.g., given

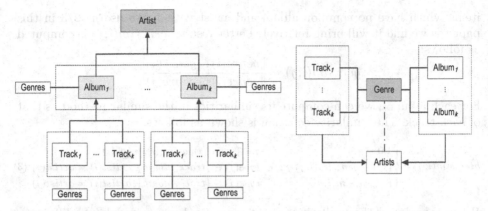

Fig. 1. Hierarchical structure among tracks, albums, artists and genres

one album, we can know its performing artist, tracks it contains and its possible genres. We can refine the hierarchical relationship among the four different types in more details, shown in Fig.1:

Based on the hierarchical tree, we can trace a node's ancestors, brothers and children, and then we could find more related items. For example, from the album 'King Of Pop' which is released by Michael Jackson, we can know its artist, albums and its genres, and find more Michael's related albums and tracks. We can also find more tracks and albums with similar styles. Therefore, we can exploit the hierarchical structure to find more reliable neighbors and make recommendations based on them. In this way, we could mitigate the negative affect which is caused by severe sparse ratings.

As there are two steps for item based neighborhood model: compute item-item similarities and make predications. We introduce the hierarchical structure based neighborhood model for each step. For the four different item types, we find their neighbors and calculate the similarities separately. We assume a node's directly linked nodes are most similar with it, and we calculate other nodes' similarity with it based on the hierarchical links as follows.

From the hierarchical structure tree, we could know that each track i's possible neighbors should contain $album(i)$, $artist(i)$, $genre(i)$ and other tracks (albums), which share the common album, artist and genres with i. Therefore, when calculating the similarities between i and other items ($track$ or $album$), there are three factors we should consider: whether they have common artist, album and genres or not. So we calculate the similarities in the following equation:

$$hier\_sim(i,j) = \begin{cases} 1.0 & j \in \{artist(i), genre(i), album(i)\} \\ (2.0 + gen\_sim(i,j))/3.0 & type(j) = track, album(i) = album(j) \\ (1.0 + gen\_sim(i,j))/2.0 & type(j) \in \{track, album\}, artist(i) = artist(j) \\ \mu \cdot gen\_sim(i,j) & type(j) \in \{track, album\}, artist(i) \neq artist(j) \end{cases} \quad (2)$$

Where $gen\_sim(i,j)$ is the similarities between $genre(i)$ and $genre(j)$, and μ is a penalty coefficient in $(0, 1]$, and it is used to adjust the similarities for the items which have no common album and artist with i. We use $\mu=2/3$ in this paper as we find it will bring relatively better results. $gen\_sim(i, j)$ is computed as follows:

$$gen\_sim(i, j) = \frac{\|genre(i) \cap genre(j)\|}{\|genre(i) \cup genre(j)\|}$$

For each album i, we can compute its similarities in the similar method as that of each track. The similarity equation is shown as follows:

$$hier\_sim(i,j) = \begin{cases} 1.0 & j \in \{artist(i), genre(i), track(i)\} \\ (1.0 + gen\_sim(i,j))/2.0 & type(j) \in \{track, album\}, artist(i) = artist(j) \\ \mu \cdot gen\_sim(i,j) & type(j) \in \{track, album\}, artist(i) \neq artist(j) \end{cases} \quad (3)$$

When considering the similarities about genres and artists, we should differentiate them from before as they are on the top layers of the hierarchical tree. The relations between artists and genres are relatively weak when we trace from the hierarchical links, so we do not consider the artists' related genres when computing their neighbors, and also ignore similarities between artists. For each artist i, we can get its related albums and tracks easily from the hierarchical, while it is a little difficult to determine their similarities, so we do not differentiate neighbors' weights and set the similarities as 1.0. Besides, we sort i's neighbors in descending order by their popularity. For the genres, we deal with them in the same way as that of artists. In addition, we also consider two genres are similar if they belong to a track's (album) genre sets together.

If we only calculate similarities based on the hierarchical links, the specific values are not very accurate. To make the similarities more reliable, we can incorporate $i \rightarrow j's$ confidence into the original equation and revise it in the form of:

$$confhier\_sim(i, j) = hier\_sim(i, j) \cdot \frac{\|R(i) \cap R(j)\|}{\|R(i)\|} \quad (4)$$

In this equation, we do not consider users' specific ratings, yet only utilize the item-item pairs' relative co-occurring frequency. So the sparse ratings have less impact on the confidence calculation. In this way, we could also compute the similarities between artists whose albums and tracks have common genres through incorporating their confidence, as sometimes a artist might be affected by other artists with similar styles .

In the stage of making predication for item candidates, we take account of other users' impact for the predicted item i and use item-mean-centered method to normalize $\hat{r}_{ui}$. We also use user-mean-centered normalization for $\hat{r}_{ui}$ in user-knn model. The rating predication framework for all the refereed item-knn models are shown as follows:

$$\hat{r}_{ui} = \bar{r}_i + \frac{\sum\limits_{j \in NH_{u(i)}} w_{ij}(r_{ij} - \bar{r}_j)}{\sum\limits_{j \in NH_{u(i)}} |w_{ij}|} \quad (5)$$

Where $NH_u(i)$ is item $i's$ neighbors which are rated by u and computed through hierarchical structure. In a further step, we could easily combine the hierarchy based neighborhood model and the rating item based neighbor into one integrated framework:

$$\hat{r}_{ui} = \alpha \cdot HierKNN(\hat{r}_{ui}) + (1 - \alpha) \cdot RateKNN(\hat{r}_{ui}) \tag{6}$$

Here α is a constant used to adjust the weight of the two parts, the value is in $(0, 1]$ and trained from the training data.

3.3 Hierarchical Structure Based Latent Factor Model

In the prior section, we utilize the hierarchical links to find more reliable neighbors for each item. Furthermore, we could explore the hierarchical structure more to model their effect on users' potential interests. Suppose the scenario that if a user likes one album very much, there might be several possible factors contributing for it: Maybe he prefers some of the tracks in the album very much, or he is a fan of the album's performing artist, or because of the associated genres. E.g., if a user loves the album 'like a virgin' released by Madonna very much, he might also prefer the song 'Angel' or 'like a virgin' in that album, or he might be a fan of Madonna, or he might loves Pop or Rock very much. On the other side, if we know one person dislikes the album, we could conclude similar reasons causing it.

We can know whether one person likes one album or not from the explicit feedback-ratings he gives for the album. However, according to our analysis above, it is not important whether he rates the albums high or low, so we just use the binary value for the albums and build the AlbumSVD model as follows:

$$\hat{r}_{ui} = b_{ui} + q_i^T \left(p_u + (||album(u)|| + 1)^{-\frac{1}{2}} \sum_{j \in album(u)} a_j \right) \tag{7}$$

Where $b_{ui} = \mu + b_i + b_u$, $album(u)$ is the items u might potentially prefer and has not rated yet, it is generated by the albums he has rated.

The $album(u)$ is generated in the following way: for each album i that u have rated, we select $top\text{-}k$ tracks with highest popularity belongs to i. We also count the frequency of related artist and genres, and choose the artists and genres whose frequent numbers no less than the threshold. The values of $top\text{-}k$ and $threshold$ are determined by validation ratings.

For the artists and genres, we could deduce in the similar way as that of albums. Although the correlated relations between artists and genres are relatively weak, yet for the users side, we could consider their mutual influence through strict restrictions. The $artist(u)$ and $genre(u)$ generating process is similar with that of the AlbumSVD. For the rated artists, we firstly select the $top\text{-}k$ support albums and tracks released by each artist whom u has rated. We also select the most frequent genres related with the tracks and albums belong to the artists he has rated.

Through the above analysis, we could easily incorporate the affect of artists, albums and tracks for users into one integrated model named AGTSVD, leading to the following framework:

$$\hat{r}_{ui} = b_{ui} + q_i^T \left(p_u + \beta_1 \cdot N_{au}^{-\frac{1}{2}} \cdot \sum_{j \in album(u)} a_j + \beta_2 \cdot N_{tu}^{-\frac{1}{2}} \cdot \sum_{j \in artist(u)} t_j \right.$$

$$\left. + \beta_3 \cdot N_{gu}^{-\frac{1}{2}} \cdot \sum_{j \in genre(u)} g_j \right) \tag{8}$$

Where $N_{au} = \|album(u)\| + 1$, $N_{tu} = \|artist(u)\| + 1$ and $N_{gu} = \|genre(u)\| + 1$. β_1, β_2 and β_3 are constants used to control the weights of three aspects, and they are determined by validation data set.

We could also take the effect of implicit feedback into account, incorporate the items that users have rated into AGTSVD model, and extend it to AGTSVD++ model :

$$\hat{r}_{ui} = b_{ui} + q_i^T \left(p_u + \beta_1 \cdot N_{au}^{-\frac{1}{2}} \cdot \sum_{j \in album(u)} a_j + \beta_2 \cdot N_{tu}^{-\frac{1}{2}} \cdot \sum_{j \in artist(u)} t_j \right.$$

$$\left. + \beta_3 \cdot N_{gu}^{-\frac{1}{2}} \cdot \sum_{j \in genre(u)} g_j + \beta_4 \cdot N_{ru}^{-\frac{1}{2}} \cdot \sum_{j \in R(u)} y_j \right) \tag{9}$$

Here $N_{ru} = \|R(u)\| + 1$. To solve this model, we only use the observed ratings and minimize the prediction errors on them. The optimization object is in the following form:

$$\min_{b_*, q_*, y_*, a_*, t_*, g_*} \sum_{(u,i) \in R} (\hat{r}_{ui} - r_{ui})^2 + \lambda \cdot \left(\sum_{j \in artist(u)} \|t_j\|^2 + \sum_{j \in album(u)} \|a_j\|^2 \right.$$

$$\left. + \sum_{j \in genre(u)} \|g_j\|^2 + \sum_{j \in R(u)} \|y_j\|^2 + b_u^2 + b_i^2 + \|p_u\|^2 + \|q_i\|^2 \right) \tag{10}$$

In this model, λ is a regulation parameter used to avoid overfitting, and it is tuned on the validation data. To solve the model, we use stochastic gradient descent technique which is very popular, effective and efficient.

3.4   Hierarchical Structure Based Integrated Model

In the prior two sections, we introduced hierarchical structure based neighborhood model and hierarchical structure based latent factor model. On one side, we utilize hierarchical links to find more reliable neighbors. On the other side, we explore hierarchical structure's impact for users' potential preferences. Furthermore, we could take account of both aspects simultaneously and incorporate them into a integrated framework in the form of :

$$\hat{r}_{ui} = b_{ui} + q_i^T \left(p_u + \beta_1 \cdot N_{au}^{-\frac{1}{2}} \sum_{j \in album(u)} a_j + \beta_2 \cdot N_{tu}^{-\frac{1}{2}} \sum_{j \in artist(u)} t_j + \beta_3 \cdot N_{gu}^{-\frac{1}{2}} \sum_{j \in genre(u)} g_j \right.$$

$$\left. + \beta_4 \cdot N_{ru}^{-\frac{1}{2}} \sum_{j \in R(u)} y_j \right) + (\|R^k(i;u)\| + 1)^{-\frac{1}{2}} \sum_{j \in R^k(i;u)} [(r_{uj} - b_{uj})w_{ij} + c_{i,j}] \qquad (11)$$

$$+ (\|H^k(i;u)\| + 1)^{-\frac{1}{2}} \cdot \sum_{j \in H^k(i;u)} (r_{uj} - b_{uj})h_{ij}$$

Here, the framework consists of two parts: latent factor model and neighborhood model. The later tier $(\|R^k(i;u)\| + 1)^{-\frac{1}{2}} \sum_{j \in R^k(i;u)} [(r_{uj} - b_{uj})w_{ij} + c_{ij}] + (\|H^k(i;u)\| + 1)^{-\frac{1}{2}} \sum_{j \in H^k(i;u)} (r_{uj} - b_{uj})h_{ij}$ can be considered as neighborhood based model's effect on the residuals of the latent factor model for the original ratings. $R^k(i;u)$ are item $i's$ top-k most similar items which have been rated by u, and they are determined by the ratings. $H^k(i;u)$ also are item $i's$ top-k most similar items while they are determined through the hierarchical links. $(\|R^k(i;u)\| + 1)^{-\frac{1}{2}} \sum_{j \in R^k(i;u)} c_{ij}$ is used to incorporate the effect of implicit feedback.

As usual, we could learn the related parameters through minimizing the regularized squared error function similarly as (10), and use gradient descent solver to modify the parameters in the opposite gradient direction until convergence, yielding:

$$b_u \leftarrow b_u + \mu_1 \cdot (e_{ui} - \lambda_1 \cdot b_u)$$
$$b_i \leftarrow b_i + \mu_2 \cdot (e_{ui} - \lambda_2 \cdot b_i)$$
$$p_u \leftarrow p_u + \mu_3 \cdot (e_{ui} \cdot q_i - \lambda_3 \cdot p_u)$$
$$q_i \leftarrow q_i + \mu_4 \cdot (e_{ui} \cdot (p_u + \beta_1 N_{au}^{-\frac{1}{2}} \sum_{j \in album(u)} a_j + \beta_2 N_{tu}^{-\frac{1}{2}} \sum_{j \in artist(u)} t_j + $$
$$\beta_3 N_{gu}^{-\frac{1}{2}} \sum_{j \in genre(u)} g_j$$
$$+ \beta_4 N_{ru}^{-\frac{1}{2}} \sum_{j \in R(u)} y_j) - q_i)$$
$$\forall j \in album(u):$$
$$a_j \leftarrow a_j + \mu_5 \cdot (e_{ui} \cdot \beta_1 \cdot N_{au}^{-\frac{1}{2}} \cdot q_i - \lambda_5 \cdot a_j)$$
$$\forall j \in H^k(i;u):$$
$$h_{ij} \leftarrow h_{ij} + \mu_6 \cdot (e_{ui} \cdot (\| H^k(i;u) \| + 1)^{-\frac{1}{2}} \cdot (r_{uj} - b_{uj}) - \lambda_6)$$
$$\forall \in R^k(i;u):$$
$$w_{ij} \leftarrow w_{ij} + \mu_6 \cdot (e_{ui} \cdot (\| R^k(i;u) \| + 1)^{-\frac{1}{2}} \cdot (r_{uj} - b_{uj}) - \lambda_6)$$
$$c_{ij} \leftarrow c_{ij} + \mu_7 \cdot (e_{ui} \cdot (\| R^k(i;u) \| + 1)^{-\frac{1}{2}} - \lambda_7)$$

Here, $e_{ui} = r_{ui} - \hat{r}_{ui}$, t_j, g_j and y_j are modified similarly as that of a_j. The learning rate (μ_*) and regularization (λ_*) could be tuned by the validation ratings.

4 Experiments

4.1 Data Set

We use yahoo! music rating dataset to evaluate our approaches in this paper. yahoo! music rating dataset is released in the track1 of the kdd cup 2011. in the

Table 1. Yahoo! Music Dataset Composition

#Users	#Items	#Ratings	#Train Ratings	#Validation Ratings	#Test Ratings
1,000,990	624,961	262,810,175	252,800,275	4,003,690	6,005,940

dataset each user has at least 20 ratings while 4 items for each user in validation ratings and 6 items for each user in the test ratings. the composition of the dataset is shown in table 1:

From Table 1, we can know the scale of ratings is very large and the sparsity is about 99.6%, even sparser than that of Netflix dataset whose sparsity is 98.9%. The rating value is integral number between **0** and **100**, thus the large range brings much bigger mistakes when compared with the Netflix rating value which is integer between 1 and 5. So it is more challenging to make predications on the Yahoo! Music dataset.

The rating items consist of 4 different types: *tracks, albums, genres and artists*. We can find the hierarchical links of the four types from the auxiliary files. For each track, it belongs to a unique album and artist, while it might have several possible genres. For each album, it also belongs to a unique artist, and it might contain tens of tracks and several possible genres. The statistics for the links and item numbers that each type includes and each type's rating numbers are shown in Table 2. From this table, we can conclude that the hierarchical links are very plentiful, and they could be modeled effectively. We could also see that tracks and artists receive more ratings compared to albums and genres, so utilizing their hierarchical information will contribute more for improving the prediction performance. The numbers of the four types and their ratings are also shown in Table 2.   We can also draw

Table 2. Hierarchical average links, Taxonomy No. and Rating percentage

	track	album	artist	genre	Item No. Percentage	Rating Percentage
track	/	0.87483	0.9167	2.2387	507172 (81.15%)	46.85%
album	5.2293	/	0.88646	2.4251	88909 (14.23%)	19.01%
artist	15.9096	2.8261	/	2.0710	27888 (4.46%)	28.84%
genre	1144.5746	217.3518	58.2218	/	992 (0.16%)	5.30%

each type rating numbers for the users with different total ratings in Figure 2. From it, we could see that for the users whose rating number is small, artists and genres take up more in the items. While as the users' rating number grows, the tracks and albums occupy more. Therefore, we could effectively utilize the hierarchical links effectively model users' potential preferences.

4.2   Evaluation Metric

We use root mean squared error (RMSE) to evaluate our approach in this paper. Compared to MAE, RMSE gives more weights for predictions with bigger errors and is used more in recent years. The evaluation rule is in the form of: $RMSE = \sqrt{\frac{\sum_{(u,i) \in T} (r_{ui} - \hat{r}_{ui})^2}{\|T\|}}$

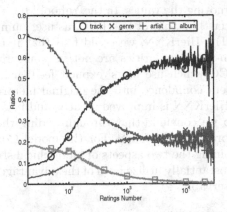

Fig. 2. The rating ratios of four types for users with different rating numbers

4.3 Experiment Results and Analysis

In this part, we evaluate the performance of our approach by comparing with the traditional models which take no account of hierarchical structure.

HierKNN. Before comparing the performance of each related neighborhood models, we tune the parameter α in the confidence hierarchical neighborhood (ConfHierKNN) and rating based item neighborhood (ItemKNN) integrated model (ConfHierInteg) on the validation data set. We tried different number of neighbors respectively to find the best α, and got $\alpha = 0.53$ for the ConfHierInteg model. The evaluation results are shown in Table 3.

Table 3. Comparison Between Neighborhood Based Models

Model \ K	5	10	15	20	50	100	200	300	500
GobalMean	38.1414	38.1414	38.1414	38.1414	38.1414	38.1414	38.1414	38.1414	38.1414
ItemMean	33.0870	33.0870	33.0870	33.0870	33.0870	33.0870	33.0870	33.0870	33.0870
UserMean	29.4216	29.4216	29.4216	29.4216	29.4216	29.4216	29.4216	29.4216	29.4216
ItemConfKNN	29.1507	28.9405	28.8703	28.8359	28.7948	28.7712	28.7415	28.7365	28.7359
UserPCC	28.8303	28.3981	28.2689	28.2092	28.1275	28.1158	28.1180	28.1201	28.1213
ItemPCC	27.0902	26.9848	27.0038	27.0299	27.0985	27.1215	27.1283	27.1289	27.1290
HierKNN	27.9769	27.9371	27.9363	27.9363	27.9369	27.9373	27.9373	27.9373	27.9373
ConfHierKNN	27.1481	26.8406	26.7950	26.7929*	26.8502	26.8868	26.9001	26.9016	26.9018
ConfHierInteg	25.9373	25.8585*	25.8804	25.9065	25.9853	26.0163	26.0266	26.0278	26.0280

From Table 3, we could conclude that neighborhood models are absolutely much better than user average, item average and global average model. When comparing between item confidence based neighborhood model (ItemConfKNN) and rating based neighborhood model (UserPCC / ItemPCC), we could easily find that utilize users' specific ratings which reflect their explicit feedback are more effective than ignoring the values. In the Yahoo! Music rating data, we also find ItemPCC is better than that of UserPCC, as user numbers are much more than item numbers. For HierKNN, we could find that it is less effective than ItemPCC as the item-item similarities are not very accurate. However, it has outperformed UserPCC which use users' explicit feedback. Moreover, through incorporating item-item confidence into the original hierarchy similarities, the performance of ConfHierKNN is improved greatly, and it is more effective than that of ItemPCC. So we conclude that through adding the items' confidence, more reliable neighbors could be found. For the model ConfHierInteg, we can also see that by combining the two aspects of hierarchical structure and ratings, ConfHierInteg will substantially utilize both of the advantages and the prediction accuracy improve observably.

HierSVD. When mentioned to the hierarchical structure related latent factor model, there are several factors we should consider: Firstly, in the generation process of $album(u)$, $artist(u)$ and $genre(u)$, two parameters need to be tuned. Through the training on the validation data, we set $top\text{-}k=10$ for AlbumSVD and GenreSVD, and $top\text{-}k=20$ for AristSVD; we choose $threshold=0.3$ for the three models. In that way the models could trade off better between performance and computation cost. Secondly, we should learn the related learning rates (μ_*) and regulation parameters (λ_*), we use APT2 method which can be classified as one two-way cross folder method, and the details can be found in [21]. The related parameters are set as: $\mu_1=0.002294$, $\lambda_1=0.005$, $\mu_2=0.08149$, $\lambda_2=0.005$, $\mu_3=0.00007$, $\lambda_3=0.015$, $\mu_4=0.003052$, $\lambda_4=0.015$, $\mu_5=0.0001$, $\lambda_5=0.015$. Lastly, we tune β_1-β_4 and set as:$\beta_1=0.15$, $\beta_2=0.4$, $\beta_3=0.3$, $\beta_4=0.3$.

Based on the conditions mentioned above, our models are compared with the original Biased-SVD and SVD++ accordingly. To show the effectiveness of our models, we execute all the models with varying factor dimensions, and the detailed comparison results can be found in Fig. 3(a) and Fig. 3(b).

As both of the figures show that our models outperform the SVD and SVD++ accordingly in each factor dimension. From Fig. 3(a) we can see that the AGTSVD decreases the RMSE by about 0.75, and as shown in Fig. 3(b), it also performs better than SVD++. Therefore, when there are not enough ratings, the AGTSVD will perform even better than the SVD++ model, as the AGTSVD only concerns about the items that users have not rated. From Fig. 3(b), we verify that through incorporating implicit feedback into AGTSVD model, predication errors will be decreased even more. We can also notice that when making comparison among the AlbumSVD, GenreSVD and ArtistSVD, or the AlbumSVD++, GenreSVD++ and ArtistSVD++, both of the artist related models work better than the other two models. It is because that artists locate in the highest level of he hierarchical structure tree, thus when considering their feedback, it has more related items to select from, which users might be interested in. So we get a conclusion that the higher level of the items, the more important they are. We can also see that the RMSE of hierarchical structure related models decrease more steadily than that of the original models as the factor dimensionality increases.

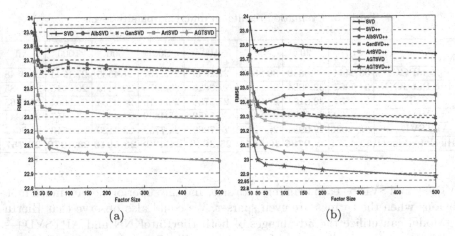

Fig. 3. RMSE of different methods on Test ratings for varying dimensionality

HierInteg. Furthermore, we conduct several experiments for the state-of-the-art model which incorporates both aspects of the hierarchical structure characteristics. When evaluating HierIntegModel, we set neighbor size with 100, as we find increase it bring little RMSE decrease, and we use the following values for the referred meta parameters: μ_6=0.000035, λ_6=0.000328, μ_7−0.0001, λ_7=0.00001. Table 4. summarize the evaluation results, we can find the RMSE decrcase by **1.03** comparing to SVD, and through the comparison with AGTSVD++, we verify that incorporating hierarchical based neighbor model into AGTSVD++ could decrease the prediction errors even more, and the integrated model could make use of the advantages of both aspects adequately.

Table 4. Performance of hierarchical structure integrated model

Factor Model	10	20	50	100	200	500
HierIntegModel	23.0842	22.8156	22.757	22.7463	22.7286	22.7032

Sparsity Analysis. To verify the hierarchical structure based model can effectively overcome the sparsity issue, we evaluate related models with different sparsity levels of training data while keeping the validation data and test data. We sample the original by the order of users' rating time with varying training data size N, and execute the related models under the same circumstance (KNN with best performance, SVD related models with f=50). The experimental results are listed in Table 5.

From Table 5, we could clearly find that compared with neighborhood models, latent factor models have much stronger ability to overcome rating sparsity problem. HierRateKNN performs much better than that of ItemPCC as it relies less on the numbers of training data. For model APTSVD++, it outperforms

Table 5. Evaluation results on different sparsity levels

Model \ N	10	20	50	100	200	500	1000	All
ItemPCC	36.2444	33.8040	30.8575	29.2649	28.1737	27.4355	27.1744	26.9848
HierRateKNN	34.5367	32.3624	29.5040	27.9626	26.9978	26.3443	26.0915	25.8585
SVD	31.7533	30.8154	29.0389	27.5738	26.2416	24.9734	24.3664	23.7663
SVD++	31.2232	30.0089	27.8409	26.5352	25.4832	24.4681	23.9457	23.3944
APTSVD++	30.52066	29.0954	26.9206	25.7069	24.7752	23.9120	23.4713	22.9642
HierIntegModel	**30.4771**	**28.9786**	**26.6651**	**25.3649**	**24.4208**	**23.5913**	**23.1979**	**22.757**

SVD++ and SVD on each training data size, and the superiorities are much more obvious when the ratings are even sparser. We could also observe that HierIntegModel can utilize the advantages of both HierRateKNN and APTSVD++, and improve the prediction performance steadily for each training data size. Through the comparison, we can get the conclusion that our approach indeed mitigate the sparsity problem effectively.

5   Conclusion

In this paper, we propose a music hierarchical structure based model to mitigate the rating sparsity problem, which heavily affects the recommendation methods' performance. We utilize the hierarchical structure in two ways. Firstly, we exploit it to find more reliable neighbors through the links between items with different types. Then, we explore the effect of hierarchical structure on users' potential preferences and find more related items that they might be interested in. Finally, we incorporate both aspects into a integrated framework. Experimental results show our model could significantly improve the recommendation performance. Furthermore, we evaluate related models on different sparsity levels of the training data, and conclude that our approach can effectively mitigate rating sparsity issue.

In the future work, we will explore rating temporal effect on improving recommendation performance, and integrate it with the framework proposed in this paper.

References

1. Amatriain, X., Bonada, J., Loscos, À., Arcos, J.L., Verfaille, V.: Content-based transformations. Journal of New Music Research 32(1), 95–114 (2003)
2. Logan, B.: Mel frequency cepstral coefficients for music modeling, pp. 723–732. ACM (2010)
3. Chen, H., Chen, A.: A music recommendation system based on music data grouping and user interests, 231–238 (2001)
4. Lamere, P.: Social tagging and music information retrieval. Journal of New Music Research 37(2), 101–114 (2008)

5. Sarwar, B., Karypis, G., Konstan, J., Reidl, J.: Item-based collaborative filtering recommendation algorithms. In: Proc. WWW 2001, pp. 285–295. ACM (2001)
6. Koren, Y.: The bellkor solution to the netflix grand prize (2009)
7. Koren, Y.: Factorization meets the neighborhood: a multifaceted collaborative filtering model. In: Proc. 14th International Conference on Knowledge Discovery and Data Mining. ACM (2008)
8. Hu, Y., Koren, Y., Volinsky, C.: Collaborative filtering for implicit feedback datasets. In: 8th IEEE International Conference on Data Mining, pp. 263–272. IEEE (2008)
9. Adomavicius, G., Tuzhilin, A.: Towards the next generation of recommender systems: a survey of the state-of-the-art and possible extensions. IEEE TKDE 17(6), 734–749 (2005)
10. Bell, R.M., Koren, Y.: Scalable collaborative filtering with jointly derived neighborhood interpolation weights. In: ICDM, pp. 43–52. IEEE Computer Society (2007)
11. Funk, S.: Netflix update (2006),
    http://sifter.org/simon/journal/20061211.html
12. Paterek, A.: Improving regularized singular value decomposition for collaborative filtering. In: Proceedings of KDD Cup and Workshop (2007)
13. Takács, G., Pilászy, I., Németh, B., Tikk, D.: Matrix factorization and neighbor based algorithms for the netflix prize problem. In: Proc. of the 2008 ACM Conference on Recommender Systems, pp. 267–274. ACM (2008)
14. Sarwar, B., Karypis, G., Konstan, J.A., Riedl, J.: Application of dimensionality reduction in recommender systemca case study (2000)
15. Shao, B., Wang, D., Li, T., Ogihara, M.: Music recommendation based on acoustic features and user access patterns, 1602–1611
16. Dror, G., Koenigstein, N., Koren, Y.: Yahoo! music recommendations: Modeling music ratings with temporal dynamics and item taxonomy. In: Proc. 5th ACM Conference on Recommender Systems, pp. 165–172 (2011)
17. Chen, T., Zheng, Z., Lu, Q.: Informative ensemble of multiresolution dynamic factorization models (2011)
18. Dror, G., Koenigstein, N., Koren, Y.: The yahoo! music dataset and kdd-cup11. In: KDD-Cup Workshop (2011)
19. Chen, P., Tsai, C.Y., et al.: A linear ensemble of individual and blended models for music rating prediction. In: KDD-Cup Workshop (2011)
20. Jahrer, M.,Töscher, A.: Collaborative filtering ensemble. In: KDD-Cup Workshop (2011)
21. Töscher, A., Jahrer, M., Bell, R.: The bigchaos solution to the netflix grand prize. Netflix Prize Documentation (2009)

Enhancing Music Information Retrieval
by Incorporating Image-Based Local Features

Leszek Kaliciak[1], Ben Horsburgh[1], Dawei Song[2,3],
Nirmalie Wiratunga[1], and Jeff Pan[4]

[1] The Robert Gordon University, Aberdeen, UK
[2] Tianjin University, Tianjin, China
[3] The Open University, Milton Keynes, UK
[4] Aberdeen University, Aberdeen, UK
{l.kaliciak,b.horsburgh,n.wiratunga}@rgu.ac.uk,
Dawei.Song@open.ac.uk, jeff.z.pan@abdn.ac.uk

Abstract. This paper presents a novel approach to music genre classification. Having represented music tracks in the form of two dimensional images, we apply the "bag of visual words" method from visual IR in order to classify the songs into 19 genres. By switching to visual domain, we can abstract from musical concepts such as melody, timbre and rhythm. We obtained classification accuracy of 46% (with 5% theoretical baseline for random classification) which is comparable with existing state-of-the-art approaches. Moreover, the novel features characterize different properties of the signal than standard methods. Therefore, the combination of them should further improve the performance of existing techniques.

The motivation behind this work was the hypothesis, that 2D images of music tracs (spectrograms) perceived as similar would correspond to the same music genres. Conversely, it is possible to treat real life images as spectrograms and utilize music-based features to represent these images in a vector form. This points to an interesting interchangeability between visual and music information retrieval.

Keywords: Local features, Co-occurrence matrix, Colour moments, K-means algorithm, Fourier transform.

1 Introduction

Almost every representation of music used in the field of Music Information Retrieval (MIR) involves extracting features from music transformed into the frequency domain. These features include chromatic, melodic, harmonic, rhythmic, and timbral measures.

Thus, [1] represents the distribution of chroma within a song as a histogram. The songs with similar chroma histogram distributions are considered similar. The temporal aspects of pitch are taken into account by [2] and [3]. In [4] authors try to capture the rhythm by constructing a self-similarity matrix based upon the similarity of each short time frequency spectra extracted from the audio. Bello [5] presents a method to describe a novelty function (a common method for identifying onsets) of a waveform inspired by Foote's [4] similarity measure. One of the challenges in MIR is how to interpret the

Y. Hou et al. (Eds.): AIRS 2012, LNCS 7675, pp. 226–237, 2012.

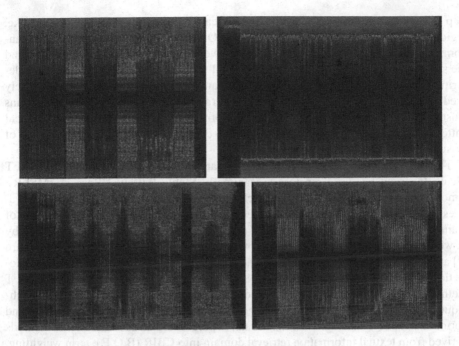

Fig. 1. Music representation in the form of 2D images

classification confusions. Some incorrectly classified instances cast doubt on whether the ground truth is correct (for example a pop song that could be labeled as funk). The solution to this problem might be the incorporation of fuzzy logic.

Our approach implements methods from both visual IR (VIR) and MIR research areas. The motivation behind this work was the hypothesis, that 2D images of music tracs (spectrograms) perceived as similar would correspond to the same music genres (perhaps even similar music tracs). Conversely, we can treat real life images as spectrograms and utilize music-based features to represent these images in a vector form. This would point to an interesting interchangeability between visual and music information retrieval. Thus, instead of extracting features directly from frequency domain, we generate an image of each song that shows how the spectral density of a signal varies with time. Two geometric dimensions represent frequency and time, and the colour of each point in the image indicate the amplitude of a particular frequency at a particular time. The advantage of such visual representation is that it does not rely on musical concepts (melody, timbre, rhythm, etc.). It is also intuitive that two songs can be compared visually based on their spectrogram representations. The next step involves the application of local features from Content Based Image Retrieval and representation of music in the form of histograms of visual words counts. Thus obtained histograms, characterizing individual songs, are used for genre music classification task.

Because global approaches find it hard to capture all the properties of an image, the implemented local features are based on the "bag of visual words" approach. The first step in the "bag of features" [1] method is to localize the points of interest (point-like, region-like) by using corner or blob detectors. Other sampling techniques include random and dense sampling. The second step involves the representation of regions around the sample points in a form of multidimensional vectors. There are various existing descriptors, the SIFT (Scale Invariant Feature Transform) being one of the most widely used. The initial extraction is performed on a training set of images and the K-means clustering is applied to it. Each cluster will correspond to one "visual word", a local pattern. Finally, each image in a data collection can be characterized by a histogram of "visual words" counts.

The most common interest points detectors are: Harris-affine, Hessian-affine, SIFT (Scale Invariant Feature Transform), Maximally Stable Extremal Regions (MSER). Among descriptors we have SIFT (detector and descriptor), local jets (image derivatives), steerable filters, generalized moment invariants. One of the variations of "bag of features" (B.O.F) method based on SIFT detector and descriptor was first proposed by Lowe in [6]. Other good sources of information about scale space and local features are [7] and [8]. In [9] authors applied the local features to nude or pornographic images detection. Instead of using the well known SIFT descriptor, they implemented Hue-SIFT method in order to take colour into account. [10] is a comparison of different techniques used in B.O.F. approach for image sampling, visual dictionary generation and normalization of the histograms. Yang et al. [11] incorporated and tested some methods derived from textual information retrieval domain into CBIR (B.O.F.): term weighting, stop word removal, feature selection. They also conducted experiments testing the influence of the vocabulary size (number of clusters) and spatial information on the retrieval performance.

In this paper we utilize fast and easy to implement method based on local features [15]. Because quite often randomly generated sample points are more discriminant than the points detected by corner detector (especially when it comes to a large number of sample points when the set of keypoints detected by corner detectors becomes saturated), we decided to implement a hybrid sampling technique. Comparison between random, dense, pure corner-based, and hybrid sampling showed the superiority of this type of sampling. Our descriptor is based on co-occurrence matrix and colour moments. For the description of the image patches around the sample points we used separately: co-occurence matrix computed at eight different orientations, and three colour moments to capture the local colour properties. The co-occurence matrix has proven to be an effective way of texture representation and by considering multiple orientations we make it invariant to rotation. Colour moments are fairly insensitive to changes in viewpoint, and their computation is trivial. Moreover, patches characterized by colour moments are also able to capture the local textural information. Despite the relative simplicity of the model, this method was able to obtain results comparable with current state-of-the-art (ImageCLEF2010 Wikipedia Retrieval Task).

One of the main advantages of our model is that by switching to visual domain we can abstract from the musical concepts and still obtain results comparable with current state-of-the-art. Moreover, the proposed approach could be used for the characterization of signal in general as an alternative to common techniques.

---

[1] Terms "bag of visual words" and "bag of features" will be used interchangeably in this paper.

The paper is organized as follows. In Section 2 we present the model and describe the developed algorithm in detail. It consists of the sub-sections introducing the local features based on the "bag of visual words", the Fast Fourier Transform and a sub-section on spectrograms generation. Section 3 is devoted to the experiments and discussion. It describes the data collection used in the experiment, the detailed experimental setup and results with their analysis. We draw conclusions in Section 4 and finally present some ideas for future research (Section 5).

2   The Proposed Model

Here, we give a detailed description of the proposed model. The framework establishes the link between MIR and VIR research areas.

Our algorithm consists of the following stages:

1. **Transformation to frequency domain:** Transform the music data from the time to frequency domain using Fast Fourier Transform (FFT). Since audio signals are periodic over time, it is convenient to represent them as a sum of infinite number of sinusoidal waves. It makes it easier to analyze sinusoidal functions than general shaped functions.

2. **Music representation, visual data generation:** Generate spectrograms in two dimensional space, where the geometric dimensions represent frequency and time, and the colour of each point in the image indicate the amplitude of a particular frequency at a particular time. These spectrograms are generated from the signal transformed by FFT. Our method of spectrograms generation was designed in such a way as to produce images containing easy to capture visual properties.

3. **Image sampling:**
   – **Keypoints detection:** Apply Shi and Tomasi (see [14]) method to find the points of interest.
   – **Random sampling:** Apply random points generator to produce another half of the sample points.
   – **Random sampling:** Alternatively, dense sample images (divide images into a number of uniform, non-overlapping rectangular sub-images).

4. **Description of local patches:** Characterize local patches in the form of co-occurence matrix or colour moments. In case of co-occurence matrix extract the meaningful statistics - energy, entropy, contrast, and homogeneity. Compute the features for individual colour channels.

5. **Feature vector construction:** Represent local patterns as 9 dimensional (moments) or 12 dimensional (co-occurence) vector.

6. **Visual dictionary generation:** Apply K-means clustering to the training set in order to obtain the codebook of visual words.

7. **Histogram computation:** Create a histogram of visual words counts by calculating the distance between image patches and cluster centroids.

8. **Music genre classification:** The classification of music data into music genres is performed by k-nearest neighbour algorithm, based on Minkowski's fractional similarity measure.

Steps 3 to 7 are related to generation of visual representations of the spectrograms. In the course of this research, global methods like colour moments, co-occurrence matrix (texture), colour correlograms were also tested. We utilize local features because of their superior performance over various global methods. The local features may also have another advantage over other models. An interesting future work would be to investigate if image patches identified by corner detectors (roughly speaking - locations of a sudden change of pixel intensities) and "visual words" correspond to some important characteristics of audio signal.

2.1 The Fast Fourier Transform

Let $x_0, ..., x_{N-1}$ be complex numbers. The Discrete Fourier Transform (DFT) is defined by

$$X_k = \sum_{n=0}^{N-1} x_n e^{-i2\pi k \frac{n}{N}} \quad k = 0, ..., N-1.$$

Computing DFT requires $O(N^2)$ operations, while FFT reduces the number to $O(N \log N)$. The implemented FFT method incorporates Cooley-Tukey algorithm, which breaks down a DFT into smaller DFTs. The audio sampled size is 65536 bytes (1.486 seconds), with sampling rate of 44100Hz.

2.2 Spectrogram Generation

The resulting spectrum is split into 512 bins (64Hz / bin). The power of each bin is converted into a pixel as follows:

```
Let colour = power / meanPower

IF {colour > 1}
   r = ((1 - (1 / colour)) / 2) + 0.5f
ELSE
   IF {colour > 0.5}
      g = (((colour-0.5)/0.5) / 2) + 0.5f
ELSE
   b = ((colour / 0.5) / 2) + 0.5f

   ENDIF
ENDIF
```

The horizontal dimension of each image represents the time (1 pixel = 1.486sec), vertical dimension represents frequency (1 pixel = 64Hz), and pixel intensities represent power.

This method of spectrogram generation produces images that varies in colour and texture. These properties make the images suitable for application of visual features. An interesting observation is that some genres are easily recognizable directly from our spectrograms. Classical music, for instance, is characterized by a presence of the blue colour joining the top and the bottom part of an image.

2.3 The Sampling Technique

Recently, an approach based on local features extraction has become quite popular in Visual Information Retrieval. Global approaches find it hard to capture all the properties of an image. The most recent state-of-the-art in Image Retrieval is based on so-called "bag of visual words". The first step in the "bag of features" method is to localize the points of interest (point-like, region-like) by using corner/blob detectors. Other sampling techniques include random and dense sampling. The second step involves the representation of regions around the sample points in a form of multidimensional vectors. There are various existing descriptors, the SIFT (Scale Invariant Feature Transform) being one of the best. The initial extraction is performed on a training set of images and the K-means clustering is applied to it. Each cluster will correspond to one "visual word", a local pattern. Finally, each image in a data collection can be characterized by a histogram of "visual words" counts. The most popular interest points detectors are: Harris-affine, Hessian-affine, Scale Invariant Feature Transform (SIFT), Maximally Stable Extremal Regions (MSER). Among descriptors we have SIFT (detector and descriptor), local jets (image derivatives), steerable filters, generalized moment invariants.

Let us now return to the local features utilized in this paper. As aforementioned, the implemented hybrid sampling method combines Shi and Tomasi [14] corner detection with a random number generator. The Shi and Tomasi method is based on the Harris corner detector. The change of pixel intensities is characterized as

$$S(x,y) = \sum_u \sum_v w(u,v)(I(u,v) - I(u+x,v+y))^2. \tag{1}$$

From Taylor approximation of the first order we get

$$I(u+x,v+y) \approx I(u,v) + I_x(u,v)x + I_y(u,v)y. \tag{2}$$

Substituting (2) in (1) we obtain

$$S(x,y) \approx \sum_u \sum_v w(u,v)(I_x(u,v)x + I_y(u,v)y)^2. \tag{3}$$

We can rewrite equation (3) in the following form

$$S(x,y) \approx \begin{pmatrix} x & y \end{pmatrix} A \begin{pmatrix} x \\ y \end{pmatrix} \tag{4}$$

where $A =$

$$\sum_u \sum_v w(u,v) \begin{bmatrix} I_x^2 & I_x I_y \\ I_x I_y & I_y^2 \end{bmatrix} = \begin{bmatrix} \langle I_x^2 \rangle & \langle I_x I_y \rangle \\ \langle I_x I_y \rangle & \langle I_y^2 \rangle \end{bmatrix}. \tag{5}$$

We define "cornerness", a measure of corner response as

$$M_c = \lambda_1\lambda_2 - \kappa(\lambda_1 + \lambda_2)^2 = det(A) - \kappa trace^2(A). \tag{6}$$

We assume that the corner was detected if M_c is sufficiently large. Shi and Tomasi found that the good corners can be obtained by setting a minimum threshold and checking if the smaller of the eigenvalues is greater than the threshold.

Another sampling technique implemented for comparison purposes was dense sampling. In this case, each image was divided into the same number of 900 identical rectangular sub-images.

2.4 Region Descriptors

Each local patch in an image was represented as

- The 8 orientational co-occurence matrix.

- Colour moments.

A simple co-occurence matrix is defined as follows

$$C_{\Delta x, \Delta y}(i, j) =$$

$$\sum_{p=1}^{n}\sum_{q=1}^{m}\begin{cases} 1, & if\ I(p,q) = i\ and\ I(p + \Delta x, q + \Delta y) = j \\ 0, & otherwise \end{cases}$$

The matrix describes the way certain grayscale pixel intensities occur in relation to other grayscale pixel intensities. It counts the number of such patterns. The most discriminating statistics extracted from co-occurence matrix are: contrast, inverse difference moment, entropy, energy, homogeneity, and variance.

The method based on three colour moments assumes that the distribution of colour can be treated as probability distribution. Three statistics extracted from individual colour channels are

- Mean $E_i = \sum_{j=1}^{n} \frac{1}{N} p_{ij}$

- Standard Deviation $\sigma_i = \sqrt{\left(\frac{1}{N}\sum_{j=1}^{N}(p_{ij} - E_i)^2\right)}$

- Skewness $s_i = \sqrt[3]{\left(\frac{1}{N}\sum_{j=1}^{N}(p_{ij} - E_i)^3\right)}$

The first moment can be interpreted as an average colour value, second as a square root of the variance of the distribution, and third as the measure of asymmetry in the distribution. One can construct the weighted similarity measure as an analogy to manhattan metric, for example:

$$d(H, I) = \sum_{i=1}^{r} w_{i1}\left|E_i^1 - E_i^2\right| + w_{i2}\left|\sigma_i^1 - \sigma_i^2\right| + w_{i3}\left|s_i^1 - s_i^2\right|.$$

Colour moments can also capture the textural properties of an image and are fairly insensitive to viewpoint changes. By computing them in HSV colour space we can make the statistics insensitive to illumination changes.

2.5 Feature Vector Construction, Visual Dictionary Generation, and Histogram Computation

The local patches are represented as multidimensional vectors constructed from different statistics, extracted from individual colour channels. By taking a sample training set consisting of collection's representative images, we can generate so-called visual vocabulary. The K-means clustering algorithm has been used for that purpose. Each cluster characterizes a local pattern, representing specific "visual word". The histogram of visual words counts is created by computing the manhattan distance between individual patches and cluster centroids, and calculating how many patches belong to specific clusters.

2.6 Music Genre Classification

The classification of music data into music genres is performed by k-nearest neighbour algorithm, using Minkowski's fractional similarity measure

$$d(x, y) = \left(\sum_{i=1}^{n} \sqrt{|x_i - y_i|} \right)^2$$

where $x = (x_i)$ and $y = (y_i)$ are the n dimensional feature vectors. It was experimentally proven (see [12]) that the fractional measures from Minkowski's family of distances yield good results in VIR.

3 Experiment and Discussion

Figure 2 shows the query by visual example retrieval based on the local features and our music representation.

For the experimental purposes we used a data collection consisting of 4759 music tracks. The genre distribution is presented in table 1.

Genre labels were extracted from iTunes. The local feature algorithm uses 900 sample points per image, for each sample point we open a square window 10 by 10 pixels wide. The dimensionality of the histograms of visual words counts is 40. The applied k-nearest neighbour algorithm uses 9-fold cross-validation, 12 nearest neighbours, distance weighting and manhattan metric. The classification accuracy we obtained with dense sampling was approximately **46 per cent** (2176 tracks) of correctly classified instances. The hybrid sampling scored lower, resulting in **43 per cent** (2051 tracks) of retrieval accuracy. The reason for this lies in the worse performance of corner detector in this domain. The local features with hybrid sampling performed better than the one with dense sampling on ImageCLEF2007 and MIRFlickr25000 collections, consisting of real-life images.

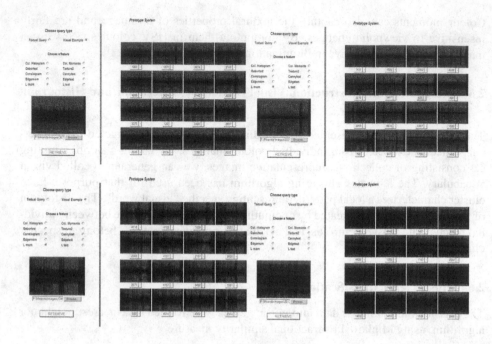

Fig. 2. Local features at work

From the confusion analysis we observed that most incorrectly classified instances were confused with similar genres, and the song-genre correspondence was arguable and subjective. That is why it is so hard to improve the retrieval performance. Good, natural solution to this problem could be the incorporation of fuzzy logic, and associate each song with certain probability of it being in one of the classes. The problem with comparisons with other methods arises because of the lack of the standardized data collections in MIR. Many data collections have unequally distributed data sets, different number of genres, more specialized or generalized classes.

All of that affects the behavior of classifier. Meng and others [13] used a multivariate autoregressive feature model, considered as current state-of-the-art in MIR, to capture the temporal information in the window. The data set used consisted of 1210 music tracks with 11 genres. The best mean classification accuracy they obtained were 44 and 40 per cent for the LM and GLM classifiers. It should be noted though that the accuracies obtained by the automatic classification need to be relative to the theoretical baseline for random classification which is 9% for [13], and 5% for our collection. It means that the performance of our method is actually much better. There are also other aspects, mentioned previously, that make the evaluation difficult.

In his PhD thesis on music genre classification, Serra presents a "non exhaustive list for the most relevant papers presented in journals and conferences for the last years" [16]. He concludes that "although accuracies are not completely comparable due to the different datasets the authors use, similar approaches have similar results. This suggest that music genre classification, as it is known today, seems to reach a "glass ceiling"". The reported accuracies were then plotted with respect to the number of genres (Figure 3).

Table 1. Genre classes

Genre	Tr.	Genre	Tr.
Pop	1024	Country	82
Alternative and Punk	919	Hip-Hop	81
Rock	862	Reggae	80
R&B	516	Easy Listening	80
Classical	293	Musicals	75
Dance	265	Latin	62
Alternative	139	Christmas	42
Folk	115	Rap	15
Metal	89	Soundtrack	11
Blues	9		

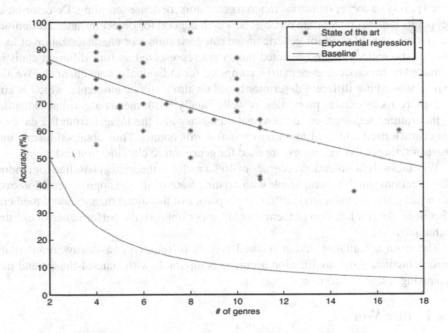

Fig. 3. State of the art in genre classification. Adapted from [16].

The human performance in classifying music genres (10 genres) is around 53% correctly cassified for 250ms samples and around 70% for samples longer than 3s [17]. Thus, the performance of current state of the art models for genre classification is comparable with the human performance.

4   Conclusions

In this paper we propose a novel approach to MIR. Having represented the music tracks in the form of two dimensional images, we apply the "bag of visual words" method from visual IR in order to classify the songs into 19 genres. The motivation behind this work was the hypothesis, that 2D images of music tracs (spectrograms) perceived as similar would correspond to the same music genres (perhaps even similar music tracs). Conversely, we can treat real life images as spectrograms and utilize music-based features to represent these images in a vector form. This would point to an interesting interchangeability between visual and music information retrieval.

First, songs are represented as images generated from Fourier frequency domain. The next step involves indexing thus obtained data collection by applying the following method, derived from content based image retrieval. Because images often consist of different patches of uniform patterns, global features find it hard to capture all the properties. Initially, when it comes to the implemented method, about half of sample points is detected by a corner detector and another half is picked at random. For relatively small number of sample points this technique proved to give better results than sampling based purely on corner detectors, random generators, or dense sampling. Detector-based keypoints tend to concentrate on objects, which is good for object instance recognition but not necessarily good for generic image categorization. For characterization of local patches the regions around selected points are represented as four different statistics extracted from co-occurrence matrix, computed for individual colour channels. We also experiment with a different descriptor based on three colour moments, which is able to capture local textural properties as well. Finally the k-means algorithm is applied to the training set to generate the visual dictionary and the images from the database are characterized with a histogram of visual words counts. Thus obtained histograms, characterizing individual songs, are used for genre music classification task.

We obtained classification accuracy of 46% (with 5% theoretical baseline for random classification) which is comparable with existing state-of-the-art approaches. Moreover, the novel features characterize different properties of the signal than standard methods. Therefore, the combination of them should further improve the performance of existing techniques.

The main advantages of our method are: more intuitive, easy way to automatic music classification, classification accuracy comparable with state-of-the-art and new promising research direction.

5   Future Work

The future work may include incorporation of the spatial information for local image patches, experimentation with different sampling techniques and incorporation of temporal information (short time Fourier transform, wavelets), which should further improve the classification accuracy. Additionally, an interesting future work would be to investigate if image patches identified by corner detectors and "visual words" correspond to some important characteristics of audio signal. In other words, new specialized visual features can be developed for this particular task.

References

1. Suyoto, I.S.H., Uitdenbogerd, A.L., Scholer, F.: Searching musical audio using symbolic queries. IEEE Transaction on Audio Speech and Language Processing 16(2), 372–381 (2008)

2. Hu, N., Dannenberg, R., Tzanetakis, G.: Polyphonic audio matching and alignment for music retrieval. In: Proc. IEEE WASPAA, New Paltz, NY (2003)
3. Collins, N.: Using a pitch detector for onset detection. In: Proc. of ISMIR 2005, pp. 100–106 (2005)
4. Foote, J., Cooper, M.: Visualizing musical structure and rhythm via self-similarity. In: Proceedings of the 2001 International Computer Music Conference, pp. 419–422. Citeseer (2001)
5. Bello, J., Daudet, L., Abdallah, L., Duxbury, S., Davies, M., Sandler, M.: A tutorial on onset detection in music signals. IEEE Transaction on Speech and Audio Processing 13(5), 1035 (2005)
6. Lowe, D.G.: Object recognition from local scale-invariant features. In: Proceedings of the International Conference on Computer Vision, vol. 2, pp. 1150–1157 (1999)
7. Mikolajczyk, K., Schmidt, C.: A performance evaluation of local descriptors. IEEE Transactions on Pattern Analysis and Machine Intelligence, 1615–1630 (2005)
8. Lindeberg, T.: Scale-space. Encyclopedia of Computer Science and Engineering 4, 2495–2504 (2009)
9. Lopes, A.P.B., De Avila, S.E.F., Peixoto, A.N.A., Oliveira, R.S., Araujo, A.A.: A bag-of-features approach based on hue-SIFT descriptor for nude detection. In: Proceedings of the 17th European Signal Processing Conference, Glasgow, Scotland (2009)
10. Nowak, E., Jurie, F., Triggs, B.: Sampling Strategies for Bag-of-Features Image Classification. In: Leonardis, A., Bischof, H., Pinz, A. (eds.) ECCV 2006. LNCS, vol. 3954, pp. 490–503. Springer, Heidelberg (2006)
11. Yang, J., Jiang, Y.G., Hauptmann, A.G., Ngo, C.W.: Evaluating bag-of-visual-words representations in scene classification. In: Proceedings of the International Workshop on Workshop on Multimedia Information Retrieval, p. 206 (2007)
12. Liu, H., Song, D., Rüger, S.M., Hu, R., Uren, V.S.: Comparing Dissimilarity Measures for Content-Based Image Retrieval. In: Li, H., Liu, T., Ma, W.-Y., Sakai, T., Wong, K.-F., Zhou, G. (eds.) AIRS 2008. LNCS, vol. 4993, pp. 44–50. Springer, Heidelberg (2008)
13. Meng, A., Ahrendt, P., Larsen, J., Hansen, L.K.: Temporal feature integration for music genre classification. IEEE Transactions on Audio, Speech and Language Processing 15(5), 1654–1664 (2007)
14. Shi, J., Tomasi, C.: Good features to track. In: IEEE Conference on Computer Vision and Pattern Recognition, pp. 593–600 (1994)
15. Kaliciak, L., Song, D., Wiratunga, N., Pan, J.: Novel local features with hybrid sampling technique for image retrieval. In: Proceedings of Conference on Information and Knowledge Management (CIKM), pp. 1557–1560 (2010)
16. Serra, X.: Audio Content Processing for Automatic Music Genre Classification: Descriptors, Databases, and Classifiers. Doctoral Dissertation (2009)
   www.tesisenred.net/bitstream/handle/10803/7559/
   tegt.pdf?sequence=1
17. Perrot, D., Gjerdigen, R.: Scanning the Dial: An Exploration of Factors in Identification of Musical Style. In: Proceedings of Social Music Perception Cognition, p. 88 (1999)

Yet Another Sorting-Based Solution to the Reassignment of Document Identifiers

Liang Shi[1,2] and Bin Wang[1]

[1] Institute of Computing Technology, Chinese Academy of Sciences, China
[2] Graduate School of the Chinese Academy of Sciences, Beijing, China
{shiliang,wangbin}@ict.ac.cn

Abstract. Inverted file is generally used in search engines such as Web Search and Library Search, etc. Previous work demonstrated that the compressed size of inverted file can be significantly reduced through the reassignment of document identifiers. There are two main state-of-the-art solutions: URL sorting-based solution, which sorts the documents by the alphabetical order of the URLs; and TSP-based solution, which considers the reassignment as Traveling Salesman Problem. These techniques achieve good compression, while have significant limitations on the URLs and data size. In this paper, we propose an efficient solution to the reassignment problem that first sorts the terms in each document by document frequency and then sorts the documents by the presence of the terms. Our approach has few restrictions on data sets and is applicable to various situations. Experimental results on four public data sets show that compared with the TSP-based approach, our approach reduces the time complexity from $O(n^2)$ to $O(\overline{|D|} \cdot n \log n)$ ($\overline{|D|}$: average length of n documents), while achieving comparative compression ratio; and compared with the URL-sorting based approach, our approach improves the compression ratio up to 10.6% with approximately the same run-time.

Keywords: Inverted File, Index Compression, Reassignment of Document Identifiers, TERM sorting-based.

1 Introduction

Web Search Engines are always facing serious performance challenges with the rapid growth of web data. First, full text index for tens of billions of pages needs to be stored in disk or main memory, which requires a huge amount of storage; second, tens of thousands of queries should be processed in hundreds of milliseconds. Inverted file has been proved to be the best way to index large and variable-length web data compared to signature file and bitmap in [1]. Inverted file is composed of the lexicon and the posting lists. The lexicon contains terms tokenized from the documents. For each term in the lexicon, a posting list is associated containing information about document frequencies, document identifiers, term frequencies and positions. The general form of inverted file is as follows:

$$t_i \rightarrow < f_{t_i}; d_{i1}, d_{i2}, \ldots, d_{if_{t_i}} >$$

(1)

Y. Hou et al. (Eds.): AIRS 2012, LNCS 7675, pp. 238–249, 2012.

where f_{t_i} represents the document frequency of term t_i (number of documents in which it appears), and $d_{i1}, d_{i2}, \ldots, d_{if_{t_i}}$ are identifiers of the documents containing t_i. For simplicity here, we leave out the frequencies and positions in the posting list.

The scale of data has caused a critical dependence on compression of the inverted file. Index compression not only reduces the storage space, but more importantly, gives chance to keep a larger fraction of inverted file in main memory to improve overall query processing throughput[2].

For efficient storage, inverted files use the d-gap representation instead of the original document identifier, because most compression regimes achieve higher compression ratio for smaller integers. We sort the document identifiers (docIDs) in each posting list by ascending order, and then take the differences among successive identifiers (d-gap). Finally, we encode the d-gaps with certain compression regimes introduced in [3,4,5,6,7,8,9].

Meanwhile, researchers propose another approach called Document Identifier Reassignment, to enhance the compression ratio of inverted files. Generally, identifiers are assigned to documents by indexed order, while the docID reassignment is concerned with reassigning docIDs to optimize the distribution of d-gaps and maximize the compression ratio of the resulting inverted file. A good docID assignment stated in [10] is to create locality by assigning consecutive docIDs to similar documents; this produces clusters of many small values interrupted by a few large values and results in better compression.

Prior work showed that compressed index size can be substantially reduced through reassignment of docIDs. The first solution to this problem named B&B is proposed in [10] by first building a similarity graph from inverted index and then recursively splits the whole graph into small subgraphs until every subgraph becomes a singleton. Finally, identifiers are reassigned to documents by depth-first traversal of the graph. Another approach proposed in [11] considers the problem as Traveling Salesman Problem (TSP). This approach tries to find a traversal from the document similarity graph that maximizes the sum of distances (document similarity) between consecutive documents. The maximal traversal gives the optimal order of docIDs.

However, the B&B and TSP-based solutions are limited on small data sets; they operate on a dense graph with $O(n^2)$ edges for n documents. Based on the assumption that similar documents share a large prefix of their URLs, [12] just simply assign docIDs to web pages according to the alphabetical order of the URLs (URL sorting-based). But for documents without URLs (e.g., FBIS or LATimes collections) or that URL similarity is irrelevant to document similarity (e.g., Wikipedia, Twitter), URL sorting-based approach does not apply.

Despite a number of recent publications on this topic, there are still many open challenges. Our goal in this paper is to illustrate a new solution to the reassignment of document identifiers. Our main contributions are as follows:

1. We present another sorting-based approach named TERM sorting that scales to tens of millions of documents. Different from the URL sorting-based approach, we first sort the terms in every document by document frequency, and then sort the documents by the presence of the terms.
2. We study three different ways of sorting the terms in our approach. Experiments show that sorting the terms by descending order of document frequency results in better compression than ascending order and original order.
3. We evaluate our approach with TSP-based approach on the small data sets like TREC FBIS and LATimes. Our experiments show significant improvements in time and space while achieving comparative compression ratio.
4. We compare our approach with URL sorting-based approach on the large data sets like Wikipedia and WT2g. Results show that for a less densely sampled set of web pages (WT2g) or documents for which URL similarity is unrelated to document similarity (Wikipedia), our approach achieves much better compression results.

The rest of this paper is organized as follows: Section 2 provides some technical background and reviews related work. In Section 3 we introduce our TERM sorting-based approach in detail. Section 4 shows the results of our evaluation. Section 5 provides conclusion and future work.

2   Background and Related Work

In this section, we first outline several known index compression regimes used in our later experiments, then we discuss related work on the reassignment of document identifiers.

2.1   Index Compression Techniques

The basic idea of index compression is to compress a sequence of integers (d-gaps) into smaller size. We now provide brief description of some compression regimes to keep the paper self-contained.

Variable-Byte Coding(VB): VB coding [4] represents an integer in variable bytes, where the highest bit of each byte called status bit indicating whether the next byte follows the current one, and the rest 7 bits in binary encoding.

Simple9(S9): The Simple9 coding [13] divides 32-bit word into 4 status bits and 28 data bits, where the 4 status bits can represent 9 different situations of the data bits: 28 1-bit numbers, 14 2-bit numbers, 9 3-bit numbers (1 bit unused) or 7 4-bit numbers and so on. Anh and Moffat extend 32-bit to 64-bit (named **S8b**) in [14] with 4 status bits indicating 16 cases of the 60 bits available for data.

PForDelta(PFD): This compression regime [6] supports fast decompression while also achieving a good compressed size. It first determines a value b so that most of the values to be encoded (say, 90%) are less than 2^b and thus can fit into a fixed bit field of b bits each. The remaining values, called exceptions which are larger than 2^b, are coded separately. If we compress 32 values each time, then we need $(32 \times b)$ bits to store ordinary values, followed by exceptions compressed in fixed-length coding. For the original positions of the exceptions, we store the offset value to the next exception. Ahead of the fixed bit field, there is a value indicating the offset of the first exception.

2.2   Document Identifier Reassignment

The compressed size of the whole inverted file is a function of all the d-gaps being compressed, which greatly depends on how we assign the docIDs.

A common assumption in prior work is that docIDs should be assigned such that similar documents (i.e., documents with a lot of common terms) are close to each other. We classify the related work into three categories: 1) cluster-based, which partitions the documents into clusters by similarity, then assigns consecutive docIDs to documents in the same cluster, 2) TSP-based, which finds the maximal distance by traversing the document similarity graph, then assigns docIDs according to the traversal order and 3) URL sorting-based, which assigns docIDs to documents by the alphabetical order of the URLs.

Cluster-Based: B&B algorithm proposed in [10] starts from a previously built inverted file, and then constructs a document similarity graph (DSG) where the vertices correspond to documents and the edges weighted by the cosine similarity between each pair of documents. The B&B algorithm recursively splits DSG into smaller subgraphs, which represent smaller subsets of the collection, until all subgraphs become singleton. The identifiers are finally reassigned according to the depth-first visit of the resulting tree.

On the basis of B&B, a lightweight k-means-like cluster algorithm is proposed in [15]. It scans the whole collection k times, and at each time chooses the longest document as the center, then adds the remaining documents to the center according to the Jaccard similarity. Finally, we get k clusters with (N/k) documents each. Identifiers are reassigned to the documents according to the order added to these clusters.

TSP-Based: In [11], Shieh et al. proposed an approach based on the Travelling Salesman Problem (TSP) which attempts to find a tour of DSG to maximize the sum of all the travelled edge weights. The experiments show a great improvement of compression ratio.

To improve the efficiency of TSP, Blanco et al. introduced SVD to reduce the dimensionality of DSG in [16]. However, SVD technique is still quadratic in the number of documents.

Ding et al. proposed a new framework in [17] for scaling TSP-based approach by Locality Sensitive Hashing, which obtains a reduced sparse graph. This technique achieves improved compression ratio while scaling to tens of millions of documents. However, for large collections it still takes a long time to hash and sample from the whole DSG.

URL Sorting-Based: This approach is proposed by Silvestri in [12], which simply sorts the documents by alphabetical order of their URLs, and then reassigns docIDs according to the new order. This is the simplest and fastest approach, and performs well for large web collections with almost unlimited scalability.

3   TERM Sorting-Based Approach

3.1   Motivation

In Section 2, we discussed three categories of approaches to the problem of docID reassignment. Of all these approaches, the TSP-based approach achieves the best compression ratio most of the time; however, it is limited to fairly small data sets. The URL sorting-based approach performs not as well, but it can scale to large collections and performs well in many commercial search engines. However, the URL sorting-based approach is only applicable to web pages where the similarity between pages can be inferred from the URLs.

Inspired by the URL sorting-based approach, we propose the TERM sorting-based approach to the reassignment of docIDs which concerns the length of the posting list. The advantages of our solution is as follows: 1) it sorts the documents by terms instead of the URLs and performs better than URL sorting-based approach, 2) it takes much less space and time than TSP-based approach while achieving comparative compression ratio and 3) it is applicable and scalable for large collections.

3.2   Methodology

Given the posting list in Equation (1), we first sort the document identifiers in ascending order and then get it's d-gap form:

$$t_i \rightarrow < f_{t_i}; d_{i1'}, d_{i2'} - d_{i1'}, \ldots, d_{if_t'} - d_{if_{t-1}'} >, d_{ik'} < d_{ij'}, \forall k < j \qquad (2)$$

Since most coding regimes encode an integer x with $O(\log(x))$ bits, let $\phi(t_i)$ be the required bits of t_i's posting list in Equation (2),

$$\phi(t_i) = O(\log(d_{i1'}) + \log(d_{i2'} - d_{i1'}) + \cdots + \log(d_{if_t'} - d_{if_{t-1}'}))$$
$$= O(\log(d_{i1'} \times (d_{i2'} - d_{i1'}) \times \cdots \times (d_{if_t'} - d_{if_{t-1}'}))) \qquad (3)$$

Our goal is to minimize $\sum_{t_i} \phi(t_i)$ by permuting all the documents and finding the best permutation which optimizes the distribution of overall d-gaps and gains most in compression ratio.

The simplest way is to exhaustively permute the N documents and find the best reassignment which achieves the best compression ratio. But the time complexity of this brute force method is $O(N!)$ which is not feasible, especially for a large N. It is proved in [18] that finding the best permutation is an NP-complete problem.

However, from Equation (3) we observe that:

- term t_i with different document frequency (DF) has different impact on the total $\sum_{t_i} \phi(t_i)$
- $\phi(t_i)$ is composed of continued product of d-gaps which would be decreased with more 1-gaps

Based on the above observations, we propose an approximate polynomial-time solution to the docID reassignment problem, named TERM sorting-based solution. At the first step, we sort the terms in every document by DF; and then sort the documents by the presence of terms. Finally, we reassign docIDs according to the new order of the sorted documents.

3.3   Choosing Order Of Terms

As discussed in previous subsection, we need to arrange the terms in every document before sorting the documents. In this paper, we study three different ways of sorting the terms. In the simplest way, we use the original order of terms (TERM-ORIGIN) in every document. Besides, we sort the terms in each document by descending order (TERM-DESC) and ascending order (TERM-ASC) of DF.

TERM-ORIGIN: Simply considered as a random order of terms.

TERM-DESC: Assuming that the terms with larger DF (i.e., longer posting list) have more impact on the compression ratio, documents should be reordered according to the presence of these terms first to generate more 1-gaps in the corresponding posting lists.

TERM-ASC: Just like the above one, but emphasizes on optimizing the shorter posting lists of terms with smaller DF.

For example, given a 4×4 matrix in Fig. 1, we have four terms t_1, t_2, t_3 and t_4 tokenized from documents d_1, d_2, d_3 and d_4. Element e_{ij} in the matrix represents presence($= 1$) or absence($= 0$) of term t_i in document d_j and numbers in brackets indicate the DF of each term. Fig. 2 illustrates the TERM-DESC (*left*) order and TERM-ASC (*right*) order of the terms.

$$
\begin{bmatrix}
 & d_1 & d_2 & d_3 & d_4 \\
t_1(2) & 1 & 1 & 0 & 0 \\
t_2(4) & 1 & 1 & 1 & 1 \\
t_3(2) & 1 & 0 & 0 & 1 \\
t_4(3) & 0 & 1 & 1 & 1
\end{bmatrix}
$$

Fig. 1. A 4×4 matrix indicates the relation between terms t_1, t_2, t_3, t_4 and d_1, d_2, d_3, d_4

$$
\begin{bmatrix}
 & d_1\ d_2\ d_3\ d_4 \\
t_2(4) & 1\ \ 1\ \ 1\ \ 1 \\
t_4(3) & 0\ \ 1\ \ 1\ \ 1 \\
t_1(2) & 1\ \ 1\ \ 0\ \ 0 \\
t_3(2) & 1\ \ 0\ \ 0\ \ 1
\end{bmatrix}
\qquad
\begin{bmatrix}
 & d_1\ d_2\ d_3\ d_4 \\
t_3(2) & 1\ \ 0\ \ 0\ \ 1 \\
t_1(2) & 1\ \ 1\ \ 0\ \ 0 \\
t_4(3) & 0\ \ 1\ \ 1\ \ 1 \\
t_2(4) & 1\ \ 1\ \ 1\ \ 1
\end{bmatrix}
$$

Fig. 2. Sort the terms in descending (*left*) and ascending (*right*) order of DF

3.4   Sorting the Documents

After sorting the terms, we then sort the documents by the presence of terms to generate more 1-gaps, following the observations in subsection 3.2. Consider the matrix in Fig. 1 and Fig. 2, the document sorting results are illustrated in Fig. 3.

$$
\begin{bmatrix}
 & d_1\ d_2\ d_4\ d_3 \\
t_1(2) & 1\ \ 1\ \ 0\ \ 0 \\
t_2(4) & 1\ \ 1\ \ 1\ \ 1 \\
t_3(2) & 1\ \ 0\ \ 1\ \ 0 \\
t_4(3) & 0\ \ 1\ \ 1\ \ 1
\end{bmatrix}
\quad
\begin{bmatrix}
 & d_2\ d_4\ d_3\ d_1 \\
t_2(4) & 1\ \ 1\ \ 1\ \ 1 \\
t_4(3) & 1\ \ 1\ \ 1\ \ 0 \\
t_1(2) & 1\ \ 0\ \ 0\ \ 1 \\
t_3(2) & 0\ \ 1\ \ 0\ \ 1
\end{bmatrix}
\quad
\begin{bmatrix}
 & d_1\ d_4\ d_2\ d_3 \\
t_3(2) & 1\ \ 1\ \ 0\ \ 0 \\
t_1(2) & 1\ \ 0\ \ 1\ \ 0 \\
t_4(3) & 0\ \ 1\ \ 1\ \ 1 \\
t_2(4) & 1\ \ 1\ \ 1\ \ 1
\end{bmatrix}
$$

Fig. 3. Sort the documents by the presence of the terms under three orders: TERM-ORIGIN (*left*), TERM-DESC (*middle*) and TERM-ASC (*right*)

After sorting the documents, we reassign the identifiers to the documents according to the new order. For example, in Fig. 3 we get the new order of documents under TERM-DESC (*middle*): d_2, d_4, d_3 and d_1. During the docID reassignment procedure, new docID $1, 2, 3, 4$ are reassigned to d_2, d_4, d_3 and d_1 accordingly.

3.5   Time Complexity Analysis

Previous studies like TSP-based and B&B approach focused on finding clustering property of the n documents through computing the similarities between documents, which takes $O(n^2)$ time and occupies large additional space for document similarity graph [15,10,11,17].

We avoid the computational bottleneck of constructing document similarity graph by just sorting the terms and the documents in place. Our TERM sorting-based approach consists of two steps: 1) sorting terms and 2) sorting documents.

It is quite difficult to calculate the exact time complexity under real situations. To estimate the time complexity, here we simply assume that all the documents are of the same length $\overline{|D|}$, i.e., average length of all the documents.

At step 1, sorting terms in n document requires $O(n \cdot \overline{|D|} \cdot \log \overline{|D|})$ time on average using quicksort. Step 2 requires $O(n \log n)$ comparisons between documents using quicksort and each comparison takes $O(\overline{|D|})$ time on average, so the time complexity of step 2 is $O(\overline{|D|} \cdot n \log n)$. As the $\overline{|D|}$ is around hundreds and n scales to thousands or millions or even billions in real data, the time complexity of our TERM sorting-based approach is $O(\overline{|D|} \cdot n \log n)$.

4 Experiments

In this section, we evaluate the performance of our TERM sorting-based approach. Our experiments are set up as follows: first, we introduce the experimental setup; then, we compare our approach with TSP-based approach and URL sorting-based approach in both compression size and run-time for the document reassignment procedure.

4.1 Experimental Setup

In our experiments, we use four data sets. Table 1 summarizes the statistics: the size of collection, number of terms, documents and postings.

Table 1. Basic statistics of our data sets

Attribute	Data Sets			
	FBIS	LATimes	WT2g	Wiki
Size	480 MB	475 MB	2 GB	17 GB
Terms	403,401	375,978	1,452,253	11,219,827
Documents	130,471	131,896	247,491	7,567,031
Postings	30,610,632	34,522,364	87,588,058	1,035,989,536

- FBIS: The TREC FBIS collection consists of $130,471$ documents from Foreign Broadcast Information Service.
- LATimes: The TREC LATimes collection consists of $131,896$ documents from the LA Times.
- WT2g: A general Web crawl, used by the TREC 1999 Web track.
- Wiki: This is a dump of English version of Wikipedia taken on April 11, 2012, which consists of about 7.3 GB zipped pages (no talk or user pages). We parse the dump and get about 17 GB raw txt from wiki format.

We ran our experiments on a Xeon 2GHz PC with 128GB of main memory. The operating system was Linux and all the experimental code was written in C++. We didn't introduce stemming or stop words in our experiments. Since our approach explicitly try to optimize docID compression which do not affect much frequency and position compression, we focus on total index size due to docID reassignment throughout this section.

4.2   Reducing in Sum of Logarithm

In this subsection, we sum up the logarithm of the d-gaps in each data set and give a quantitative method-independent evaluation of the benefits that accrue from the docID reassignment. Fig. 4 shows the results in four data sets mentioned above. It is showed that docID reassignment significantly reduces the sum of the logs of the d-gaps when the terms are sorted in descending order. It is worth noticing that docID reassignment does not achieve improvements in FBIS. We think this is because the optimization through reassignment is limited compared to the scale of posting lists.

4.3   Comparing with TSP-Based Approach

In this subsection, we compare five reassignment methods: an original order of docIDs without reassignment (ORIGIN), TSP-based approach (TSP), TERM-ORIGIN, TERM-DESC and TERM-ASC based on our TERM sorting approach.

Table 2 and Table 3 show the compression size and run-times of the five approaches in four compression regimes on FBIS and LATimes.

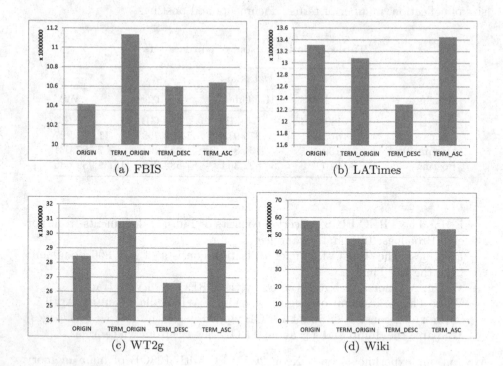

(a) FBIS   (b) LATimes

(c) WT2g   (d) Wiki

Fig. 4. Sum of the logs of the d-gaps on four data sets: (a) FBIS (b) LATimes (c) WT2g (d) Wiki. (ORIGIN - the original docID order without reassignment; TERM-ORIGIN - the sorted DocID order with the terms in original order; TERM-DESC: the sorted DocID order with the terms in descending order; TERM-ASC: the sorted DocID order with the terms in ascending order.)

We show the number of bits per docID in VB, S9, S8b and PFD. The last column shows the run-time to do the reassignment required in each approach. The best reassignment for each compression regime is bolded.

From the experimental results we see that: 1) TERM sorting-based approach performs better with TERM-DESC, 2) TSP gets better compression ratio in VB, S9 while TERM-DESC performs better in S8b and PFD and 3) TSP takes about 22 hours for FBIS and 24 hours for LATimes, while our TERM sorting-based approaches only need about 5 seconds and 8 seconds respectively.

Table 2. Bits per docID and run-times of ORIGIN, TSP, TERM-ORIGIN, TERM-DESC and TERM-ASC reassignments in VB, S9, S8b and PFD. (FBIS)

	Method				
	VB	S9	S8b	PFD	Run-time
ORIGIN	9.43	7.60	7.68	7.54	/
TSP	**9.31**	**7.28**	7.49	7.31	22h19m
TERM-ORIGIN	9.49	7.69	7.65	7.39	4.92s
TERM-DESC	9.43	7.39	**7.40**	**7.30**	5.70s
TERM-ASC	9.39	7.63	7.62	7.41	5.78s

Table 3. Bits per docID and run-times of ORIGIN, TSP, TERM-ORIGIN, TERM-DESC and TERM-ASC reassignments in VB, S9, S8b and PFD. (LATimes)

	Method				
	VB	S9	S8b	PFD	Run-time
ORIGIN	9.69	8.06	7.89	7.63	/
TSP	**9.42**	**7.39**	7.48	7.39	24h13m
TERM-ORIGIN	9.61	7.79	7.64	7.44	7.62s
TERM-DESC	9.48	7.44	**7.36**	**7.32**	8.83s
TERM-ASC	9.64	7.94	7.76	7.49	8.80s

To the best of our knowledge, PFD is widely used in many commercial search engines for its fast decoding speed and high compression ratio. Considering the compression ratio and run-time in both reassignment and decoding, we think TERM-DESC is a better choice for FBIS and LATimes.

4.4 Comparing with URL Sorting Approach

In this subsection, we compare our TERM sorting-based approach with URL sorting-based approach (URL) on WT2g and Wiki data sets. (The TSP-based approach is not referred here because it is limited to small data sets.)

Table 4. Bits per docID and run-times of ORIGIN, URL, TERM-ORIGIN, TERM-DESC and TERM-ASC reassignments in VB, S9, S8b and PFD. (WT2g)

	Method				
	VB	S9	S8b	PFD	Run-time
ORIGIN	**9.55**	7.77	8.07	8.21	/
URL	9.57	**7.75**	**8.03**	8.22	29.46s
TERM-ORIGIN	9.69	8.12	8.29	8.19	19.0s
TERM-DESC	9.59	7.86	8.06	**7.91**	26.3s
TERM-ASC	9.71	8.55	8.72	8.31	23.7s

Table 4 shows the results on WT2g data set. As discussed in [17], URL-based approach performs not so well for the less densely sampled set of web pages like WT2g and outperforms the others only in S9 and S8b. Again in PFD, TERM-DESC improves the compression ratio by 3.7%.

Table 5. Bits per docID and run-times of ORIGIN, TSP, TERM-ORIGIN, TERM-DESC and TERM-ASC reassignments in VB, S9, S8b and PFD. (Wiki)

	Method				
	VB	S9	S8b	PFD	Run-time
ORIGIN	11.11	10.66	10.01	9.75	/
URL	10.57	9.59	9.20	9.09	10m15s
TERM-ORIGIN	10.49	9.27	8.84	8.70	9m46s
TERM-DESC	**10.31**	**8.67**	**8.32**	8.32	10m12s
TERM-ASC	10.78	10.11	9.51	9.13	10m16s

Table 5 shows the results on Wiki data set. From this table, we see that TERM-DESC improves the compression ratio by 2.5%, 9.6%, 10.6% and 8.5% in VB, S9, S8b and PFD respectively, compared to the URL-sorting based method.

5   Conclusion and Future Work

In this paper, we propose the TERM sorting-based approach to the reassignment of document identifiers. We compare our approach with the TSP-based approach and show that the compression ratio of our approach is comparative with the TSP, with run-time significantly decreased. Compared with the URL sorting-based approach, our approach improves the compression ratio up to 10.6%.

However, there are still several open questions for future research. First, the order of the terms still needs further study. Second, current studies focus on maximizing the number of small gaps, and it is an open question on how to minimize the compressed size of inverted file which also contains term frequencies and positions. We leave this to our future work.

Acknowledgment. This work is supported by the National Science Foundation of China under Grant No. 61070111 and the Strategic Priority Research Program of the Chinese Academy of Sciences under Grant No. XDA06030200.

References

1. Witten, I.H., Moffat, A., Bell, T.C.: Managing Gigabytes: Compressing and Indexing Documents and Images, 2nd edn. Morgan Kaufmann, San Francisco (1999)
2. Scholer, F., Williams, H.E., Yiannis, J., Zobel, J.: Compression of inverted indexes for fast query evaluation. In: SIGIR, pp. 222–229. ACM (2002)
3. Moffat, A., Stuiver, L.: Binary interpolative coding for effective index compression. Inf. Retr. 3(1), 25–47 (2000)
4. Williams, H.E., Zobel, J.: Compressing integers for fast file access. Comput. J. 42(3), 193–201 (1999)
5. Yan, H., Ding, S., Suel, T.: Inverted index compression and query processing with optimized document ordering. In: WWW, pp. 401–410. ACM (2009)
6. Zukowski, M., Héman, S., Nes, N., Boncz, P.: Super-scalar ram-cpu cache compression. In: ICDE, p. 59. IEEE Computer Society Press (2006)
7. Elias, P.: Universal codeword sets and representations of the integers. IEEE Transactions Information Theory 21(2), 194–203 (1975)
8. Silvestri, F., Venturini, R.: Vsencoding: efficient coding and fast decoding of integer lists via dynamic programming. In: CIKM, pp. 1219–1228. ACM (2010)
9. Rice, R., Plaunt, J.: Adaptive variable-length coding for efficient compression of spacecraft television data. IEEE Transactions Communication Technology 19(6), 889–897 (1971)
10. Blandford, D.K., Blelloch, G.E.: Index compression through document reordering. In: DCC, pp. 342–351. IEEE Computer Society (2002)
11. Shieh, W.-Y., Chen, T.-F., Shann, J.J.-J., Chung, C.-P.: Inverted file compression through document identifier reassignment. Inf. Process. Manage. 39(1), 117–131 (2003)
12. Silvestri, F.: Sorting Out the Document Identifier Assignment Problem. In: Amati, G., Carpineto, C., Romano, G. (eds.) ECIR 2007. LNCS, vol. 4425, pp. 101–112. Springer, Heidelberg (2007)
13. Anh, V.N., Moffat, A.: Inverted index compression using word-aligned binary codes. Inf. Retr. 8(1), 151–166 (2005)
14. Anh, V.N., Moffat, A.: Index compression using 64-bit words. Software: Practice and Experience 40(2), 131–147 (2010)
15. Silvestri, F., Orlando, S., Perego, R.: Assigning identifiers to documents to enhance the clustering property of fulltext indexes. In: SIGIR, pp. 305–312. ACM (2004)
16. Blanco, R., Barreiro, A.: Document Identifier Reassignment Through Dimensionality Reduction. In: Losada, D.E., Fernández-Luna, J.M. (eds.) ECIR 2005. LNCS, vol. 3408, pp. 375–387. Springer, Heidelberg (2005)
17. Ding, S., Attenberg, J., Suel, T.: Scalable techniques for document identifier assignment ininverted indexes. In: WWW, pp. 311–320. ACM (2010)
18. Blanco, R., Barreiro, A.: Characterization of a simple case of the reassignment of document identifiers as a pattern sequencing problem. In: SIGIR, pp. 587–588. ACM (2005)

What Snippets Say about Pages in Federated Web Search

Thomas Demeester[1], Dong Nguyen[2],
Dolf Trieschnigg[2], Chris Develder[1], and Djoerd Hiemstra[2]

[1] Ghent University, Ghent, Belgium
{tdmeeste,cdvelder}@intec.ugent.be
[2] University of Twente, Enschede, The Netherlands
{d.nguyen,d.trieschnigg,d.hiemstra}@utwente.nl

Abstract. What is the likelihood that a Web page is considered relevant to a query, given the relevance assessment of the corresponding snippet? Using a new federated IR test collection that contains search results from over a hundred search engines on the internet, we are able to investigate such research questions from a global perspective. Our test collection covers the main Web search engines like Google, Yahoo!, and Bing, as well as a number of smaller search engines dedicated to multimedia, shopping, etc., and as such reflects a realistic Web environment. Using a large set of relevance assessments, we are able to investigate the connection between snippet quality and page relevance. The dataset is strongly inhomogeneous, and although the assessors' consistency is shown to be satisfying, care is required when comparing resources. To this end, a number of probabilistic quantities, based on snippet and page relevance, are introduced and evaluated.

Keywords: Web search, test collection, relevance judgments, federated information retrieval, evaluation, snippet.

1 Introduction

Finding our way around among the vast quantities of data on the Web would be unthinkable without the use of Web search engines. Apart from a limited number of very large search engines that constantly crawl the Web for publicly available data, a large amount of smaller and more focused search engines exist, specialized in specific information goals or data types (e.g., online shopping, news, multimedia, social media).

The goal of Federated Information Retrieval (FIR) [1] is to combine multiple existing search engines into a single search system. With the wide variety of existing resources, including those that are not directly accessible by Web crawlers, federated search on the Web has an enormous potential, but is a huge research challenge all the same. A number of FIR research collections have been created in the past, but they are mostly artificial and do not represent the heterogeneous Web environment, i.e., search engines with different retrieval methods,

Y. Hou et al. (Eds.): AIRS 2012, LNCS 7675, pp. 250–261, 2012.

highly skewed sizes, many types of content, and various ways of composing result snippets.

We created a large dataset for this setting, introduced in Nguyen et al. [2], containing sampled results from 108 search engines on the internet, and relevance judgements for both snippets and pages on a number of test topics. We are convinced that such a dataset can stimulate research on federated Web search, and have therefore made our dataset available to researchers[1].

The goal of this paper is threefold. First, we discuss the relevance judgments for the new dataset. Second, we point out some potential difficulties in Web FIR evaluation due to the non-homogeneous character of the resources. Third, we provide a probabilistic analysis of the relationship between the indicative snippet relevance and the relevance of pages.

After a brief overview of related work, several aspects of the relevance judgments are described, starting with the setup of the test collection and the user study, followed by an analysis of the relevance judgments, and focusing on the relationship between snippets and pages. Finally, some conclusions are formulated.

2 Related Work

FIR has been under investigation for many years. The early work is described in detail by Callan [3], identifying three main tasks. *Resource description* is the task of gathering knowledge about the different collections, *resource selection* deals with selecting a subset of collections that are most likely to return relevant results, and *results merging* deals with integrating the ranked lists from different resources into a single ranked list. An extensive overview of more recent work, including FIR in the Web context, is given by Shokouhi and Li in [1].

Reference data in the form of manually annotated relevance judgments are essential in IR evaluation. These are often being created by means of dedicated assessor panels, e.g., for the Text Retrieval Conference (TREC) collections [4]. An important issue for a reliable evaluation is the consistency of the relevance judgments. Voorhees [5] studied the consistency among different assessors, whereas Scholar [6] looked at the self-consistency of assessors for standard TREC collections. Carterette [7] argued that the evaluation accuracy may be affected by assessor errors, which in turn are more likely to occur when judgments are gathered by means of crowdsourcing. Because of the unknown properties of our collection, we only used reliable assessors.

Most online search engines present their results as a ranked list of snippets, giving a sneak preview of the document's content. Nowadays, these snippets are mostly *query-biased*, i.e., they are extracted from the result document, based on the query terms. The fast generation of result snippets has been studied in the past, e.g. by Turpin [8]. For this paper, we will however look at the snippets from a user's perspective, regardless of the different methods used to create them. To our knowledge, no recent work has been published that investigates the relation between results snippets from various origins, and page relevance.

---

[1] http://www.snipdex.org/datasets

3   Relevance Assessments

3.1   Test Collection Setup

Our primary goal was to create a test collection with similar properties as the actual Web. Therefore, our data was crawled from the Web, using results from actual search engines, each with their own document collection and retrieval algorithms. Between December 2011 and January 2012, we collected data from 108 resources, ranging from large general Web search engines to search engines over small, specific collections. These resources cover a broad range of application domains and data formats, and we divided them into 12 categories. Besides General Web Search engines, we have Multimedia, Academic, News, Shopping, Encyclopedia, Jobs, Blogs, Books... and several other types of resources. Due to space limitations, we have to refer to Table 4 for a complete listing, and to Nguyen et al. [2] for some example resources per category.

The collection consists of a large amount of web data, obtained by query-based sampling and intended for resource selection experiments, and also a large set of relevance judgments for a number of test topics, to be used for the evaluation of retrieval algorithms. This paper focuses on the relevance judgment analysis, but more detailed properties of the complete dataset are given in [2].

3.2   User Study Setup

We decided to obtain relevance judgments for the top 10 results that each search engine returned in response to a number of test topics. With over a hundred search engines, more than a thousand pages would need to be judged for a single query. Alternative strategies requiring less judgments have been studied, e.g., by Carterette [9], but we wanted a reliable overview of the relevance for this highly inhomogeneous collection. Hence, we decided for independent graded relevance judgments. Also, the setting of many different real-life search engines provided an excellent opportunity to study, besides the pages, the snippets as generated by these resources. When assessors were presented with snippet and page simultaneously, a preliminary experiment showed that the page influenced their snippet judgment. Hence, a snippet and the corresponding page had to be judged separately and in that order, but still by the same assessor to guarantee an unbiased analysis. We decided to first gather all the snippet judgments, and then the page judgments, as it allowed to minimize the page annotation effort, as explained below.[2]

In the following, we give an overview of the different aspects of the relevance judgments.

Topics. The judged topics are those from the 2010 TREC Web Track [10], as we preferred to use existing topics designed for the Web context, to ensure an

---

[2] This however does not allow us to investigate how the knowledge of a snippet influences the corresponding page judgement. In fact, based on the insights gained in this study, we will adapt the design of a future user study in several ways, to be able to study such issues.

objective characterization of our collection. These topics are divided into two categories (*ambiguous* and *faceted*), and provided with one general information need, as well as several specific descriptions (for the Web Track diversity task). We only presented the assessors with the query terms and the general information need. Most of the topics are especially suited for general Web search engines, and therefore we pay extra attention to these in our analysis. However, the goal of the test collection is to include the other resource categories as well. Therefore, the judges were instructed to interprete the information need in a broad ('multimedia') sense. Consider, e.g., the query *joints* (a search for information about joints in the human body). A picture or video fragment of human joints, even without further textual data, could be highly relevant. Of course, it is often not possible to interpret a topic from the point of view of all types of search engines. For example, a job offer for an orthopedic surgeon, although related, could not be considered relevant to the query *joints*.

Snippets. For each topic, the snippets were shown one by one and in a random order to the assessors. The title, snippet text, preview (if present), and page url, were displayed for each snippet in a uniform manner, albeit provided by the different web search engines. The goal of the snippet annotation task was to predict whether the corresponding page would be relevant. The following labels were used: No, Unlikely, Maybe, and Sure. The corresponding guidelines are summarized in Table 1. We will call these the *snippet relevance* levels, although strictly speaking they only represent the judge's estimation of the page's relevance given the snippet. Especially the smaller resources often provided less than 10 results per query, and in total only 35.651 snippets had to be judged. About 71% of the snippets were judged only once, but for the snippets of 14 topics we obtained between 2 and 5 judgments, with in total over 53000 snippet judgments being collected.

Pages. For the page judgments, the Web Track 2010 relevance levels and descriptions were used: Non, Rel, HRel, Key, Nav, see Table 1. We presented the judges with a snapshot of each page, as well as the html content. Judging pages appeared much more time-consuming than judging snippets, but we were able to reduce the page annotation effort by two thirds, based on the following assumption. In case none of the annotators had labeled a particular snippet higher than No, the corresponding page was not judged, and by default given the label Non. As such, we are not able to estimate the probability of page relevance for totally non-relevant snippets. Yet, even if one of those snippets would correspond to a relevant page, the user would not have clicked the snippet, and therefore not have visited the page.

Assessors. From the 10 assessors that contributed to the relevance judgments, 3 were external students, who created 55% of the snippet judgments, and 57% of the page judgments. One of them was especially hired for one month, and created almost as many judgments as all the other judges together. Apart from

Table 1. Relevance levels and descriptions for snippets and pages

Snippet judgment guidelines: this snippet's page ...	
No	... is definitely not relevant; you would not click the link
Unlikely	... is probably not relevant; you would not click the link
Maybe	... is probably relevant; you would click the link
Sure	... is definitely relevant; you would click the link
Page judgment guidelines: this page ...	
Non	... provides no useful information on the topic
Rel	... provides minimal information on the topic
HRel	... provides substantial information on the topic
Key	... is dedicated to the topic and is worthy of being a top result
Nav	... represents the intended home page of the query entity

the external students, we had 4 junior IR researchers judge 39% of snippets and pages, and 3 senior IR researchers, who did 6% of the snippets and 4% of the pages. We decided to limit the annotation effort for the pages to 39 topics to be judged once and 11 topics twice. The judgments that will be used throughout this paper, are those for which one and the same person has judged all snippets and all required pages of a full topic, i.e., those 39 topics with a single judge, and 11 topics with two judges. Except for the experiments where 2 assessors are compared, we will include those 11 topics in our analysis by randomly selecting the judgments of only one assessor.

3.3   Relevance Judgment Consistency

In the following paragraphs, we will take a closer look at the consistency of the relevance judgments. For clarity, we introduce the random variables S and P, indicating the snippet label, respectively, page label, taking the values listed in Table 1. In some experiments, we reduced our graded relevance levels to a binary relevance, using different cut-offs. For example, the relevance cut-off $S \geq$ Maybe means all snippets with labels Maybe and Sure are considered relevant, and the cut-off $P \geq$ Key indicates that only pages with labels Key and Nav are considered relevant.

Consistency of Page Judgments per Assessor. Some of the resources (especially general Web search engines) often returned the same urls. This allows to study how consistently each assessor performed the page judgments. In [6], assessor consistency on standard TREC collections was investigated by comparing the relevance for pairs of duplicate documents where at least one document had been judged relevant. There were fractions of 15% to 19% of inconsistent judgments for binary relevance, and between 19% and 24% for ternary relevance. We repeated this experiment, for all (4644) pairs of duplicate urls judged by the same user and with at least one judgment level of Rel or higher. For our five graded relevance levels, we found 24% of differently judged pairs, and for binary relevance judgments (i.e., relevant for labels HRel and above) we found 13%, comparable with the TREC judgments.

In order to obtain consistent page judgments for our analysis further on, we grouped all the judgments for a particular url by the same user, and replaced them with the average judgment level.

Inter-Assessor Consistency. The consistency among pairs of assessors has been investigated for a standard TREC collection [5]. It appeared that despite a considerable disagreement among assessors, the ranking of different IR systems mostly remains independent of which set of relevance judgments is used.

We compared our assessors by splitting up our double assessments into two sets (called set 1 and set 2), using binary relevance to allow comparing with [5]. The results are shown in Table 2. The first shown parameter is the *overlap*, calculated as the size of the intersection of the relevant documents in set 1 and set 2, divided by the size of its union, and averaged over the 11 considered topics. In the case of binary page relevance for labels P≥HRel, the resulting overlap of 0.43 (taking into account all resources) is similar to [5], where mean overlap values from 0.42 to 0.49 were reported. From the overlap at different relevance levels, it appears to be more difficult for an assessor to choose between two higher labels (e.g., HRel and Key), than between two lower labels (such as Non and Rel).

We also calculated the fraction of relevant documents from set 1 that are also labeled as relevant in set 2, and vice versa. The average of these two values is shown in Table 2, for convenience called the *precision*[3]. The result at relevance level P≥HRel, including all resources, is 0.62. This imposes a practical upper bound of 62% precision at a recall of 62% on the performance of retrieval systems that are evaluated with these data, because the assessors only agree up to that level. In [5], a value of 65% was reported over the TREC-4 topics. However, our results are strongly influenced by the different types of resources. For the general Web search engines (WSE) on their own, the consistency between judges is much higher than without WSE, as shown in Table 2. This might be due to the fact that the information needs become less well-defined when interpreted in a broader sense than only for general Web search. Also, for the weaker relevance criterion P≥Rel, comparable to the relevance criterion for old TREC collections, we find a much higher precision of 78%.

In order to get an idea of the consistency in judging snippets, we included the snippet overlap and precision in Table 2. Comparing the snippet level S=Sure with the page level P≥HRel, it appears that for the non-WSE, judging snippets is much more difficult than judging pages. This already became clear during the annotation phase, because of the properties of the content (e.g., videos were reduced to small static previews or not shown at all in the snippets), and because the amount of data the decision is based on, is much smaller in the case of

---

[3] These values can indeed be interpreted as the average of the precision when considering the relevant entries in set 1 as the retrieved set with set 2 as the reference set and vice versa, or, equivalently the average of the recall, by exchanging set 1 and set 2.

Table 2. Average overlap and precision between assessors over pairwise judged queries, for different binary relevance levels, shown for (1) all resources, (2) only general Web search engines (WSE), and (3) without WSE

	relevance	all resources		only WSE		all but WSE	
		overlap	precision	overlap	precision	overlap	precision
pages	≥Rel	0.64	0.78	0.81	0.90	0.56	0.71
	≥HRel	**0.43**	**0.62**	**0.59**	**0.74**	**0.31**	**0.51**
	≥Key	0.34	0.46	0.36	0.48	0.25	0.33
snippets	≥Maybe	0.53	0.72	0.70	0.83	0.45	0.65
	Sure	**0.37**	**0.59**	**0.60**	**0.75**	**0.22**	**0.42**

snippets. These are difficulties that make evaluation of FIR in a realistic Web context a difficult task.

3.4   Distribution and Uniqueness of Relevant Results

Relevance Distribution. Figure 1 gives an overview of the average number of resources per topic that returned a given number of relevant pages, split up into ambiguous and faceted topics. The Nav judgments were merged with the Key judgments, because only a few topics were apt for navigational relevance. On average 8 resources returned at least one Key+Nav result for the ambiguous topics, and 9 resources for the faceted topics. However, only for half of the ambiguous topics there was at most one resource that returned 4 or more Key+Nav results, whereas three resources returned 4 or more Key+Nav results for each of the faceted topics. With respect to the total amount of resources (108), the number of resources that returned relevant results is very small for both types of topics; the relevance distribution is highly skewed, which was to be expected, due to the large variation in size and nature of the resources.

An important characteristic of the test collection is also the distribution of the relevant results among the different types of search engines, see Figure 2. The types of search engines that are best suited for the Web Track topics, are in the first place the WSE, but also search engines that provide information from encyclopedia, books, and blogs, and multimedia search engines.

Uniqueness of Results. Table 3 gives the number of snippets and pages with a high relevance level, returned by the WSE and the multimedia resources (as these provide by far the most relevant results), and in total. The 10 WSE together provide more highly relevant results than the 21 multimedia search engines, and even more than the total of 98 non-WSE. However, after normalizing the urls to a uniform form (by omitting search-engine-specific additions etc.) and counting duplicates only once, the contribution of WSE and multimedia engines is comparable, and in total the non-WSE produced about twice the number of unique highly relevant pages as the WSE. This is a strong argument in favour of FIR on a Web scale. However, when considering the highest relevance level,

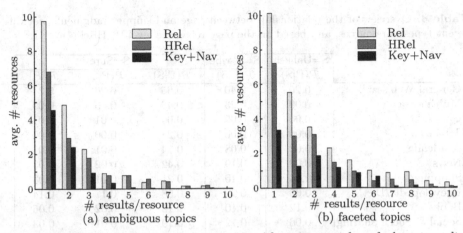

Fig. 1. Average number of resources per topic with a given number of relevant results for the shown relevance levels, for different types of topics

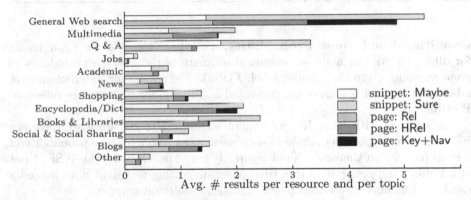

Fig. 2. Average number of resources per topic with a given number of relevant results for the shown relevance levels, for different types of topics

Table 3. Total number of highly relevant snippets and pages

Resources	# snipp. Sure	# pag. ≥HRel	# unique pag. ≥HRel
General Web Search	1932	1626	675
Multimedia	1038	707	701
Total	4237	2973	1950

P≥Key, the WSE engines provide about twice as many unique relevant results compared to the multimedia engines.

3.5 Snippet vs. Page Judgments

This section describes the relation between the relevance labels of snippets and their corresponding pages, judged independently but by the same assessor(s).

Table 4. Overview of the relationship between page and snippet judgments, for different types of resources, and based on the page relevance level P≥HRel

	S=Unlikely	S=Maybe	S=Sure		
	$\mathcal{P}(P\|S)$	$\mathcal{P}(P\|S)$	$\mathcal{P}(P\|S)$	$\mathcal{P}(P,S)$	$\mathcal{P}(P)$
General Web search	0.20	0.40	0.65	0.26	0.34
Multimedia	0.09	0.23	0.48	0.06	0.09
Q & A	0.00	0.00	0.06	0.01	0.01
Jobs	0.00	0.06	0.24	0.00	0.00
Academic	0.03	0.08	0.14	0.01	0.01
News	0.09	0.19	0.42	0.02	0.03
Shopping	0.06	0.10	0.21	0.01	0.03
Encyclopedia/Dict	0.05	0.23	0.58	0.11	0.14
Books	0.12	0.10	0.18	0.02	0.05
Social & Social Sharing	0.06	0.12	0.19	0.01	0.03
Blogs	0.12	0.23	0.40	0.05	0.07
Other	0.04	0.08	0.34	0.01	0.01
All	0.09	0.21	0.50	0.06	0.08

Conditional and Joint Probabilities. Based on the relevance judgments for all topics, we can make an empirical estimate of the average probability of page relevance given the snippet label, $\mathcal{P}(P|S)$. The results for different snippet labels and types of resources are presented in Table 4, for binary page relevance P≥HRel.

For snippets with label No, we cannot estimate $\mathcal{P}(P|S)$, since we have not judged the corresponding pages. However, for most resources, the probability of a HRel page for an Unlikely snippet is already very low. Only for the WSE, 1 out of 5 Unlikely snippets points to a HRel page, suggesting we might have missed a significant number of relevant pages behind non-relevant snippets.

Comparing $\mathcal{P}(P|S)$ for the snippet labels Maybe and Sure shows that for the most suited types of resources, a relatively large amount of HRel pages are behind snippets which were judged only Maybe. This shows that often a HRel page's snippet cannot convince the user that the page is indeed highly relevant.

For snippets labeled Sure, we showed $\mathcal{P}(P|S)$, $\mathcal{P}(P,S)$, and $\mathcal{P}(P)$. The page relevance probability $\mathcal{P}(P)$ by itself gives a false impression, as it does not encorporate any effects of the snippet quality. Instead, we could use $\mathcal{P}(P,S)$, denoting for S=Sure and P≥HRel the joint probability that a snippet is judged as Sure *and* its page is at least HRel, but for most of the listed resource categories it is very low, just like $\mathcal{P}(P)$. We can conclude no more than that these collections probably contain little relevant information on our test topics. This is why we instead consider $\mathcal{P}(P|S)$. It only allows studying how accurately the snippets reflect the page relevance, but is less dependent on the scope of the test topics. For example, the News resources display a relatively high $\mathcal{P}(P|S)$, against a very low $\mathcal{P}(P,S)$. In other words, they returned only very few relevant results for our topics, but if a snippet is found relevant, 4 out of 10 times it points to one of those few relevant results (see Table 4).

Table 5. Average $\mathcal{P}(\text{P}|\text{S})$, for page and snippet judged by different assessors (cross), vs. the same (self)

		S≤Maybe		S=Sure	
		$\mathcal{P}_{\text{self}}(\text{P}|\text{S})$	$\mathcal{P}_{\text{cross}}(\text{P}|\text{S})$	$\mathcal{P}_{\text{self}}(\text{P}|\text{S})$	$\mathcal{P}_{\text{cross}}(\text{P}|\text{S})$
General Web Search	P≥HRel	0.16	0.17	0.67	0.65
	P≥Key	0.04	0.08	0.37	0.29
Multimedia	P≥HRel	0.05	0.05	0.49	0.45
	P≥Key	0.00	0.00	0.07	0.04

Assessor Dependency. An important question is how the relationship $\mathcal{P}(\text{P}|\text{S})$ between snippets and pages would generalize among different users. For example, in a federated search scenario where snippet lists are cached locally and shared among peers, what is the probability that, if a snippet is considered relevant by one user, the corresponding page is relevant to other users? Using the 11 topics with double judgments, we calculated $\mathcal{P}_{\text{cross}}(\text{P}|\text{S})$, the estimated probability of page label P from one assessor given snippet label S from the other assessor (averaged over both directions). In Table 5, these values are shown for different levels of page relevance, together with the average self-probability $\mathcal{P}_{self}$ per assessor. The estimated probability that the page of a relevant snippet is also relevant, would be expected to decrease, if both are not judged by the same assessor. Alternatively, the probability that a page is still relevant, despite a less relevant snippet, is likely to increase in that case. This phenomenon is indeed observed, especially for the stricter page relevance level (P≥Key). This confirms that our precaution, to have one and the same annotator judge corresponding snippets and pages, was necessary.

Comparison between Largest Web Search Engines. As the test topics are best suited for the general Web search engines, we can explicitly compare the performance of four of the largest general Web search engines in our collection, i.e., Google, Yahoo!, Bing, and Baidu, as well as Mamma.com, which is actually a metasearch engine. Table 6 presents the results.

For the snippet label S=Sure and two page relevance levels (P≥HRel and P≥Key), we show $\mathcal{P}(\text{S})$ and $\mathcal{P}(\text{P}|\text{S})$, the estimate that a user makes about the page's relevance based on the snippet alone, and how well the page relevance is linked with that estimate. A search engine should however try to optimize the joint probability $\mathcal{P}(\text{P,S})$. For a better comparison between these resources, we therefore explicitly report $\mathcal{P}(\text{P,S})$ as well. For these resources, we can use $\mathcal{P}(\text{P,S})$ as a measure of comparison, because they have a similar target area (i.e., general Web search). We also showed the more traditional $\mathcal{P}(\text{P})$, which is actually the averaged precision@10 of page relevance, and that is consistently higher than $\mathcal{P}(\text{P,S})$, as it does not take the snippet into account.

The metasearch engine outperforms the others, as it aggregates results from a number of resources, such as Google, Yahoo!, and Bing. We want to stress that the considered test topics are still no representative collection of, for example,

Table 6. Comparison of the largest general Web search engines

	$\mathcal{P}$(S=Sure)	P≥HRel and S=Sure			P≥Key and S=Sure		
		$\mathcal{P}$(P|S)	$\mathcal{P}$(P,S)	$\mathcal{P}$(P)	$\mathcal{P}$(P|S)	$\mathcal{P}$(P,S)	$\mathcal{P}$(P)
Google	0.42	0.68	0.28	0.38	0.39	0.16	0.19
Yahoo!	**0.47**	0.69	**0.32**	**0.44**	0.38	0.18	**0.22**
Bing	0.41	0.60	0.24	0.28	0.30	0.12	0.13
Baidu	0.21	0.43	0.09	0.12	0.23	0.05	0.06
Mamma.com	0.43	**0.73**	0.31	0.41	**0.44**	**0.19**	**0.22**

popular Web queries, and therefore we cannot draw any further conclusions about these search engines beyond the scope of our test collection. Yet, here is another example of how the table might be interpreted, with that in mind. Considering only Key results, we could compare Yahoo! and Bing. Yahoo! seems to score higher for all reported parameters, so either Bing's collection contains a smaller number of relevant results, or Yahoo!'s retrieval algorithms are better tuned for our topics. The lower value of $\mathcal{P}$(P|S) for Bing shows that it has a slightly increased chance that the page for a promising snippet appears less relevant. However, the ratio of $\mathcal{P}$(P,S) and $\mathcal{P}$(P) is higher for Bing than for Yahoo!, indicating that for Yahoo!, its own recall on Key pages will be decreased more due to the quality of the snippets, than for Bing. In fact, we found that $\mathcal{P}$(S=Sure|P≥Key) is 79% for Yahoo!, but 91% for Bing.[4]

4   Conclusions and Future Work

In this paper, we analyzed the relevance judgments for a new federated IR test collection, containing data from over a hundred online search engines with various purposes, such as general Web search, multimedia, academic, books, shopping, etc. It appeared that on average the judgments were created with a level of consistency comparable to that of standard TREC collections, with however a higher consistency for the general Web search engines, as compared to the others. We found the judged test topics from the 2010 Web Track to be biased towards general Web search, leading to a large variation in the fraction of relevant results per resource category. Yet even then, the total number of different relevant results from the non general Web search resources was large enough to confirm the potential interest of federated Web search.

Due to the different scope of the resource categories, an absolute comparison in terms of the empirical probability of relevance for our specific test topics would yield little information. Instead, we discussed the conditional probability of page relevance given the snippet judgment, which allowed a limited but less biased means of comparison. Within the more homogeneous category of general Web

---

[4] We did not elsewhere elaborate on the parameter $\mathcal{P}$(S|P), due to length constraints. However, the reader could approximate it (due to the low number of digits shown), dividing $\mathcal{P}$(P,S) by $\mathcal{P}$(P) in Tables 4 and 6.

search engines, we were able to use the joint probability of snippet and page relevance to compare between the resources. The effect of the snippets varied among the resources, but the joint probability appeared consistently lower than the probability of relevance for only the pages, hence showing that the effect of the snippets cannot be left out when characterizing search engines.

In the future, new query sets will be used, designed to make more use of all the resources under study, including the more specialized ones. The user study will be redesigned, to allow investigating issues like how bad snippets damage recall. We will also investigate the effect of the type and number of topics on the quality of evaluation data for a federated Web search test collection, as well as the potential of resource selection and results merging algorithms.

Acknowledgments. This research was partly supported by the Netherlands Organization for Scientific Research, NWO, grants 639.022.809 and 640.005.002, and partly by the IBBT (Interdisciplinary Institute for Broadband Technology) in Flanders.

References

1. Shokouhi, M., Li, L.: Federated Search. Foundations and Trends in Information Retrieval 5(1), 1–102 (2011)
2. Nguyen, D., Demeester, T., Trieschnigg, D., Hiemstra, D.: Federated Search in the Wild: the Combined Power of over a Hundred Search Engines. In: CIKM 2012. ACM (2012)
3. Callan, J.: Distributed information retrieval. Advances in Information Retrieval 7, 127–150 (2002)
4. Voorhees, E.: TREC: Experiment and Evaluation in Information Retrieval. The MIT Press, Cambridge (2005)
5. Voorhees, E.: Variations in relevance judgments and the measurement of retrieval effectiveness. Information Processing and Management 36, 697–716 (2000)
6. Scholer, F., Turpin, A., Sanderson, M.: Quantifying test collection quality based on the consistency of relevance judgements. In: SIGIR 2011, pp. 1063–1072. ACM (2011)
7. Carterette, B., Soboroff, I.: The effect of assessor error on IR system evaluation. In: SIGIR 2010, pp. 539–546. ACM (2010)
8. Turpin, A., Tsegay, Y., Hawking, D., Williams, H.E.: Fast generation of result snippets in web search. In: SIGIR 2007, pp. 127–134. ACM (2007)
9. Carterette, B., Allan, J., Sitaraman, R.: Minimal test collections for retrieval evaluation. In: SIGIR 2006, pp. 268–275. ACM (2006)
10. Clarke, C.L.A., Craswell, N., Soboroff, I., Cormack, G.V.: Overview of the TREC 2010 Web Track. In: TREC, pp. 1–9 (2010)

CoFox: A Synchronous Collaborative Browser

Jesus Rodriguez Perez[1], Teerapong Leelanupab[2], and Joemon M. Jose[1]

[1] School of Computing Science
University of Glasgow
Glasgow, G12 8RZ, United Kingdom
j.rodriguez-perez.1@research.gla.ac.uk, jj@dcs.gla.ac.uk
[2] Faculty of Information Technology
King Mongkut's Institute of Technology Ladkrabang
Bangkok, 10520, Thailand
teerapong@it.kmitl.ac.th

Abstract. Search activities are evolving to new ways and many search activities are conducted collaboratively. This paper introduces the development and evaluation of a synchronous collaborative web browsing system called CoFox. CoFox provides a platform to allow a pair of users to tackle collaborative search tasks. We introduce the architecture of the system and the important components of such a system. A user-centred evaluation was conducted and the results are presented here. We hypothesise that using CoFox to perform collaborative tasks will result in higher user satisfaction, effectiveness and in gaining access to a higher number of relevant results at an earlier state when compared to systems designed for individual web search. The results of a user evaluation carried out by eighteen pairs of users, showed that the hypotheses hold, thus confirming that the system enhances the results achieved during web searching.

1   Introduction

During recent years the Internet has evolved into an increasingly social place, where more and more activities need of the collaboration and views of multiple parties. There are numerous activities which can benefit from the collaboration between users, including: trip planning, document making, researching, shopping, etc. However, web searching has remained largely an individual activity more than a collaborative task.

Most research in collaborative search focuses on implicit collaboration (i.e. users are not explicitly aware of each other). Examples include implicit query suggestion and collaborative filtering. Data extracted from other users' searching sessions can improve future searches and help current users with similar search patterns and information needs. On the contrary, other collaborative systems have opted for an explicit and synchronous approach[2][6][8].

This paper describes the implementation and evaluation of a system for synchronous collaborative web searching, named CoFox. CoFox enables awareness of all the actions performed by a remote search partner through the use of a remote view window. This window shows a live visualisation of the remote user's

Y. Hou et al. (Eds.): AIRS 2012, LNCS 7675, pp. 262–274, 2012.

browser window, which enables the local user to keep track of the other user's steps and also to evaluate together a web page as if they were co-located, through the built-in communication tools. Whilst this functionality is available in other systems such as Coagmento [11], its effects have not yet been properly studied. It has been considered as an extra feature, and not an opportunity to support awareness in a more natural setting. Furthermore CoFox supports effective division of labour by letting the users undertake their own individual searches allowing them to put their views together at any time whilst being aware of the remote user's actions.

A user centred experiment was carried out to evaluate the performance achieved by 18 pairs of users using CoFox in three different tasks and settings. The results show that the collaborative search experience and constant awareness of the friend's actions, provided by CoFox, helps users to work more efficiently when compared to the baseline systems when performing a collaborative task.

1.1 Motivation

There are many existing systems facilitating collaboration through algorithmic mediation [10]. Alternatively, some systems allow users to manually organise results within different containers [4] which are then shared with a group of users. Although these systems have been proven to help users collaborate effectively, a common factor to many is the steep learning curve of new concepts and working practices in order to maximise collaboration effectiveness. The motivation of this study is to explore other alternatives for collaboration. By adapting and combining everyday elements, such as video players, instant messaging clients and a browser history, we wish to faciliate a more natural approach to collaboration, as it is observed in the group information immersion and sharing in physically co-located environments. We propose using video as the main awareness supporting mechanism along with a set of tools to allow for a collaborative activity to take place and we evaluate its effectiveness when compared with individual and co-located web searching.

2 Related Work

Awareness in collaborative browsing refers to the extent that a user is aware of what the other user's thoughts and actions are, so that collaboration can take place. Liechti & Sumi [5] identify four different types of awareness: Group awareness; Workspace awareness; Contextual awareness and Peripheral awareness.

The effects of awareness were studied by Villa et al [12] in the context of multimedia retrieval. They concluded that, while a positive trend was observed, awareness did not necessarily lead to better performance.

Collaborative systems, such as SearchTogether [6], focus on the group dynamics of collaboration. It does so by keeping a common history of the activity of all users. Each user is provided with tools to store, sort and access queries or URLs contained in the history provided by any of the group members. SearchTogether also provides a tool called "split search", which divides the results of a search amongst the users for them to evaluate. A similar system is Coagmento [11], which additionally allows for the sharing of snippets and comments amongst collaborating users within the same project.

CoSense [9] was intended to help ease the sensemaking process when users are engaging in collaborative information seeking activity. CoSense organises all data related to the collaborative session, such as chat messages, URLs, comments associated with the URL, etc. CoSense uses novel information visualization technique to reduce the complexity of data presented to the collaborating users.

Collaborative systems have been also explored in other contexts such as collaborative multimedia retrieval [4]. Adcock and Pickens [1] designed a system for collocated, synchronous collaboration between searchers for video search.

Alternative input and output devices have been extensively studied in collaborative search tools. For example, CoSearch [2] is a system developed for co-located search by exploiting the use of many pointing devices or mobile phones, in order to explore results or to suggest new queries. Other systems such as WeSearch [8] use a table-top multi-touch display to allow a group of co-located users to collaboratively search, share results, discuss and evaluate information.

Whilst some systems allow the view of a remote user's screen, it is not part of the user's search process and it is regarded as a supplemental feature (as in the case of SearchTogether). We believe that a system built around the natural feedback of a video stream taking a primarily visual approach to collaboration will greatly benefit users. Thus we will be supporting mutual awareness through the bidirectional visualization of user screens as a live video stream, building the rest of the system around this visual context.

3   Interface Description

In this section, the collaborative interface is introduced. To support awareness, division of labour and independence, the system includes a series of specialised tools. They have been tailored to resemble standard tools commonly to enhance the user familiarity, diminishing the learning curve. As a result, we expect the users to feel comfortable with the look and feel of the GUI when performing collaborative search tasks.

The Local User Area (LUA. See figure 1, dot 1), The right area of the interface comprehends the work area for the local user which contains a full browser featuring the Gecko rendering engine. It allows the user to perform tab-browsing, search for text on a given page, etc. At the bottom of this window there is a progress bar which indicates the loading of web pages.

The Remote User Area (RUA. See figure 1, dot 2), located at the left side of the interface, it comprehends the tools to enable synchronous collaborative searching. These tools included are: the live video feed of the remote user's window, the shared links tool and the instant messaging tool. The tools provided by the system are now introduced paying attention to the core concepts of collaborative search awareness, division of labour and independence: The remote user's window located in the RUA represents our strategy to support awareness in our

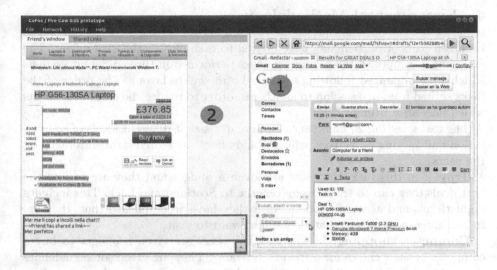

Fig. 1. Screen shot of the system during one session

system. The remote user's window shows what the remote user is currently look-ing at all times as a video stream. This allows the users to know at a glance, what the other user is doing in a straightforward and natural way. For example, this is manifested as the users share information just by highlighting snippets within a web page for the remote user to see, providing a natural way to referring to on-screen information, which resembles finger-pointing on a co-located scenario.

Division of labour is a core concept of collaborative browsing. It refers to the ability of dividing a problem into smaller units so that it can be approached by a number of collaborating users at the same time. Division of labour is supported in CoFox's as follows: In contrast with other systems, where two or more users use a single computer [2] with one or multiple pointing devices, CoFox allows each user to interact independently with their own computer, whilst sharing their findings and ideas in real-time in a similar fashion to SearchTogether [6]. The system has been designed to work specifically in pairs because as stated by [3], over 70% of collaborative searches are carried out in pairs. Moreover, users can use the instant messaging tool to communicate, organise, and evaluate results at the same time that they contrast information through the remote user window. Furthermore, the shared links tool included in the RUA allows users to keep a common history of relevant URLs they consider. Independence is an important factor in CoFox, and it strives to make the collaborative interaction as unobtrusive as possible, whilst still keeping an optimal level of awareness. CoFox allows local users to perform their own searches without interfering or being interfered by the remote user by providing a flexible interface. The users are able to hide the tools provided when they are not using them or resizing the different components to occupy as much of the screen as desired.

4 Experimental Methodology

Eighteen pairs of participants were invited to take part in the experiment. The experiment comprised three different tasks performed in three different systems which were organized following a Greco-Latin square layout. The order of the tasks shifted from one session to another in order to reduce the statistical effects which may arise from learning.

The tasks, lasting 20 minutes each, were devised at different levels of collaboration and varying objectives, which would affect the search strategies taken by the users.

Travel: In this task, the users are given a budget and they are to find the best deals they can to travel from Tenerife to Scotland and back. This includes flight tickets and accommodation. After they have accomplished that, they are to find interesting places that they might want to visit.

Computer shopping: In this task the users are given a budget to buy a computer of their choice for a friend's birthday. They are given some guidelines as to what their friend might like the most but they are free to make any choices.

Egypt research: In this tasks, the users are given some questions about ancient Egypt to which they will have to find the answers on the internet. This task is mostly exploratory since the users are likely not to be familiar with the domain and questions can be tricky to answer.

4.1 Baseline Systems

Two systems (or scenarios) were use as baselines, namely Solo and Together. When working with the Solo baseline system, a pair of users are given the same task to complete it on their own without interaction between them. This baseline system simulates the case in which two people, interested in the same topic, would perform the same search on their respective machines to find relevant information to later discuss about it.

In the case of the Together baseline system, the users are sitting in front of a single machine (co-located). One of the users is interacting with the computer whilst the other is assisting him by suggesting new queries, helping to evaluate results and so on. This baseline system simulates the most common collaborative situation [3], which may take place in contexts such as, trip planning or group study.

4.2 Subjects

The average subject is between 25 and 35 years old, with good knowledge of IT, good web searching skills, which frequently uses instant messaging clients to communicate with his/her friends. Sometimes he/she needs to perform searches along with his friend, in order to fulfil a common information need and he/she feels that current tools do not properly support his collaborative activity, and therefore he/she would have liked a system to assist him.

4.3 Hypotheses

Building upon the expected benefits introduced in the previous section, the following hypotheses have been formulated to evaluate our system:

(H1) CoFox helps users to effectively collaborate when working together to fulfil a common information need;

(H2) CoFox helps users to find more relevant information sooner, when compared to the proposed baseline systems; and

(H3) Users experience higher satisfaction using CoFox for collaborative searching, when compared to the baseline systems.

4.4 Data Collection

To evaluate our hypotheses we collected both qualitative and quantitative data during the experiments. Qualitative information was gathered through the use of questionnaires. For the quantitative data, all the interactions with the system were logged, paying special attention to these events:

LocalBrowser_UrlLoad_Completed: It represents the load of a webpage, as a result from the user clicking on a link. The URLs are stored to be later processed.

LocalBrowser_Query_Captured: This event represents every query issued to a search engine such as Google, Bing, Yahoo, etc. The text submitted as a query is stored as its value.

LocalBrowser_folder_changeTab: This event stores clicking events of switching between tabs.

ChatSent: It stores the text sent through the instant messaging tool to another user.

Finally, a list of relevant URLs was built by using the results submitted by the users in the evaluation.

5 Results and Discussion

This section presents the results obtained during the experiments described in section 4, which are discussed with respect to the hypotheses introduced in section 4.3.

5.1 Effective Collaboration

To compare the effectiveness across systems we can measure the differences with respect to the number relevant pages visited, the overall number of pages visited and the time spent. H1 states that the effectiveness achieved by the users when using CoFox outperforms that of the baseline systems. Table 1 shows a series of summary values per each of the systems averaging over all 18 sessions. By looking at the values of the first column (Rel. Pages), we can see that CoFox outperforms the other systems in terms of total number of relevant pages visited during the search sessions. In the other hand when looking at the total number of pages visited (Tot. Pages), CoFox also leads with over the two baseline systems. Furthermore if we compare the Together baseline against the others it is easy to argue that they are in disadvantage since they only have access to a single machine whereas the other scenarios involve two users in two sepparate machines. In the other hand one could also argue that the time invested by the users performing a task in the Together scenario has yielded lower results in terms of relevant pages and total pages visited, therefore with respect to our metrics the users have wasted more time that in the other cases. In the other hand, when comparing CoFox to the Solo scenario in terms of total and relevant pages, we can appreciate that even though the number to total pages remains very similar, the total number of relevant pages visited is approximately 1.5 times greater than what was achieved in the Solo scenario. This can be attributed to the users being able to organize their search, decreasing the replication of results, and efficiently focusing their efforts to yield better results in the same amount of invested time.

Table 1. Global performance descriptive data

System	Rel. Pages	Tot. Pages	Time	RelPerMin	%
Solo	247	1567	19	13.00	15.76
Cofox	393	1643	17	23.12	23.92
Tog	126	744	12	10.50	16.94

The column "%" of table 1 represents the percentage of relevant pages visited. When using CoFox 23.92% of the pages have been assessed as relevant which is superior to both baseline systems. The average ending times of CoFox and Solo are very similar, whereas Together sessions are noticeably shorter. The apparent reason for this, based on direct observation of the subjects, is related to the roles that the users play in such a situation. One of the users seems to lead in the search as the other user remains mostly passive, which negatively affects the search as the active user gets bored and rushes to finish the task. When looking at the relevant pages visited per minute (RelPerMin), CoFox takes the lead with practically double the relevant pages per minute compared to the Solo system and slightly more than double compared to the Together system. This reaffirms that users are finding more relevant results per unit of time when using CoFox when compared to the other systems.

Table 2 holds qualitative data gathered from the questionnaries. Users where asked to rate their agreement with the statement "CoFox helps me and my search partner to be more effective". The rating average was 6.43 with a verly low standard deviation showing a very high level of agreement. Similarly users rated the organizational and synchronization levels they perceive with their search partners, as well as the awareness about their search partner's actions and thoughts. The results results were very positive with a mean value of 6.14 and 5.86 respectively.

Finally, the users were asked to rank the systems by preference order to perform collaborative tasks. A majority of 64% of the subjects chose CoFox as their

Table 2. Descriptive qualitative data for effectiveness. Range of values: 1 ->Strongly disagree ; 7 ->Strongly agree

Effectiveness perceived			
	Mean	Std Dev.	Std Err.
1. CoFox helps me and my search partner to be more effective.	6.43	0.79	0.15
2. More organised and syncronised when using Co-Fox	6.14	1.25	0.23
3. When using CoFox I am more aware of my partner's actions and thoughts	5.86	1.51	0.28

first option, followed by the baselines systems Together(25%) and Solo(11%). Furthermore, users were asked about their experiences using CoFox, a common argument being that CoFox, allowed them to effectively search on their own machines, whilst still being able to communicate with the other user when needed. Moreover, when performing the task with the Together baseline, they felt that they spent valuable time discussing resulting in a less productive search and exploring less results, compared to CoFox.

Based on the data presented and the users' statements we can assert that CoFox can increase the effectiveness of the search, in terms of effectiveness perceived by the user and overall system performance when carrying out collaborative searches, compared to the baselines.

5.2   Finding Relevant Information

This subsection looks at how the users find the relevant information that they are seeking. H2 states that users find more relevant information at a earlier stage when using CoFox when compared to the baselines. To evaluate this hyphoteses we divide the data in proportional chunks of data with respect to time. Figure 2(a) shows the frequency of url loads per time range. In this figure it can be observed that the differences between CoFox and Solo systems with respect to the quantity of web pages visited is not very significant. This is not the case when we observe the values for the Together baseline system, which yields substantially lower page loads. The reasoning behind is the limitation of this collaborative setting of having only one terminal at the users disposal. Therefore it is intuitive that given any time range, this system scenario would result in significantly less visited pages, whilst the other two systems remain similar with respect to this metric.

Moreover figure 2(b) shows the number of relevant web pages visited in equivalent time ranges. To build this figure we crossreferenced the list mentioned in section 4.4 containing the urls of relevant web pages with all the visited pages for each of the systems. We can observe that the differences become more acute as the time passes. CoFox outperforms the other systems following a bell shaped distribution which falls down after 7.5 minutes as the users compile their answers in the emails. The distributions for Together and Solo baselines are behave similarly. It is imporant to note that even though more queries are issued during the Solo scenario than Together, the performance achieved is very similar.

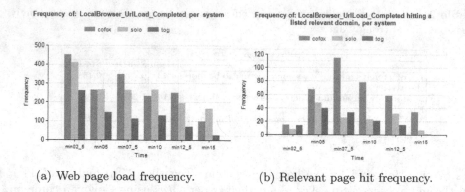

(a) Web page load frequency. (b) Relevant page hit frequency.

Fig. 2. Frenquencies of URL loads and relevant pages visited

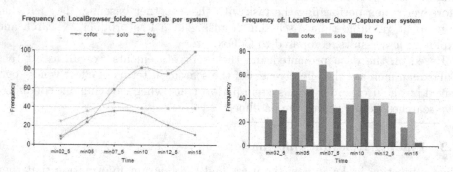

(a) Frequency of changing between tabs (b) Frequencies of queries issued per system

Fig. 3. Frequencies of query issuing and tab changing

5.3 User Satisfaction

Figure 3(b) shows the frequencies of queries issued, divided in ranges of time of 2 minutes and a half. We can observe that the number of queries issued using CoFox during the first 7:30 can be compared to the Together system, except for the first 2 minutes where the users searching with CoFox have issued fewer queries. This is very likely to be due to the initial divison of labour phase by which users organize themselves since it correlates with the Together system.

After minute 7:30 the number of queries issued using CoFox drops significantly, and remains lower than the other systems for the rest of the session. If we also consider figure 2(b), the peak representing the range of time when the higher number of relevant pages are visited correlates with the time before the number of queries drops in figure 3(b). This is evidence of the earlier fulfilment of the information seeking activity by many users when using CoFox, which confirms what the users have qualitatively expressed in the questionnaires.

Another observation is that even as the number of queries issued by users when searching with CoFox drops and becomes lower than in other systems, as

Table 3. User satisfaction descriptive data. Range of values: 1 ->Strongly disagree ; 7 ->Strongly agree

User Satisfaction			
	Mean	Std Dev.	Std Err.
1. Using CoFox is effortless	6.11	1.57	0.30
2. CoFox can be used effectively without instruction	5.96	1.40	0.26
3. CoFox is easy to learn	6.43	0.84	0.16
4. CoFox works the way I want	5.89	1.50	0.28
5. Comfortable and in control using CoFox	6.28	0.88	0.16
6. Comfortable and in control using Together	6.00	1.41	0.27
7. Comfortable and in control using Solo	5.81	1.39	0.27

seen in figure 3(b), the frequency of relevant pages being visited remains higher than any of the baseline systems.

To further evaluate H2 we examine figure 3(a). This figure shows a plot of the frequency associated with changing from own tab to another. Intuitively the pages in the tabs can be considered as relevant as users do not want to get rid of, and they partially fullfil their information need. Therefore the more switching amongst tabs, the more possibly relevant pages are being considered as solutions to the task at hand. In figure 3(a) CoFox follows an incremental distribution where the other systems behave more linearly, suggesting that users are considering more documents towards their final answer. Further evidence can be found near the ending of the tasks the switching between tabs frequency increases rapidly as users revisit the tabs to copy and include the URLs into the solutions they propose for the task. Based on these observations we can evaluate that CoFox helps collaborating users to be more effective when performing a collaborative information seeking task, thus confirming H2.

User satisfaction is measured through the feedback gained from the questionnaries, after the users have performed each of the tasks. Table 3 shows descriptive values extracted from the answers to those questionnaires. When users were asked to rate their level of agreement to the statement: "Using CoFox is effortless" in a likert scale of 7 points, 7 being strongly agree and 1 strongly disagree, the mean of their answers was 6.11 with a std deviation of 1.57. This shows that the users felt really comfortable using CoFox, probably due to the familiarity brought by its user interface. When the users rated the statements: "CoFox can be used effectively without instruction" and "CoFox is easy to learn", the results were also very positive, which shows that CoFox has a very small learning curve. Likewise users also shown a very high level of agreement when asked to rate the statement: "CoFox works the way I want".

Furthermore, the users were asked to rate their agreement with the statement: "I was comfortable and in control of the system at all times" for each of the systems right after they performed one of the given tasks. The means are very high in all the systems, but looking closely to the data, we can observe that the rating for CoFox has a much lower Std Deviation, showing a higher level of agreement amongts users.

To further evaluate the user satisfaction the users were optionally asked to state what they liked and what they disliked about the different systems. For the Solo system the users stated "You can do what you want" and "It is most efficient for most cases ".

The comments for attributed to the Together system include "Slower because there is only one computer"; "The ideas of both users are not developed completely"; "Prefer to speak rather than chat"; "Faster to evaluate results"; "Cannot be as objective as with CoFox"; "Generates more conflict and discussion". Finally the comments arised by the use of CoFox are "Saves time because of division of labour"; "No conflict to use the computer"; "CoFox is more fun to search with"; "Liberty of being able to work yourself and at the same time contributing to the task"; "Allows me to discuss information with someone who is not co-located, so distance is not a problem anymore"; "Better organization than Together"; "It is comfortable to see what the other user is doing" and "Keep control over the computer"

CoFox receives most of the positive comments which can be considered as evidence of a higher user satisfaction. It was observed that CoFox may be less efficient in non-collaborative tasks than the baselines, as some users stated "Solo is Most efficient for most cases", as the system might be an annoyance for non-collaborative tasks. Furthermore a study carried out by Morris [7] shows that a very low number of people engage in collaborative searches in a daily or weekly basis (0.9% and 25.7% respectively), therefore this user is right in stating that the Solo baseline system will be better for most cases.

Assuming a collaborative task is to be carried out by a pair of users, they experience a higher satisfaction using CoFox and the collaborative tools it provides, thus confirming H3.

6   Conclusions

In this paper, we propose a system for paired synchronous collaborative web browsing namely CoFox. CoFox has been designed to support awareness, division of labour and independence. An evaluation was carried out to assess CoFox's effectiveness, overall user satisfaction and capability to support users in finding relevant information, by performing a series of experiments and then comparing the acquired data against the two baseline systems. In the other hand, we also explore the differences in search strategies between the baseline systems and the proposed collaborative system. Through the analysis of the quantitative and qualitative data extracted from the experiments, we demonstrate that CoFox is an effective tool for collaborative searching, assuming that users have a common information need and they are trying to complete a collaborative task. Users manage to build more efficient queries which allows them to find more relevant documents in a smaller timeframe, even thought there is a communication overhead. This study have important practical implications as users over the internet are evolving to be everyday more sociable. Internet browsing is slowly changing from an activity performed individually to a more social and collaborative activity. A possible way to extend this study is to introduce explicit collaborative query suggestion. A pair of users with a common information need, could benefit from being able to keep a summary of the queries that each other have issued to the search engine as well as the pages visited as a result of each of the queries. Suggestions could be drawn from the comparison of query histories so users can benefit from their partner's search or perform a more efficient division of labour. Another possible way to extend this study is to create the infrastructure to allow more users to effectively take part in the search. This represents a challenge since awareness supported through a video stream between more than 2 users becomes tricky due to space constrains. Therefore a balance in the amount of information shown needs to be achieved to avoid cluttering the interface and unnecessarily distracting the users.

Acknowledgments. This research is partially supported by the EU funded project LiMoSINe (288024)

References

1. Adcock, J., Pickens, J.: Fxpal collaborative exploratory video search system. In: Proceedings of the 2008 International Conference on Content-Based Image and Video Retrieval, pp. 551–552. ACM (2008)
2. Amershi, S., Morris, M.: Cosearch: a system for co-located collaborative web search. In: Proceeding of the Twenty-Sixth Annual SIGCHI Conference on Human Factors in Computing Systems, pp. 1647–1656. ACM (2008)
3. Amershi, S., Morris, M.: Co-located collaborative web search: understanding status quo practices. In: Proceedings of the 27th International Conference Extended Abstracts on Human Factors in Computing Systems, pp. 3637–3642. ACM (2009)
4. Halvey, M., Vallet, D., Hannah, D., Jose, J.: Vigor: a grouping oriented interface for search and retrieval in video libraries. In: Proceedings of the 9th ACM/IEEE-CS Joint Conference on Digital Libraries, pp. 87–96. ACM (2009)
5. Liechti, O.: Awareness and the www: an overview. ACM SIGGROUP Bulletin 21(3), 3–12 (2000)
6. Morris, M., Horvitz, E.: Searchtogether: an interface for collaborative web search. In: Proceedings of the 20th Annual ACM Symposium on User Interface Software and Technology, pp. 3–12. ACM (2007)
7. Morris, M.R.: A survey of collaborative web search practices. In: Proceeding of the Twenty-Sixth Annual SIGCHI Conference on Human Factors in Computing Systems, CHI 2008, pp. 1657–1660. ACM, New York (2008)
8. Morris, M.R., Lombardo, J., Wigdor, D.: Wesearch: supporting collaborative search and sensemaking on a tabletop display. In: Proceedings of the 2010 ACM Conference on Computer Supported Cooperative Work, CSCW 2010, pp. 401–410. ACM, New York (2010)
9. Paul, S., Morris, M.: Cosense: enhancing sensemaking for collaborative web search. In: Proceedings of the 27th International Conference on Human Factors in Computing Systems, pp. 1771–1780. ACM (2009)

10. Pickens, J., Golovchinsky, G., Shah, C., Qvarfordt, P., Back, M.: Algorithmic mediation for collaborative exploratory search. In: 31st Annual International ACM SIGIR Conference on Research and Development in Information Retrieval, SIGIR 2008, pp. 315–322. ACM, New York (2008)
11. Shah, C.: Coagmento-a collaborative information seeking, synthesis and sensemaking framework. In: Integrated demo at CSCW, pp. 6–11 (2010)
12. Villa, R., Gildea, N., Jose, J.M.: A study of awareness in multimedia search. In: Proceedings of the 8th ACM/IEEE-CS Joint Conference on Digital Libraries (2008)

FISER: An Effective Method for
Detecting Interactions between Topic Persons

Yung-Chun Chang[1,2], Pi-Hua Chuang[1], Chien Chin Chen[1], and Wen-Lian Hsu[2]

[1] Department of Information Management, National Taiwan University
No. 1, Sec. 4, Roosevelt Rd., Taipei City 10617, Taiwan (R.O.C)
[2] Institute of Information Science, Academia Simica
No. 128, Sec. 2, Academia Rd., Taipei City 11529, Taiwan (R.O.C)
{changyc,hsu}@iis.sinica.edu.tw, r99725008@ntu.edu.tw,
paton@im.ntu.edu.tw

Abstract. Discovering the interactions between the persons mentioned in a set of topic documents can help readers construct the background of the topic and facilitate document comprehension. To discover person interactions, we need a detection method that can identify text segments containing information about the interactions. Information extraction algorithms then analyze the segments to extract interaction tuples and construct an interaction network of topic persons. In this paper, we define interaction detection as a classification problem. The proposed interaction detection method, called FISER, exploits nineteen features covering syntactic, context-dependent, and semantic information in text to detect interactive segments in topic documents. Empirical evaluations demonstrate the efficacy of FISER, and show that it significantly outperforms many well-known Open IE methods.

Keywords: Topic Person Interaction, Information Extraction, Interactive Segment.

1   Introduction

The Web has become an abundant source of information because of the prevalence of Web2.0, and Internet users can express their opinions about topics easily through various collaborative tools, such as weblogs. Published documents provide a comprehensive view of a topic, but readers are often overwhelmed by large number of topic documents. To help readers comprehend numerous topic documents, several topic mining methods have been proposed. For instance, Chen and Chen [4] summarized the incidents of a topic timeline to help readers understand the story of a topic quickly. Basically, a topic is associated with specific times, places, and persons [11]. Discovering the interactions between the persons can help readers construct the background of the topic and facilitate document comprehension. According to [12], interaction is a kind of human behavior that makes people take each other into account or have a reciprocal influence on each other. Examples of person interactions include compliment, criticism, collaboration, and competition. The discovery of topic person interactions involves two key tasks, namely *interaction detection* and

Y. Hou et al. (Eds.): AIRS 2012, LNCS 7675, pp. 275–285, 2012.
© Springer-Verlag Berlin Heidelberg 2012

interaction extraction. Interaction detection first partitions topic documents into segments and identifies the segments that convey possible interactions between persons. Then, interaction extraction applies an information extraction algorithm to extract interaction tuples from the interactive segments. In this paper, we investigate interaction detection. In contrast to Open IE research [1], which focuses on discovering static and permanent relations (e.g., *capital-of*) between entities, the interaction relations we investigate are dynamic and topic-dependent. To identify dynamic interactions between persons, we define interaction detection as a classification problem. We also propose an effective interaction detection method, called FISER (Feature-based Interactive SEgment Recognizer), which employs nineteen features cover syntactic, context-dependent, and semantic information in text to detect interactive segments in topic documents. Our experiment results show that FISER can identify interactive segments accurately and the proposed features outperform those of well-known Open IE systems dramatically.

The remainder of this article is organized as follows. In Section 2, we discuss Open IE and explain how it differs from our research. We describe the proposed FISER method in Section 3, and evaluate its performance in Section 4. Then, in Section 5, we present our conclusions.

2   Related Work

Our research is closely related to Open IE, which is a novel information extraction paradigm proposed by Banko et al. [1]. In Open IE, the objective is to recognize the relations between entities without providing any relation-specific human input. Like our approach, Open IE involves two tasks, namely, *relation detection* and *relation extraction* [5, 9]. In [9], Li et al. demonstrated that relation detection is critical to outputting reliable relation tuples. However, our survey of Open IE literature revealed that most Open IE approaches omit relation detection, or they exploit simple heuristics to detect relation segments. For example, TEXTRUNNER [1] employs six syntactic features to detect relation segments in a text corpus. The drawback with the approaches is that they do not consider text semantics, so they may not perform well in terms of relation detection [5]. In [2, 14], the authors view relation extraction as a sequence labeling problem, and employ conditional random fields (CRFs) [8] to recognize relation expressions. Because the models are trained and tested with relation segments, in practice, a relation detection component is needed to achieve a good relation extraction performance [5, 9]. Zhu et al. [14] proposed a statistical framework, called StatSnowball, to conduct both traditional information extraction and Open IE. The framework employs discriminative Markov logic networks (MLNs) to learn the weights of relation extraction patterns, which are generally linguistic-structure rules and keyword-matching rules.

Our method differs from existing Open IE approaches in a number of respects. First, to the best of our knowledge, all existing Open IE approaches detect static and permanent relations. By contrast, our method detects interactive segments and the interactions between persons are dynamic and topic-dependent. Second, in addition to syntactic features, we devise useful context-dependent and semantic features to detect interactive segments effectively. Finally, most Open IE approaches analyze the text between entities. Our method further considers the contexts before and after person names to enhance the relation detection performance.

3   FISER System

The process of FISER is comprised of three key components, namely, *candidate segment generation, feature extraction,* and *interactive segment recognition.* At present, FISER is designed for Chinese topics. Candidate segment generation extracts important person names from a set of topic documents, and then partitions the documents into candidate segments that may contain information about the interactions between the topic persons. Next, the feature extraction component extracts representative text features from each candidate segment. The features are used by the interactive segment recognition component to classify interactive segments. We discuss each component in detail in the following sub-sections.

3.1   Candidate Segment Generation

Given a topic document d, we first apply the Chinese language parser CKIP AutoTag[1] to decompose the document into a sequence of sentences $S = \{s_1,...,s_k\}$. The parser also breaks a sentence into tokens and tags their parts-of-speech. It also labels the tokens that represent a person's name. In our experiment, we observed that many of the labeled person names rarely occurred in the topic documents and the rank-frequency distribution of person names followed Zipf's law [10]. To discover the interactions between important topic persons, the low frequency person names are excluded and $P = \{p_1,...,p_e\}$ denotes the set of important topic person names. In Chinese, the main constituent of a sentence is a simple phrase [13]. Therefore, two or more consecutive sentences may express a coherent discourse; and an interactive segment may include a number of sentences. In our candidate segment generation algorithm, we consider two types of candidate segments: an *intra-sentential* segment in which the person names appear in the same sentence, and an *inter-sentential* segment in which the person names are distributed among consecutive sentences. Given a topic person name pair (p_i, p_j), the algorithm processes document sentences one by one and considers a sentence as the initial sentence of a candidate segment if it contains person name p_i (p_j). Then, it examines the initial sentence and subsequent sentences until it reaches an end sentence that contains person name p_j (p_i). If the initial sentence is identical to the end sentence, the algorithm generates an intra-sentential candidate segment; otherwise, it generates an inter-sentential candidate segment. However, if a period appears in between the initial and end sentences, we drop the segment because a period indicates the end of a discourse in Chinese. In addition, if p_i (p_j) appears more than once in a candidate segment, we truncate all the sentences before the last p_i (p_j) to make the candidate segment concise. By running all person name pairs over the topic documents, we obtain a candidate segment set $CS = \{cs_1,...,cs_m\}$.

---

[1] http://ckipsvr.iis.sinica.edu.tw/

```
Candidate Segment Generation
INPUT: (pᵢ, pⱼ) – a topic person name pair; S = {s₁,...,sₖ} – a sequence of sentences from a topic
document d.
BEGIN
inCandidate = false
cs = {}
FOR l = 1 TO l = k
   IF sₗ contains pᵢ (pⱼ) && inCandidate == false
      add sₗ into cs and set inCandidate = true
   ELSE IF sₗ contains pᵢ (pⱼ) && inCandidate == true
      cs = {} and add sₗ into csₙ
   ELSE IF sₗ contains pⱼ (pᵢ) && inCandidate == true
      add sₗ into cs, save cs into candidate segment set CS
      set inCandidate = false and csₙ = {}
   ELSE IF inCandidate == true && sₗ has a period
      cs = {} and set inCandidate = false
END FOR
END
```

Fig. 1. Candidate segment generation algorithm

3.2   Interactive Segment Recognizer and Feature Extraction

To recognize interactive segments in CS, we treat interaction detection as a binary classification problem. In this work, we utilize the maximum entropy (ME) classification method [3], which is a logistic regression-based statistical model. Let IS denote that a segment is interactive. ME classifies a candidate segment in terms of the following conditional probability:

$$P(IS \mid cs_l) = \frac{1}{Z(cs_l)} \exp(\sum_j w_j * f_j(IS, cs_l)) \tag{1}$$

$$Z(cs_l) = \exp(\sum_j w_j * f_j(IS, cs_l) + \sum_k w_k * f_k(\neg IS, cs_l)), \tag{2}$$

where f_j is a feature function and w_j is its weight. A feature function indicates a specific condition between IS and cs_l. $Z(cs_l)$ is a smoothing factor that is used to normalize $P(IS|cs_l)$ within the range [0,1]. Given a training dataset, the weights of the feature functions can be derived appropriately by the conditional maximum likelihood estimation method. The learned weights are then used by Eq. 1 to detect interactive segments. Generally, interactions between entities are described by verbs [7], but not all verbs express interactions. For instance, the candidate segment cs_1 shown below contains more than one verb; however, the interaction between the given person names 胡錦濤(Hu Jintao) and 歐巴馬(Barack Obama) is not described by a verb. While the verb 審問(interrogated) in cs_2 indicates repulsion between the given person names 馬英九(Ma Ying-jeou) and 蔡英文(Tsai Ing-wen), the segment contains another important topic person name, 宋楚瑜(James Soong), who is irrelevant to the interaction. Because detecting interactions is so difficult, in addition to syntactic properties, the semantic and context information should be considered to ensure that interaction detection is successful. The following presents the proposed features including syntactic, context-dependent, and semantic information in text.

[cs_1] 中國(Nc)$_{China}$ 國家主席(Na)$_{paramount\ leader}$ 胡錦濤(Nb)$_{Hu\ Jintao}$ 即將(D)$_{will}$ 訪問(VC)$_{visit}$ 美國(Nc)$_{United\ States}$ ，(COMMACATEGORY) 將(D)$_{will}$ 有與(V_2)$_{have}$ 美國(Nc)$_{United\ States}$ 總統(Na)$_{president}$ 歐巴馬(Nb)$_{Barack\ Obama}$ 見面(VA)$_{meet}$ 的(DE) 機會(Na)$_{chance}$ ，(COMMACATEGORY)

(*Hu Jintao, the current Paramount Leader of the People's Republic of China, will visit the U.S., and have a chance to meet Barack Obama, the President of the United States.*)

[cs_2] 馬英九(Nb)$_{Ma\ Ying\text{-}jeou}$ 總統(Na)$_{president}$ 在(P)$_{during}$ 辯論會(Na)$_{debate}$ ，(COMMACATEGORY) 不但(Cbb)$_{not\ only}$ 反駁(VC)$_{retorted}$宋楚瑜(Nb)$_{James\ Soong's}$ 的(DE) 質疑(VE)$_{questions}$，(COMMACATEGORY) 且(Cbb)$_{but\ also}$審問(VC)$_{interrogated}$ 宇昌案爭議(Nb)$_{Yu\text{-}Chang\ controvers}$ ，(COMMACATEGORY) 指責(VC)$_{challenge}$ 蔡英文(Nb)$_{Tsai\ Ing\text{-}wen}$ 的(DE) 道德標準(Na)$_{moral\ standards}$ 。(PERIODCATEGORY)

(*Taiwan President Ma Ying-Jeou not only retorted James Soong's questions during the debate, but also interrogated Tsai Ing-Wen on the Yu-Chang controversy, challenging her moral standards.*)

Syntactic feature set:
- **VERB RATIO (*vr*)**: The ratio of transitive verbs to intransitive verbs for the given person names in a candidate segment.
- **VERB COUNT (*vc*)**: The number of verbs in a candidate segment.
- **VERB COUNT BETWEEN TOPIC PERSONS (*vcp*)**: The number of verbs for the given person names in a candidate segment.
- **SEGMENT LENGTH (*sl*)**: The length of a candidate segment (i.e., the number of tokens).
- **VERB DENSITY (*vd*)**: The ratio of verbs to the length of a candidate segment.
- **SPECIFIC PUNCTUATION (*sp*)**: It is equal to 1 if the punctuation {: ; 、} appears in a candidate segment; otherwise, it is 0.
- **DISTANCE OF TOPIC PERSONS (*dp*)**: The number of tokens in the given person names of a candidate segment; that is, the distance of the given person names.
- **MIDDLE TOPIC PERSON (*mp*)**: It is equal to 1 if person names other than the given person names occur in a candidate segment. For instance, *mp* is 1 for cs_2.
- **INTRA-SENTENTIAL SEGMENT (*iss*)**: It is equal to 1 if a candidate segment is intra-sentential; otherwise, it is 0. For instance, *iss* is 0 for cs_2.
- **FIRST POSITION (*fp*)**: The first position of the given person names in a candidate segment. For instance, *fp* is 1 for cs_2.
- **LAST POSITION (*lp*)**: The last position of the given person names in a candidate segment. For instance, *lp* is 16 for cs_2.

Context-dependent feature set:
- **TRI-WINDOW COUNT (*tc*)**: The number of verbs in the tri-window (i.e., three consecutive tokens) before and after the given person names. For instance, *tc* is 1 for cs_2.
- **INTERACTIVE VERB (*iv*)**: It is equal to 1 if a candidate segment contains a verb on an interactive verb list; otherwise, it is 0. The verb list is compiled by using the log

likelihood ratio (LLR) [10], which is an effective feature selection method. Given a training dataset comprised of interactive and non-interactive segments, LLR calculates the likelihood that the occurrence of a verb in the interactive segments is not random. A verb with a large LLR value is closely associated with the interactive segments. We rank the verbs in the training dataset in terms of their LLR values, and select the top 150 verbs to compile the interactive verb list.

- **INTERACTIVE BIGRAM (*ib*):** It is equal to 1 if a candidate segment contains a bigram of an interactive bigram list; otherwise, it is 0. The bigram list is compiled in a similar way to the verb list by selecting the top 150 bigrams in the training dataset based on their LLR values.

Semantic feature set:

- **SENTIMENT VERB COUNT (*svc*):** This is the number of sentiment verbs in a candidate segment. Intuitively, interactions can occur with positive or negative semantics. For instance, the verb 審問 (interrogated) in cs_2 describes criticism between the given person names, and it is a sentiment verb with negative semantics. Here, we employ the NTU Chinese Sentiment Dictionary (NTUS)[2], which contains 2812 positive and 8276 negative Chinese sentiment verbs compiled by linguistic experts.

- **NEGATIVE ADVERB COUNT (*nac*):** The number of negative adverbs (e.g., 未曾 (have never)) in a candidate segment.

- **INTERACTIVE SEMEME (*is*):** It is equal to 1 if a sememe of a verb in a candidate segment is on an interactive sememe list; otherwise, it is 0. A sememe is a semantic primitive of a word defined by E-HowNet [6], which is a Chinese lexicon compiled by Chinese linguistic experts. Basically, an interaction can be described by different synonyms. By considering the sememes of the verbs in a candidate segment, we may increase the chances of detecting interactions. For each sememe in E-HowNet, we compute its information gain [10] in discriminating the interactive and non-interactive segments of the training dataset. However, a sememe with a high information gain can be an indicator of non-interactive segments. Therefore, we process sememes one by one according to the order of their information gains. We compute the frequency that a sememe occurs in the interactive segments. If the sememe tends to occur in the interactive segments, we regard it as an interactive sememe; otherwise, it is a non-interactive sememe. We compile the interactive sememe list by selecting the first 150 interaction sememes.

- **NON-INTERACTIVE SEMEME (*ns*):** It is equal to 1 if a sememe of the verbs in a candidate segment is on a non-interactive sememe list; otherwise, it is 0. Similar to *is* the non-interactive sememe list is compiled by selecting the first 150 non-interactive sememes in the training dataset.

- **FREQUENT SEMEME (*fs*):** It is equal to 1 if a sememe of the verbs in a candidate segment is on a frequent sememe list; otherwise, it is 0. We rank the sememes of verbs according to their occurrences in the interactive segments of the training dataset. The frequent sememe list is compiled by selecting the first 150 frequent sememes.

---

[2] http://nlg18.csie.ntu.edu.tw:8080/opinion/publ.html

4   Performance Evaluation

In this section, we present the data corpus used for the performance evaluation; examine the effects of the proposed features; and compare the proposed feature set with those of well-known Open IE methods.

4.1   Data Collection

In information extraction, evaluations are normally based on official corpora. Most previous information extraction studies used the Automatic Context Extraction (ACE) datasets[3] to evaluate system performance. However, the relations (e.g., *capital of*) defined in the datasets are static and therefore irrelevant to interactions between persons. To the best of our knowledge, there is no official corpus for interaction detection; therefore, we compiled our own data corpus for the performance evaluations. Table 1 shows the statistics of the data corpus, which comprises ten political topics in Taiwan from 2004 to 2010. Each topic consists of 50 news documents (all longer than 250 words) downloaded from Google News. As mentioned in Sec. 3, many of the person names labeled by CKIP rarely occur in topic documents. Hence, we selected the first frequent person names whose accumulated frequencies reached 70% of the total person name frequency count in the topic documents. We extracted 1747 candidate segments from the topic documents by using the candidate segment generation algorithm. Then, three experts labeled 455 segments as interactive, and the Kappa statistic of the labeling process was 0.615. The statistic indicates that our annotated data corpus is substantial. It is noteworthy that approximately 45% of the interactive segments are inter-sentential. In other words, expressions of person interactions often cross sentences. Meanwhile, 77% of the intra-sentential segments are non-interactive. Since interaction expresses are rare and sentence-crossing, detecting interactions is difficult.

Table 1. The statistics of data corpus

# of topics	10
# of topic documents	500
# of tagged person names	436
# of evaluated person names	85
# of person name pairs	432
# of interactive segments (intra)	266
# of interactive segments (inter)	189
# of non-interactive segments (intra)	905
# of non-interactive segment (inter)	387

4.2   Effects of the Features

We use 10-fold cross validation to examine the effects of the proposed features. For each evaluation run, a topic and the corresponding candidate segments are selected as

---

[3] http://www.itl.nist.gov/iad/mig/tests/ace/

test data, and the remaining topics are used to train FISER. The results of the 10 evaluation runs are averaged to obtain the global performance. The evaluation metrics are the precision, recall, and F1-score. We use F1 to determine the superiority of the features because it balances the precision and recall scores.

Table 2 shows the performance of the syntactic, context-dependent, and semantic features, denoted as $FISER_{syntactic}$, $FISER_{context-dependent}$, and $FISER_{semantic}$ respectively. As shown in the table, the syntactic features cannot detect interactive segments correctly. This is because they are incapable of discriminating between interactive segments. Since both interactive and non-interactive segments are comprised of grammatical sentences, they have similar syntactic feature values. For instance, the averages of VERB RATIO (vr) for the interactive and non-interactive segments are 2.11 and 2.33 respectively; and there is no significant difference in terms of t-testing with a 99% confidence level. Besides, as mentioned in Sec. 4.1, both intra-sentential and inter-sentential segments can be non-interactive. Therefore, the syntactic INTRA-SENTENTIAL SEGMENT (iss) and SPECIFIC PUNCTUATION (sp) features, which are used to judge inter-sentential segments, are indiscriminative. By contrast, the context-dependent features detect interactive segments successfully. We observe that the compiled interactive verb list and interactive bigram list are closely associated with person interactions, so the INTERACTIVE VERB (iv) and INTERACTIVE BIGRAM (ib) features discriminate interactive segments effectively. Meanwhile, the verbs used to describe person interactions tend to occur immediately before or after the given person names. Thus, the context-dependent TRI-WINDOW COUNT (tc) feature is useful for filtering out non-interactive segments. It is noteworthy that the semantic features produce a high precision rate, but a low recall rate. Our analysis of the experimental data showed that segments containing positive or negative verbs generally reveal person interactions; hence, the semantic feature SENTIMENT VERB COUNT (svc) yields high detection precision. However, a significant proportion of the interactive segments do not have sentimental semantics, so the feature cannot increase the detection recall rate. While the semantic features INTERACTIVE SEMEME (is), NON-INTERACTIVE SEMEME (ns), and FREQUENT SEMEME (fs) try to increase the detection of interactive segments by considering the sememes of verbs, the expert-compiled E-HowNet is not comprehensive enough to identify various person interactions. Notably, FISER achieves its best performance when all the features are applied together (denoted as $FISER_{all}$). In other words, the context-dependent features and semantic features do not conflict with each other.

Table 2. Experimental result of each feature category

Features Category	Precision	Recall	F1-score
$FISER_{syntactic}$	19.4%	1.1%	2.0%
$FISER_{context-dependent}$	66.3%	41.5%	49.3%
$FISER_{semantic}$	66.7%	16.9%	26.2%
$FISER_{all}$	70.5%	51.9%	58.6%

4.3   Comparison with Open IE Methods

Since interaction detection is an innovative research issue, we hardly find systems for comparisons. Nevertheless, our research is closely related to the relation detection task of Open IE. Hence, we compare FISER with three well-known Open IE methods, namely, TEXTRUNNER [1], O-CRF [2], and StatSnowball [14]. TEXTRUNNER, the first Open IE system, employs six syntactic features to extract the relations between entities. O-CRF considers syntactic features, including POS tags, conjunctions, and regular expressions of syntax. It also uses context words to identify relation keywords between entities. StatSnowball also adopts syntactic features to identify relation keywords between entities. The selected features include POS tags and occurrences of non-stop words. Notably, O-CRF and StatSnowball, which are designed for relation extraction, extract interaction keywords from a candidate segment in our experiment. Hence, a candidate segment is classified as non-interactive if no interactive keyword is extracted from it. Additionally, to examine the effectiveness of the proposed features in a fair manner, we also train a ME classifier using the features of each compared method and employ the 10-fold cross validation to obtain its global performance.

Table 3. The interaction detection result of compared methods

Features	Precision	Recall	F1-score
TEXTRUNNER	32.7%	2.2%	4.0%
O-CRF	42.1%	8.8%	14.6%
StatSnowball	48.1%	5.5%	9.9%
TEXTRUNNER$_F$	48.8%	34.3%	38.9 %
O-CRF$_F$	53.2%	39.8%	43.5%
StatSnowball$_F$	52.6%	25.2%	32.1%
FISER$_{all}$	70.5%	51.9%	58.6%

As shown in the Table 3, FISER outperforms all the compared methods and feature sets. As the compared methods and feature sets simply use syntactic features, they cannot sense the semantics of person interactions in candidate segments successfully. By contrast, FISER incorporates semantic and context-dependent features, and thus achieves the best precision, recall, and F1 score. O-CRF outperforms StatSnowball and TEXTRUNNER because its feature set considers the context information of a candidate segment. It is interesting to note that O-CRF and StatSnowball are inferior to O-CRF$_F$ and StatSnowball$_F$, and the recall rates of O-CRF and StatSnowball are very low. Basically, O-CRF and StatSnowball employ the CRF model to learn the extraction patterns of interaction keywords. Since the non-interactive segments have no interaction keywords, only the interactive segments of the training data are useful for pattern learning. As shown in Table 1, most of the candidate segments are non-interactive. Thus, the learned extraction patterns cannot detect interactive segments completely, and the recall rates of the methods deteriorate. The outcome corresponds well with the observation in [9] that detecting relation segments is necessary to ensure that extractions of relation keywords are reliable. Based on the experimental results, we conclude that syntactic features cannot detect interactive segments correctly. Existing Open IE studies focus on discovering static and permanent relations between entities. In [2], Banko and Etzioni claim that 86% of relation expressions are in the given entities. However, according to our data corpus, only 56% of the interaction expressions are in the given person names in Chinese. Therefore, the compared methods are inferior in terms of detecting interactive segments.

5   Conclusion and Feature Works

A topic is associated with specific times, places, and persons. Discovering the interactions between the persons would help readers construct the background of the topic and facilitate document comprehension. In this paper, we have proposed an interaction detection method called FISER, which employs nineteen features covering syntactic, context-dependent, and semantic information in text to detect interactive segments in topic documents. Our experiment results demonstrate the efficacy of FISER and show that it outperforms well-known Open IE methods.

In our future work, we will employ sophisticated syntactic features, such as the dependency tree of a sentence, to enhance FISER's syntactic features. We will also investigate using information extraction algorithms to extract interaction tuples from the detected interactive segments and construct an interaction network of topic persons.

References

1. Banko, M., Cafarella, M.J., Soderland, S., Broadhead, M., Etzioni, O.: Open information extraction from the web. In: Proceedings of the 20th International Joint Conference on Artifical Intelligence, pp. 2670–2676 (2007)
2. Banko, M., Etzioni, O.: The tradeoffs between open and traditional relation extraction. In: Proceedings of the 46th Annual Meeting on Association for Computational Linguistics on Human Language Technologies, pp. 28–36 (2008)
3. Berger, A.L., Pietra, V.J.D., Pietra, S.A.D.: A maximum entropy approach to natural language processing. Computational Linguistics 22, 39–71 (1996)
4. Chen, C.C., Chen, M.C.: TSCAN: A content anatomy approach to temporal topic summarization. IEEE Transactions on Knowledge and Data Engineering 24, 170–183 (2012)
5. Hirano, T., Asano, H., Matsuo, Y., Kikui, G.: Recognizing relation expression between named entities based on inherent and context-dependent features of relational words. In: Proceedings of the 23rd International Conference on Computational Linguistics: Posters, pp. 409–417 (2010)
6. Huang, S.L., Chung, Y.S., Chen, K.J.: E-HowNet: the expansion of HowNet. In: Proceedings of the 1st National HowNet Workshop, pp. 10–22 (2008)
7. Kim, M.Y.: Detection of gene interactions based on syntactic relations. Journal of Biomedicine and Biotechnology 2008, 371–380 (2008)
8. Lafferty, J.D., McCallum, A., Pereira, F.C.N.: Conditional random fields: probabilistic models for segmenting and labeling sequence data. In: Proceedings of the 18th International Conference on Machine Learning, pp. 282–289 (2001)

9. Li, W., Zhang, P., Wei, F., Hou, Y., Lu, Q.: A novel feature-based approach to Chinese entity relation extraction. In: Proceedings of the 46th Annual Meeting of the Association for Computational Linguistics on Human Language Technologies, pp. 89–92 (2008)
10. Manning, C.D., Schütze, H.: Foundations of statistical natural language processing, 1st edn. MIT Press, Cambridge (1999)
11. Nallapati, R., Feng, A., Peng, F., Allan, J.: Event threading within news topics. In: Proceedings of the 13th ACM International Conference on Information and Knowledge Management, pp. 446–453 (2004)
12. Vernon, G.M.: Human interaction: an introduction to sociology, 1st edn. Ronald Press Co., New York (1965)
13. Wang, Y.K., Chen, Y.S., Hsu, W.L.: Empirical study of Mandarin Chinese discourse analysis: an event-based approach. In: Proceedings of 10th IEEE International Conference on Tools with Artificial Intelligence, pp. 466–473 (1998)
14. Zhu, J., Nie, Z., Liu, X., Zhang, B., Wen, J.R.: StatSnowball: a statistical approach to extracting entity relationships. In: Proceedings of the 18th International Conference on World Wide Web, pp. 101–110 (2009)

Cross-Lingual Knowledge Discovery: Chinese-to-English Article Linking in Wikipedia

Ling-Xiang Tang[1], Andrew Trotman[2], Shlomo Geva[1], and Yue Xu[1]

[1] Science and Engineering Faculty, Queensland University of Technology,
Brisbane, Australia
{l4.tang,s.geva,yue.xu}@qut.edu.au
[2] Department of Computer Science, University of Otago,
Dunedin, New Zealand
andrew@cs.otago.ac.nz

Abstract. In this paper we examine automated Chinese to English link discovery in Wikipedia and the effects of Chinese segmentation and Chinese to English translation on the hyperlink recommendation. Our experimental results show that the implemented link discovery framework can effectively recommend Chinese-to-English cross-lingual links. The techniques described here can assist bi-lingual users where a particular topic is not covered in Chinese, is not equally covered in both languages, or is biased in one language; as well as for language learning.

Keywords: Wikipedia, Cross-lingual Link Discovery, Link Mining, Anchor Identification, Link Recommendation, Chinese Segmentation, Translation.

1 Introduction

Wikipedia is the largest multi-lingual encyclopaedia online with over ten million articles in almost every written language. However, knowledge in Wikipedia could have boundaries because of language barriers. The anchored links in Wikipedia articles are mainly created within the same language domain. Knowledge sharing and discovery are impeded by the absence of links between different language domains. Users are forced to use one language version of the resource and are not easily able to switch languages where appropriate. A user may prefer multiple explanations, or just the one in their preferred language, or the richer content, or to extend their understanding of a language through reading translations.

For example, in Hong Kong the word 花蟹 ("flower crab") is colloquial for the ten-dollar note. There are, indeed, 花蟹 entries in both Chinese and English Wikipedia but they are not linked to each other. Fig. 1 shows English and Chinese Wikipedia pages on the Hong Kong ten-dollar note. From the figure, it can be seen that there should be bi-directional language links, but that they have not yet been created. The boxed texts in the Chinese page could be used to further generate anchored links for bi-lingual users to explore those anchors' English counterparts.

Y. Hou et al. (Eds.): AIRS 2012, LNCS 7675, pp. 286–295, 2012.

Previous studies of link discovery between documents in different languages include the followings. Sorg & Cimiano [1] tackle the German and English Wikipedia language-link problem using a classification-based approach. Their study particularly examines missing language-links between Wikipedia articles on the same topic. Melo & Weikum [2] do the opposite, they examine incorrect Wikipedia language-links between articles on the same topic. In the NTICR Crosslink task, Fahrni *et al.* [3] implemented a CLLD system using a graph-based method for disambiguation and achieved very good results; the Kim & Gurevych approach performed the best in linking English documents to Chinese when measured with manual assessment results.

In this paper, we focus on the realisation of efficient and effective automated Chinese to English link discovery in Wikipedia and study the effects of Chinese segmentation and Chinese to English translation on the hyperlink recommendation.

Fig. 1. The Wikipedia pages on "flower crab"

2   Chinese / English Wikipedia

2.1   Corpora Information

Dumps of the Chinese and English Wikipedia taken in June 2010 were converted into files marked up using the YAWN system [1]. After conversion, there were 3,484,250 properly formatted English articles and 316,251 properly formatted Chinese articles. In the collection, just over half of the Chinese articles (170,637), but only 5% of the English articles (169,974), were language cross-linked.

2.2   Links In Wikipedia

Language Link. Wikipedia of different languages is connected through links between articles on the same topic with a single page-to-page language link. Those language links can be used to produce Chinese / English title mapping table T_{lang}. This table can be utilised as a dictionary for translation which will be discussed in section 4.2.

Anchored Mono-Lingual Links. There are nearly 8 million (mostly mono-lingual) links in the Chinese corpus; and around 90 million links in the English corpus. From each corpus, a link table T_{link} ($T_{link\text{-}chinese}$ for Chinese and $T_{link\text{-}english}$ for English) can be mined. All T_{link} tables contain a list of linked documents each with a unique id, a link frequency (lf), and a document frequency (df). The usage of link information mined from the corpora will be discussed in the next section. Several entries taken from T_{lang} and $T_{link\text{-}chinese}$ are showed in Table 1.

Table 1. Extracts from T_{lang} and $T_{link\text{-}chinese}$

T_{lang}		$T_{link\text{-}chinese}$			
Title (zh)	Title (en)	Title (zh)	ID	lf	df
花旗银行	Citibank	花旗银行	53090	42	46
椰子蟹	Coconut crab	椰子蟹	536691	10	10
米高佐敦	Michael Jordan	英国	39793	5212	6866

3   Linking Chinese to English

3.1   Chinese Natural Language Processing

Study of both English mono-lingual and English-to-Chinese document linking has been covered by the recent research on link discovery [4, 5]. However, linking Chinese documents to English still has certain unique problems that need to be addressed. To the best of our knowledge, there are no published research papers that address this.

Chinese Wikipedia is a collaborative effort of contributors from different Chinese spoken geographic areas with different knowledge backgrounds and language variations. They cite modern and ancient sources combining simplified and traditional Chinese text, as well as regional variants. Therefore, in order to link Chinese documents to English documents while considering the linguistic complexity in the Chinese Wikipedia articles, it is necessary to break the Chinese text into separate words (to segment the text). Chinese segmentation breaks long strings of characters into n-gram words. It is presumed that this is a particularly critical step in Chinese-to-English cross-lingual link discovery because it affects not only the identification of the anchors but also the ability to translate them into English. The error rate of anchor translation, and translation in general, is dependent on the quality of the segmentation [6].

3.2   Article Linking

The state-of-the-art techniques for document linking have been seen in past studies. For mono-lingual link discovery there are: the Link mining (ML) method [7] and the Page Name Matching (PNM) method [8]. In this paper, it was intended to make use of these two techniques to build a comprehensive, effective Chinese-to-English link discovery framework that can recommend high quality link efficiently between two knowledge domains in different languages.

Link Mining. Link mining method mines the existing links in a single language version of Wikipedia to create a *link table*, T_{link}, of mono-lingual anchor-to-target ($a \rightarrow d$) pairs. From this link table, the probability of any sequence of terms being an anchor can be computed (for pre-existing anchors). Based on the existing link information that is extracted during the mining phase, the best target for an anchor can also be computed. Note that the same anchor text may be linked to different destinations in different instances where it appears and so it is necessary to identify the most likely link.

Itakura & Clarke [7] trawl English Wikipedia and extract all anchor target pairs. They then re-trawl the collection looking for the frequency of the anchor phrases used either as a link or in plain text. From this they compute an anchor weight, γ, the probability that a given phrase is an anchor and linked to a specific target document as follows:

$$\gamma = \frac{number\ of\ pages\ that\ have\ link(a \rightarrow d)}{number\ of\ pages\ that\ have\ text\ of\ anchor(a)} \tag{1}$$

where the numerator is the link frequency, *lf*, of anchor a pointing to document d; and the denominator is the document frequency (*df*) of anchor a in the corpus.

To link documents within Wikipedia or any documents with Wikipedia, Mihalcea & Csomai [9] and Milne & Witten [10] also use a similar method to weight phrases.

Page Name Matching. An alternative approach for link discovery is title matching (also known as name-matching, and entity matching). For mono-lingual link discovery Geva [8] builds a *page title table*, a list of titles of all documents in Wikipedia. For a given document, a list of all possible n-gram substrings are built and then from the list the longest that are also in the page title table are chosen as the anchors. The targets are the documents with the given title.

To use this in Chinese to English link discovery, it is necessary to first construct a table of corresponding English and Chinese documents. Then, for a new Chinese document, identify all substrings that match Chinese document titles as the anchors. The targets are the corresponding English documents.

4   The Proposed Approach

4.1   Anchor Identification

In this work, we use both the anchor weight [7] and the page name matching [8] methods to identify anchors. The reasons are: first, they are very efficient methods, and anchors can be created easily on-the-fly because the title mapping table T_{lang}, and anchor weights (γ scores) of all possible anchor candidates can be pre-mined and pre-computed; second, the recommended anchors are mainly from the anchor pool that they are either article titles that readers might look up or ones that were previously linked by the human editors.

With the anchor weighting method, for a new previously unlinked document all possible n-gram substrings from the document are first computed. For each of these the γ score is looked-up and the anchors sorted by these values. An arbitrary number (based on a threshold, or alternatively a density) of highly ranked links are then chosen. In the case of overlapping anchors, the longest anchor is chosen.

Page name matching has a similar anchor identification process that from the document all possible n-grams that can be found in the Chinese title table are extracted and but then sorted based on the length of title. The rationale for choosing the longer titles – which also proves correct in experiments – is that longer phrase matches are less likely to be coincidental, and longer phrases in text are generally more specific than shorter ones.

The issue with these two anchor identification methods is that without Chinese segmentation anchors may be created for unrelated topics. For example, the following two sentences contain non-Chinese words (underscored) that could be mistakenly linked to the unrelated Chinese articles:

"胸甲骑兵在腓特烈大帝和拿破仑的军队中都扮演过非常重要的角色。"— taken from the Chinese Cuirassier article[1]. In this sentences, the two adjoining characters—中 and 都 means "in" and "both" separately, but together they (中都) are often used as place names (e.g. an old name for *Beijing* city).

4.2   Anchor Translation

Triangulation. One way to use page name matching and link mining approaches is to mine in one language and to identify target documents translated into the second language.

To do this, a table of documents existing in both languages could be used. Such a table, T_{lang}, can be generated from the page-to-page language links present in Wikipedia. This is similar to the translation memory approach that is commonly used in Machine Translation. This is a form of triangulation. An English page is a good target to a Chinese anchor if there exists a link from the anchor to the Chinese document and from the Chinese document to the English document. The relationship of the triangulation is illustrated in Fig. 2.

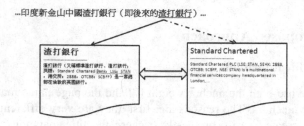

Fig. 2. Cross-lingual triangulation

---

[1] http://zh.wikipedia.org/wiki/胸甲骑兵

Machine Translation. As an addition to the translation with triangulation, Candidate anchors can be translated into English using Google Translate API[2]. Machine translation will be particularly helpful when triangulation fails to provide a proper translation for a high valued anchor and this will be often the case because table T_{lang} is an incomplete set of the mapping of Chinese / English article titles in Wikipedia.

4.3   Link Recommendation

Link recommendation is the final step of our link discovery approach. As in link mining and page name matching methods, all anchor candidates (either from T_{lang} or $T_{link\text{-}chinese}$) already have been associated with a specific target document. So with these two methods, once an anchor is identified, the target document is also determined. So Different anchor identification, translation and final document linking methods will lead to different discovered link sets.

5   Experiments

5.1   Anchor / Link Specification

Although there is no hard limit to the number of anchors that may be inserted into a document, a user will become overwhelmed if almost every term in an article is also an anchor. For evaluation purposes we impose a limit of 50 links per document.

5.2   Evaluation

To simulate Chinese-English cross-lingual linking, we create a set of 36 topics[3] (including 香港十元紙幣 (*Hong Kong ten-dollar note*)), then mine the remaining corpus to generate the two kinds of tables, T_{link} and T_{lang}. With the Wikipedia groundtruth, the Precision-at-N and Link Mean Average Precision (LMAP) metrics employed in NTCIR-9 Crosslink task [5, 11] are used to quantify the performance of the different cross-linking methods.

5.3   Experimental Runs

By combining different translation methods (either triangulation or machine translation) and different anchor weighting strategy (γ score computed using either $T_{link\text{-}chinese}$ or $T_{link\text{-}english}$), the resulting discovered link sets are also different. The runs are outlines in **Table 2**. The segmentation approach proposed by Tang *et al.* [12] was used to complete the segmentation task.

---

[2] http://code.google.com/apis/language/translate/overview.html
[3] http://crosslink.googlecode.com/files/zh-topics-36.zip

Table 2. Experimental runs information

Run Name	Description
LinkProb	Anchor identified with the link table $T_{link\text{-}chinese}$ with link mining method, γ computed with $T_{link\text{-}chinese}$, and target links were identified trough triangulation
PNM	Page name matching through triangulation with T_{lang}
LinkProbEn	Anchor identified with the link table $T_{link\text{-}chinese}$, then with machine translation link probability taken from $T_{link\text{-}english}$
LinkProbEn2	Similar to *LinkProbEn* but final ranking with $T_{link\text{-}chinese}$
LinkProb_S	*LinkProb* run with segmentation
LinkProbEn_S	*LinkProbEn* run with segmentation

Table 3. Performance of experimental runs

Run ID	LMAP	P@5	P@10	P@20	P@50
LinkProb	0.168	0.800	0.694	0.546	0.386
PNM	0.123	0.667	0.567	0.499	0.351
LinkProbEn2	0.095	0.456	0.428	0.338	0.247
LinkProbEn	0.085	0.489	0.394	0.315	0.211
LinkProb_S	0.059	0.411	0.322	0.268	0.201
LinkProbEn_S	0.033	0.233	0.186	0.144	0.118

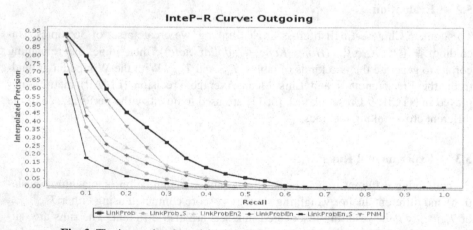

Fig. 3. The interpolated precision/recall curves for the different methods

6   Results and Discussion

The LMAP and P@N scores for the different runs are given in Table 3. Runs are scored on the extracted Wikipedia ground-truth and sorted on LMAP. Precision and recall curves are given in Fig. 3. All runs except *PNM* use the same anchor identification strategy. So, the difference in the performance of those runs can be attributed to the segmentation and translation. Overall, the best performing run, *LinkProb* has the best combination of strategies (and not a different method of choosing anchors).

6.1   Segmentation in CELD

In all cases non-segmented runs out performed the segmented variant of the run. Contrary to intuition, segmentation interferes with anchor identification. This reflects both the non-perfect performance of any segmentation algorithm, and the links themselves being unlikely to be ambiguous in context (because they are named-entities).

There is no doubt that segmentation can increase the accuracy of Chinese text processing, but for link discovery the problem lies in the difficultly of controlling the segmentation granularity for perfect anchor identification. Small granularity will result in small size of words and may help reducing errors of matching the anchors to unrelated topics but may miss out the named entities with compound words; large granularity in segmentation will however cause the exact opposite problem. The extra step for Chinese segmentation in link discovery will increase the computational complexity. Therefore, Chinese segmentation is not absolutely required for Chinese-to-English link discovery if the goal is set to achieve ultimate linking performance.

6.2   Translation in CELD

All runs that used machine translation performed worse than *LinkProb* and *PNM*. Run *LinkProbEn2* and run *LinkProb* indentified the same set of initial candidate Chinese anchors and used the same link ranking strategy. *LinkProbEn2*, however, performs worse than *LinkProb*. This suggests that the performance deteriorates as a consequence of the translation process. A failure analysis of the runs suggests that the problem is caused by translation error. Table 4 lists some of the anchor candidates (column 1) that were incorrectly machine translated (column 2) and the preferred target document seen through link mining (column 3). The failure in translation is similar to that caused by segmentation. Without perfect knowledge of all entities the translation software cannot produce perfect results. Such results cannot be expected because the entity list cannot be closed.

Table 4. Example translation errors in the runs

Anchor	MT	Wiki
資治通鑑	Mirror	Zizhi Tongjian
社稷	Boat	Soil and grain
白骨精	White-Boned Demon	Bai Gu Jing

The result suggests that the mined mapping table T_{lang} used in runs *LinkProb* and *PNM*, is a better translation table than classical machine translation. This is hardly surprising as it is domain specific, and entity list (rather than phrasal text). An alternative we did not test was a combination of the two approaches – using machine translation if an entity could not be translated.

6.3   Chinese-to-English Document Linking

As can be seen from both Table 3 and Fig. 3, run *LinkProb* performed best when scored using LMAP and P@N. Given that the number of candidate links in T_{lang} used by cross-lingual page name matching algorithm is much smaller than $T_{link\text{-}chinese}$ used by link mining method the good performance of *PNM* is surprising but encouraging.

Run *LinkProbEn2* ranked third performing better than *LinkProbEn*. The difference between the two runs was the source of the link probability γ score. In the former the probability came from the Chinese language corpus, but in the latter it came from the English corpus. This suggests that Chinese is a better predictor of which English documents to link to than is English. So the link mining was the best algorithm we tested for Chinese-English cross-language link discovery. As the experiments are the first reported for solving the Chinese-to-English document linking problem, the LMAP and P@N scores of run *LinkProb* are the best results to date.

7   Conclusion and Future Work

In this paper we presented a Chinese to English link discovery framework for automatically identifying anchors in Chinese document that should target documents in English. The experimented Chinese-to-English Cross-linking approach included the use of Chinese word segmentation, Chinese to English translation, and link mining.

Although Chinese segmentation and machine translation are two essential steps in Chinese to other-language information retrieval, our results suggest that they are not needed for link discovery. This is because segmentation is implicit in the anchor mining and the translation is implicit in cross-language triangulation.

The experimental results show that the implemented link discovery framework can effectively recommend Chinese-to-English cross-lingual links. This CELD framework can also be used as a Wikipedia article recommendation system to suggest articles for further reading. In future, to further improve our system performance we would like to explore other techniques such as linkage factor graph model used by Wang et al. [13] in their work of linking English Wikipedia to other online Chinese encyclopaedias.

References

1. Sorg, P., Cimiano, P.: Enriching the Crosslingual Link Structure of Wikipedia - A Classification-Based Approach. In: AAAI 2008 Workshop on Wikipedia and Artifical Intelligence (2008)
2. Melo, G.D., Weikum, G.: Untangling the cross-lingual link structure of Wikipedia. In: Proceedings of the 48th Annual Meeting of the Association for Computational Linguistics, pp. 844–853 (2010)

3. Fahrni, A., Nastase, V., Strube, M.: HITS' Graph-based System at the NTCIR-9 Cross-lingual Link Discovery Task. In: Proceedings of NTCIR-9, pp. 473–480 (2011)
4. Huang, W.C(D.), Geva, S., Trotman, A.: Overview of the INEX 2009 Link the Wiki Track. In: Geva, S., Kamps, J., Trotman, A. (eds.) INEX 2009. LNCS, vol. 6203, pp. 312–323. Springer, Heidelberg (2010)
5. Tang, L.-X., Geva, S., Trotman, A., Xu, Y., Itakura, K.Y.: Overview of the NTCIR-9 Crosslink Task: Cross-lingual Link Discovery. In: Proceedings of NTCIR-9, pp. 437–463 (2011)
6. Chang, P.-C., Galley, M., Manning, C.D.: Optimizing Chinese word segmentation for machine translation performance. In: Proceedings of the Third Workshop on Statistical Machine Translation (2008)
7. Itakura, K.Y., Clarke, C.L.A.: University of Waterloo at INEX2007: Adhoc and Link-the-Wiki Tracks. In: Fuhr, N., Kamps, J., Lalmas, M., Trotman, A. (eds.) INEX 2007. LNCS, vol. 4862, pp. 417–425. Springer, Heidelberg (2008)
8. Geva, S.: GPX: Ad-Hoc Queries and Automated Link Discovery in the Wikipedia. In: Fuhr, N., Kamps, J., Lalmas, M., Trotman, A. (eds.) INEX 2007. LNCS, vol. 4862, pp. 404–416. Springer, Heidelberg (2008)
9. Mihalcea, R., Csomai, A.: Wikify!: linking documents to encyclopedic knowledge. In: Proceedings of CIKM 2007, pp. 233–242 (2007)
10. Milne, D., Witten, I.H.: Learning to link with wikipedia. In: Proceeding of CIKM 2008 pp. 509–518 (2008)
11. Tang, L.-X., Itakura, K.Y., Geva, S., Trotman, A., Xu, Y.: The Effectiveness of Cross-lingual Link Discovery. In: Proceedings of The Fourth International Workshop on Evaluating Information Access (EVIA), pp. 1–8 (2011)
12. Tang, L.-X., Geva, S., Trotman, A., Xu, Y.: A Boundary-Oriented Chinese Segmentation Method Using N-Gram Mutual Information. In: Proceedings of CIPS-SIGHAN Joint Conference on Chinese Language Processing, pp. 234–239 (2010)
13. Wang, Z., Li, J., Wang, Z., Tang, J.: Cross-lingual knowledge linking across wiki knowledge bases. In: Proceedings of the 21st international conference on World Wide Web, pp. 459–468 (2012)

Unified Recommendation and Search in E-Commerce

Jian Wang[1], Yi Zhang[1], and Tao Chen[2]

[1] School of Engineering, University of California Santa Cruz
Santa Cruz, CA, 95060, USA
{jwang30,yiz}@soe.ucsc.edu
[2] Robert H. Smith School of Business, University of Maryland
College Park, Maryland, 20742 USA
taochen@rhsmith.umd.edu

Abstract. In an e-commerce website, a recommender system and a search engine are usually developed and used separately. In fact they serve a similar goal: helping potential consumers find products to purchase. Recommender systems use a user's prior purchase history to learn the user's preferences and recommend products that the user might like. Search engines use a user's query information, which tells much about a user's purchase intention in the current search session, to find matching products. This paper explores how to integrate the complementary information to build a **unified recommendation and search system**. We propose three approaches. The first one is using a multinomial logistic regression model to integrate a rich set of search features and recommendation features. The second one is using a gradient boosted tree based ranking model. The third one is a new model that explicitly models the user's categorical choice, purchase state (repurchase, variety seeking or new purchase) in addition to the final product choice. Experiments on a data from an e-commerce web site (shop.com) show that unified models work better than the basic search or recommendation systems on average, particularly for the repeated purchase situations. The new model predicts a user's categorical choice and purchase state reasonably well. The insight and predicted purchase state may be useful for implementing the user-state specific marketing and advertising strategies.

Keywords: Recommender System, E-commerce, Purchase Intention.

1 Introduction

Recommender systems and search engines in e-commerce sites help consumers locate products to purchase. Although they have similar goals, they are usually developed and used separately in an e-commerce site. In the research community, they are also studied as two separate problems.

Traditional recommendation approaches, including collaborative filtering and content-based filtering models, learn a user's long-term preferences from the user's purchase or rating history. They work well in domains such as movie and

Y. Hou et al. (Eds.): AIRS 2012, LNCS 7675, pp. 296–305, 2012.

music recommendations. Yet their effectiveness in the e-commerce domain is limited [3] since the high data sparsity is a serious problem. More importantly, a consumer often makes purchase decisions based on various short-term or contextual reasons that are unknown to the e-commerce system. It is extremely challenging for recommender systems to make prediction under this circumstance. Fortunately an e-commerce website is not limited to the user purchase history. This is especially true for a search session, in which a user has issued one or more queries before she makes a purchase decision. These queries provide much information about the consumer's current purchase intention, and could benefit the recommender system if used appropriately. On the other hand, a typical e-commerce search engine focuses on the short-term need of a user and returns products that match the consumer's current queries. The retrieval ranking is usually based on the similarity between the product and the queries. However, different users have different preferences such as brand loyalty and so on. A consumer's prior purchase history indicates her general purchase preferences and characteristics, and could benefit the search engine if used appropriately.

In this work, we intend to design a **unified recommendation and search system** to capture both the long-term preferences and short-term needs of a user. To do so, the system incorporates the user's purchase history and her search information. We can roughly break down different pieces of information into three groups: the user's history-related information, the query-related information and the product's marketing-related information. We can use different regression model (multinomial logistic regression, gradient boosting tree (GBT), etc.) to integrate all three groups of information mentioned above. In order to better understand and explain the user's purchase intention, we propose a new unified model that jointly models a user's unobservable categorical choice, purchase state (repurchase, new purchase, or variety seeking) and product choice. The new model provides more insights about a user's purchase intention and decision process. We compare our proposed models with a variety of baselines on an e-commerce dataset collected from shop.com. The experimental results demonstrate the value of a **unified recommendation and search system** with a better prediction power.

The major contributions of this work include the following:

- Demonstrate the value of a **unified recommendation and search system** to predict a user's purchase intention in e-commerce sites.
- Propose and evaluate three unified modeling approaches in the paper. The first one is multinomial logistic regression. The second one is gradient boosting trees. The third one is a three-stage model with explicit user purchase state which better understands the user's purchase intention. It jointly models the user's unobservable categorical choice, purchase state and product purchase.
- Compare the unified models with a variety of recommendation and search algorithms. They all achieve better prediction performance. In addition, the new model with user purchase state has better explanatory power.

2   Related Work

Research on recommender systems puts much effort on understanding and modeling the user's long-term preferences. The algorithms powering a recommender system could be simple lookups, or more complex solutions such as content-based filtering, collaborative filtering, or hybrid algorithms. Recommendation in the e-commerce domain has received much attention. Several methods have been studied, including item-based filtering algorithms [14], MDP-based methods [13], utility theory-based methods [15] and so on. The work presented in this paper can be viewed as recommendation augmented with search queries.

Search engine is a major research focus of the information retrieval community. In information retrieval field, the state-of-the-art learning algorithms are based on **gradient boosting trees** [5]. Although search engines in e-commerce are widely used, the research is very limited. Guo and Agichtein [6] presented search models to capture a user's fine-grained interaction with the search results. Another recent work by Li et al. [9] studied the problem of using recommender models to enhance the basic search performance with the linear sum of two models' results. The work presented in this paper can be viewed as personalized search with long-term user history. Besides, we analyze the user's behavior intention in more details by designing a model with different user states, and learning state-dependent ranking models.

3   Problem Analysis

3.1   Problem Definition

If a user issues one or more queries before she purchases some products, we call the user session a *search session*. A search session may contain one or more orders. This paper focuses on generating a good recommendation list for a user in a search session. After a user issues one query and before she makes a purchase, our system generates a ranked recommendation list and presents it to the user to choose from. In this paper, the following notations are used:

- u: the index of user.
- i or j: the index of product.
- c: the index of product's category. Examples of categories are "digital cameras", "camera filters", "dental/oral care", "office supplies" and so on.
- t: time. Many features are dependent on the time t. For example, the similarity between product i and the user's previous purchases depends on the user's purchase history at time t.
- Q: the user's current query/queries at time t. It contains all queries issued in the current session after the last purchase order.
- $\mathbf{x}_{u,i,t}$: the feature vector of the candidate product i for user u at time t. It includes all features about user u's history (including queries issued) at time t, features of product i, and features capturing interaction between user u and product i at time t.
- $y_{u,t}$: user u's purchase decision at time t. If $y_{u,t} = i$, it indicates that user u purchases product i at time t.

4   Unified Models

4.1   Information for the Unified Model

On the high level, we classify all available information of the unified system into three groups: $x_{u,i,t} = \{x^H, x^Q, x^P\}$

[**User's History-related Information x^H**] The **first** type of information reflects the user's general characteristics in the current session. The **second** type of information captures the user's general characteristics in each candidate category c. For instance, a user might be a loyal consumer in the "food" category yet not in the "clothing" category. The **third** type of information captures the user's interaction with each candidate product i.

[**Query/Search-related Information x^Q**] This group of features is entirely based on the user's queries Q. The queries issued before purchasing are highly related to the user's purchase intention in that session.

[**Product Marketing-related Information x^P**] This group of features contains the product's marketing-related information, including the product's category, popularity, price, and the corresponding category's popularity, the re-purchase tendency of the product/category and so on.

4.2   Basic Learning Model 1: Multinomial Logistic Regression

A straightforward unified approach is to use a learning algorithm to predict the user purchase based on all features $x_{u,i,t} = \{x^H, x^Q, x^P\}$. One approach is using multinomial logistic regression to model the probability of user u purchasing product i at time t as follows:

$$P(y_{u,t} = i) = \frac{exp\{w^T x_{u,i,t}\}}{\sum_j exp\{w^T x_{u,j,t}\}} \tag{1}$$

where w is the model parameter to be learned from the training data.

4.3   Basic Learning Model 2: Gradient Boosted Tree

An alternative model to combine all features is using **Gradient Boosted Trees**. This model has been proven to be very successful in the web search community. Most of the top performing algorithms in the *Learning To Rank Challenge*[1] are variations of GBT [2]. We use it as a unified model to solve a two-class classification problem: predicting whether a user will purchase a product or not. The gradient tree boosting learns a weighted additive model of simple *trees* from the labeled data. It is a very general and powerful machine learning algorithm that performs well on learning-to-rank tasks.

Given users' purchase history, we can generate the training data set $(x_{u,i,t}, r)$, where $x_{u,i,t}$ contains all information in Section 4.1 and $r = 1$ if the user purchases the product i. The learned GBT is used to predict the purchase likelihood for each product and all products are ranked accordingly.

---

[1] http://learningtorankchallenge.yahoo.com/workshop.php

4.4   Model with Categorical Choice and Purchase State

Marketing researchers found that a user's purchase state can be classified into the repeated purchase, the variety-seeking purchase and the new purchase [7,10,1]. In a repurchase state, the user is more likely to stick to her favorite products. In a variety-seeking state, the user would consider new options in the category that she purchased before. In a new purchase state, the user would purchase some new products in the category that she did not purchase before. Understanding these purchase intentions will help better design recommendation, marketing or advertising strategy. Before a user makes a purchase, e-commerce websites don't observe the user's categorical choice or purchase state a priori. To model this, we introduce two random variables into our system: a consumer u's categorical choice $c_{u,t}$ and the purchase state $s_{u,t}$ at time t.

We use the following multinomial logistic regression function to predict a user's categorical choice $c_{u,t}$.

$$P(c_{u,t} = k) = \frac{\sum_{i \in \psi_k} exp\{\theta_1^T(\mathbf{x_{u,i,t}})\}}{\sum_{i \in \psi_{all}} exp\{\theta_1^T(\mathbf{x_{u,i,t}})\}} \tag{2}$$

where ψ_k contains all products in category k and ψ_{all} contains all products.

Depending on the categorical choice $c_{u,t}$, we can predict the consumer's purchase state $s_{u,t}$, which can take three possible values: repurchase, variety-seeking or new purchase state. If category $c_{u,t}$ was never purchased by u before, the purchase state is always a new purchase (i.e. $s_{u,t} = 0$). Otherwise, the purchase state $s_{u,t}$ could either be repurchase $s_{u,t} = 1$ or variety-seeking purchase $s_{u,t} = -1$. In the second case, the state can be estimated based on the user's history and the user's categorical choice, and we use the following standard binomial logistic regression to model $P(s_{u,t}|c_{u,t})$:

$$P(s_{u,t}|c_{u,t}) = \frac{1}{1 + exp\{-s_{u,t}(\theta_2^T \mathbf{x_{u,i,t}})\}} \tag{3}$$

Given that a user is in the purchase state $s_{u,t}$ and would choose from category $c_{u,t}$, we model how she decides which product to purchase at time t using a multinomial logistic regression. The probability of product choice $y_{u,t}$ depends on the categorical choice, the purchase state, and all other features as follows:

$$P(y_{u,t} = i|s_{u,t}, c_{u,t}) = \frac{I_{i \in \psi_{s,c,u,t}} exp\{\mathbf{w_{s_{u,t}}}^T \mathbf{x_{u,i,t}}\}}{\sum_{j \in \psi_{s,c,u,t}} exp\{\mathbf{w_{s_{u,t}}}^T \mathbf{x_{u,j,t}}\}} \tag{4}$$

where I_* is an indicator function. $I_* = 1$ if $*$ is true, otherwise 0. $\psi_{s,c,u,t}$ is a set of all products that match the category $c_{u,t}$ and purchase state $s_{u,t}$ at time t. Depending on the value of $s_{u,t}$, the model parameter $\mathbf{w_{s_{u,t}}}$ is either $\mathbf{w_1}$, $\mathbf{w_{-1}}$ or $\mathbf{w_0}$. By having different $\mathbf{w_s}$ parameters, this model learns state-dependent multinomial logistic regression function for each purchase state.

The parameters we need to estimates are $\Theta = (\theta_1, \theta_2, w_1, w_0, w_{-1})$. Each purchase decision/action of one product can be treated as a training/testing data point. Assuming the prior distribution of each model parameter is a Gaussian centered on zero, the optimal parameters can be learned from the training data using the *maximum a posteriori probability (MAP)* estimation:

$$(\mathbf{w_1}, \mathbf{w_{-1}}, \mathbf{w_0}, \theta_1, \theta_2)$$

$$= \arg maxP(\Theta) \prod_{u,t} P(y_{u,t}, s_{u,t}, c_{u,t})$$

$$= \arg maxP(\theta_1)P(\theta_2)P(\mathbf{w_1})P(\mathbf{w_{-1}})P(\mathbf{w_0}) \prod_{u,t} [P(y_{u,t} = i|s_{u,t}, c_{u,t})P(s_{u,t}|c_{u,t})P(c_{u,t})]$$

To use the model for recommendation, we can estimate the probabilistic distribution of $y_{u,t}$ by summing over all possible values of the unobservable state $s_{u,t}$ and categorical choice $c_{u,t}$. $P(y_{u,t} = j) = \sum_{c_{u,t}} \sum_{s_{u,t}} P(y_{u,t}, s_{u,t}, c_{u,t})$. Then we can rank products based on the purchase probability $P(y_{u,t} = j)$.

5   Experimental Setup

5.1   Shop.com Dataset

We create an evaluation dataset based on the purchase record that was collected from 2004-01 to 2009-03 on an e-commerce website (shop.com). The Apache log that includes users' search history is collected from 2007-06 to 2008-11. In the entire dataset, there are 124,813 products, 244,226 users, 296,145 unique orders, i.e. (user, order time) and 341,598 unique purchases, i.e. (user, product). There are 378 secondary-level categories for all products in total. For sessions with the Apache log information, the first 80% is used as the training data to calculate feature values, the following 10% is used as the training data to build models and the last 10% is used for testing. Sessions with no Apache log are also used as the training data to calculate feature values. For the unified model, we select top $k\%$ products($k = 0.01$ in our experiment) in each basic recommender solution and search solution's own ranked list to put into the ranking candidate pool. In addition, products that the user purchased before the order time are also added into the pool. The unified model learns the ranking function from the training data and ranks all products in the pool in the testing step.

5.2   Algorithms to be Compared

Our experiments compare the following three groups of system solutions:

[**Basic Recommendation Solutions**] We implement a collaborative filtering model, a content-based filtering model, and two variations of a hybrid filtering model. *PureSVD* [4] is a representative collaborative filtering model with a decent performance on standard datasets. *ContentRec* is a content-based filtering model. It estimates the score of each product as its total similarity to the user's previous purchase(s). *GBT.Rec* and *GBT.Rec.P* are variations of a hybrid filtering model. *GBT.Rec* use Gradient Boosting Tree(GBT) [5] to integrate all user's history-related features $\mathbf{x^{Hp}}$ that are generated from the user's prior purchase history. *GBT.Rec.P* further integrates the product's marketing-related information to *GBT.Rec.*

[**Basic Search Solutions**] For basic search-based solutions, we implement three methods. The first one is the well-known BM25 ranking algorithm which uses the most recent query and all terms of the candidate product to calculate a relevance score. The second one is *GBT.Search* which integrates features of $\mathbf{x^Q}$. The third one is *GBT.Search.P* which further integrates the product's marketing-related features $\mathbf{x^P}$.

[**Unified Solutions**] We implemented and compared three unified models in this paper. *Multi.U* is to use the multinomial regression model in Section 4.2 to combine all features. *GBT.U* is to use the gradient boosting tree algorithm in Section 4.3. *State.U* is to use the model in Section 4.4. It incorporates the consumer's categorical choice and purchase state. We focus on evaluating the ability of unified models in incorporating the user history related features $\mathbf{x^H}$ with other search/product related features. Thus we synthesize features of $\mathbf{x^Q}$ and $\mathbf{x^P}$ into a single feature: the output of the *GBT.Search.P* algorithm.

5.3   Evaluation Metric

In a typical e-commerce website, the ranking of all recommendations is more important than rating predictions. Instead of using some common rating prediction accuracy measures, we evaluate all algorithms in the context of a ranking task [4]. Each product purchasing order corresponds to a testing point, which is uniquely identified by a (user, order time) pair. Let $S_{purchased}$ be the set that contains all products in this order. We rank each product in the candidate pool according to its predicted score for user u at time t, where t is right before an order time. Let $S_{K,recommended}$ be the set that contains top K products. Conversion rate is used as the metric. If a user purchases at least one product from the recommended top K list, we consider that the user converts from a browser into a buyer. Thus *conversion rate@K* = 1 if $S_{purchased}$ and $S_{K,recommended}$ has overlap product(s), otherwise *conversion rate@K* = 0.

6   Experimental Results

6.1   Overall Performance Analysis

The basic recommendation solutions, the basic search solutions, and the unified solutions are compared for user search sessions with at least one query and prior purchase (Figure 1). The recommendation solutions without integrating the query-related information are the least effective. On the other hand, the basic search solutions are more effective. After integrating a user's prior purchase/query history and her current query/queries, unified solutions are better. Based on the paired two-tailed t-test, the basic unified model *Multi.U* is better than the best search solution(*GBT.Search.P*) and statistically significantly better than the best recommendation solution(*GBT.Rec.P*). *GBT.U* and *State.U*

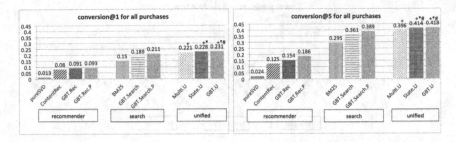

Fig. 1. Comparison of different system's performance for conversion rate@1 and conversion rate@5. $Value^+$ is significantly better than $GBT.Rec.P$. $Value^*$ is significantly better than $GBT.Search.P$. $Value^\#$ is significantly better than $Multi.U$.

are statistically significantly better than the best recommendation solution, the best search solution and the basic unified model $Multi.U$. There is no significant difference between $GBT.U$ and $State.U$ in all top 5 positions.

Why does the traditional recommender system perform not so well in the e-commerce system? We compare the density of this dataset with the other commonly used datasets in the research of recommender system. The shop.com dataset is significantly more sparse and the huge sparsity leads to the poor performance of the traditional recommender system. This serious data sparsity issue also exists in other e-commerce websites [3]. In a search session with the query information available, it is essential to incorporate the information into a unified recommender and search system.

6.2   Performance of Different Purchase Types

We further examine the performance of several algorithms on different purchase types: repurchase orders(14.03% of all orders), new purchase orders(73.08% of all orders) and variety-seeking orders(20.54% of all orders). The algorithms to compare include the best recommendation solution ($GBT.Rec.P$), the best search solution ($GBT.Search.P$) and three unified models. Conversion rate for top 5 recommendations is shown in Figure 2. Here are some observations: 1) The recommendation solution performs better than the search solution for repurchases, but worse for new purchases and variety-seeking purchases. 2) Unified models perform significantly better than the recommendation solution in new purchase and variety-seeking orders. The search queries in the current session largely reflect the consumer's short-term purchase intention. Without a user's query, it is hard for the recommender system to learn the consumer purchase intention purely from her prior purchase history. 3) Unified models perform significantly better than the best search solution ($GBT.Search.P$) in repurchase orders, but worse in new purchase and variety-seeking orders. The unified models try to integrate the consumer's prior history at the cost of higher model complexity. Improving the performance of repurchase orders but hurting the performance of

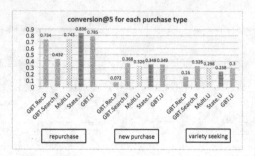

Fig. 2. Performance of representative algorithms in conversion rate@5 for three purchase types: repurchase, new purchase and variety-seeking

the other two is a trade-off for the unified model. All three unified models are trained to optimize the overall performance, not in a specific purchase type.

6.3   User States in the New Model

Although two unified models *GBT.U* and *State.U* perform similarly to each other, the new model *State.U* has its own contribution. It explicitly models a user's purchase state and may tell us more about the user intention. Explicit predictions of the categorical choice and purchase state allow great opportunities for e-commerce sites to implement various marketing strategies. Here are some potential use cases for each purchase state: 1) Marketers could provide related display ads to consumers in the repurchase state. Marketing research indicates that the advertising may impact high loyalty consumer segments while have no impact on low loyalty consumer segments [12]. 2) Marketers could implement different promotion frequencies for prominent brands and less prominent brands for variety-seeking consumers [8]. 3) If a consumer is likely to purchase in a new category, marketers can better categorize the recommendation list which would potentially increase the user's satisfaction with their choices [11].

7   Conclusion and Future Work

Instead of building a recommendation system and a search engine separately, we incorporate the user's prior purchases and queries into a unified model for e-commerce websites. We propose to use three unified models: multinomial regression, gradient boosting tree algorithm, and a new model with user states. Experimental results on shop.com dataset show that the unified models perform better than the basic search or recommendation models.

Acknowledgments. We would like to thank shop.com for sharing the data. This work was funded by National Science Foundation IIS-0953908 and CCF-1101741. Any opinions, findings, conclusions or recommendations expressed in this paper are the authors, and do not necessarily reflect those of the sponsors.

References

1. Berning, C.A.K., Jacoby, J.: Patterns of information acquisition in new product purchases. The Journal of Consumer Research 1(2), 18–22 (1974)
2. Burges, C.J.C., Svore, K.M., Bennett, P.N., Pastusiak, A., Wu, Q.: Learning to rank using an ensemble of lambda-gradient models. Journal of Machine Learning Research - Proceedings Track 14, 25–35 (2011)
3. Chen, Y., Canny, J.F.: Recommending ephemeral items at web scale. In: Proceedings of the 34th International ACM SIGIR 2011, pp. 1013–1022 (2011)
4. Cremonesi, P., Koren, Y., Turrin, R.: Performance of recommender algorithms on top-n recommendation tasks. In: Proceedings of the Fourth ACM RecSys, pp. 39–46. ACM, New York (2010)
5. Friedman, J.H.: Greedy function approximation: A gradient boosting machine. Annals of Statistics 29, 1189–1232 (2001)
6. Guo, Q., Agichtein, E.: Ready to buy or just browsing?: detecting web searcher goals from interaction data. In: Proceeding of the 33rd International ACM SIGIR 2010, pp. 130–137. ACM (2010)
7. Hoyer, W.D.: An examination of consumer decision making for a common repeat purchase product. Journal of Consumer Research: An Interdisciplinary Quarterly 11(3), 822–829 (1984)
8. Kahn, B.E., Raju, J.S.: Effects of price promotions on variety-seeking and reinforcement behavior. Marketing Science 10(4), 316–337 (1991)
9. Li, W., Ganguly, D., Jones, G.J.F.: Enhanced Information Retrieval Using Domain-Specific Recommender Models. In: Amati, G., Crestani, F. (eds.) ICTIR 2011. LNCS, vol. 6931, pp. 201–212. Springer, Heidelberg (2011)
10. McAlister, L., Pessemier, E.: Variety seeking behavior: An interdisciplinary review. Journal of Consumer Research 9(3), 311–322 (1982)
11. Mogilner, C., Rudnick, T., Iyengar, S.S.: The mere categorization effect: How the presence of categories increases choosers' perceptions of assortment variety and outcome satisfaction. Journal of Consumer Research 35(2), 202–215 (2008)
12. Raj, S.P.: The effects of advertising on high and low loyalty consumer segments. Journal of Consumer Research 9(1), 77–89 (1982)
13. Shani, G., Heckerman, D., Brafman, R.I.: An mdp-based recommender system. J. Mach. Learn. Res. 6, 1265–1295 (2005)
14. Wang, J., Sarwar, B., Sundaresan, N.: Utilizing related products for post-purchase recommendation in e-commerce. In: Proceedings of the Fifth ACM RecSys 2011, pp. 329–332. ACM (2011)
15. Wang, J., Zhang, Y.: Utilizing marginal net utility for recommendation in e-commerce. In: Proceedings of the 34th International ACM SIGIR 2011, pp. 1003–1012. ACM (2011)

Identifying Local Questions in Community Question Answering

Long Chen, Dell Zhang, and Mark Levene

DCSIS, Birkbeck, University of London
Malet Street, London WC1E 7HX, UK
{long,mark}@dcs.bbk.ac.uk, dell.z@ieee.org

Abstract. Community Question Answering (CQA) services, such as Yahoo! Answers, WikiAnswers, and Quora, have recently proliferated on the Internet. A large number of questions in CQA ask for information about a certain place (e.g., a city). Answering such local questions requires some local knowledge; therefore, it is probably beneficial to treat them differently from global questions for answer retrieval and answerer recommendation etc. In this paper, we address the problem of automatically identifying local questions in CQA through machine learning. The challenge is that manually labelling questions as local or global for training would be costly. Realising that we could find many local questions reliably from a few location-related categories (e.g., "Travel"), we propose to build local/global question classifiers in the framework of PU learning (i.e., learning from positive and unlabelled examples), and thus remove the need of manually labelling questions. In addition to standard text features of questions, we also make use of locality features which are extracted by the geo-parsing tool Yahoo! Placemaker. Our experiments on real-world datasets (collected from Yahoo! Answers and WikiAnswers) show that the probability estimation approach to PU learning outperforms S-EM (spy EM) and Biased-SVM for this task. Furthermore, we demonstrate that the spatial scope of a local question can be inferred accurately even if it does not mention any place name. This is particularly helpful in a mobile environment as users would be able to ask local questions via their GPS-equipped mobile phones without explicitly mentioning their current location and intended search radius.

Keywords: Community Question Answering, Local Search, PU Learning.

1 Introduction

Community Question Answering (CQA) services, such as Yahoo! Answers[1], Baidu Zhidao[2], Quora[3], Facebook Questions[4], have been booming recently.

---

[1] http://answers.yahoo.com/
[2] http://zhidao.baidu.com/
[3] http://www.quora.com/
[4] http://www.facebook.com/questions/

Y. Hou et al. (Eds.): AIRS 2012, LNCS 7675, pp. 306–315, 2012.
© Springer-Verlag Berlin Heidelberg 2012

Such platforms enable users to ask and answer questions in natural language, and thus satisfy their information need with pertinent user-generated content tailored to their complex intent.

In Broder's [1] seminal work, user's intent is defined as "the need behind the query". Unfortunately, most CQA portals do not consider user's locality intent, therefore user's information need is not satisfied geographically. For example, at the time of writing this paper, the question "Whats the best restaurant to watch fireworks from in Hongkong?" attracts only one response from Yahoo!Answers, leaving a large margin space for the system to attract more Hongkong based users to answer it.

To shed light on user's locality intent, we propose to classify questions into two categories according to the locality intent: local and global. By considering the question, for instance, "What's the best restaurant to watch fireworks from in Hongkong?" as a local one, a CQA system can route the question directly to some specific local answerers by identifying the corresponding spatial scope. On the other hand, by identifying "Where is a good place I can chat to people about money making ideas?" as a global question, we can highlight the question on the home page to attract more people to answer it, regardless of their locality background. After performing the classification over local and global, we further pinpointed spatial scope of the question by analyzing its thematic features so as to enable the search radius to vary depending on user's information need. For example, users querying for a coffee shop are probably looking for one within walking distance. If they are consulting the local tax rate, they will expect a distance of vicinity council. If they want to buy cheap ticket for travelling, however, distance may not be important as tickets can be bought from the Internet. CQA systems can then automatically pinpoint the specific locality scope by combining one's GPS and the spatial scope of the topic. This is a tempting scenario for mobile environment: one can ask local question without explicitly mentioning their current location and intended search radius, producing a significantly enhanced user-experience in terms of simplicity and flexibility.

In this paper, we build a predictive model through machine learning based on both text and metadata features to identify the locality intent. Our investigation reveals that the Probability Estimation model achieves a superior performance than S-EM and Biased-SVM — two state-of-the-arts PU learning models. In addition to revealing the general locality intent, the spatial scope of the question is also further targeted and exploited. Our experiment shows that the F_1 score of 0.738 and 0.754 can be achieved on Yahoo!Answers and WikiAnswers datasets respectively.

The rest of this paper is organised as follows. In Section 2, we review the related work. In Section 3, we define our taxonomy of user's locality intent in CQA. In Section 4, we introduce the PU approach to question classification with only positive and unlabelled examples. In Section 5, we describe the experimental setup and present our findings. In Section 6, we conclude our work and contributions.

2   Related Work

The problem of understanding user's locality intent is first proposed in the context of Web search engines. Luis et al. [4] classify the locality intent of Web search queries into two categories: global and local. However, this taxonomy is not that suitable for CQA, because web search engines aim to retrieve the most relevant web pages while CQA services strive to find the most appropriate people with the matching knowledge.

Tom et al. [12] proposed a classification-based approach for question routing, which directs questions to answerers who are most likely to provide answers. They propose to use local and global features to enhance classifier's performance. Baichuan et al. [5] provide a question routing framework, which comprehensively considers user's expertise, availability and answerer rank by having these features integrated into a single language model. These works are somewhat similar to our motivations, but none of them leverage user's geographical information to make inference.

With regard to the task of semi-supervised learning in CQA, our previous work [2] has already revealed that unlabelled questions can be made useful to improve the performance of question classification. In that work, we employ a Co-training framework to identify subjective and social questions in CQA. But different from the Co-training framework in which both positive and negative labelled examples are compulsory for training, in this task we take advantage of the PU learning framework that only requires positive ones to use. To the best of our knowledge, this is the first CQA work that integrates PU learning framework in the designing of the system.

3   Taxonomy

Taking into account the special locality characteristics of CQA, we propose the following taxonomy that classifies questions into two categories in terms of their underlying geographical locality: local and global.

Local Questions. The intent of such questions is to get information regarding a certain geographical locality, the best answers are likely to be produced by local answerers. For example, the question "Which country in Africa that was colonized by France did assimilation policy succeed?" asks for specific details of a particular location. Usually, local questions include one or more location names, as in the case of the question "What's the best restaurant to watch fireworks from in Hongkong?", the asker tries to set up a connection with the Hong Kong community, from whom the user can learn more details about a particular entity afterwards.

Global Questions. The intent of such questions is to get information irrespective of the geographical locality, the best matches are usually general answerers. For example, the question "Why I cannot block someone on YouTube when there's a new channel design update?" asks information from a general trouble shooter regarding web site configuration, regardless of their locality background. But more implicitly, the question "Is the Aurora Borealis phenomenon found anywhere else in the world?" may appear to be a local question (notice that Aurora Borealis always corresponds to the pole areas), until one comes to realize that there is no answerer available in such austerity region.

Notice that our taxonomy is a two-level hierarchy, in which local category are further broken down into subcategories to pinpoint question's spatial scope. We inherit the administrative place types of Yahoo! Placemaker namely, *Country, State, County, Town, and Local Administrative Area* to further break down local questions into the second level of spacial scope. More-detailed information regarding different Places vs. Place Names can be found at the Yahoo! Placemaker Key Concepts page[5].

4   Approach

In the locality classification task, dozens of local questions can be automatically detected from location-based categories. For example, in the Dining Out category of Yahoo! Answers, questions have been broken down into cities subcategories scattered around the world. On the other hand, however, it's impractical to label large amounts of global examples manually. Traditional supervised learning models are thus not helpful to the building of an automated training model — they require training in both local and global examples. Naturally we think PU learning framework can fit in to this context quite well.

Basically, PU learning is a semi-supervised learning framework, which builds a classifier with only positive and unlabelled training examples, to predict both positive and negative examples in unlabelled or test dataset. A short introduction, which describes PU learning models, is given in the following sub-sections.

4.1   S-EM

S-EM model is first proposed in [7] and can break down into two steps. The first step is to identify reliable negative examples from the unlabelled set U, which works by sending some *spy* examples from the positive set P to the U. The reliable negative examples is found through multiple iterations by running the first step couple of times. The second step is to use EM algorithm to build the final classifier. But EM algorithm makes some mixture model assumptions [9] of the datasets which can not be guaranteed to take hold, and thus it is often suffered from the problem of mismatch.

---

[5] http://developer.yahoo.com/geo/placemaker/guide/
concepts.html#placesandplacenames

4.2   Biased-SVM

The Biased-SVM [6] approach modifies the SVM formulation to make it fit in to the setting of PU learning, which can be described in the following SVM reformulations.

$$Minimize: \frac{<w \cdot w>}{2} + C_+ \sum_{i=1}^{k-1} \xi_i + C_- \sum_{i=k}^{n} \xi_i$$
$$Subjectto: y_i(< w \cdot x_i > +b) \geq 1 - \xi_i, i = 1, 2, ..., n \tag{1}$$
$$\xi_i \geq 0, i = 1, 2, ..., n$$

In the Equation 1, x_i is the input vector of the training example and y_i is its class label, $y_i \in \{1, -1\}$. The first $k - 1$ examples are positive examples labelled 1, while the rest are unlabelled examples, which are treated as negative labelled -1. C_+ and C_- are parameters to weight positive errors and negative errors differently. We give a bigger value for C_+ and a smaller value for C_- because unlabelled examples, which assumed as negative, contains positive examples. The C_+ and C_- values are chosen by using a separate validation set to verify the performance of the resulting classifier.

4.3   Probability Estimation

Probability Estimation approach is famous for its prominent accuracy and distinctive computational simplicity. This approach is first proposed in [3] and utilizes some probabilistic formulas.

Denote x an example and y the binary label (local and global); let local questions be positive examples and global questions be negative ones. Let $s = 1$ if the example x is labelled, and $s = 0$ if otherwise. Thus, the condition that only positive examples are labelled can be described as:

$$Pr(s = 1|x, y = -1) = 0 \tag{2}$$

The formula informs us that when $y = -1$, the probability of x being labelled is zero. So the objective now is to learn the classification function $f(x) = Pr(y = 1|x)$. To start with, *selected completely at random* assumption has to be satisfied: the labelled positive examples are chosen completely at random from all the positive examples, and thus,

$$Pr(s = 1|x, y = 1) = Pr(s = 1|y = 1) \tag{3}$$

The training set consists of two parts: the labelled dataset P (when $s = 1$) and the unlabelled dataset U (when $s = 0$). Let $g(x) = Pr(s = 1|x)$ be the function that estimates the probability of an example being labelled, $f(x) = Pr(y = 1|x)$ be the function that estimates the probability of an example belonging to positive category. Then the following lemma shows how to derive $f(x)$ from $g(x)$.

Lemma 1: suppose the "selected completely at random" assumption holds. Consequently,

$$f(x) = \frac{g(x)}{c} \tag{4}$$

The above equation suggests that we can attain a positive-negative classifier (this is exactly what we need) by having a positive-unlabelled classifier divided by c – the probability that a random positive example being labelled. Notice that in Equation (4), $c = Pr(s = 1|y = 1)$ is a constant that represents the probability of positive examples being labelled. So the problem now lies in how to estimate the constant c by using a trained classifier g and a validation dataset. Three estimators are proposed in [3] namely, $e_1 = \frac{1}{n} \sum_{x \in P} g(x)$, $e_2 = \sum_{x \in P} g(x) / \sum_{x \in V} g(x)$, and $e_3 = max_{x \in V} g(x)$. In the above formulas, V is the validation datasets, P consists of all the labelled examples of V, n is the cardinality of P.

5 Experiments

5.1 Dataset

The Yahoo! Answers dataset is derived from Yahoo! Answers Comprehensive Questions and Answers (v1.0), a dataset kindly provided by Yahoo Research Group[6]. It consists of 4483032 questions and respect answers from 2005/01/01 to 2006/01/01. An unlabelled set and test set are randomly selected across all 26 Yahoo main categories. Note that as we leverage a PU learning framework in our task, the training set will only involve local questions. The training set is automatically extracted from Dining Out, Travel, and Local Business categories with questions of a city name being assigned as the subcategory, whereas test set is manually labelled for both local and global examples.

The WikiAnswers dataset is collected from WikiAnswer[7] dating from 2012/01/01 to 2012/05/01 contains a total of 824320 questions (note that this is only a subset and cannot cover all the questions during that period of time). All the local questions are derived from WikiAnswers Local category as we find this is the only category in WikiAnswers that completely devoted to locality intent. We present the detailed statistics regarding the test and training sets, and validation set in Table 1. Acronym YA and WA represent Yahoo! Answers and WikiAnswers respectively.

With respect to the second-level classification, we use the same dataset by selecting all the questions containing at least one location reference(which is tagged by using Yahoo! Placemaker). There are 324537 and 12401 such questions available in Yahoo! Answers and WikiAnswres datasets, respectively, which directly serves as the second-level datasets for classification. What's more, all the location references in the training set are hidden to emulate the scenario when mobile users forget to type in the specific localities.

5.2 Performance Measure

Since the class sizes are imbalanced in this problem, we use the F_1 score [8] instead of accuracy to measure the performance of question classification.

---

[6] http://webscope.sandbox.yahoo.com/
[7] http://wiki.answers.com/

Table 1. Summary of CQA datasets

data	local	global	*total*
YA training	1000	0	1000
YA test	256	844	1100
YA validation	1000	0	1000
WA training	1000	0	1000
WA test	172	928	1100
WA validation	1000	0	1000
all	4428	1772	6200

The F_1 score is the harmonic mean of precision P and recall R: $F_1 = \frac{2PR}{P+R}$, where $P = \frac{\text{true positive}}{\text{true positive + false positive}}$, $R = \frac{\text{true positive}}{\text{true positive + false negative}}$. Furthermore, both micro-averaged F_1 (miF_1) and macro-averaged F_1 (maF_1) [11] will be reported in the next section. The former carries out averaging over all test questions while the latter over all question categories, therefore the former is dominated by performance on major question categories while the latter treats all question categories equally.

5.3 Experimental Results

A number of machine learning algorithms implemented in Weka[8], including C4.5, Random Forest, Naive Bayes, k-Nearest-Neighbours, and Linear Support Vector Machine (SVM), have been tested for semi-supervised learning (PU learning). What we find is that SVM can constantly outperform other schemes so we use it as the basic learning scheme in the following subsections.

Semi-Supervised Learning. We exploit several locality features that can help detecting the locality intent within the question, namely location frequency and location level, in addition to the textual features. But the original question datasets are not geographically annotated and contain no locality information. In order to extract location references and assign geographical scope to each question, Yahoo! PlaceMaker is employed to equipped original datasets with the location-specific explanation. There are two versions of scopes available in Place-maker, namely the geographical scope and the administrative scope. Geographic Scope is the place that best describes the document and may be of any place type. Administrative Scope is the place that best describes the document and has an administrative place type (which refers to Country, State, County, Local Administrative Area and Town). We use the geographical scope in this paper because we find this version provide more details than administrative scope[9].

---

[8] http://www.cs.waikato.ac.nz/ml/weka/

[9] http://developer.yahoo.com/geo/placemaker/guide/concepts.html

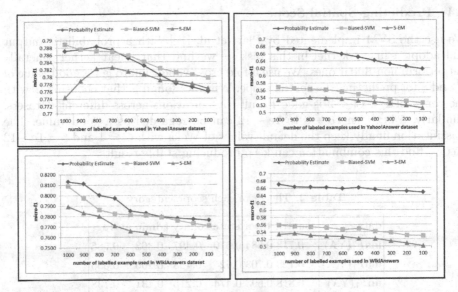

Fig. 1. The performance of PU learning with decreasing number of training examples used over Yahoo! Answers (up) and WikiAnswers (down)

Figure 1 is the learning curve of PU learning schemes given a varying number of positive labelled examples (the unlabelled examples are fixed at 5000). We employ the S-EM scheme to serve as baseline and the Biased-SVM as the state-of-the-art. As far as we can tell in the miF_1 figures, two datasets have a similar result, in which Probability Estimation and Biased-SVM perform significantly better than S-EM given sufficient amounts of labelled examples, but the gap starts to decrease when we shrink the labelled size. All three approaches give a comparable performance when providing only 500 labelled examples or less.

As for maF_1 figure over the YA dataset, Probability Estimation can consistently outperform the other two schemes – an approximately 23% error reduction on the basis of Biased SVM – irrespective of the labelled data size; At the same time, Biased-SVM is slightly better than S-EM approach with an average 2% improvement. The result generated on the WA dataset for maF_1 is quite similar and we do not mention it here. We propose that the probability approach can overwhelm the other two due to the maldistributional nature of the test set: 20% positive examples vs. 80% negative ones. In Probability Estimation model, having the non-traditional classifier divided by a constant c enables the classifier to be more tolerant towards the positive classifying errors by sacrificing some negative examples — we believe that is why Probability Estimation, in some cases, is even slightly worse than Biased SVM under miF_1 – producing a superior result for maF_1 by picking up the minority class in general.

5.4 Predicting Spatial Scope

We use the SVM implemented by Platt et al. [10] with a probabilistic output and adopt a linear kernel in this task — we find that linear kernel generally outperform non-linear ones. We use 5-fold cross-validation to get a good value of C and ξ for predicting the spatial scope. What we found is that choosing C with the range of $1 < C < 2$ is good for all configurations across different datasets. Similarly, ξ between $1e-8$ to $1e-12$ can be regarded as a good value. The classification finally keep the default values of Weka with $C = 1$ and $\xi = 1e-12$, as we find this combination can keep producing stable results.

Table 2. The F_1 of each scope category

data	country	town	state	county	admin	average
maF_1(YA)	0.713	0.684	0.670	0.497	0.363	0.585
maF_1(WA)	0.729	0.703	0.625	0.458	0.260	0.555
miF_1(YA)	0.818	0.693	0.474	0.215	0.131	0.738
miF_1(WA)	0.833	0.703	0.324	0.183	0.177	0.754

Table 2 gives the result of the maF_1 and miF_1 comparison over each scope level. Under the evaluation of maF_1, the prediction on country, town and state scopes have a superior performance than the rest, this suggest that these three scopes are relatively easier to identify by inferring the question's topic (for both Yahoo! Answers and WikiAnswers). But the system only display a mediocre performance regarding county and local administrative area scopes, which leads to our speculation that the questions in a higher scope level may have more discriminative power than questions in lower scope level. This is quite explainable: questions with a larger scope tend to have generalization behaviour whereas questions with smaller scope are liable to have uniqueness behaviour. Under the evaluation of miF_1, the performance over Yahoo! Answers and WikiAnswers are 0.738 and 0.754 respectively, which suggests that majority of the local questions' scope can be accurately predicted even if user does not mention the place names.

6 Conclusions

The main contribution of this paper is threefold. First, we identify several metadata features which can be used together with standard text features by machine learning algorithms to classify questions according to their geographical locality. Second, we prove that Probability Estimation approach can consistently outperform the S-EM and Biased-SVM on the evaluation of maF_1 and miF_1. Third, we prove that the spatial scope of a local question can be inferred accurately even if it does not mention any place name.

Acknowledgements. We are grateful to anonymous reviewers for their scrupulous review and constructive criticism, improving the quality of this work. We would like to thank Martyn Harris and Tom Ue for their meticulous proofreading and beneficial discussions.

References

1. Broder, A.: A taxonomy of web search. SIGIR Forum 36, 3–10 (2002)
2. Chen, L., Zhang, D., Levene, M.: Understanding user intent in community question answering. In: Proceedings of the 21st International Conference Companion on World Wide Web, WWW 2012 Companion, pp. 823–828. ACM, New York (2012)
3. Elkan, C., Noto, K.: Learning classifiers from only positive and unlabeled data. In: Proceedings of the 14th ACM SIGKDD International Conference on Knowledge Discovery and Data Mining, KDD 2008, pp. 213–220. ACM, New York (2008)
4. Gravano, L., Hatzivassiloglou, V., Lichtenstein, R.: Categorizing web queries according to geographical locality. In: Proceedings of the Twelfth International Conference on Information and Knowledge Management, CIKM 2003, pp. 325–333. ACM, New York (2003)
5. Li, B., King, I., Lyu, M.R.: Question routing in community question answering: putting category in its place. In: Proceedings of the 20th ACM International Conference on Information and Knowledge Management, CIKM 2011, pp. 2041–2044. ACM, New York (2011)
6. Liu, B., Dai, Y., Li, X., Lee, W.S., Yu, P.S.: Building text classifiers using positive and unlabeled examples. In: Proceedings of the Third IEEE International Conference on Data Mining, ICDM 2003, pp. 179–188. IEEE Computer Society, Washington, DC, USA (2003), http://dl.acm.org/citation.cfm?id=951949.952139
7. Liu, B., Lee, W.S., Yu, P.S., Li, X.: Partially supervised classification of text documents. In: Proceedings of the Nineteenth International Conference on Machine Learning, ICML 2002, pp. 387–394. Morgan Kaufmann Publishers Inc., San Francisco (2002)
8. Manning, C.D., Raghavan, P., Schtze, H.: Introduction to Information Retrieval. Cambridge University Press, New York (2008)
9. McCallum, A., Nigam, K.: A comparison of event models for naive bayes text classification. In: AAAI 1998 Workshop on Learning for Text Categorization, pp. 41–48. AAAI Press (1998)
10. Platt, J.C.: Fast training of support vector machines using sequential minimal optimization. In: Advances in Kernel Methods, pp. 185–208. MIT Press, Cambridge (1999)
11. Yang, Y., Liu, X.: A re-examination of text categorization methods. In: Proceedings of the 22nd Annual International ACM SIGIR Conference on Research and Development in Information Retrieval, SIGIR 1999, pp. 42–49. ACM, New York (1999)
12. Zhou, T.C., Lyu, M.R., King, I.: A classification-based approach to question routing in community question answering. In: Proceedings of the 21st International Conference Companion on World Wide Web, WWW 2012 Companion, pp. 783–790. ACM, New York (2012)

Improving Content-Based Image Retrieval
by Identifying Least and Most Correlated Visual Words

Leszek Kaliciak[1], Dawei Song[2,3], Nirmalie Wiratunga[1], and Jeff Pan[4]

[1] The Robert Gordon University, Aberdeen, UK
[2] Tianjin University, Tianjin, China
[3] The Open University, Milton Keynes, UK
[4] Aberdeen University, Aberdeen, UK
{l.kaliciak,n.wiratunga}@rgu.ac.uk, Dawei.Song@open.ac.uk,
jeff.z.pan@abdn.ac.uk

Abstract. In this paper, we propose a model for direct incorporation of image content into a (short-term) user profile based on correlations between visual words and adaptation of the similarity measure. The relationships between visual words at different contextual levels are explored. We introduce and compare various notions of correlation, which in general we will refer to as image-level and proximity-based. The information about the most and the least correlated visual words can be exploited in order to adapt the similarity measure. The evaluation, preceding an experiment involving real users (future work), is performed within the Pseudo Relevance Feedback framework. We test our new method on three large data collections, namely MIRFlickr, ImageCLEF, and a collection from British National Geological Survey (BGS). The proposed model is computationally cheap and scalable to large image collections.

Keywords: content-based image retrieval and representation, local features, correlation, pseudo relevance feedback, similarity measure.

1 Introduction

Recently, content-based image retrieval (CBIR) based on local feature extraction has attracted a lot of attention. One of the widely used approaches is based on so-called "bag of visual words" or "bag of features" (BOF) [1]. This model was inspired by the "bag of words" (BOW) framework from text information retrieval. The BOW represents documents as orderless "bag" of terms containing some words from the dictionary. In CBIR, terms from the text retrieval correspond to groups of local image patches (called visual words). A BOF representation of an image is a histogram of the visual words' counts in the image. The BOF approach is a mid-level representation that helps to reduce semantic gap between human perception and machine representation of images.

The local features based on BOF disregard the information about correlations between visual words. However, when the vocabulary size (the number of clusters) is small, the BOF's coefficients tend to be highly correlated. Such correlations can be exploited in order to improve the BOF performance.

---

[1] Terms "bag of visual words" and "bag of features" will be used interchangeably in this paper.

Y. Hou et al. (Eds.): AIRS 2012, LNCS 7675, pp. 316–325, 2012.

Proximity-based correlations are often utilized to capture the spatial relative information between instances of visual words and enhance the visual representations. Here, we will utilize both proximity-based and image-level correlations to adapt the similarity measure and re-rank the top images returned in the first round retrieval. To the best of our knowledge, no systematic comparison has been conducted between image-level and proximity based notions of correlation in the context of query expansion in image retrieval.

Existing approaches (query expansion like frameworks) often modify the current query, which leads to the normalization of histograms. This may not be desirable, since the (mid-level) semantic meaning of bins may be lost and the representations may become less discriminative due to the varied complexity of images. Moreover, many researchers incorporate the $tf \cdot idf$ weighting scheme from text retrieval although some experiments suggest that even the most frequent visual words are important to the retrieval (see [1]). However, others ([13]) report performance improvement for $tf \cdot idf$ and thus the results are not conclusive. We believe that this may be domain specific. $tf \cdot idf$ may work better in case when the precise object matching is important. In this paper we are concerned with generic image retrieval only and our model avoids the re-normalization by modifying the similarity measure.

Current approaches are also data storage and computationally expensive which makes them less suitable for real user oriented applications, for example, to incorporate content into a user profile.

To tackle the aforementioned problems, we propose a novel approach to exploit the inter-relationships between the visual words. We introduce and test a few notions of correlation. First, we generate a matrix of correlations between visual words for each top image returned in the first round retrieval. Second, we aggregate the matrices and identify the dominant and least correlated coefficients. Thus obtained information, along with the visual words' frequencies from the current query, is then utilized to weight the similarity measure. Certain coefficients in the similarity measure corresponding to highly correlated terms are then increased, while the coefficients related to least correlated visual words are deemphasized. The images returned in the first round retrieval are then re-ranked according to the modified similarity measure. The improved performance, observed on three different data collections, is in our opinion a promising indicator for the real user evaluation. The proposed approach should let us directly incorporate image content into user profiles, where each profile would be represented in a form of a matrix of correlations between visual words obtained from the query history. Thus obtained user profile, which would store user visual preferences, could be utilized to adjust the similarity measure with respect to each individual user.

2 Related Work

Readers interested in local features and the "bag of visual words" approach are referred to [1,2,3] for the detailed description and application of aforementioned methods.

An interactive image retrieval model with adaptive similarity measure is introduced in [14]. The weights for adjusting the similarity measure (with respect to the image content representation - global features) are calculated according to the consistency of the vectors' components representing images collected from user relevance feedback. First, the representations of images deemed relevant by the user are stacked to form a matrix. Next, if a column contains elements with similar values then this particular dimension is considered to be a good indicator of the user's information need and the weight is calculated as an inverse of standard deviation across this dimension.

Liu et. al. ([6]) exploit co-occurrence information in spatial domain. Authors make an assumption that the related visual words would appear in a certain neighbourhood. They utilize the equivalent of $tf \cdot idf$ weighting scheme from text retrieval. Having obtained the information about the relationships, they use it to update the current query by weighting all the coefficients in the histogram. This leads to the normalization process which may hamper the performance [1].

Another approach [7] tries to capture the spatial relationships between pairs of visual words by building a visual word tree. The tree is generated by clustering interest points that co-occurred within some spatial distance. Latent Semantic Analysis (LSA) is then applied to compute the importance of each visual word to the given query, and the most important ones become so-called topic words. The $tf \cdot idf$ weighting scheme and the topic words are then utilized to re-rank the images. This approach, although quite efficient in comparison with others, is not applicable to real user evaluation because of the computational cost (high dimensional Scale Invariant Feature Transform descriptor, costly LSA).

Model proposed in [8] utilizes data mining techniques to discover spatially co-occurrent patterns of visual words. Authors report limitations of standard codebook generation techniques (related to synonymy and polysemy of visual words) and propose a novel approach, which constructs a higher-level visual phrase lexicon consisting of groups of co-located visual words.

Spatial correlations are also exploited in [9] where they are represented by correlograms. Experimental results show, that the joint models (B.O.F and correlogram) outperform standard appearance-only models. However, models based only on correlograms perform worse than standard B.O.F approach.

Jamieson et. al. [10] propose to group features that exist within a local neighbourhood, claiming that arrangements or structures of local features are more discriminative. Such groups of visual words are then associated with annotation words.

Trigram model is proposed in [11] to help in image classification. The method captures spatial correlations between image patches. Comparison between unigram and trigram models shows that the latter one improves the classification accuracy.

Another model [12] defines visual phrase-based image similarity. First, they count occurrences of each visual word. Then the occurrences of adjacent patch pairs formed by frequent visual words are counted and finally, the visual phrases are generated by selecting the adjacent patch pairs whose occurrences are higher than the threshold. The similarity between two images is measured by cosine metric with $tf \cdot idf$ weighting scheme adapted from text retrieval.

In general, methods that utilize information about correlations between visual words try to group semantically similar visual words' together. They usually consider co-occurrences at one contextual level and are computationally expensive and not scalable.

Our method, in contrast, is computationally and data storage cheap, utilizes co-occurrences at various contextual levels, and avoids the normalization of histograms of visual words' counts. These properties make it suitable for a real user evaluation, where a user profile would represent user visual preferences. Such type of user profile would be utilized to put a query into the right context.

3 Notions of Correlation Between Visual Words

Here, we introduce image-level and proximity-based notions of correlation. In text retrieval (see [5]) document-level correlations seem to be stronger. A document may contain correlated terms not because of their proximity, but because they refer to the same topic.

Because our histograms of visual words' counts can be classified as a mid-level representation (the BOF reduces the semantic gap), we can introduce the correlations in a relatively intuitive way. Let us first focus on the correlations at the image level.

Correlation 1 can be regarded as the number of all pairs between the instances of different visual words (see Figure 1). Here, for instance, the squares denoted as A represent different instances of the same visual word (image patches) that appears within an individual image. When dealing with a set of images, we would aggregate the correlation matrices generated for each image. In case of Correlation 1, this would be equivalent to putting histograms of visual words counts as rows in a matrix and multplying the transposition of this matrix by itself. This is an analogy to document-level correlation in text IR. Correlation 2 is a normalized version of Correlation 1, where the denominator is a total number of all possible pairs between occurrences of visual words (Figure 2).

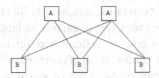 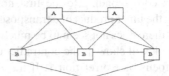

Fig. 1. Interpretation of Correlation 1. This is the common document/image level correlation. Here, squares denote instances of visual words (image patches) and the links the relationships between them.

Fig. 2. Normalization factor in Correlation 2. Here, squares denote instances of visual words (image patches) and the links the relationships between them.

Correlation 3 (Figure 3) can also be regarded as the number of pairs between the occurrences of different visual words, but this time the correspondence is as follows (see Figure 3).

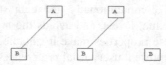

Fig. 3. Interpretation of Correlation 3. Here, squares denote instances of visual words (image patches) and the links the relationships between them.

1. $corr(vt_i, vt_j) = vt_i f \cdot vt_j f$

2. $corr(vt_i, vt_j) = \dfrac{2 \cdot vt_i f \cdot vt_j f}{(vt_i f + vt_j f) \cdot (vt_i f + vt_j f - 1)} = \dfrac{vt_i f \cdot vt_j f}{\binom{vt_i f + vt_j f}{2}}$

3. $corr(vt_i, vt_j) = \min(vt_i f, vt_j f)$

4. $corr(vt_i, vt_j) = \dfrac{vt_i f \cdot vt_j f}{\binom{vt_i f + vt_j f}{2}} + \min(vt_i f, vt_j f)$

where vt_i, vt_j denote the ith and jth visual term respectively, and $vt_i f$, $vt_j f$ denote the frequencies (number of occurrences) of the terms. By calculating the correlations between all visual words in a particular image, we will obtain a matrix of correlations:

$$\begin{pmatrix} corr(vt_1, vt_1) & corr(vt_1, vt_2) & \dots & corr(vt_1, vt_n) \\ corr(vt_2, vt_1) & corr(vt_2, vt_2) & \dots & corr(vt_2, vt_n) \\ \vdots & \vdots & \dots & \vdots \\ corr(vt_n, vt_1) & corr(vt_n, vt_2) & \dots & corr(vt_n, vt_n) \end{pmatrix}$$

The matrix corresponding to the first notion of correlation can also be obtained by calculating the inner product of a transposed vector image representation and itself $h^t \cdot h$.

At first, there does not seem to be much difference between these three relationships. A closer look will show us the contradictions with our intuition of correlation.

Let us focus on Correlation 1. If the frequencies of two pairs of visual words are $\{5, 10\}$ and $\{5, 100\}$ then the latter will be assigned higher correlation value. We would, however, expect the former pair to be at least equally correlated.

Normalization (Correlation 2) helps to overcome the above issue. However, if the frequencies are proportional, for example $\{10, 20\}$ and $\{40, 80\}$ then the former will score higher. But, intuitively, the latter is more correlated.

Correlation 3 seems to be intuitively right, but will ignore the additional information from the frequencies (see example for Correlation 1). Normalization of correlation 3 will produce similar side effects to Correlation 2. Therefore, we introduce the Correlation 4, which does not seem to contradict our intuition. Experimental results confirm the superiority of this notion of correlation in the user simulation.

Above notions of correlation consider two instances of visual words to co-occur if they appear somewhere within an image (visual context - the whole image). Let us now introduce, by analogy to text retrieval, what we will refer to as proximity-based correlation. Two instances of visual words will be considered correlated if they appear

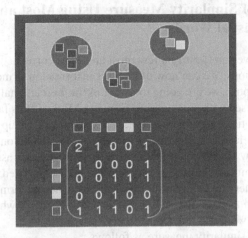

Fig. 4. Proximity-based correlation. For the clarity of presentation, the matrix corresponds to only three instances of visual words (circles' centres)

together within a certain neighbourhood (visual context - "sliding window"). In case of dense sampling this is rather straightforward. When dealing with sparse sampling, however, we need to shift the window (square, circular) from one instance of visual word to another. Figure 4 shows an example of proximity-based correlation. Here, the squares denote instances of various visual words. Now we can show how to incorporate the information about correlations into the Pseudo Relevance Feedback (PRF). PRF assumes that the top documents from the first round retrieval are all relevant to the query. Then, the additional information from the top documents is usually utilized to expand the query.

Initially, the first round retrieval is performed. Then, for each image from the top returned images, the matrix of correlations will be created. We aggregate all the matrices in order to obtain the final matrix from which the most and least dominant correlations will be identified (in terms of values). Notice, that in case of Correlation 1, this approach would be equivalent to constructing a matrix with rows corresponding to each image representation (from the top returned images)

$$M = \begin{pmatrix} vt_1^1 f & vt_2^1 f & \dots & vt_n^1 f \\ vt_1^2 f & vt_2^2 f & \dots & vt_n^2 f \\ \vdots & \vdots & \dots & \vdots \\ vt_1^m f & vt_2^m f & \dots & vt_n^m f \end{pmatrix}$$

and multiplying $M^t * M$, where t denotes the transpose operation. The advantage of our method is that it does not restrict us to one notion of correlation and we can define it in a more intuitive way.

4   Adaptation of Similarity Measure Using Most and Least Correlated Visual Words

As aforementioned, we can identify a few most and least correlated visual words from the matrix of correlations. We can now utilize this information to modify the similarity measure. For this purpose, we are going to use Minkowski fractional similarity measure (the method may be used with any measure from the Minkowski's family of distances).

First, we must identify a certain number (see Experimental Setup section for details) of most and least correlated visual terms by looking at the correlation matrix's elements' values above or below the diagonal (symmetrical matrix). Let's assume that we have identified the dominant correlation $\{v_k, v_l\}$ and the least correlated pair $\{v_n, v_m\}$. Let us now look at the query and extract the frequencies of visual terms corresponding to v_k, v_l, v_n, v_m. We can assume that $v_k f \geq v_l f$ and $v_n f \geq v_m f$, where vf denotes the frequency of a visual word taken from the query.

We can weight the similarity measure as follows

$$
\begin{aligned}
d(Q, I) = {}& \left(\sum_{i=1}^{N} \sqrt{|v_{Q_i} f - v_{I_i} f|} \right)^2 = \\
= {}& \left(\sqrt{|v_{Q_1} f - v_{I_1} f|} \right)^2 + \left(\sqrt{|v_{Q_2} f - v_{I_2} f|} \right)^2 \\
& + \ldots + \left(\sqrt{\frac{v_{Q_k} f}{v_{Q_l} f \cdot \log_b c(v_k, v_l)} |v_{Q_l} f - v_{I_l} f|} \right)^2 \\
& + \ldots + \left(\sqrt{\frac{v_{Q_m} f}{v_{Q_n} f \cdot \log_b c(v_n, v_m)} |v_{Q_n} f - v_{I_n} f|} \right)^2 \\
& + \ldots + \left(\sqrt{|v_{Q_N} f - v_{I_N} f|} \right)^2
\end{aligned}
$$

where Q denotes the query representation, I is an image representation from the data collection, and $c(v_k, v_l)$ and $c(v_n, v_m)$ are the correlation values taken from the correlation matrix..

Thus, we increase the elements corresponding to the visual word in the query with lower frequency value (dominant correlations), and decrease the elements corresponding to the visual word in the query with higher frequency value (least correlated pairs). Having done the similarity measure weighting, we re-rank the top images by calculating the new distance between the query and the images returned in the first round retrieval.

5   Experiments and Discussion

We evaluate the proposed method on three large data collections: ImageCLEFphoto 2007 (20000 images), MIRFlickr 25000 (25000 images) and a collection from British Geological Survey (BGS, 7432 images). The collections differ significantly in size and content. For each of the 100 query topics (60 for ImageCLEF) we retrieve 16 images and calculate Mean Average Precision (MAP). To test the influence of the correlations on the retrieval performance, we generate the correlation matrix from these 16 images, weight the similarity measure, and re-rank the top images. Next, we compute the Mean Average Precision and compare it with the baseline (which does not take the correlations into account).

Some images belong to a few categories. The evaluation on MIRFlickr and BGS collections was the "lenient" one. We assumed that an image is relevant if it shares at least one category with the query image (based on the ground truth data provided).

5.1 Experimental Setup

The implemented local features utilize the random sampling technique. We set the number of sample points to 900. A large number of sample points (in random sampling) is expected to give better results than other sampling methods (see [3]). For each sample point of an image, a 10×10 square patch around it was characterized as multidimensional vector by applying a local descriptor. Each image patch has 9 dimensions (3 for each colour channel), and the codebook size is 40. The visual features, despite using low dimensional vectors and small vocabulary, are comparable with more sophisticated approaches (ImageCLEF2010 Wikipedia Retrieval Task). For a detailed description of the local features used, the reader is referred to [15].

When exploiting the correlations between visual words, we identify 5 dominant and 1 least correlated pair. The p and c parameters' values in the similarity measure are set to 0.5 and 1.31 for all three data collections and were determined experimentally. In case of proximity-based correlation, we will consider two instances of visual words to be correlated if they both appear within a circle of radius 14.15. This is approximately the sum of two diagonals of square sub-images (image patches may overlap).

5.2 Experimental Results and Discussion

Tables 1, 2, and 3 show the experimental results. They present the results for the case when no Pseudo Relevance Feedback was incorporated (NP), when only the dominant correlations were taken into account (D), and the MAPs for both dominant and least correlated pairs (DL). The performance of five notions of correlation is also depicted in the tables. Labels C1, C2, C3 and C4 correspond to correlations 1, 2, 3 and 4 accordingly (see Correlation Between Visual Words section, image-level) whereas C0 denotes proximity-based correlation. The computation of correlation 1 and then addition of matrices is equivalent to commonly used multiplication of the transpose of an image representation matrix by itself. It is one of the standard ways for capturing correlation. Therefore, correlation 1 can also be considered as another baseline. Results presented in bold font are significantly different (two-tailed t-test, 0.05) from the baseline.

It can be seen that C4 and C3 correlations obtained the best results on all three data collections. The addition of information about the least correlated visual words often further improves the performance. Moreover, image level correlations outperformed proximity based one. This may be due to the notion that an image may contain correlated visual words not because of their proximity but because they refer to the same topic.

Table 1. ImageCLEF2007 results (MAP)

	C4	C3	C2	C1	C0
NP	0.0204	0.0204	0.0204	0.0204	0.0204
D	0.0211	0.0211	0.0210	0.0208	0.0206
DL	0.0213	0.0213	0.0211	0.0209	0.0207

Table 2. MIRFlickr results (MAP)

	C4	C3	C2	C1	C0
NP	0.6794	0.6794	0.6794	0.6794	0.6794
D	**0.6938**	**0.6936**	0.6859	0.6869	0.6802
DL	**0.6951**	**0.6936**	0.6854	0.6871	0.6807

Table 3. BGS results (MAP)

	C4	C3	C2	C1	C0
NP	0.3158	0.3158	0.3158	0.3158	0.3158
D	**0.3286**	**0.3286**	0.3187	0.3199	0.3172
DL	0.3268	0.3265	0.3194	0.3193	0.3176

We should be aware, however, that the assumption in PRF framework that all the top documents are relevant to the query may produce a number of false correlations. The process will therefore depend on the adequacy (the ability to capture relevant properties) of the image representation and the retrieval performance of the implemented methods. The real user evaluation should, however, be able to overcome these limitations because all the queries will be selected by the user.

6 Conclusions and Future Work

In this paper we propose a new approach for identifying and utilizing the information about correlations between visual words. We implement and test various notions of correlation at different contextual levels (we refer to them as image-level and proximity based). To the best of our knowledge, this is the first time these two were compared within this type of framework in image retrieval.

Experimental results show the superiority of two notions of correlation, C4 and C3, which are image level correlations. For these two correlations, we report significant improvement in terms of Mean Average Precision on two data collections within PRF evaluation framework. Moreover, the addition of information about the least correlated visual words often further improves the performance. Proximity based notion of correlation does not show a significant improvement in the context of this model.

The proposed method is computationally and data storage cheap, utilizes correlation at different contextual levels, and avoids the normalization of histograms. We believe that the our approach can be successfully incorporated into the experiment involving real users. Thus, a user profile (correlation matrix generated from the query history) could be stored for each individual user, and the information from the profile would be utilized to put the query in the right visual context.

We are planning to extend our evaluation to other various weighting schemes and similarity measures. The ultimate goal, however, would be the aforementioned real user evaluation. The proposed method was developed for this purpose. We will try to take into account the ranking of the top retrieved images and the order of the queries in the query history, as the current query should be given more importance than others. When it comes to the automated methods, like Pseudo Relevance Feedback for example, the assumption that all the top documents are relevant to the query may produce a number of false correlations. The process will therefore depend on the adequacy (the ability to capture relevant properties) of the image representation and the retrieval performance of the implemented methods. The real user evaluation should, however, be able to overcome these limitations and the promising results encourage us to pursue the proposed approach.

References

1. Yang, J., Jiang, Y.G., Hauptmann, A.G., Ngo, C.W.: Evaluating bag-of-visual-words representations in scene classification. In: Proceedings of the International Workshop on Multimedia Information Retrieval, p. 206 (2007)
2. Nowak, E., Jurie, F.: Learning visual similarity measures for comparing never seen objects. In: IEEE Conference on Computer Vision and Pattern Recognition, pp. 1–8 (2007)
3. Nowak, E., Jurie, F., Triggs, B.: Sampling Strategies for Bag-of-Features Image Classification. In: Leonardis, A., Bischof, H., Pinz, A. (eds.) ECCV 2006. LNCS, vol. 3954, pp. 490–503. Springer, Heidelberg (2006)
4. Grubinger, M., Clough, P., Hanbury, A., Müller, H.: Overview of the ImageCLEFphoto 2007 Photographic Retrieval Task. In: Peters, C., Jijkoun, V., Mandl, T., Müller, H., Oard, D.W., Peñas, A., Petras, V., Santos, D. (eds.) CLEF 2007. LNCS, vol. 5152, pp. 433–444. Springer, Heidelberg (2008)
5. Biancalana, C., Lapolla, A., Micarelli, A.: Personalized Web Search Using Correlation Matrix for Query Expansion. In: Cordeiro, J., Hammoudi, S., Filipe, J. (eds.) Web Information Systems and Technologies. LNBIP, vol. 18, pp. 186–198. Springer, Heidelberg (2009)
6. Liu, T., Liu, J., Liu, Q., Lu, H.: Expanded bag of words representation for object classification. In: 16th IEEE International Conference on Image Processing (ICIP), pp. 297–300 (2010)
7. Zhang, S., Huang, Q., Lu, Y., Wen, G., Tian, Q.: Building pair-wise visual word tree for efficient image re-ranking. In: IEEE International Conference on Acoustics Speech and Signal Processing (ICASSP), pp. 794–797 (2010)
8. Yuan, J., Wu, Y., Yang, M.: Discovery of collocation patterns: from visual words to visual phrases. In: IEEE Conference on Computer Vision and Pattern Recognition, CVPR 2007, pp. 1–8 (2007)
9. Savarese, S., Winn, J., Criminisi, A.: Discriminative object class models of appearance and shape by correlatons, pp. 1063–6919. IEEE Computer Society (2006)
10. Jamieson, M., Dickinson, S., Stevenson, S., Wachsmuth, S.: Using language to drive the perceptual grouping of local image features. In: IEEE Computer Society Conference on Computer Vision and Pattern Recognition, vol. 2, pp. 2102–2109 (2006)
11. Wu, L., Li, M., Li, Z., Ma, W.Y., Yu, N.: Visual language modeling for image classification. In: Proceedings of the International Workshop on Workshop on Multimedia Information Retrieval, pp. 115–124 (2007)
12. Zheng, Q.F., Wang, W.Q., Gao, W.: Effective and efficient object-based image retrieval using visual phrases. In: Proceedings of the 14th Annual ACM International Conference on Multimedia, pp. 77–80 (2006)
13. Sivic, J., Zisserman, A.: Video Google: Video Google: A Text Retrieval Approach to Object Matching in Videos. In: Proceedings of the 9th IEEE International Conference on Computer Vision, vol. 2, pp. 1470–1477 (2003)
14. Rui, Y., Huang, T.S., Ortega, M., Mehrotra, S.: Relevance Feedback: A Power Tool in Interactive Content-Based Image Retrieval. IEEE Transactions on Circuits and Systems for Video Technology 8, 644–655 (1998)
15. Kaliciak, L., Song, D., Wiratunga, N., Pan, J.: Novel local features with hybrid sampling technique for image retrieval. In: Proceedings of Conference on Information and Knowledge Management (CIKM), pp. 1557–1560 (2010)

Topical Relevance Model

Debasis Ganguly, Johannes Leveling, and Gareth J.F. Jones

CNGL, School of Computing, Dublin City University, Ireland
{dganguly,jleveling,gjones}@computing.dcu.ie

Abstract. We introduce the topical relevance model (TRLM) as a generalization of the standard relevance model (RLM). The TRLM alleviates the limitations of the RLM by exploiting the multi-topical structure of pseudo-relevant documents. In TRLM, intra-topical document and query term co-occurrences are favoured, whereas the inter-topical ones are down-weighted. The multi-topical nature of pseudo-relevant documents results from the multi-faceted nature of the information need typically expressed in a query. The TRLM provides a framework to estimate a set of underlying hypothetical relevance models for each such aspect of the information need. Experimental results show that the TRLM significantly outperforms the RLM for ad-hoc and patent prior art search.

1 Introduction

An information need expressed in a short query can encompass a wide range of more focused sub-information needs. For instance, the query *Poliomyelitis and Post-Polio* may seek relevant information on polio disease, its outbreaks, on medical protection against the disease, and on post-polio problems. This multi-faceted nature of the information need expressed in a query is manifested in the retrieved documents, as they tend to form clusters of topics [1]. The model proposed in this paper estimates multiple relevance models, each pertaining to a single aspect of the overall information need expressed in a query, as opposed to estimating only one relevance model [2]. Thus, for the example query *Poliomyelitis*, our model would estimate multiple relevance models, one catering for the disease information, one associated with the prevention of the disease, one pertaining to the post-polio problems and so on. Sometimes, the expression of multiple aspects of an information need can be explicit, such as in queries of associative document search [3], where full documents are used as queries to retrieve related documents from the collection, e.g. patent prior art search [4], where each *claim* field of a patent query expresses an individual information need for prior art related to a particular claim. These very long queries describe diverse, sometimes orthogonal information needs, in contrast to short queries. To cater for the different characteristics of the two types of queries, i.e. short queries with implicit multi-topical information needs, and explicitly multi-faceted long queries, we propose two variants of our model: one with the assumption that terms in a query are generated by sampling from a number of relevance models each pertaining to a specific aspect of the information need; and the other with the assumption that each relevance model generates a subset of query terms. We call the two variants unifaceted topical relevance model (uTRLM) and multifaceted topical relevance model (mTRLM), respectively. We provide a formal description for the two

Y. Hou et al. (Eds.): AIRS 2012, LNCS 7675, pp. 326–335, 2012.

variants of TRLM and evaluate both on standard datasets. The remainder of this paper is organized as follows. In Section 2, we describe related work in PRF and topic models. Section 3 introduces the topical relevance model and provides estimation details for the model. Section 4 describes the experimental setup, followed by Section 5 presenting the evaluation results. Section 6 concludes the paper with directions for future work.

2   Related Work

Pseudo-Relevance Feedback. Pseudo-relevance feedback (PRF) is a standard automatic technique in IR which seeks to improve retrieval effectiveness in the absence of explicit user feedback [5]. PRF assumes that top ranked initially retrieved documents are relevant, which are then used to identify terms that can be added to the original query performing an additional retrieval run with the expanded query [5]. PRF can also involve re-weighting of query terms [5,6] and re-ranking initially retrieved documents by recomputing similarity scores, such as the relevance model estimation [2].

Relevance Model (RLM). RLM is a statistical generative model utilizing the co-occurrence between terms in the query and pseudo-relevant documents to improve retrieval quality [2]. However, a limitation of RLM is that it uses whole documents for co-occurrence statistics. This shortcoming of the RLM was addressed by the positional relevance model (PRLM) [7], which assigns higher weights to co-occurrences within close proximity to better estimate the relevance model. We hypothesize that proximity with query terms alone do not adequately identify relevant topics in a document, and that it would be better to apply techniques of topic modelling on the set of pseudo-relevant documents. To this effect, we propose the topical relevance model (TRLM) as an extension to the RLM.

Topic Models. The most widely used topic modelling technique is the latent Dirichlet allocation (LDA) which treats every document as a mixture of multinomial distributions with Dirichlet priors [8]. IR with LDA based document models (LBDM) involves estimating LDA model for the whole collection by Gibbs sampling and then linearly combining the standard LM term weighting with LDA-based term weighting [9]. Linear combination was done because LDA itself may be too coarse to be used as the only representation for IR. In fact they report that optimal results are obtained by setting the proportion of LDA to 0.3 as a complementary proportion of 0.7 for standard LM weighting. Our proposed method overcomes the coarseness of the topic representation limitation by restricting LDA to only the top ranked pseudo-relevant set of documents. This also makes the estimation a lot faster. Another major difference to [9] is that we do not linearly combine document language model scores and the KL divergence scores. We simply calculate the KL divergence between the estimated topical relevance model and the document language model to re-rank each document. Thus, our model does not require an extra parameter for a linear combination, which makes optimization easier.

3   Topical Relevance Model

Overview of the Relevance Model. The key idea in RLM-based retrieval is that both relevant documents and query terms are assumed to be sampled from an underlying

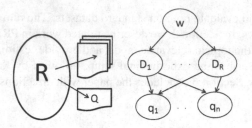

Fig. 1. a) Schematic representation (left) and b) dependence graph (right) of a relevance model

hypothetical model of relevance R pertaining to the information need expressed in the query. In the absence of training data for the relevant set of documents, the only observable variables are the query terms assumed to be generated from the relevance model. Thus, the estimation of the probability of a word w being generated from the relevance model is approximated by the conditional probability of observing w given the observed query terms. This is illustrated in Figure 1a.

Given the query $Q = \{q_i\}_{i=1}^n$ of n independent terms, the probability of generating a word w from an underlying relevance model R is thus estimated as follows.

$$P(w|\text{R}) \approx P(w|q_1, \ldots, q_n) \propto \prod_{i=1}^n P(w|q_i) \qquad (1)$$

An assumption that the query terms are conditionally sampled from multinomial document models $\{D_j\}_{j=1}^R$, where R is the number of top ranked documents obtained after initial retrieval as shown in Figure 1b, leads to Equation (2).

$$P(w|q_i) = \sum_{j=1}^R P(w|D_j)P(D_j|q_i) \propto \frac{1}{R} \sum_{j=1}^R P(w|D_j)P(q_i|D_j) \qquad (2)$$

The last step of Equation (2) has been obtained by discarding the uniform prior for $P(q_i)$, and taking the uniform prior of $P(D_j) = \frac{1}{R}$ outside the summation. Equation (2) has an intuitive explanation in the sense that the likelihood of generating a word w from the relevance model R will increase if the numerator $P(w|D_j)P(q_i|D_j)$ increases, or in other words if w co-occurs frequently with a query term q_i in a pseudo-relevant document D_j. RLM thus utilizes co-occurrence of a non-query term with the given query terms to boost the retrieval scores of documents, which otherwise would get a lower language model similarity score due to vocabulary mismatch. For more details on the RLM, the reader is referred to [2].

Motivation for TRLM. Since co-occurrences in the RLM are computed at the level of whole documents, the co-occurrence of a word belonging to a topic different from the query topics, is not down-weighted as it should be. Thus, it is potentially helpful to compute co-occurrence evidences at the sub-document level, as done in the PRLM using proximity [7]. However, instead of relying on proximity we generalize the RLM by introducing the notion of topics. The RLM has an oversimplified assumption that

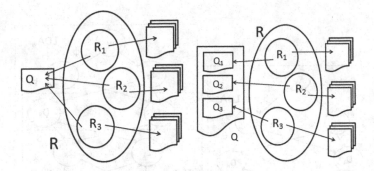

Fig. 2. Generalizations of the RLM a) unifaceted (left) and b) multifaceted (right)

all relevant documents are generated from a single generative model. A query typically encompasses multiple aspects of the overall information need expressed in it. Thus in a more general case, it would be reasonable to assume that the query terms are sampled from a number of relevance models instead of one. This is illustrated in Figure 2a, where it is assumed that the query words are sampled from three different RLMs R_1, R_2 and R_3, and that each RLM R_i generates its own set of relevant documents. Broadly speaking, the sub-relevant models can be thought of addressing each separate topic of the overall information need shown by the encompassing RLM R. We call this model *unifaceted*, because the query terms themselves are assumed to belong to a single topic, whereas the underlying information need might be broad and pertain to different topics. The prefix "uni" in the name thus relates to the query characteristic.

Queries can also be explicitly multifaceted, i.e. structured into diverse information needs, e.g. patent applications in patent prior art search are structured into claims and the requirement is to retrieve prior articles for each such claim. In such a case, we can hypothesize that a query essentially is comprised of a set of sub-queries each of which is sampled from a separate relevance model, as shown in Figure 2b.

TRLM Description. Let R represent the underlying relevance model that we are trying to estimate. In the standard RLM, it is assumed that words, both from the relevant documents and the query, are sampled from R as shown in Figure 1a. In contrast to this, the unifaceted topical relevance model (uTRLM) assumes that a query expresses a single overall information need, which in turn encapsulates a set of sub-information needs. This is shown in Figure 2a where R_1, R_2 and R_3 are specific sub-information needs encapsulated within the more global and general information need R. This is particularly true when the query is broad and is comprised of a wide range of underspecified information needs. The uTRLM thus assumes that the relevance model is a mixture model, where each model R_i generates the words in relevant documents addressing a particular topic, and in addition the query terms as well.

Another generalization which can be made to the RLM is for the case where a query explicitly conveys a set of largely different information needs. Queries in the associative document search domain fall under this category. Segmenting a query into a set of non-overlapping blocks of text and then using each block as a separate query has successfully been applied for associative document search [3], which illustrates that

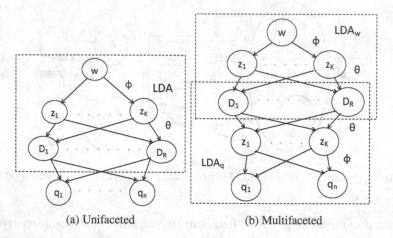

(a) Unifaceted   (b) Multifaceted

Fig. 3. Graphical representation of the two variants of a topical relevance model (TRLM)

such long queries are comprised of multiple information needs. Speaking in terms of the TRLM, it is reasonable to assume that each topical relevance model thus generates its own set of relevant documents and its own subset of query terms. This is illustrated in Figure 2b, which shows that R_1 generates its own set of relevant documents with the subsets of query terms, and leads to our definition of the multifaceted TRLM (mTRLM).

Estimation of the TRLMs. The only observable variables in a TRLM are the query terms. Hence one needs to approximate the probability of generating a non-query term w from the RLM R, by the probability of generating w given that the model has already generated $q_1, \ldots, q_n$. This probability is $P(w|q_1, \ldots, q_n)$, which is thus used as the approximated probability of generating a term from the RLM R, similar to Equation (1). Let us assume that a word w can be generated from a finite universe of topics $z = \{z_1, \ldots, z_K\}$, where each topic z_i addresses the relevance criterion expressed in the sub-relevance model R_i, as shown in Figure 2. Assuming $z \in \mathbb{R}^K$ follows a multinomial distribution $\phi \in \mathbb{R}^K$, with Dirichlet prior β for each ϕ_i, each document $d \in \{D_j\}_{j=1}^R$ in turn comprises of a number of topics, where it is assumed that a topic $z \in \{z_k\}_{k=1}^K$ is chosen by a multinomial distribution $\theta \in \mathbb{R}^K$ with the Dirichlet prior α. With this terminology, we derive the estimation equations for the two variants of TRLM.

The dependence graph of a unifaceted TRLM is shown in Figure 3a. Let us assume that the query terms $\{q_i\}_{i=1}^n$ are conditionally sampled from multinomial unigram document models $\{D_j\}_{j=1}^R$, where R is the number of top ranked documents obtained after an initial retrieval step. Every query term q_i is generated from a document D_j with $P(q_i|D_j)$. Each $P(w|q_i)$ in turn is given by

$$P(w|q_i) = \sum_{j=1}^R P(w|D_j)P(D_j|q_i) \tag{3}$$

Due to the addition of a layer of latent topic nodes, there is no longer a direct dependency of w on D_j, as in the RLM (see Figure 1b and Equation (2)). Hence to estimate

$P(w|D_j)$, we need to marginalize this probability over the latent topic variables z_k. Thus, we have

$$P(w|D_j) = \sum_{k=1}^{K} P(w|z_k)P(z_k|D_j) \tag{4}$$

Substituting Equation (4) in Equation (3) and applying Bayes rule, we obtain

$$P(w|q_i) = \sum_{j=1}^{R} \frac{P(q_i|D_j)P(D_j)}{P(q_i)} \sum_{k=1}^{K} P(w|z_k)P(z_k|D_j)$$

$$\propto \frac{1}{R} \sum_{j=1}^{R} P(q_i|D_j) \sum_{k=1}^{K} P(w|z_k)P(z_k|D_j) = \frac{1}{R} \sum_{j=1}^{R} P(q_i|D_j)P_{LDA}(w|D_j, \hat{\theta}, \hat{\phi}) \tag{5}$$

The last step in Equation (5) is obtained by discarding the uniform prior $P(q_i)$ and replacing the inner summation with the LDA document model. This is shown by the box labelled "LDA" in the dependence graph of Figure 3a. $P(q_i|D_j)$ is the standard probability of generating a term q_i from a smoothed unigram multinomial document model D_j. Equation (5) has a very simple interpretation in the sense that a word w is more likely to belong to the TRLM if it i) co-occurs frequently with a query term q_i in the top ranked documents; and ii) w has a consistent topical class across the set of pseudo-relevant documents. It is also seen from Equation (5) that the uTRLM uses a document model $P_{LDA}(w|D)$, different to the standard unigram LM document probability $P_{LM}(w, D)$ for a document D. This may be interpreted as smoothing of word distributions over topics, somewhat similar to [9], the difference being that the smoothing is done over the set of top ranked documents instead of the whole collection. Using marginalized probabilities $P(w|z_k)$ in Equation (4) leads to a different maximum likelihood estimate in comparison to $P(w|D)$, which is the standard maximum likelihood of a word w as computed over the whole document D. It also ensures that each topic is estimated separately with variable weights as per the prior for each topic i.e. $P(z_k|D_j)$.

The difference between the mTRLM and the uTRLM is the way in which query terms are sampled from document models. While in the uTRLM, a query term is directly generated from a document model, in the mTRLM the query term generation probability is marginalized over the latent topic models, as shown in Figure 3b. Thus it models the fact that not only the pseudo-relevant documents but also a query comprises multiple topics. This is shown by the additional layer of latent topic nodes inserted between the document nodes and the query term nodes. Taking into account the latent topics in a query, $P(q_i|D_j)$ of Equation (5) has to be marginalized over the topic nodes as shown in Equation (6).

$$P(q_i|D_j) = \sum_{k=1}^{K} P(q_i|z_k)P(z_k|D_j) \tag{6}$$

Substituting Equation (6) in Equation (5) and ignoring the denominator $P(D_j)$ by assuming uniform priors, leads to the modified TRLM equation for the mTRLM.

$$P(w|q_i) = \frac{1}{R} \sum_{j=1}^{R} \left(\sum_{k=1}^{K} P(q_i|z_k)P(z_k|D_j) \right) P_{LDA}(w|D_j, \hat{\theta}, \hat{\phi})$$

$$= \frac{1}{R} \sum_{j=1}^{R} P_{LDA}(q_i|D_j, \hat{\theta}, \hat{\phi}) P_{LDA}(w|D_j, \hat{\theta}, \hat{\phi}) \tag{7}$$

Equation (7) thus involves two levels of LDA estimated term generation probabilities, one for the words in pseudo-relevant documents and the other for the query terms. This is shown by the two boxes LDA_w and LDA_q respectively in Figure 3b. Equation (7) ensures that it assigns higher probability to a term being generated from the relevance model, if the term co-occurs with a query term in pseudo-relevant documents and is also likely to belong to the same topic as that of the query term.

4   Experimental Setup

Dataset. We evaluate the uTRLM on the TREC 6, 7, 8 and Robust adhoc test collections using the title field of these queries typically comprising of a few keywords. In addition to testing it on these queries, we also use longer queries in the form of the TREC Robust TDN (Title, Description, Narrative) topics to test both uTRLM and mTRLM. The rationale behind using these longer queries is to examine how the two variants of the model perform for queries which have an intermediate length between the two extremes of either being very short comprising of a few keywords, or being very long as in associative document search. For evaluating TRLM on very long queries we use the CLEF-IP[1] 2010 dataset, which comprises of a collection of patents from the European patent office, where the queries are themselves full patent documents.

Selecting Baselines. Since the evaluation objective is to examine whether the TRLM improves on the RLM, we used the RLM as one of our baselines. Additional term-based query expansion with query re-weighting on top of RLM estimation (denoted as RLM+QE) has been found to improve its effectiveness further [10]. We thus use RLM+QE as a stronger baseline for comparison with the TRLM. To compare RLM and TRLM on the same platform, we implemented both in SMART[2]. GibbsLDA++[3] was used for Gibbs sampling for LDA inference in TRLM. The reason for not using the LBDM approach as a baseline for our experiments is that according to the experiments described in [9], it could not outperform RLM, which in turn implies that our choice of RLM and RLM+QE as baselines is stronger than LBDM.

Parameters. The reported results for our experiments were obtained after tuning the parameters through a series of initial retrieval experiments. The smoothing parameter of initial retrieval (LM) i.e. λ, was optimized empirically to 0.4 and 0.6 respectively for the TREC and CLEF-IP collections. The hyper-parameters α and β which control the Dirichlet distributions for TRLM, were set to $\frac{50}{K}$ and 0.1 respectively as suggested

---

[1] http://www.ir-facility.org/clef-ip/
[2] ftp://ftp.cs.cornell.edu/pub/smart/
[3] http://gibbslda.sourceforge.net/

Table 1. Mean average precision (MAP) values obtained by applying uTRLM on TREC title queries and mTRLM on CLEF-IP queries. * and + indicates statistically significant improvement of TRLM over RLM and RLM+QE respectively.

Topic Set	LM	RLM	RLM+QE	TRLM
TREC-6	0.2075	0.2146	0.2244	**0.2484***+
TREC-7	0.1614	0.1789	0.1805	**0.1816**
TREC-8	0.2409	0.2380	0.2612	**0.2631***
TREC-Robust	0.2618	0.3052	0.3064	**0.3351***+
CLEF-IP	0.0960	0.1081	0.0947	**0.1095**

in [11]. The number of iterations for Gibbs sampling i.e. N, was set to 1000 for all TRLM experiments. We tuned the common parameter R, i.e. the number of top ranked documents used for pseudo-relevance, within the range of $[5, 50]$ so as to obtain the best settings for both the RLM and the TRLM. We did not split up the topic sets into separate training and test sets, but rather the parameters were tuned separately for each individual dataset.

5   Results

Short Queries. The results in Table 1 show that the uTRLM significantly[4] outperforms the RLM for three query sets viz. TREC-6, 8 and Robust. The uTRLM also outperforms RLM+QE, i.e. the RLM with explicit term-based query expansion, even though the latter performs a second retrieval run with additional expansion terms. The limitation of RLM can particularly be seen on TREC-8 where re-ranking documents by RLM in fact decreases MAP with respect to the initial retrieval, whereas RLM+QE increases MAP significantly. By outperforming RLM+QE, the TRLM, which relies only on re-ranking, provides empirical evidence to a more accurate and more robust estimation of the relevance model compared to the RLM.

Very Long Queries. It can be seen from the last row of Table 1, that the mTRLM performs better than the RLM on CLEF-IP dataset. The mTRLM achieves significantly higher MAP over the initial retrieval result LM result, whereas the RLM's improvement over LM is not significant. RLM+QE which performed well for short queries, gives poor results for these long queries. This conforms to previous findings that query expansion is of little or no use for patent search, due to the fact that expansion terms tend to add more noise to the already very long and noisy queries [12]. The mTRLM overcomes the necessity to add expansion terms, thus outperforming RLM+QE, and also marginalizes co-occurrence computation over individual topics in a query instead of the whole query, thus outperforming the RLM.

Sensitivity to the Number of Topics. An important parameter of the TRLM is the number of topics K, which was optimized empirically in the range of $[2, 50]$ for the values reported in Table 1. Figure 4 shows the effect of variations in the number of

---

[4] *Significance* refers to statistical significance by Wilcoxon test with 95% confidence measure.

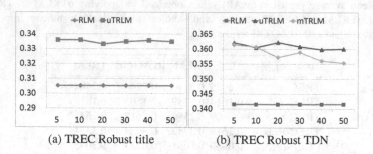

(a) TREC Robust title   (b) TREC Robust TDN

Fig. 4. Effect of varying the number of topics K on MAP

topics on retrieval as measured by MAP. It can be seen from the figures that the retrieval effectiveness is relatively insensitive to the choice of the number of topics. The justification of using a much smaller value range of K in comparison to the global LDA based approach [9], which used much higher values of K in the range of 100 to 1500, is that LDA estimation in the TRLM is done on only a small number of documents in contrast to the full corpus. To see the effect of the parameter K on individual queries, we looked at the MAP values for TREC 6, 7, 8 and Robust queries for different K values in the range of $[2, 50]$ and found that only 24 of 250 queries register a standard deviation higher than 0.02 in MAP, which suggests that the MAP is fairly insensitive to the choice of K and performance is stable for a majority of queries.

Figure 5 highlights the observations for three queries with the highest variances in MAP values. Three patterns of MAP variations for different values of K can be observed in Figure 5: i) a sharp increase, ii) a peak, and iii) a sharp decrease, with increasing K. The first case is illustrated by query *Gulf War Syndrome*, where we note a sharp increase in the MAP with an increase of K, which intuitively suggests that this query is of a very generic nature and the pseudo-relevant documents are associated with a high number of diverse topics. A wide range of symptoms occurring in different

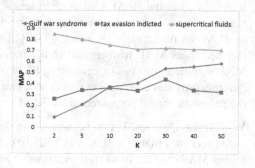

Fig. 5. Effect of K (number of topics) on MAP for three example queries

individuals tend to form separate topics, as a result of which the model is optimized for a high value of K. The case of a distinct peak in MAP is illustrated by the query *tax evasion indicted*. The peak is suggestive of the ideal number of relevant topics for this particular query. This query encapsulates expresses two broad information needs: firstly about tax evasion, and secondly about the people who lost money. Both of these can in turn address individual sub-topics, e.g. there can be many different types of organizations involved in tax evasion. The third case is shown by the query *supercritical fluids*, which is suggestive of a very specific and precise information need. The TRLM for this query thus yields the optimal result with only 2 topics, and the MAP decreases with an increase in the number of topics.

6   Conclusions and Future Work

This paper has presented the TRLM, a novel framework for exploiting the topical association of terms in pseudo-relevant documents. The key contributions of this paper are: i) a theoretical justification of the use of topic models in local context analysis thus addressing aspects of relevance; ii) investigating the use of LDA smoothed document and query models for relevance model estimation; iii) proposing an effective technique for associative document retrieval in a single retrieval step; and iv) outperforming the standard RLM, RLM+QE on queries of diverse types and lengths.The work presented in this paper treats an entire pseudo-relevant document as a unit in the LDA estimation. A possible extension to this approach, which will be investigated as part of our future work, is to use smaller textual units, i.e. sentences or paragraphs as document units in the LDA estimation. This would naturally take into account proximity evidence as well, in addition to the topical distribution of terms.

References

1. Xu, J., Croft, W.B.: Query expansion using local and global document analysis. In: SIGIR 1996, pp. 4–11. ACM (1996)
2. Lavrenko, V., Croft, B.W.: Relevance based language models. In: Proceedings of SIGIR 2001, pp. 120–127. ACM (2001)
3. Takaki, T., Fujii, A., Ishikawa, T.: Associative document retrieval by query subtopic analysis and its application to invalidity patent search. In: Proceedings of CIKM 2004, pp. 399–405 (2004)
4. Piroi, F., Lupu, M., Hanbury, A., Zenz, V.: CLEF-IP 2011: Retrieval in the intellectual property domain. In: Proceedings of CLEF 2011 (2011)
5. Robertson, S.E., Sparck Jones, K.: Relevance weighting of search terms, pp. 143–160. Taylor Graham Publishing (1988)
6. Hiemstra, D.: Using Language Models for Information Retrieval. PhD thesis, Center of Telematics and Information Technology, AE Enschede (2000)
7. Lv, Y., Zhai, C.: Positional relevance model for pseudo-relevance feedback. In: Proceedings of SIGIR 2010, pp. 579–586. ACM (2010)
8. Blei, D.M., Ng, A.Y., Jordan, M.I.: Latent Dirichlet allocation. J. Mach. Learn. Res. 3, 993–1022 (2003)
9. Wei, X., Croft, W.B.: LDA-based document models for ad-hoc retrieval. In: Proceedings of SIGIR 2006, pp. 178–185. ACM (2006)
10. Lang, H., Metzler, D., Wang, B., Li, J.T.: Improved latent concept expansion using hierarchical markov random fields. In: Proceedings of CIKM 2010, pp. 249–258 (2010)
11. Griffiths, T.L., Steyvers, M.: Finding scientific topics. PNAS 101, 5228–5235 (2004)
12. Magdy, W., Leveling, J., Jones, G.J.F.: Exploring Structured Documents and Query Formulation Techniques for Patent Retrieval. In: Peters, C., Di Nunzio, G.M., Kurimo, M., Mandl, T., Mostefa, D., Peñas, A., Roda, G. (eds.) CLEF 2009. LNCS, vol. 6241, pp. 410–417. Springer, Heidelberg (2010)

Adaptive Data Fusion Methods
for Dynamic Search Environments

Shengli Wu[1], Yuping Xing[1], Jieyu Li[1], and Yaxin Bi[2]

[1] School of Computer Science and Telecommunication Engineering
Jiangsu University, Zhenjiang, China
{swu,xingyuping,jli}@ujs.edu.cn
[2] School of Computing and Mathematics
University of Ulster, Northern Ireland, UK
{y.bi}@ulster.ac.uk

1 Introduction

In the web age, publishing information and opinions online is very easy and fast. Since the web is reachable by a huge number of grassroots people, the number and scale of social networking sites [1] are growing at a tremendous speed. It is an interesting thing to find out information, news & events, opinions, etc., exchanged in these sites. Thus quite a few researchers focus on this and some related issues. One major characteristic of these social networking sites is their dynamic nature. When new things or themes appear, they are discussed in quirk, and then forgotten very quickly. It is also true that the thriving and decline of such sites may happen very quickly. How to cope with this dynamic environment is a challenging issue for the information/opinion search services.

Data fusion has been demonstrated as a useful technique to improving retrieval effectiveness. Different data fusion methods, such as CombSum [8,9], CombMNZ [8,9], the linear combination method [2,13,14,16], the correlation method [17], the Borda count [1], the Condercet fusion [11], the probabilistic fusion method [10], Markov chain-based methods [6,12], the multiple criteria approach [7], and others, have been proposed and investigated.

Among all the data fusion methods, the linear combination method is very flexible since different weights can be assigned to different component systems. When choosing component systems, it is not as picky as other data fusion methods such as CombSum, CombMNZ, and Borda count, since poor component systems are not harmful to the linear combination method (if weights are assigned properly) but can be disastrous to some others. Therefore, the linear combination method is very good when component systems vary in effectiveness. A recent investigation [15] finds that if weights can be assigned by proper training using such as multiple linear regression, then the linear combination method is superior to other data fusion methods involved and is able to beat the best component systems by a large margin. For the weights obtained by training, they do not need to change if the condition (such as document collection,

---

[1] http://en.wikipedia.org/wiki/List_of_social_networking_websites

Y. Hou et al. (Eds.): AIRS 2012, LNCS 7675, pp. 336–345, 2012.

query topics, and component retrieval systems involved) is not changed. However, this may cause serious effectiveness deterioration in a dynamic environment in which some or all above three aspects change rapidly. Rather than using a static weighting scheme, we consider that an incremental weighting scheme is more desirable. Therefore, we try to find out effective weighting methods which are applicable in a dynamic environment.

Up to now, there is not much research on dynamic data fusion methods. A few papers such as [3,5] investigated query-specific data fusion methods. Such methods may be useful for routing tasks in which the underneath document collection is updated regularly but the same group of queries are used again and again. In this paper, we focus on the aspect of upgrading of component retrieval systems, which is an important dimension of dynamic retrieval environment. To our knowledge, this has not been explored before.

2   Generating a Benchmark for Testing Dynamic Data Fusion Methods

There are two classes of data fusion methods: equally-treated and biased methods. Equally-treated methods treat all component results equally, while biased methods treat different component results in different ways. CombSum, Comb-MNZ, Borda, Condorcet belong to the first category; while the linear combination, weighted Borda, weighted Condorcet go to the second category. Comparing these two types of methods, equally-treated methods are usually more efficient, but may not perform well when the performance of all component results or the similarity between different pairs of component results vary considerably. On the other hand, biased methods are more able to deal with different situations, but training is needed so as to obtain proper weights for all the results involved.

In the real world, things keep changing. This is especially true for information retrieval/web search. First, the document collection may always under change, for example, in the case of web search. It is also true for many different types of digital libraries, online information services, blogs, and so on. Second, queries issued by users vary over time. Third, the information retrieval system (search engine) may be upgraded regularly. Therefore, the information retrieval environment can be very dynamic if all or some of the above three aspects change over time considerably.

Although quite a few information retrieval events such as TREC, NTCIR [2], CLEF [3] have been held annually for some time, it is difficult for them to take dynamic search environment into much consideration. In these events, a task consists of a group of queries. However, when running queries one by one, the same collection of documents is used and the information retrieval system involved does not evolve over those different queries. That is to say, only one of the three aspects (query) is dynamic. Therefore, the data sets being used in

---

[2] http://research.nii.ac.jp/ntcir/index-en.html
[3] http://www.clef-initiative.eu/

these events are not ideal for testing adaptive data fusion methods. In the first place, it is desirable to generate benchmarks for the testing of adaptive data fusion methods.

Considering the huge cost of generating new data sets, it is reasonable to reuse data sets in TREC or other events with some necessary change. One good candidate for this is the data set of the TREC 2008 blog opinion task. In most TREC tasks, 50 queries are used. However, in the TREC 2008 blog opinion task, it includes 150 queries (queries 851-950 and queries 1001-1050). Later in this paper we re-number them from 1 to 150. As a matter of fact, it has the second largest number of queries in all past years' TREC tasks, only after the TREC 2004 robust track in which 250 queries are used. When more queries are involved, it is more likely to generate some runs with various level of performance over different queries. It is also possible to use more queries for training and/or testing. Thus the experimental results should be more reliable.

What we would do is to make some individual runs, each of which performs quite differently across different block of queries. This can be done by generating some "artificial" runs, and any of which is a mixture of several different original runs. We use the following procedure: first of all, we divide all 150 queries into 3 groups and each comprises 50 queries (1-50, 51-100, or 101-150), where 3 is chosen arbitrarily. 2, 4, 5, or 6 might be reasonable options as well. Then from all 191 original runs, we randomly pick up three different runs r_1, r_2, and r_3. We take the result for the first 50 queries from r_1, result for the second 50 queries from r_2, and result for the third 50 queries from r_3, and mix them together to generate a new run r_m. The above process is repeated until we obtain 191 generated runs, the same number as that of the original ones.

The generated runs have increased the dynamic property of the component results to some degree, mainly in the respect of information retrieval systems. There is not much difference between the generated and original runs on the aspect of average performance. In such a scenario, we can expect that those data fusion methods that work well in the standard TREC setting can still work reasonably well.

3   Adaptive Data Fusion Methods

Suppose that we have a group of component systems, and for any query issued each of them will provide a ranked list of documents. These ranked lists are fused by some data fusion methods. We also assume that the result from any component system for a single query will be evaluated and its effectiveness will be known immediately. Thus we are able to apply adaptive data fusion methods and to adjust weights dynamically. It may be argued that such a condition is quite difficult to satisfy. Anyway, it is possible for us to use some form of feedback information provided by users or just some click-through data from users as a kind of pseudo-relevance feedback. Then we can reckon the performance of the information retrieval system approximately. Another situation is that the user interaction is included in the search process.

The adaptive data fusion methods work in the following way: at the very beginning, since no knowledge about any of the component systems/results is available, we just treat all component systems equally. Afterwards, When the queries are processed, we have more knowledge about the effectiveness of those results involved. Thus we can update the weight for the linear combination method accordingly.

In this study, we propose and investigate two methods of updating weights. One is the simple performance-square updating (referred to as PSU later in this paper), the other is a mixture of performance-square updating and linear regression analysis updating (referred to as the mixed updating method). Both methods update the weights of component systems per query.

PSU is related to the performance-square weighting, which is investigated in [16] for the linear combination method. PSU uses the following equation to update the weight of any component system

$$w_i' = c * w_i + (1 - c) * p^2 \tag{1}$$

where w_i and w_i' are the weights before and after the updating, respectively. p is the performance (measured by average precision over all relevant document levels, and $0 \le p \le 1$) of the system in question on the current query, and c is a parameter that needs to be set in the range of 0-1. Two extreme situations are $c = 0$ and $c = 1$. If $c = 0$, then w_i' is only decided by p^2 of the current query; if $c = 1$, then w_i' is only decided by w_i, and no adaptive updating is allowed.

Multiple linear regression is found to be an effective technique for determining the weights of component systems for the linear combination method [15]. In this study, we apply multiple linear regression to the data from a single query, not from a large number of queries as in [15]. This decision mainly based on the following two considerations: first, because adaptive data fusion methods are used in a dynamic environment and weights assignment and fusion need to be done in running time, storing and processing data generated from a large number of queries may be too costly; second, the search environment may change very rapidly, then it is not good to use too many old historical data.[4]

Suppose for one query q, n documents $d_1,...,d_n$ are retrieved by all m component systems. Every component system IRS_i assigns a score s_{ij} to document d_j. The multiple linear regression tries to minimize u in the following equation

$$u = \sum_{j=1}^{n} [y_j - (\beta_1' * s_{1j} + \beta_2' * s_{2j} + ... + \beta_m' * s_{mj})]^2 \tag{2}$$

where y_j is the judged score of document d_j to the query. If binary relevance judgment is used, it is 1 for a relevant document and 0 for an irrelevant document. When y_j for $(1 \le j \le n)$ and s_{ij} for $(1 \le i \le m, 1 \le j \le n)$ are known, β_1', $\beta_2',..., \beta_m'$ can be calculated out for a group of training data.

---

[4] Using a few more queries for the multiple linear regression is also tried but there is no significant improvement over using a single query.

Further those coefficients (β') can be normalized by

$$\beta_i = \frac{m * \beta'_i}{\sum_{i=1}^{m} \beta'_i} \tag{3}$$

where m is the number of systems involved. After normalization, the average of all βs is 1.

Finally, for the mixed method, we may use the following equation to update weight for every information retrieval system involved:

$$w'_i = c * w_i + 0.5(1 - c) * p^2 + 0.5(1 - c) * c_1 * \beta_i \tag{4}$$

where c and c_1 are two parameters that need to be set empirically. As in Equation 1, w_i and w'_i are the weights before and after the updating, respectively. c represents the rate of weight inherited from previous queries and $1 - c$ the rate of weight updated by the current query. c_1 is another parameter which is used to normalize the value from multiple linear regression so as to make it comparable with the value from performance-square updating.

Theoretically, the multiple linear regression can be used alone. However, when running the multiple linear regression, quite often (about 14% of all the cases tested in this study) no valid coefficients can be found. We guess the reason for this is the data records used in this study are too few (only from one query). This phenomenon never happen if using the data records for 25 queries or more. [5] If no valid coefficients can be found, we just assign 1 to every system involved. The advantage of the mixed method is, when the multiple linear regression does not work, we can still use the performance-square weighting to perform the task.

4   Experimental Setting and Results

It has been found that the logistic model is good for score normalization of information retrieval results [4,15]. The logistic model is also used in this study.

$$s(t) = \frac{e^{a+b*ln(t)}}{1 + e^{a+b*ln(t)}} = \frac{1}{1 + e^{-a-b*ln(t)}} \tag{5}$$

In Equation 5, $s(t)$ is the normalized score of the document at rank t. After using the data in all generated 191 runs, we obtain the values of the two coefficients: $a=0.718$, and $b=-2.183$. Note that the logistic model used here is far from optimum because we treat all the runs and all the queries equally. No matter in which run and for which query, a document's score is only decided by its ranking position.

From all available (generated) runs, we randomly choose 3, 4, 5, 6, 7 ,8 ,9, or 10 of them to perform the data fusion experiment. Apart from the two adaptive methods, CombSum is also tested for comparison.

---

[5] We also tried data records from 2 to 10 queries, this phenomenon still happen from time to time.

For both adaptive methods, every result is given a equal weight of 0.2 initially. The consideration for using such a value is because the average of all 191 runs is about 0.4. After a few steps, the average weight for all the systems will be close to 0.2 ($0.4*0.4=0.16$). For both methods, c is set to 0.05. Thus at each step, the weight generated takes 95% of the old value and takes 5% of the update. c can be set to different values so as to meet different application requirements. The general principle is: the more dynamic the search environment is, the larger value we should set c. For the mixed method, c_1 is set to 0.2. Note in Equation 4, because the average p^2 is about 0.2 and the average of α_i is 1, we set c_1 to 0.2 so as to let both components affect the final result equally. The purpose of such treatment is to make the weights of all the systems as stable as possible throughout the whole process.

4.1   Average Performance of Data Fusion Methods

For each given number (3-10), 200 randomly selected combinations are tested. The experimental results are shown in Figures 1-2, in which each data point is the average of 200 combinations and 150 queries in each combination.

From Figures 1-2, we can see that PSU and the mixed method are a little better (about 1% on AP and 0.5%-0.6% on P@10) than CombSum, no matter which measure is used. Although the difference is small, it is significant at a level of 1% (two-tailed T test). PSU and the mixed method are very close, the difference between them are sometimes statistically significant, but sometimes not. Note that all component runs are close in performance is a favorable condition

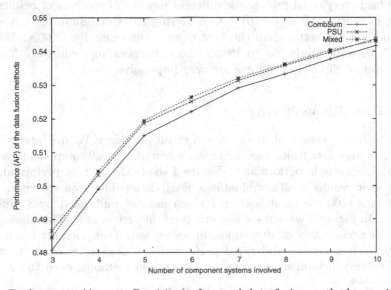

Fig. 1. Performance (Average Precision) of several data fusion methods over 191 runs generated from TREC 2008 blog opinion task

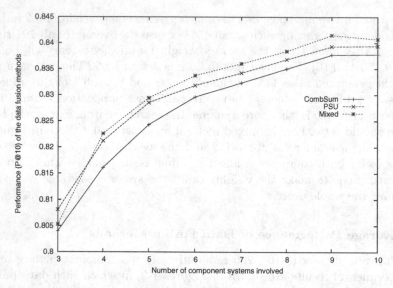

Fig. 2. Performance (10 Document-level Precision) of several data fused methods over 191 runs generated from TREC 2008 blog opinion task

for CombSum to achieve good fusion performance, this is a major reason that is why in this experiment, all three data fusion methods are close in performance.

We also compare all three data fusion methods with the best component result. We find that all three data fusion methods are better than the best component result. This is very consistent across different number of component results and different measures (AP or P@10). [6] Considering AP, CombSum, PSU, and the mixed method are better than the best component result by 42.60%, 43.68%, and 43.74%, respectively. As to P@10, the corresponding figures are 33.64%, 34.09%, and 34.25%. These figures are very impressive.

4.2   Further Discussion

Next let us have a close look at the fusion result per query. We are interested to see how adaptive data fusion methods react when some or all component systems change significantly in performance. Figure 3 shows the average performance of all component results in all combinations. Each data point is the average of 1600 combinations (200 combinations for a given number multiplied by 8 different numbers). In Figure 3 we can see the saw teeth-like curve of it. This shows that the average performance of all component results varies considerably from query to query. Such uncertain and rapid performance change from query to query can not be very helpful in any way to data fusion methods, even to adaptive

---

[6] Several other measures including P@5, P@20, P@100, RP and RR are also used for performance evaluation. The results on them are not presented because they are very analogous to the presented.

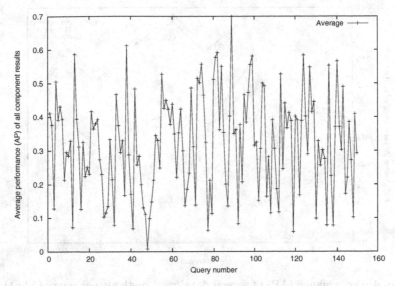

Fig. 3. Average performance (average precision) of all component results per query (1-150)

data fusion methods. If there is a more systematic change, then it is possible for adaptive data fusion methods to make certain changes to the fusion model so as to achieve better performance.

Because each generated run is assembled by three equal-sized blocks (comprising 50 queries) from three different original runs, those queries at the beginning of the second and third blocks are worth attention. Such an effect is just like that the information retrieval system has been changed considerably after 50 queries and after 100 queries again. It is interesting to see how data fusion methods perform when such a considerable change has been made to information retrieval systems. Figure 4 shows the performance of the three data fusion methods per query (query numbers 52-61). The result is very similar for another group of queries 102-111. Here we do not distinguish the number of component results involved. Each data point in Figure 4 is the average of 1600 combinations (200 combinations for a given number multiplied by 8 different numbers). The average performance of all component results are also shown.

From Figure 4, we can see that both PSU and the mixed method are better than CombSum, and all three data fusion methods are better than the average of all component results. For queries 52-61, both PSU and the mixed method are better than CombSum by a little over 2%; for queries 102-111, the improvement rate is 1.5%. In both cases, the difference is significant at a level of 1% (two-tailed T test). It shows that both adaptive methods are doing well when there is a radical change on the implementation of the information retrieval system. Comparing two sub-groups of queries (52-61, 102-111)

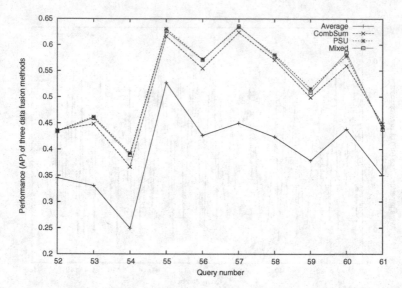

Fig. 4. Performance (AP) of several data fused methods per query (52-61)

with all the queries , we can see that adaptive methods do better for queries 52-61 and queries 102-111 than the average of all 150 queries. The corresponding figures are 2%, 1.5%, and 1% respectively. This shows that the two adaptive data fusion methods are able to cope with systematic changes from component retrieval systems.

5   Conclusions

The purpose of this paper is to investigate adaptive data fusion methods in a dynamic environment. In order to test them, a benchmark has been generated from 191 runs submitted to the 2008 opinion retrieval task. In this benchmark, any run is a mixture of partial results from three original runs submitted. The effect is very much like that the information retrieval systems have been made radical changes twice when running those 150 queries. It happens for the first time after 50 queries have been performed, and happen for the second time after another 50 queries have been performed.

Two adaptive methods, PSU and the mixed method, are presented. Experiments on the benchmark show that both PSU and the mixed method are better than CombSum. Although on average the improvement is small, the difference is statistically significant. More importantly, the proposed methods do better when there is radical changes to some or all of the information retrieval systems.

References

1. Aslam, J.A., Montague, M.: Models for metasearch. In: Proceedings of the 24th Annual International ACM SIGIR Conference, New Orleans, Louisiana, USA, pp. 276–284 (September 2001)
2. Bartell, B.T., Cottrell, G.W., Belew, R.K.: Automatic combination of multiple ranked retrieval systems. In: Proceedings of ACM SIGIR 1994, Dublin, Ireland, pp. 173–184 (July 1994)
3. Bigot, A., Chrisment, C., Dkaki, T., Hubert, G., Mothe, J.: Fusing different information retrieval systems according to query-topics: a study based on correlation in information retrieval systems and trec topics. Information. Retrieval 14(6), 617–648 (2011)
4. Calvé, A.L., Savoy, J.: Database merging strategy based on logistic regression. Information Processing & Management 36(3), 341–359 (2000)
5. Diamond, T., Liddy, E.D.: Dynamic data fusion. In: Proceedings of TIPSTER 1998 workshop, Baltimore, USA, pp. 123–128 (October 1998)
6. Dwork, C., Kumar, R., Naor, M., Sivakumar, D.: Rank aggregation methods for the web. In: Proceedings of the Tenth International World Wide Web Conference, pp. 613–622, Hong Kong, China (May 2001)
7. Farah, M., Vanderpooten, D.: An outranking approach for rank aggregation in information retrieval. In: Proceedings of the 30th ACM SIGIR Conference, Amsterdam, The Netherlands, pp. 591–598 (July 2007)
8. Fox, E.A., Koushik, M.P., Shaw, J., Modlin, R., Rao, D.: Combining evidence from multiple searches. In: The First Text REtrieval Conference (TREC-1), Gaitherburg, MD, USA, March 1993, pp. 319–328 (March 1993)
9. Fox, E.A., Shaw, J.: Combination of multiple searches. In: The Second Text REtrieval Conference (TREC-2), Gaitherburg, MD, USA, pp. 243–252 (August 1994)
10. Lillis, D., Toolan, F., Collier, R., Dunnion, J.: Probfuse: a probabilistic approach to data fusion. In: Proceedings of the 29th Annual International ACM SIGIR Conference, Seattle, Washington, USA, pp. 139–146 (August 2006)
11. Montague, M., Aslam, J.A.: Condorcet fusion for improved retrieval. In: Proceedings of ACM CIKM Conference, McLean, VA, USA, pp. 538–548 (November 2002)
12. Renda, M.E., Straccia, U.: Web metasearch: rank vs. score based rank aggregation methods. In: Proceedings of ACM 2003 Symposium of Applied Computing, Melbourne, USA, pp. 841–846 (April 2003)
13. Vogt, C.C., Cottrell, G.W.: Predicting the performance of linearly combined IR systems. In: Proceedings of the 21st Annual ACM SIGIR Conference, Melbourne, Australia, pp. 190–196 (August 1998)
14. Vogt, C.C., Cottrell, G.W.: Fusion via a linear combination of scores. Information Retrieval 1(3), 151–173 (1999)
15. Wu, S.: Linear combination of component results in information retrieval. Data & Knowledge Engineering 71(1), 114–126 (2012)
16. Wu, S., Bi, Y., Zeng, X., Han, L.: Assigning appropriate weights for the linear combination data fusion method in information retrieval. Information Processing & Management 45(4), 413–426 (2009)
17. Wu, S., McClean, S.: Improving high accuracy retrieval by eliminating the uneven correlation effect in data fusion. Journal of American Society for Information Science and Technology 57(14), 1962–1973 (2006)

NMiner: A System for Finding Related Entities by Mining a Bimodal Network

VenkataSwamy Martha[1], Stephen Wallace[1], Halil Bisgin[1], Xiaowei Xu[1],
Nitin Agarwal[1], and Hemant Joshi[2]

[1] University of Arkansas at Little Rock, Little Rock, AR, USA
{vxmartha,sxwallace,hxbisgin,xwxu,nxagarwal}@ualr.edu
[2] DataMinr Inc., NY, USA
hemant@dataminr.com

Abstract. Motivated from related entity finding problem, in this paper, we introduce a novel approach to query answering called "NMiner." NMiner takes advantage of heuristics to find answers to complex semantic queries. It uses a combination of natural language processing techniques to parse sentences and extract entities, hypertext structure of the documents to derive relational information, and semantic web data to extract relevant entities as search result candidates. Further, a bimodal network of sentences and entities is created from the search result candidates. Content Centric Ranking (CCR) and Cumulative Structural Similarity (CSS), are proposed to rank the candidate entities. Our empirical study on the ClueWeb09 corpus (with approximately 25 terabytes of web documents) shows that both CSS and CCR outperform PageRank and HITS. Moreover, NMiner proved to be significant in solving the problem of answering complex queries performed against a largely unstructured corpus of text documents.

Keywords: Related entity finding, Question answering, Network mining.

1 Introduction

There are several tools available that are designed to browse the large amount of data available on the Internet via basic text searches. Through the use of standard logical and relational operators, users formulate queries that are applied to the corpus of web data, returning a subset of results that meet their search criteria. However, these tools are limited and are typically performing keyword matching. In an attempt to further research along these lines, the Text Retrieval Conference (TREC) [1] introduced a track called Related Entity Finding (REF), designed to extend the knowledge base in the areas of question answering, entity relationships and dedicated entity home pages. Motivated by the REF track, our research has led to the development of a search engine called NMiner.

The proposed system and the underlying algorithms depend on the extraction of entities and their relationships. The Stanford NER tagger [3] is capable of

Y. Hou et al. (Eds.): AIRS 2012, LNCS 7675, pp. 346–355, 2012.

identifying persons, organizations, and locations, and is used extensively by the research community for entity extraction purposes. In addition to relationships among documents in a corpus, the entities in documents exhibit relationships among themselves. Modeling relationships among entities eases the task of finding related entities.

NMiner allows end users to enter a query along with a target entity class and returns related entities for the given query. A sample query might ask for "universities with an NCAA Division I men's lacrosse team". NMiner uses two major components. One component indexes the unstructured web documents corpus from which candidate documents for question satisfaction would later be derived. The other component of the system processes the identified documents to obtain a ranked list of entity answers for a given query. The documents are processed using various third party tools for entity recognition purposes. The recognized entities are then ranked using principles of network analysis. We propose two algorithms to achieve best ranking, they are Content-Centric Ranking (CCR) and Cumulative Structural Similarity (CSS). Experiments are run on the system using various queries and configurations. The results are compared against existing algorithms such as PageRank [5] and Hyperlink Induced Topic Search (HITS) [17], it is evident that NMiner offers a performance enhancement over current search techniques.

Key contributions of the work are as follows:

– Developed a mechanism to identify valid entities using a consensus approach from various entity recognition techniques.
– Proposed a method to model sentence and entity relationships as a network.
– Proposed new ranking methods, Content-Centric Ranking (CCR) and Cumulative Structural Similarity (CSS) for ranking entities in a network.
– Evaluated the new ranking methods against existing ranking methods such as PageRank and HITS.
– Implemented a complete system for finding related entities of a query

Current state of art is presented in Section II, design details for NMiner are discussed in Section III, Network modeling and mining approaches are presented in Section IV, experiments followed by results and discussions in Section V and VI respectively.

2   Related Work

There have been several lines of research concerning Related Entity Finding - Linking Open Data (REF-LOD) in recent years. The application of data mining techniques to web-based data sources [2] have resulted in data collections useful in REF-LOD tasks. Ontologies such as DBpedia, EntityCube, KnowItAll, ReadTheWeb, and YAGO-NAGA have become integral to these efforts by providing comprehensive repositories concerning entities [26]. These repositories consist of reliable information that can be easily accessed and extracted. Machine learning based Natural Language Processing (NLP) tools have long been studied as part of Information Retrieval (IR) systems [23]. Named Entity Recognition (NER) is a subset of this research [24]. Meij et al. [21] used a combination of machine learning and IR tools to tie query-based mechanisms to the DBpedia data repository. Bonnefoy et al. [6] investigated using only web resources to determine with the finest granularity possible the type to which a given entity belongs to. Without a context, the relationship between any two entities can be ambiguous [22]. Graphs can be used to help visualize these relationships [25]. Graphs also allow for the transmission of contextual information through directionality and weighting. Li and Cunningham [19] explained the advantages of using vector space models (VSMs) as a tool for information retrieval (IR) tasks. Mehler et al. [20] and Dehmer et al. [8] expanded on the concept of using VSMs for data representations by incorporating contextual information derived from the hypertext portion of a web document. Several tools have been developed [18] that act as interfaces for extraction of information about Universal Resource Identifiers (URIs) of entities. Typically, IR systems must sort through a list of possible answers and provide a ranked list [10]. Elbassuoni et al. [9] used a Kullback-Leibler divergence calculation as a method for ranking the results returned from such a retrieval tool. A hybrid approach incorporating both context dependent and context independent data is proposed by Mirizzi et al. [7].

Besides the semantic based techniques, there are several network mining algorithms to rank vertices in a network. There is no participant (at least for our knowledge) who used network mining algorithms to find related entities in Clueweb09 dataset. PageRank [5] and HITS [17] are two of them and are used in this work to evaluate NMiner. The PageRank algorithm is widely recognized as being an effective tool for discovering and ranking the relationships among web pages and is the basis for the Google search engine. However, its primary method of ranking pages relies on the hyperlinks between pages and not the content of the pages themselves. This lack of contextual reference to the page content limits the effectiveness of the algorithm in searches requiring an examination of the semantic association of the query terms. The HITS algorithm is a variation on PageRank. Like PageRank, HITS examines the hyperlinks between web pages. Unlike PageRank, the HITS algorithm attempts to identify authoritative pages for a given search term by looking for hubs within the network that "join" other pages together. While this variation does provide a more refined picture of the network structure, it still does not give enough detail to reliably answer queries involving semantic content. Unlike PageRank and HITS, in this paper we propose two algorithms in which one bases on content and other ranks a vertex based on structural information of the vertex. Following section provides illustrative details of NMiner.

3   Nminer Design Details

There are two critical components in NMiner, one for candidate document retrieval and the other is for processing the candidate documents. The first component indexes the unstructured web documents corpus from which candidate documents for question satisfaction would later be derived. The other component examines the structure of each of the candidate documents and identified entities contained therein. The sentences in candidate documents and entities in the sentences are identified using various third party tools. A bi-modal network is constructed to represent the sentence and entity relationships.

Two novel networking mining algorithms, Content-Centric Ranking (CCR) and Cumulative Structural Similarity (CSS), are proposed to rank the entity vertices in the network. The algorithms proved to be significant in solving the problem of answering complex queries performed against a largely unstructured corpus of web documents.

3.1 Documents Retrieval

The objective of this component is to retrieve relevant documents for a given topic/query. The implementation details of the component involves both corpus indexing and candidate documents retrieval.

Corpus Indexing. The Clueweb09 dataset [12] is indexed using Indri software [16] for efficient retrieval. Since the dataset is too large to implement on a single machine, the dataset was segmented into 46 partitions (each partition with a set of raw web documents) and distributed across 46 processing platforms.

Candidate Documents Retrieval. The candidate documents retrieval phase leverages the index built in the previous phase to retrieve query related documents. It involves two steps, topic analysis and document retrieval.

Topics Analysis. A given query typically includes the query narrative and the class of the target entity and the given query is reformulated using topic analysis techniques. Our experiments included both manual and automated topic analysis techniques.

Manual Query Reformulation. Given topic is converted into Indri Query Language by hand. Human intervention is considered advantageous for the following reasons;1. Refinement of generic references, 2. To maintain the integrity of multiword phrases, 3. To introduce synonymic terms.

Automated Query Reformulation. The automated approach involves complex processing of given topic narrative. The processing consists of the following steps:

- Named entities are identified within the topic narrative.
- Stop words are removed except for the term 'of'. The resultant phrase contains the *query keywords.*
- The query keywords are translated to form a weighted query in Indri Query Language where weight is the length of the specific keyword.

3.2 Documents Processing

The documents obtained from the search are processed in an attempt to find answers to given topic. Figure 1 shows the steps involved in the document processing. The documents are parsed using a HTML parser obtained from [13] and to obtain text segments of the documents. A text segment is split into sentences using Apache Incubators OpenNLP sentence detection tool [14]. The lack of individual performance of NER systems motivated us to use more than one entity identification technique. Rather than running the entity taggers in parallel, here they run in sequence. The sequence processing improved accuracy by feeding discovered knowledge about entities to subsequent taggers. The Stanford NER tagger [3] is used for the first phase of entity recognition and the classes assigned by the Stanford NER tagger are ignored. Assuming all nouns in a sentence are entities, Apache Incubator's Parser [14] is used to recognize all noun phrases (including NP, NN, NNP, NNS, NNPS) in other words entities in given virtual sentence.

The entity recognition systems in previous steps introduce noise in terms of false positives (invalid entities) into the dataset. A simple noise cancellation is implemented by using an entity validation process. DBPedia [15] is chosen as our validation dataset. For faster access, the dataset is indexed using Apache Lucene. The entities that are found in DBPedia and matches with target entity class are potential answers for given query. Each sentence is compared to given query and the comparison is measured using the similarity measure called *relevance score*.

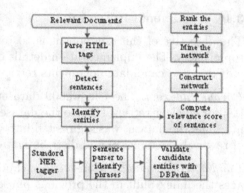

Fig. 1. Document Processing

The score of a sentence is computed as the sum of the length of the query key words found in the sentence.

By the end of document processing, we have the following characteristics for each sentence collected from all of the related documents: *sentence_id, relevance_score, target_entities*.

4   Network Processing

4.1   Network Construction

This phase builds a sentence-entity network based on the relationships inherent in the document structure from which the data is derived. The network is bimodal but not bipartite, as there were relationships among sentences. The network is constructed using the following criteria.

- Each sentence is a vertex.
- Each entity is a vertex.
- A sentence and an entity have an edge if the entity appeared in the sentence.
- The weight of a sentence vertex is its associated relevance score.
- Two sentences are connected if they co-occurred closely in the HTML tree of a document.

4.2   Network Mining

We propose two novel network mining algorithms for ranking entity vertices in sentence-entity network.

CCR Algorithm. The Content Centric Ranking (CCR) algorithm is specifically tailored for the entity ranking purposes. As the name suggests, it leverages content of the network to find critical vertices. Domain knowledge such as relevance score is used for this purpose. As it is a bi-modal network, the algorithm follows different approaches to compute CCR score for each type of vertex. A CCR score of a sentence varies with various parameters. The parameters are as follows.

- The score of a sentence varies in proportion to its relevance score.
- A large number of entities in a sentence leads to less chance of becoming an answer entity. The score of a sentence reduces as the number of neighbor entity vertices rise.
- More generic terms appear in the beginning and end of a document and a lower number of neighbor sentence vertices represent a sentence at a deeper level of the HTML tree and gets in a higher score.

In summary, the score of a sentence vertex of the network can be written as
    CCR score of a sentence = (relevance_score / (countneighbor entities * count neighbor sentences)

A sentence with a high score is assumed to contain answer entities. However, an entity may occur in more than one sentence and can inherit scores from multiple sentence vertices. When there is more than one neighbor sentence for an entity vertex, the maximum value of all neighbor sentences is the entitys CCR score.
    CCR Score of an entity = max score of neighbor sentences

The sentence from which the score is inherited is called a supporting sentence. The supporting sentence acts as evidence to support the validity of the entity as a potential answer for the query.

The CCR algorithm computes a CCR score for each entity and sentence in the network. The scores for entities are sorted to produce ranked answer entities.

Cumulative Structural Similarity (CSS) Algorithm. This is the other algorithm proposed in this paper and can be applied to any undirected network. Similarity between any two network nodes measures how likely they are to share a neighbor [26,4]. There are various similarity measures; cosine, jaccard and minimum. The robustness of cosine similarity inspired us to consider it in this work. Given two vertices i and j, their similarity can be expressed as $S(i,j) = \frac{N_i \cdot N_j}{|N_i||N_j|}$. The Cumulative Structural Similarity (CSS) of a vertex is the sum of similarity scores with its neighbors. $CSS(i) = \sum_{j \in N_i} S(i,j)$ The CSS of a vertex represents an accumulated similarity with its neighbors. A CSS value was calculated for each vertex in the sentence-entity network by considering it as a single-mode network. There was no difference in processing CSS values for sentence and entity nodes. The CSS values from all the entities are sorted to return a ranked list of entities.

5   Experiments

Queries from the TREC 2011 competition for the REF/REF-LOD track are used as input for our experiments. There are 50 queries in all and they are from different domains such as medical, political, educational, etc. For each query, various configurations in the system includes different search engines, different query reformulation techniques, and different result ranking algorithms are tested.

In addition to the Indri search engine over clueweb09 dataset, a third party search engine (Microsofts Bing) is also used as an alternative document searcher. Extensive variations in system configurations helps in analysis and to achieve the improved accuracy.

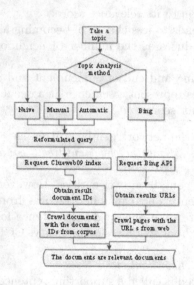

Fig. 2. Topic analysis techniques in implementation

Topic Analysis. Four topic analysis techniques are tested. A flow chart for applying the four topic analysis techniques is presented in Figure 2.

Naive. No changes are made to the original query narrative. The purpose of this configuration is to obtain a baseline.

Manual. Queries are manually reformulated by our team to obtain the best possible related document matches.

Automatic. This configuration is completely automated and programmed to perform query reformulation.

Bing. This approach is similar to the nave configuration with the addition of the Microsoft's Bing search engine as a replacement for Indri.

Results Ranking. In addition to the various search strategies, we also tested four different ranking algorithms in each configuration. The four ranking algorithms are: CCR, CSS, PageRank, and HITS. Therefore, for each query, we have 16 experiments. In each experiment, the outcome is a list of answer entities sorted by their ranks. It is not feasible to manually evaluate the outcome from the experiments, so widely use evaluation measures such as nDCG (Distributed Cumulative Gain at n) and MAP (Mean Average Precision) and are computed on the result sets.

6   Evaluation and Discussion

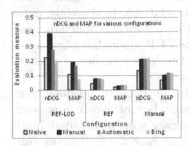

Fig. 3. Plot of nDCG and MAP for four topic analysis techniques

Evaluation of our results needs true answer sets (ground truths) for each query and we prepared three such separate sets. One set is from the TREC REF track organizers, another is from the TREC REF-LOD track organizers, and a third one is manually prepared by browsing through the Internet [1]. In each of our experiments, the outcome of the NMiner system is compared with the three true answer sets to obtain nDCG and Mean Average Precision (MAP) measurements. There are 16 experiments and in each experiment we obtain 6 evaluation measures. This process is repeated for all 50 queries. It is not feasible to plot charts for all 50 queries for this paper, so the evaluation measures are summarized. The summary is in two phases. One is to find the best topic analysis mechanism and the other is to find the appropriate network algorithm.

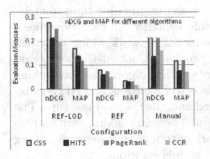

Fig. 4. nDCG and MAP plot for different network mining algorithms

From figure 3, it is clear that manual topic analysis outperforms the other techniques. However, it is not feasible for human intervention to be used for each query in a real-time system. Alternately, we chose the Automatic topic analysis technique as our standard. High nDCG and MAP for Bing run suggests that the search engine plays a critical role in our system.

The nDCG and MAP measures are summarized for each of the ranking algorithms tested and Figure 4 plots the results. From the plot, it is clear that CSS outperforms the other ranking methods. CSS gives best results because it derives the scores from topology of the network. In contrast to PageRank and HITS, CSS is a local structural measure thus finding best answers around given query keywords. The robustness of the cosine similarity adds stability to CSS. Also, the CSS value of a vertex (entity or sentence) is obtained from 2-hop structural information of the network starting from the vertex where PageRank and HITS values are computed from global structural information. The increased granularity of this scoring method provides a more accurate picture of the entitys relationship with its neighbors.

While PageRank performs well on the manual ground truth, the CSS algorithm exhibits the best performance in each case. However, the CSS algorithm

---

[1] There are no enough submissions in REF-LOD track to compare our results and REF track evaluates entity home pages rather than entities.

is computationally costly due to the similarity calculations (i.e. the dot product of two vertex vectors) when compared to CCR. Even so, is typically faster than PageRank or HITS. CCR, on the other hand, is a relatively light-weight data mining tool. Its performance is lower than the other algorithms, but not prohibitively so. From these tests, we surmise that the CCR algorithm is good for limited computational resource environments and CSS performs best if enough computational resources are available.

The running time of the NMiner for a query highly depends on document search system. In our experiments, we used a Linux box of 8 core processor with 16GB RAM. A query took 25 seconds on average with distributed Indri search engine while the same took 45 seconds on average using Bing as this configuration involves crawling the documents from web. The query processing time can be reduced by keeping NLP models in memory as most of the query processing time involved NLP models loading.

7   Conclusion

The primary aim of this research is to answer a given question with a ranked list of entities. This paper has detailed a novel methodology for identifying entities and ranking them in context to a given query. With the advantage of third party parsers like HTML parser and sentence parser, we are able to extract semantic relationships from among the entities and sentences in each document. These relationships are used to construct a bi-modal network which was then used to discover and rank target entities. The proposed Content-Centric Ranking (CCR) and Cumulative Structural Similarity (CSS) algorithms for ranking entities exhibited the best accuracy in comparison to PageRank and HITS ranking techniques because prior one is domain based and later is local measure. The experiments also proved that the CSS algorithm is best in ranking entities provided sufficient resources are available. This work reiterates that network mining algorithms can address complex problems with simple solutions.

Acknowledgments. This work was supported in part by the National Science Foundation under Grant CRI CNS-0855248, EPS-0701890, EPS-0918970, MRI CNS-0619069, and OISE-0729792, US Office of Naval Research (Award: N000141010091) and the US NSF (Awards: IIS-1110868 and IIS-1110649).

References

1. NIST Special Publication 500-207: The 1st Text REtrieval Conference (TREC-1), http://trec.nist.gov/pubs/trec1/t1_proceedings.html
2. Chakrabarti, S.: Mining the Web. Discovering Knowledge from Hypertext Data. Morgen and Kaufmann Publishers (2003)
3. Finkel, J.R., Grenager, T., Manning, C.: Incorporating Non-local Information into Information Extraction Systems by Gibbs Sampling. In: 43rd Annual Meeting of the ACL, pp. 363–370 (2005)

4. Xu, X., Yuruk, N., Feng, Z., Schweiger, T.A.J.: SCAN: a structural clustering algorithm for networks. In: KDD 2007, pp. 824–833 (2007)
5. Brin, S., Page, L.: The anatomy of a large-scale hypertextual Web search engine. In: The Seventh Int'l Conference on WWW7, pp. 107–117 (1998)
6. Bonnefoy, L., Bellot, P., Benoit, M.: The Web as a source of evidence for filtering candidate answers to natural language questions. In: 2011 IEEE/WIC/ACM, WI-IAT 2011, vol. 1, pp. 63–66 (2011)
7. Mirizzi, R., Ragone, A., Di Noia, T., Di Sciascio, E.: Ranking the Linked Data: The Case of DBpedia. In: Benatallah, B., Casati, F., Kappel, G., Rossi, G. (eds.) ICWE 2010. LNCS, vol. 6189, pp. 337–354. Springer, Heidelberg (2010)
8. Dehmer, M., Streib, F.E., Mehler, A., Kilian, J.: Measuring the Structural Similarity of Web-based Documents: A novel Approach. World Academy of Science, Engineering and Technology (2007)
9. Elbassuoni, S., Ramanath, M., Schenkel, R., Sydow, M., Weikum, G.: Language-model-based Ranking for Queries on RDF-Graphs. In: CIKM 2009, pp. 977–986 (2009)
10. Goker, A., McCluskey, T.L.: Towards an Adaptive Information Retrieval System. In: Raś, Z.W., Zemankova, M. (eds.) ISMIS 1991. LNCS, vol. 542, pp. 348–357. Springer, Heidelberg (1991)
11. Grishman, R., Sundheim, B.: Message Understanding Conference-6: A Brief History. In: Proceedings of the Int'l Conference on Computational Linguistics (1996)
12. http://boston.lti.cs.cmu.edu/Data/clueweb09/
13. http://htmlparser.sourceforge.net/javadoc/overview-summary.html
14. http://incubator.apache.org/opennlp/documentation/manual/opennlp.html
15. http://wiki.DBpedia.org/Downloads37
16. http://www.lemurproject.org/
17. Kleinberg, J.: Authoritative sources in a hyperlinked environment. In: 9th ACM-SIAM Symposium on Discrete Algorithms and IBM Research Report RJ 10076 (1998)
18. Mendes, P.N., Jakob, M., García-Silva, A., Bizer, C.: DBpedia Spotlight: Shedding Light on the Web of Documents. In: I-SEMANTICS (2011)
19. Li, Y., Cunningham, H.: Geometric and Quantum Methods for Information Retrieval. ACM SIGIR Forum 42, 2 (2008)
20. Mehler, A., Dehmer, M., Gleim, R.: Towards Logical Hypertext Structure: A Graph-Theoretic Perspective. In: Böhme, T., Larios Rosillo, V.M., Unger, H., Unger, H. (eds.) IICS 2004. LNCS, vol. 3473, pp. 136–150. Springer, Heidelberg (2006)
21. Meij, E., Bron, M., Hollink, L., Huurnink, B., Rijke, M.: Mapping queries to the Linking Open Data cloud: A case study using DBpedia. In: Web Semantics: Science, Services and Agents on the WWW, vol. 9(4), pp. 418–433.
22. Minno, M., Palmisano, D., Mostarda, M.: Slicing Linked Data by Extracting Significant, Self-describing Subsets: The DBpedia Case. In: Daniel, F., Facca, F.M. (eds.) ICWE 2010. LNCS, vol. 6385, pp. 223–231. Springer, Heidelberg (2010)
23. Moffat, A., Zobel, J., Hawking, D.: Recommended Reading for IR Research Students. ACM SIGIR Forum 39(2) (2005)
24. Nadeau, D., Sekine, S.: A survey of named entity recognition and classification. Lingvisticae Investigationes 30(1)
25. Navarro, E., Sajous, F., Gaume, B., Prévot, L., ShuKai, H., Tzu-Yi, K., Magistry, P., Chu-Ren, H.: Wiktionary and NLP: Improving synonymy networks. In: Workshop on the People Web Meets NLP, ACL-IJCNLP, pp. 19–27 (2009)
26. Weikem, G., Theobald, M.: From Information to Knowledge: Harvesting Entities and Relationships from Web Sources. ACM SIGMOD PODS 2010 (2010)

Combining Signals for Cross-Lingual Relevance Feedback

Kristen Parton[1] and Jianfeng Gao[2]

[1] Columbia University, New York, NY, USA
kristen@cs.columbia.edu
[2] Microsoft Research, Redmond, WA, USA
jfgao@microsoft.com

Abstract. We present a new cross-lingual relevance feedback model that improves a machine-learned ranker for a language with few training resources, using feedback from a better ranker for a language that has more training resources. The model focuses on linguistically non-local queries, such as [world cup] and [copa mundial], that have similar user intent in different languages, thus allowing the low-resource ranker to get direct relevance feedback from the high-resource ranker. Our model extends prior work by combining both query-and document-level relevance signals using a machine-learned ranker. On an evaluation with web data sampled from a real-world search engine, the proposed cross-lingual feedback model outperforms two state-of-the-art models across two different low-resource languages.

1   Introduction

Modern web search engines reply heavily on data-driven approaches that go beyond traditional information retrieval (IR) models by incorporating additional features into machine-learned rankers. Typical ranking features include static link analysis features like PageRank, click-through data and document classifiers [2, 6, 10]. The quality of a learned ranker depends to a large degree upon the amount of training data such as human relevance judgments, user feedback and the size of the index or web-graph.

The web is a global resource, serving users in hundreds of regions who speak hundreds of different languages. Optimizing a web search ranker for each of these language/region settings, or markets, is an expensive process, requiring a great deal of annotated data. Even after collecting annotations, ranking features derived from click-through data may not be available for markets with small numbers of users, while link analysis features such as PageRank may not be as helpful for nascent markets with fewer documents and links. Rather than collecting expensive annotated data for each new low-resource market, several strategies have been applied to exploit existing data or models. One approach is to exploit a market with more training data, such as English/US, via model adaptation (e.g. [1, 7]). Another approach, which we explore in this paper, is to use cross-lingual feedback from a high-resource market.

In this study we focus on linguistically non-local (LNL) queries, defined by [8] as concepts that are searched for by users in different markets. For instance, the concept [world cup] [copa mundial] and [coupe du monde] are LNL since they are all about the world cup. In contrast, [brooklyn beaches] is a local query. In practice, a query in

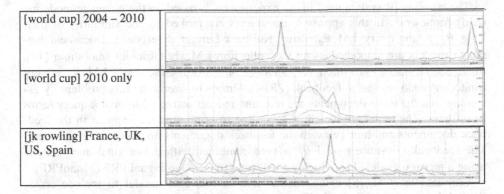

[world cup] 2004 – 2010	
[world cup] 2010 only	
[jk rowling] France, UK, US, Spain	

Fig. 1. Query volume over time for three LNL queries

language L1 is considered LNL if it has a high-confidence translation into a language L2 and the query translation is found in L2 query logs. By this metric, [8] found that at least 11.5% of Chinese queries from their dataset were LNL. Figure 1 shows query volume over time for several LNL queries. Queries in Korean, Russian, Arabic and Spanish for the concept [world cup] follow similar distributions over time, both on a large scale (2004 - 2010) and a much shorter time frame (summer 2010). Although this is not surprising, since the world cup is a time-constrained event that occurs every 4 years, it does suggest that these queries share similar user intent, even though they are in different languages. The query [jk rowling] also shows similar query volume over time in four European markets (France, UK, US, Spain), again indicating related user intent, possibly related to the publication of Harry Potter books or news headlines.

Cross-lingual relevance feedback works by retrieving results for a LNL query in the original language, L1, as well as an assisting language, L2, with a better ranker. Results from L2 are assumed to be better than L1 results, and can be used to improve L1 results, with the help of a translation dictionary. Note that this is not an instance of cross-lingual information retrieval, since the goal is still to return results in L1 only.

Our model, called the unified model, generalizes the existing cross-lingual relevance feedback models by incorporating both query expansion and document re-ranking to further amplify the signal from the high-resource ranker. We use a learning to rank approach, which requires labeled training data. Unfortunately, there are no publicly available corpora with relevance judgments for non-English web search. The datasets used in our experiments are sampled from real-world datasets indexed by a commercial search engine. We present experiments on datasets from two markets, Korean/South Korea and Russian/Russia, using English/US as the assisting market. Our evaluation shows that the proposed unified model outperforms two previous cross-lingual relevance feedback models across two different domains.

2   Cross-Lingual Relevance Feedback

Our unified model extends two previous models, MultiPRF, by Chinnakotla et al. [5], and the model proposed by Gao et al. [8], which we will refer to as DocSim. In this section, we review these models before presenting the unified model.

The baseline IR system used in our experiments is based on the language modeling (LM) framework. In this approach, documents are ranked by the similarity of their LMs θ_d to the query LM θ_q, using Kullback-Leibler divergence. Document language models are smoothed using the collection LM via Dirichlet smoothing [13]. Since search queries are often very short (2 or 3 words), the query LM θ_q is very limited. Pseudo relevance feedback (PRF) attempts to overcome this problem by assuming that the top n documents are relevant and extracting additional k query terms from them. The terms are weighted according to how often they appear in the feedback documents and how relevant the feedback documents are to the original query. The feedback relevance model θ_f is then combined with the original query model using a mixture model [11], which we will refer to as monolingual PRF (MonoPRF).

Chinnakotla et al. [5] extended the monolingual PRF model to include cross-lingual documents. Given an LNL query and search results in L1 and L2, PRF is performed in both languages. PRF terms from L2 are translated back into L1 using a probabilistic translation dictionary $P(f|e)$: $P(f|\theta_{trans}) = \sum_e P(f|e)P(e|\theta_f)$. The final model is called multilingual PRF (MultiPRF) because it does PRF in the query language (L1, θ_f) as well as in the assisting language (L2, θ_{trans}), combining them with a mixture model of Equation (1)

$$P(w|\theta_q'') = (1 - \lambda - \gamma)P(w|\theta_q) + \lambda P(w|\theta_f) + \gamma P(w|\theta_{trans}) \qquad (1)$$

The intuition behind the MultiPRF model is that the L2 corpus is larger than the original L1 corpus, so there are likely to be more relevant documents in L2. Doing query translation (to retrieve the assisting language feedback documents) and then back translation (to translate back the PRF model) also yields a query expansion effect, i.e., synonyms and related terms are added to the query.

Gao et al. [8] introduced the concept of LNL queries, and presented a document retrieval model for cross-lingual relevance feedback, which we will refer to as DocSim. Given an LNL query and search results in L1 and L2, a weighted bipartite graph is created over the documents, connecting L1 documents with L2 documents. The weight of each edge is the cross-lingual document similarity, which is calculated via cross-lingual cosine similarity. Finally, a relational ranking support vector machine is applied so that the ranks of L1 documents move closer to the ranks of similar L2 documents. For example, the official world cup webpage in Arabic is very similar to the official world cup webpage in English, so if the English page is ranked highly, the DocSim model will re-rank the Arabic page to also have a high rank.

3   The Unified Model

Both the MultiPRF model and the DocSim model are motivated by the same observation that there is a high-resource ranker in L2 that has better monolingual accuracy than the L1 ranker. But they are developed based on two different, yet complementary, assumptions. MultiPRF exploits the fact that L1 and L2 queries have shared *query intent*, and works via cross-lingual query expansion. In contrast, the DocSim model assumes that the documents that are relevant to an LNL query contain *related document content*, albeit in different languages.

Our unified model is intended to build on both complementary assumptions, and is a significant extension of the previous research in two aspects. First, we extend both the MultiPRF model and the DocSim model to handle web document structure. Second, the unified model takes the learning to rank framework to which a wide variety of features based on cross-lingual relevance feedback, e.g., those derived from both MultiPRF models and DocSim models, rather than being just limited to query expansion (MultiPRF) or document similarity (DocSim), are incorporated. In our implementation, we used a neural net ranker, called LambdaRank [4], which has been shown empirically to optimize NDCG (Normalized Discounted Cumulative Gain [9]). In the next section we describe the web document structural aspect of the features used by the model, and in the following section we explain the features used in the ranker in detail.

3.1   Web Document Structure

Web documents consists of multiple streams, or fields, which can be divided into content streams, such as url, title and body, and popularity streams, such as anchor text and queries used to access the page. [6] analyzed cross-stream perplexity and found that different language styles are used for composing the document body, title, anchor text, and queries. For instance, the anchor language model is more similar to the query language model than the body language model is. Therefore, each stream should be modeled separately and combined, rather than modeling the document as a single bag of words extracted from different streams. Similarly, BM25F combines weighted term frequencies from different fields, recognizing that some fields are more salient than others [12].

For cross-lingual relevance, document structure is important because popularity fields are the most useful for estimating relevance, but are also more likely to be missing for low-resource languages. Cross-lingual relevance feedback can project popularity fields from the richer market back onto the low-resource market. One potential pitfall could be translation, since popularity fields (such as anchor texts and user queries) are short and have very little context, so they are harder to translate accurately with machine translation systems than body text (or even title text), which usually consists of full sentences.

Another major advantage of incorporating web document structure into the model is speed. If relevance can be approximated by shorter document fields (such as anchor text or title), then doing cross-lingual document similarity is much cheaper. Each document similarity calculation involves word-by-word translation and then cosine similarity, and for example the model of Gao et al. [8] does $n \times m$ similarity calculations.

3.2   Ranking Features

The features used in the ranker can be grouped into three categories, monolingual features, MultiPRF features, and DocSim features.

Monolingual features include baseline ranking features that are used in almost all web search ranking models, such as PageRank (which is query-independent) and

BM25F (which is query-dependent). A baseline retrieval model and monolingual PRF model were built for each document stream. In our experimental dataset, documents have four streams: body, title, url and anchor text, as well as a bag-of-words stream "allfields". Ranking scores for each stream were defined as monolingual features.

MultiPRF features are derived as follows. A MultiPRF model was built for each document stream, and ranker scores for each stream were used as a feature. Overall, there are 5 MultiPRF features.

DocSim features are derived as follows. A single L1 document is compared to each L2 document, and then the cross-lingual similarity score, defined as a DocSim feature, is normalized and combined using a weighted average. Intuitively, this score represents the rank of similar L2 documents. This DocSim feature is computed for each of the 4 document streams. Two standard similarity functions were applied to each document-feedback document pair: Jaccard similarity and cosine similarity. As in [8], cross-lingual similarity was calculated using a translation dictionary. In addition to the cross-lingual similarity functions, similarity functions without translation were also used as features. The goal was to capture transliterations, translations and Latin spellings, which were particularly important for the url field, since the URLs were all in Latin. Certain words may also appear in Latin, even when the document is in another language (e.g., "windows"). Overall, there were two monolingual and two cross-lingual similarity functions for each stream, and 5 streams, for a total of 20 DocSim features.

4   Experimental Setup

4.1   Data

The unified model targets monolingual search in languages with few training resources. We used a re-ranking experimental paradigm where we try to improve the web search results by re-ranking documents retrieved from the entire web using a commercial search engine. We used data from two language/region settings that are linguistically and culturally different from the English/US setting, to see how well feedback from a better ranker from an unrelated domain can improve results. The domains we selected were Korean/South Korea and Russian/Russia. They are quite different from each other in order to see how well the model generalizes across domains.

Since we are only interested in linguistically non-local (LNL) queries, as defined above, we further filtered the data by selecting queries with high confidence translations and queries whose translations were present in English/US query logs. All queries were translated into English with the Bing translator public API[1]. Translations were considered high-confidence if back-translation produced a fuzzy match to the original query. The high-confidence English query translations that occurred in a large set of English/US queries were selected as LNL queries.

Given a LNL query, the English query translation was passed to the public Bing API[2] and the top 50 results were retrieved. Each result consists of a URL, title and

---

[1] http://www.microsofttranslator.com
[2] http://www.bing.com/developers

snippet. For each query, all URLs that were annotated for relevance were crawled, and their anchor texts were also retrieved. Many documents could not be crawled, due to dead pages or errors. Only queries with 10 or more judged documents and non-empty feedback results were kept. Table 1 shows the statistics of the final evaluation dataset used in our experiments.

Table 1. Evaluation datasets

Domain	Queries	Documents	Avg. docs/query
Korean/South Korea	134	1,986	14.8
Russian/Russia	102	1,257	12.3

4.2   IR Setup

Each crawled document was parsed and split into different streams (fields): url, title, body, anchor text and everything, which included all the other streams. Each feedback result was also split into these streams4, and the body field was replaced with the snippet. Since the snippet contains a small amount of text highly relevant to the query, using the snippet instead of the full document retains the signal from the feedback documents, while greatly speeding up the document comparison calculation.

A unigram index was built for each document stream. Each stream was tokenized according to the document language. We used a Viterbi decoder based on a unigram model to break a URL string into tokens. As in the World Wide Web, documents from different languages exist in the same global corpus, although real search engines have more sophisticated techniques for region and language matching.

5   Results

5.1   PRF Baselines

The baseline IR model is defined in Section 2 and has only one tunable parameter, the Dirichlet parameter. The monolingual PRF model (MonoPRF) does PRF based on documents returned by the baseline IR model, and has three additional parameters: the number of feedback documents, the number of feedback terms and the mixture model parameter λ, as in Equation (1) where $\gamma = 0$. The MultiPRF model is a mixture model over the MonoPRF model and the cross-lingual PRF model, as defined in Equation (1). It has four additional parameters: the number of cross-lingual feedback documents and terms, the number of translations per feedback term, and the mixture model parameter γ.

In our experiments, the model parameters were tuned using leave-one-out cross-validation and grid search. Each model's parameters were tuned separately, so for instance, the Dirichlet parameter could end up being different for baseline, MonoPRF and MultiPRF.

Results comparing the baseline IR model with monolingual and multilingual PRF on the two LNL web datasets are shown in Figure 2. MultiPRF outperforms the baseline and the monolingual PRF model in most cases (except for NDCG at 1 for Korean). The improvements for the Russian domain are all statistically significant.

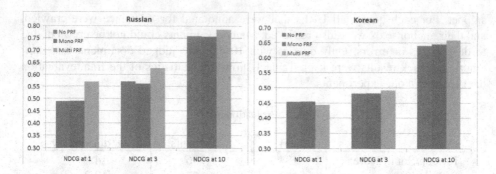

Fig. 2. Results of PRF baselines

5.2   DocSim Baseline

The DocSim model uses LambdaRank to learn a ranking model, based on the 4 document similarity features computed only on the "allfields" stream of each document, as described in Section 4.2. The LambdaRank model was a single layer neural network with 200 iterations. All reported results are from 5-fold cross-validation.

In contrast with the MultiPRF baseline, which uses query expansion for feedback, the DocSim model learns to rank L1 documents similar to the rank of similar L2 documents, based on the assumption that the L2 ranker is better. If too few feedback documents are used, there may be no similar documents to learn from. However, if too many are used, there may be too much noise in the features for the model to learn a coherent ranker. Results of applying the DocSim model with different numbers of feedback documents are shown in Figure 2. With a small number of feedback documents, the results are often worse than the baseline. However, as the number of feedback documents increases, the NDCGs improve. For the Korean domain, 25 feedback documents performed best, while for the Russian domain, 50 feedback documents were best.

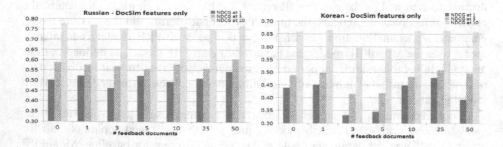

Fig. 3. Results of DocSim baseline

Table 2. Machine-learned rankers with differen features

Models	Russian NDCG at 1	NDCG at 3	NDCG at 10	Korean NDCG at 1	NDCG at 3	NDCG at 10
Baseline	50.13	58.80	77.56	43.99	46.94	65.47
MultiPRF	54.67	61.88	79.41	48.61	51.31	67.89
DocSim	55.59	59.79	77.14	47.88	50.83	67.67
Unified	56.27	62.80	79.35	47.30	51.74	68.46

5.3   Unified Model

For the final comparison, machine-learned rankers were built using baseline features (PageRank and monolingual BM25F) plus features from each model. The MultiPRF ranker had as features ranker scores from the MultiPRF rankers, while the DocSim ranker had all the DocSim features. The unified ranker had all features. Results in Table 2 shows that cross-lingual relevance feedback almost always outperforms the monolingual baseline. For the Russian domain, the unified model outperforms both the MultiPRF model and the DocSim model, except at NDCG at 10, where MultiPRF does slightly better. For the Korean domain, the unified model outperforms both DocSim and MultiPRF at NDCG at 3 and 10, but does worse at NDCG at 1. Overall, the unified model is the best performer in our experiments across both datasets.

6   Discussion and Examples

The motivating hypotheses behind the cross-lingual relevance feedback model was that linguistically non-local (LNL) queries in different languages have similar query intent and relevant documents have related content, so a poor ranker should be able to get direct feedback from a better ranker in another language. The original experiments with MultiPRF used CLEF collections, where the queries were compared over the different sites to ensure that a high percentage of them will find some relevant documents in all [language/domain] collections [3]. In contrast, extracted LNL queries were simply those that had been searched for in both languages, so there may not be as much relevant content in L2.

Surprisingly, queries that skewed heavily towards L1 were not always harmed by MultiPRF. For instance, the queries [환율] (naver) and [в контакте] (in touch) are both navigational queries to popular local websites. In the first case, the L2 ranker already knows that [naver] is a navigational query to the Korean website, so the feedback only helps. In the second case, the highest weighted English terms are still relevant, and the irrelevant terms have much lower weight.

As expected, MultiPRF did harm queries when the query translation was bad. Although query translation used a state-of-the-art MT system and the translations were filtered for "high confidence" translations via back translation, some queries were still translated poorly: [한국 일보 미국] was "hankook ilbo usa" (a partial transliteration) instead of "korea times usa"; [гадание] was "divination", instead of the more colloquial "fortune telling" or "palm reading".

Table 3. LNL queries and their retrieval results

LNL query (translation)	MonoPRF (translation)	L2 PRF	
живая природа (wildlife)	это (this), автор (author), 2010, раздел (section), alexey, далее (more), читать (read), природы (nature)	wildlife, animals, fish, service, www, utah, society, colorado, us, texas	Topic drift towards irrelevant L2 domain
работа во франции (work in France)	франции (France), туры (tours), франции (France), франция (France), работа (work), ru, отдых (relax), www, au, pair	france, work, french, employment, working, visa, travel, living, abroad, visas	L2 focuses more on work than tourism

However, even queries that are truly LNL and correctly translated can be harmed by MultiPRF. For instance, in the first example in Table 3, searching for [wildlife] in the English/US domain brings up many US-specific wildlife associations, which harm the Russian results. In the second example, the English results help re-focus the query towards working and living in France (and getting visas) instead of visiting and touring France.

7   Conclusions and Future Work

We presented a cross-lingual feedback model that aims to improve a ranker from a market with few training resources using feedback from a better ranker with richer training resources. Focusing on linguistically non-local queries allows the model to use direct feedback from the better ranker, rather than just using domain adaptation. Our model extends and generalizes prior work by incorporating both query- and document-level features. Query expansion using multilingual pseudo-relevance feedback exploits the similar *intent* of the original query and the translated query, while the document similarity features leverage related content in both languages, using translation dictionaries to bridge the cross-lingual gap. The model incorporates web document structure to further amplify the noisy signal from the better ranker. The cross-lingual unified relevance feedback model outperformed the monolingual baseline across two different domains.

While the results of this pilot study are promising, the biggest hurdle we faced was data size, and in future work we would like to apply our model to a much larger dataset. Unfortunately, we are unaware of any publicly available web search relevance judgments for languages other than English.

Another promising direction is to exploit more cross-lingual web features. For example, there are many cross-lingual anchor texts (e.g., English links pointing to Chinese pages) and user clicks (e.g., Russian queries that lead to English pages). These types of features would give stronger evidence of shared content across cross-lingual documents, or shared cross-lingual query intent. The English/US domain is also richer in popularity fields (such as anchor text and clickstream features) than some other domains. Exploiting this structural asymmetry should improve the feedback model even more.

References

1. Bai, J., Zhou, K., Xue, G., Zha, H., Sun, G., Tseng, B., Zheng, Z., Chang, Y.: Multitask learning for learning to rank in web search. In: CIKM (2009)
2. Bennett, P.N., Svore, K.M., Dumais, S.T.: Classification-enhanced ranking. In: WWW, pp. 111–120 (2010)
3. Braschler, M., Peters, C.: Cross-language evaluation forum: Objectives, results, achievements. Information Retrieval 7, 7–31 (2004)
4. Burges, C.J.C., Ragno, R., Le, Q.V.: Learning to rank with nonsmooth cost functions. In: NIPS, pp. 193–200 (2006)
5. Chinnakotla, M.K., Raman, K., Bhattacharyya, P.: Multilingual prf: english lends a helping hand. In: SIGIR, pp. 659–666 (2010)
6. Gao, J., He, X., Nie, J.-Y.: Click through-based translation models for web search: from word models to phrase models. In: CIKM, pp. 1139–1148 (2010)
7. Gao, J., Wu, Q., Burges, C., Svore, K.M., Su, Y., Khan, N., Shah, S., Zhou, H.: Model adaptation via model interpolation and boosting for web search ranking. In: EMNLP, pp. 505–513 (2009)
8. Gao, W., Blitzer, J., Zhou, M.: Using english information in non-english web search. In: Proceeding of the 2nd ACM workshop on Improving Non English Web Searching, iNEWS, pp. 17–24 (2008)
9. Jarvelin, K., Kekalainen, J.: Ir evaluation methods for retrieving highly rele-vant documents. In: SIGIR, pp. 41–48 (2000)
10. Joachims, T.: Optimizing search engines using clickthrough data. In: KDD, pp. 133–142 (2002)
11. Lavrenko, V., Croft, W.B.: Relevance-based language models. In: SIGIR, pp. 120–127 (2001)
12. Robertson, S.E., Zaragoza, H., Taylor, M.J.: Simple bm25 extension to mul-tiple weighted fields. In: CIKM, pp. 42–49 (2004)
13. Zhai, C., Lafferty, J.D.: A study of smoothing methods for language models applied to ad hoc information retrieval. In: SIGIR, pp. 334–342 (2001)

A Conceptual Model for Retrieval of Chinese Frequently Asked Questions in Healthcare

Rey-Long Liu and Shu-Ling Lin

Department of Medical Informatics
Tzu Chi University, Hualien, Taiwan
rlliutcu@mail.tcu.edu.tw

Abstract. Frequently asked questions (FAQs) in healthcare provide general readers with both reliable and readable healthcare information. In this paper, we present a conceptual retrieval technique that serves as a supplement to enhance existing FAQ retrievers to find Chinese healthcare FAQs for each input query. By analyzing the structures and goals of Chinese healthcare FAQs, we identify three types of essential concepts in healthcare FAQs: *event*, *condition*, and *aspect*, as a Chinese healthcare FAQ often cares about some aspects (e.g., cause) of some events (e.g., cardiovascular disease) under some condition (e.g., patients of the periodontal disease). The proposed conceptual retrieval technique is thus named ECA (Event, Condition, and Aspect). Given healthcare FAQs annotated by the three types of concepts, ECA can measure the conceptual similarities between an input query and the FAQs. Empirical evaluation on real-world Chinese healthcare FAQs shows that the conceptual similarity information provided by ECA is helpful for an FAQ retriever to have significantly better performance in identifying relevant FAQs for input queries.

Keywords: Frequently Asked Questions, Healthcare, Conceptual Retrieval.

1 Introduction

Healthcare information should be both reliable and readable. Frequently asked questions (FAQs) in healthcare provide such reliable and readable healthcare information as they are often written and compiled by healthcare professionals for general reader. Many healthcare information providers thus collect and maintain a large number of healthcare FAQs for general readers. Given an input query in natural language form, the retrieval of relevant FAQs is thus essential for the utility of the readable and reliable healthcare information for health promotion and disease management.

Many FAQ retrieval techniques have been developed in previous studies. Given a database of FAQs and a natural language question as an input query, an FAQ retriever ranks the FAQs in the database based on the relevancy of the FAQs to the query. The FAQ retrieval task is challenging as both the query and the FAQs are often quite short, making it difficult to collect much information to identify relevant FAQs.

In this paper, we analyze the conceptual structure of Chinese healthcare FAQs and employ the conceptual structure to provide helpful information to existing FAQ

Y. Hou et al. (Eds.): AIRS 2012, LNCS 7675, pp. 366–375, 2012.

retrieves so that relevant FAQs can be ranked higher. More specifically, we present a conceptual retrieval technique that serves as a supplement to enhance existing FAQ retrievers to find Chinese healthcare FAQs relevant to input queries. In the conceptual model, we identify three types of essential concepts: *event, condition,* and *aspect,* as a Chinese healthcare FAQ often cares about some aspects (e.g., cause) of some events (e.g., cardiovascular disease) under some condition (e.g., patients of the periodontal disease). For example, a Chinese healthcare FAQ "兒童常吃山藥會不會引發性早熟?" (For children, will frequently eating yams cause precocious puberty?) has two event concepts "山藥" (yams) and "性早熟" (precocious puberty); a condition concept "兒童" (children); and an aspect concept "引發" (cause). Obviously, to identify relevant FAQs for an input query, the FAQ retriever should consider the similarities on the three types of concepts. Therefore, our conceptual retrieval technique for healthcare FAQs is named ECA (Event, Condition, and Aspect), which aims at providing such conceptual similarity information to the retriever so that more relevant FAQs may be ranked higher for the input query.

Main contributions of ECA include (1) practically, retrieval of healthcare FAQs is a key to share reliable and readable healthcare information; (2) technically, previous studies have developed many FAQ retrievers but none of them have considered the above conceptual structures of Chinese healthcare FAQs, and hence the collaboration between ECA and the previous retrievers can further enhance the retrievers by conceptual similarity information. An empirical evaluation on thousands of Chinese healthcare FAQs shows that, by collaborating with ECA, an FAQ retriever can perform significantly better in ranking relevant FAQs higher for input queries.

2   Related Work

An FAQ often consists of two parts: a question part and an answer part. Several previous studies considered both parts for FAQ retrieval (e.g., [6][7][8]). In this paper, the proposed technique ECA focuses on the question part, which has been shown to be the most important part in FAQ retrieval [4]. ECA provides the conceptual similarity on the question part as additional information for FAQ retrievers.

Given q and f as an input query (question) and the question part of an FAQ respectively, previous techniques have employed several types of information to estimate the similarity between q and f: (1) the overlap of the words in q and f [1]; (2) the cosine similarity based on the vectors of q and f [1–2][7]; (3) the relatedness of the words in q and f (e.g., measured by the distance of the words on an ontology [2][6] [7] or the correction of the spelling errors of words in q [1][3]); (4) the correlation between the question types of q and f [7]; (5) the mapping or translation between the words in q and f (for tackling the problem of word mismatch between q and f [4][8]); and (6) the similarity between the syntactic or semantic structures of q and f (by deeper analysis such as parsing [6]).

However, none of the similarity estimation techniques considered essential concepts (*event, condition,* and *aspect*) in Chinese healthcare FAQs. ECA is the first framework aiming at the conceptual retrieval of FAQs for healthcare consumers.

Technically, the essential concepts considered by ECA can indicate a part of semantics of the query and FAQs, and hence parsing is perhaps a way to recognize the essential concepts in the query and FAQs. However, a good parser for Chinese healthcare queries is often unavailable since (1) parsing Chinese questions is a challenging task [5], and (2) a Chinese healthcare query is not always well-formed for parsing and it may even consist of multiple sentences or fragments, which are difficult to parse. ECA thus does not rely on parsing.

Table 1. Essential concepts in a healthcare FAQ

Type	Concept	Definition
Event	E1	The first target event (in the FAQ) under discussion
	E2	The second target event (in the FAQ) under discussion (may be none if the question has only one target event)
Condition	C	The condition or the context (in the FAQ) of the discussion
Aspect	A	A_{cause}: The causal aspect of the discussion (e.g., risk factors and prevention of diseases)
		$A_{process}$: The processing aspect of the discussion (e.g., treatment and management of diseases)
		$A_{diagnosis}$: The diagnosis aspect of the discussion (e.g., symptoms and diagnosis of diseases)

3   Conceptual Retrieval for Chinese Healthcare FAQs

3.1   Conceptual Structure of Chinese Healthcare FAQs

We hypothesize that core intentions of healthcare FAQs should be about health promotion and disease management. Analysis on thousands of healthcare FAQs on KingNet[1] justifies the hypothesis and provides the evidences of identifying the fundamental types of concepts in Chinese healthcare FAQs: (1) *event*: the target event under discussion, (2) *condition*: the condition of the discussion, and (3) *aspect*: the information aspect of the discussion.

More specifically, Table 1 defines the three types of concepts. As healthcare questions are often quite short and specific to one to two target events, we consider at most two target events E1 and E2. All concepts governing the context or condition of the discussion are the condition concept (C). To facilitate the recognition of the aspect concept (A), we define three high-level categories of aspects: A_{cause}, $A_{process}$, and $A_{diagnosis}$ that are about the *causal*, *processing*, and *diagnosis* aspects of the health topics respectively[2]. Note that a healthcare FAQ does not necessarily have all types of concepts, but it should have at least one concept, which is E1 (e.g., an FAQ may be simply a disease name, and hence it is asking for all information about the disease).

The expressive power of the concept types is justified by a manual annotation of the three types of concepts to thousands of Chinese healthcare FAQs. It is interesting

---

[1] Available at http://www.kingnet.com.tw
[2] The terms corresponding to each aspect should consist of at least two Chinese characters.

to note that the annotation of the concepts to each FAQ is both *feasible* and *helpful* for FAQ retrieval, based on two reasons: (1) concept annotation to an FAQ is conducted only once (e.g., conducted when the FAQ is edited and entered to the database); and (2) the annotation is not a heavy burden and can be done by most people with basic understanding of the definition of the concepts.

3.2   Measurement of Conceptual Similarity between Queries and FAQs

Given q as an input query and f as the question of an FAQ annotated with concepts, ECA estimates the conceptual similarity (CR) between q and f by Equation 1, where $S_{E1}(q,f)$, $S_{E2}(q,f)$, $S_C(q,f)$, and $S_A(q,f)$ are the similarity values on E1, E2, C, and A, respectively.

$$CR(q,f) = \begin{cases} 0, & \text{if } f \text{ has no } E2, \text{ and } S_{E1} = 0; \\ 0, & \text{if } f \text{ has both } E1 \text{ and } E2, \text{ and } S_{E1} = S_{E2} = 0; \\ average \{S_x(q,f) \mid x \text{ is } E1, E2, C, \text{ or } A \text{ in } f\}, & \text{otherwise}. \end{cases} \quad (1)$$

ECA assigns 0 to the similarity between q and f if they talk about totally different target events; otherwise the conceptual similarity is the average of the similarity values on those concepts that are annotated to f (recall that an FAQ does not necessarily have E2, A, and C). When ECA works with an FAQ retriever, the final score of f for q is computed by Equation 2, where $Score(q,f)$ is the score from the FAQ retriever.

$$FAQscore(q,f) = Score(q,f) \times 2^{CR(q,f)} \quad (2)$$

The estimation of S_{E1}, S_{E2}, S_C, and S_A can be approached in different ways (e.g., intelligent string matching or employing an ontology of synonym terms). In this paper, ECA employs Equation 3 as a string matching method to estimate the similarity (*StringS*) between two strings t_1 and t_2, where $idf(w)$ is the inverse document frequency[3] of the Chinese character w. *StringS* is higher if there are more matched characters with higher idf values.

$$StringS(t_1, t_2) = \frac{\sum\limits_{y \in t_1 \text{ and } y \in t_2} idf(y)}{\sum\limits_{z \in t_1 \text{ or } z \in t_2} idf(z)} \times \log_2 |\{y \mid y \in t_1 \text{ and } y \in t_2\}| \quad (3)$$

Similarity Measurement for *Event* (S_{E1} and S_{E2}). Suppose $e1_f$ is the string in f that is annotated with concept E1. ECA employs Equation 4 to find the string $e1_q$ in q that has the largest similarity with $e1_f$, and accordingly sets $S_{E1}(q,f)$ to the similarity between $e1_q$ and $e1_f$ (by Equation 5).

$$e1_q = \arg\max_{t \in q} \{StringS(t, e1_f)\} \quad (4)$$

---

[3] The *idf* value of a Chinese character w is calculated by treating each FAQ as a document, and hence no additional training data is required.

$$S_{E1}(q,f) = StringS(e1_q, e1_f) \tag{5}$$

If f has another event concept E2 (with $e2_f$ as the corresponding string), ECA constructs a query q' by deleting $e1_q$ from q. ECA uses $e2_f$ to find $e2_q$ from q' (by Equation 6) and then estimates $S_{E2}(q,f)$ by Equation 7.

$$e2_q = \arg\max_{t \in q - e1_q}\{StringS(t, e2_f)\} \tag{6}$$

$$S_{E2}(q,f) = StringS(e2_q, e2_f) \tag{7}$$

Similarity Measurement for *Condition* (S_C). If f is annotated with a condition concept and the corresponding string is c_f, ECA constructs a query q' by deleting $e1_q$ and $e2_q$ (if any) from q. ECA uses c_f to find c_q from q' (by Equation 8) and then estimates $S_C(q,f)$ by Equation 9.

$$c_q = \arg\max_{t \in q - e1_q - e2_q}\{StringS(t, c_f)\} \tag{8}$$

$$S_C(q,f) = StringS(c_q, c_f) \tag{9}$$

Similarity Measurement for *Aspect* (S_A). As information aspects are actually *categories* of interest (rather than specific events and conditions), an aspect may be indicated by many different strings. We thus collect the strings annotated for A_{cause}, $A_{process}$, and $A_{diagnosis}$, and accordingly construct TA_{cause}, $TA_{process}$, and $TA_{diagnosis}$ as their sets of corresponding strings respectively. The strengths of correlating q to A_{cause}, $A_{process}$, and $A_{diagnosis}$ are thus estimated by Equation 10~12, respectively.

$$AS(q, A_{cause}) = \max_{t \in TA_{cause}} \max_{c \in q - e1_q - e2_q - c_q}\{StringS(t, c)\} \tag{10}$$

$$AS(q, A_{process}) = \max_{t \in TA_{process}} \max_{c \in q - e1_q - e2_q - c_q}\{StringS(t, c)\} \tag{11}$$

$$AS(q, A_{diagnosis}) = \max_{t \in TA_{diagnosis}} \max_{c \in q - e1_q - e2_q - c_q}\{StringS(t, c)\} \tag{12}$$

We employ 0.5 as the threshold for the strength: q is said to ask for an aspect if its strength corresponding to the aspect is higher than 0.5; otherwise it does not ask for the aspect. It is interesting to note that if q does not ask for any aspect (e.g., q consists of a disease name only), a *'don't-care'* is assigned to each aspect for q, indicating that *all* aspects may be related to the intention of q. Similarly, if no aspect is annotated to f, a *'don't-care'* is assigned to each aspect for f as well. Based on the aspect category recognition, Table 2 defines a way to estimate the similarity value on an aspect (A_{cause}, $A_{process}$, or $A_{diagnosis}$). Finally $S_A(q,f)$ is the average of the similarity values on the three aspects.

Table 2. Similarity $S_m(q,f)$, where $m \in \{A_{cause}, A_{process}, A_{diagnosis}\}$, and 'O' denotes 'asks for m'; 'X' denotes 'does not ask for m'; and '?' denotes *don't-care* on aspect m

q	f	S_m	q	f	S_m	q	f	S_m
O	O	1	X	O	0	?	O	1/2
O	X	0	X	X	1	?	X	1/2
O	?	1/2	X	?	1/2	?	?	1

4   Empirical Evaluation

4.1   Healthcare FAQs

The experimental FAQs were from KingNet[4], which is a Chinese healthcare information provider. All FAQs in KingNet were collected, and we thus got 3517 FAQs. Following the definition of the three types of essential concepts (ref. Section 3.1), each FAQ was manually annotated with its concepts. The sets of strings annotated for A_{cause}, $A_{process}$, and $A_{diagnosis}$ (i.e., TA_{cause}, $TA_{process}$, and $TA_{diagnosis}$, respectively) have 74, 124, and 35 strings respectively.

4.2   Test Queries

Test queries were collected from other five healthcare information providers (not from KingNet, which was the source of experimental FAQs). We totally got 200 test queries[5]: (1) from the first provider[6], 90 test queries were collected by selecting the top-5 most popular FAQs in each category of FAQs; (2) from the second provider[7], 22 test queries were collected by selecting all FAQs of the categories about physical fitness and nutrition; (3) from the third provider[8], 25 test queries were collected by selecting five FAQs from each category; (4) from the fourth provider[9], 60 test queries were collected by selecting five FAQs from each category; and (5) from the fifth provider[10], 3 test queries were collected by selecting top-3 FAQs.

Each of the 200 test queries was manually checked to identify relevant FAQs from the 3517 experimental FAQs. For each pair of a query and an FAQ, a relevancy level was tagged based on the question parts of the query and the FAQ: *definitely relevant*, *partially relevant*, and *non-relevant*. Among the 200 queries, 129 queries had relevant (definitely relevant or partially relevant) FAQs and the average number of relevant FAQs of a query was 3.87.

---

[4] All FAQs on http://www.kingnet.com.tw were collected in February 2012.
[5] The test queries were collected in June 2012.
[6] Available at http://www.healthcare.com.tw/healthcare-front/
[7] Available at http://www.ch.com.tw/index.asp?title=1
[8] Available at http://www.tmn.idv.tw/
[9] Available at http://olddoc.tmu.edu.tw/pinging/index.htm
[10] Available at http://cisc.twbbs.org/index.php

4.3   Underlying FAQ Retriever

We employed FAQFinder [2] as the underlying FAQ retriever. ECA collaborated with FAQFinder, and we aimed at investigating the extent to which the conceptual similarity information provided by ECA was helpful for FAQFinder to have significantly better performance in identifying relevant FAQs for input queries.

The selection of FAQFinder as the underlying retriever can be justified by two reasons: (1) FAQFinder served as a baseline in many previous studies as well (e.g., [7]), (2) the similarity measures employed by FAQFinder were employed by many previous FAQ retrievers as well (e.g., the cosine similarity based on the vectors of input queries and FAQs [1][7] and the relatedness of the words in input queries and FAQs using an ontology [6][7]). Given that no previous retrievers considered conceptual similarity as ECA and FAQFinder shared technical designs with many previous retrievers, contribution of ECA to FAQFinder is of technical significance.

Given q and f as an input query and an FAQ respectively, FAQFinder linearly combined three parts of similarities between q and f: (1) cosine-based term-vector similarity, (2) relatedness of terms measured by the distance of the terms in an ontology, and (3) percentage of the terms in q that matched the terms in f. To make FAQ-Finder able to process Chinese healthcare questions more properly, a sequence of steps were conducted to q and f: (1) segmenting the question into Chinese terms by the CKIP system[11]; (2) translating the Chinese terms into English terms by the Google translation system[12]; (3) identifying the concept ID for each term on a medical ontology UMLS[13] (Unified Medical Language System) by the MMTx terminology matching system[14]. The concept IDs were used to locate the concepts on the ontology MRREL[15] from UMLS so that the distance between two concepts could be measured (for measuring the second part of similarity values considered by FAQFinder).

FAQFinder has several system parameters, including the one that controls the maximum distance between two terms on the ontology and the weights to linearly combine the three parts of similarity values. To tune the parameters, we conducted 4-fold cross validation using the 129 test queries that have relevant FAQs.

4.4   Evaluation Criteria

We employed two evaluation criteria: *mean average precision* (MAP) and *normalized discount cumulative gain at x*, which are popular criteria in measuring the performance of ranking systems. MAP is defined to be

---

[11] Available at http://ckipsvr.iis.sinica.edu.tw/.
[12] Available at http://translate.google.com.tw/.
[13] Available at http://www.nlm.nih.gov/research/umls/.
[14] Available at http://metamap.nlm.nih.gov/.
[15] Available at http://www.nlm.nih.gov/research/umls/licensedcontent/umlsknowledgesources.html

$$MAP = \frac{\sum_{i=1}^{|Q|} P(i)}{|Q|}, \quad AP(i) = \frac{\sum_{j=1}^{k} \frac{j}{FAQ_i(j)}}{k}$$

where $|Q|$ is the number of queries, k is number of relevant FAQs for the ith query, and $FAQ_i(j)$ is the number of FAQs whose ranks are higher than or equal to that of the jth relevant FAQ for the ith query. That is, $AP(i)$ is actually the average precision of the ith query, and MAP is the average of the AP values of all queries. On the other hand, NDCG@x is defined to be

$$NDCG @ x = \frac{\sum_{i=1}^{|Q|} N_i(x)}{|Q|}, \quad N_i(x) = Z_x \sum_{j=1}^{x} \frac{2^{r_i(j)} - 1}{\log_2(1+j)}$$

where $r_i(j)$ is the relevance level of the FAQ whose rank is j with respect to the ith query, and Z_x is set to some value so that a perfect ranking for the ith query gets an $N_i(x)$ value of 1.0. NDCG@x is the average of the $N_i(x)$ values of all queries. In the experiment, we report results when x ranges from 1 to 10.

For NDCG@x, relevance levels of definitely relevant, partially relevant, and non-relevant FAQs were 2, 1, 0, respectively. Also note that MAP only considers binary relevance levels (i.e., relevent and non-relevant), and hence when computing AP for a query, those FAQs that are definitely relevant or partially relevant to the query were considered to be relevant to the query.

4.5   Result and Analysis

Fig. 1 shows the average results of the 4-fold experiment. The results showed that the conceptual similarity information provided by ECA was helpful for FAQFinder, which has employed different kinds of similarity information that were employed by many other FAQ retrievers as well. To verify whether the performance improvements were statistically significant, we conducted two-sided and paired t-test with 95% confidence level. The results showed that ECA helped FAQFinder to achieve significantly better performance in both MAP and NDCG@2~10.

Table 4 shows an example to illustrate the contribution of ECA. There are two FAQs: one is relevant and the other is non-relevant for an input query. The essential concepts of the two FAQs were annotated manually, while E1 of the query was recognized by matching E1 of each FAQ in the query (in the example, for both FAQs, the identified E1 of the query is the same: 'non-stable food'). FAQFinder assigns scores to both FAQs by linearly combining its three parts of similarities (the weights for the three parts are tuned to be 0.8, 0.2, and 0 respectively), but it assigns a larger score to the non-relevant FAQ. ECA successfully assigns a smaller CR score to the non-relevant FAQ as E2 and C of the FAQ cannot be identified in the query.

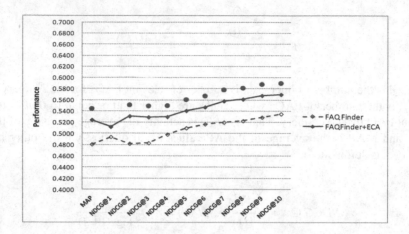

Fig. 1. Average results of the 4-fold experiment: FAQFinder significantly performs better when it collaborates with ECA ('•' indicates statistically significant performance difference with 95% confidence level)

Table 3. An example to illustrate the contribution of ECA: Given a query q, ECA gives the non-relevant FAQ a smaller similarity score (CR) as E2 and C cannot be identified in q

	Essential Concepts	Score by FAQ-Finder	Score by ECA
[Query] 嬰兒副食品 是否可以添加調味料? (Can seasonings be added to non-staple food for infants?)	Concept recognition (by ECA): (E) E1=副食品 (non-staple food); (C) None; (A) A_{cause}=True.	None	None
[Relevant FAQ] 餵食 副食品可能會遇到的 問題? (What problems should be considered when eating non-staple food?)	Concept annotation: (E) E1=餵食副食品 (eating non-staple food); (C) None; (A) A_{cause} = $A_{process}$ = $A_{diagnosis}$ = don't-care.	Part 1: 0.2445 Part 2: 0.5 Part 3: 0.5 → Score = 0.2956 (= 0.8 * 0.2445 + 0.2 * 0.5)	S_{E1}=0.77; S_A=0.5; → CR = **0.635** (= 0.77 + 0.5) / 2).
[Non-relevant FAQ] 我的小baby 厭奶,可 以補充副食品嗎? (Can non-staple food be given to my little baby who is tired of milk?)	Concept annotation: (E) E1=副食品(non-staple food); E2=厭奶 (tired of milk) (C) C=小baby (little baby); (A) A_{cause} = $A_{process}$ = $A_{diagnosis}$ = don't-care.	Part 1: 0.2705 Part 2: 0.5 Part 3: 0.5 → Score = **0.3164** (= 0.8 * 0.2705 + 0.2 * 0.5).	S_{E1}=1; S_{E2}=0; S_C=0; S_A=0.5; → CR = 0.375 (= 1 + 0 + 0 + 0.5) / 4).

5   Conclusion and Future Work

In this paper, we analyze the conceptual structure of Chinese healthcare FAQs, and based on the analysis, present a technique ECA that provides conceptual similarity information to enhance existing FAQ retrievers in ranking Chinese healthcare FAQs for input queries. Empirical evaluation on thousands of Chinese healthcare FAQs justifies the expressive power of the conceptual model and the significant contribution of ECA to FAQ ranking. The result is of practical significance to the utility of healthcare FAQs for general readers as well as theoretical significant to the studies of FAQ retrieval in the healthcare domain. We are exploring two interesting extensions for ECA: (1) developing machine learning methods to more properly integrate the conceptual similarity information from ECA with other kinds of similarity information (i.e. refining Equation 2), and (2) employing more resources (e.g., domain-specific ontology of terms) to estimate the similarity between the concepts in input queries and FAQs (i.e., refining Equation 3).

Acknowledgments. This research was supported by Tzu Chi University under the grant TCRPP101007.

References

1. Bernhard, D., Gurevych, I.: Answering Learners' Questions by Retrieving Question Paraphrases from Social Q&A Sites. In: Proceedings of the Third ACL Workshop on Innovative Use of NLP for Building Educational Applications, Columbus, Ohio, USA, pp. 44–52 (2008)
2. Burke, R.D., Hammond, K.J., Kulyukin, V., Lytinen, S.L., Tomuro, N., Schoenberg, S.: Question Answering from Frequently Asked Question Files Experiences with the FAQ FINDER System. AI Magazine 18(2) (1997)
3. Contractor, D., Kothari, G., Faruquie, T.A., Subramaniam, L.V., Negi, S.: Handling Noisy Queries In Cross Language FAQ Retrieval. In: Proceedings of the 2010 Conference on Empirical Methods in Natural Language Processing, pp. 87–96. MIT, Massachusetts (2010)
4. Jeon, J., Croft, W.B., Lee, J.H.: Finding Similar Questions in Large Question and Answer Archives. In: CIKM 2005, Bremen, Germany (2005)
5. Lee, C.-W., Day, M.-Y., Sung, C.-L., Lee, Y.-H., Jiang, T.-J., Wu, C.-W., Shih, C.-W., Chen, Y.-R., Hsu, W.-L.: Boosting Chinese Question Answering with Two Lightweight Methods: ABSPs and SCO-QAT. ACM Trans. Asian Lang. Inform. Process. 7(4), Article 12 (2008)
6. Wu, C.-H., Yeh, J.-F., Lai, Y.-S.: Semantic Segment Extraction and Matching for Internet FAQ Retrieval. IEEE Transactions on Knowledge and Data Engineering 18(7) (2006)
7. Wu, C.-H., Yeh, J.-F., Chen, M.-J.: Domain-Specific FAQ Retrieval Using Independent Aspects. ACM Transactions on Asian Language Information Processing 4(1), 1–17 (2005)
8. Xue, X., Jeon, J., Croft, W.B.: Retrieval Models for Question and Answer Archives. In: Proceedings of SIGIR 2008, Singapore (2008)

LDA-Based Topic Formation and Topic-Sentence Reinforcement for Graph-Based Multi-document Summarization

Dehong Gao[1], Wenjie Li[1], You Ouyang[2], and Renxian Zhang[1]

[1] Department of Computing, The Hong Kong Polytechnic University, Hong Kong
[2] Miaozhen Systems, Beijing, China
{csdgao,csrzhang,cswjli}@comp.polyu.edu.hk
ouyangyou@miaozhen.com

Abstract. In recent years graph-based ranking algorithms have attracted much attention in document summarization. This paper introduces our recent work on applying a topic model, namely LDA, in graph-based summarization. In the proposed approach, LDA is used to automatically identify a set of semantic topics from the documents to be summarized. The identified topics are then used to construct a bipartite graph to represent the documents. Topic-sentence reinforcement is implemented to calculate the salience scores of topics and sentences simultaneously. By incorporating the information embedded in the topics, the sentence ranking result can be improved. Experiments are conducted on the DUC 2004 data set to evaluate the effectiveness of the proposed approach.

Keywords: Multi-document summarization, Graph-based sentence ranking, Latent Dirichlet Allocation.

1 Introduction

Document summarization requires producing a short summary for one or more documents that conveys the most important information in the original document(s). With the explosive increase of documents on the Internet, document summarization has proved to be an essential task for accessing the overloaded information. Generally, document summarization approaches can be divided into abstractive or extractive. Abstractive summarization approaches usually rewrite the original documents to generate a concise abstract that requires sophisticated nature language generation techniques. In contrast, extractive summarization approaches just select a number of indicative text fragments from the original documents to form a summary and thus are much easier to implement. Up to the present, most summarization approaches are still built upon the extractive summarization framework in which sentence ranking is the issue of most concern.

Various sentence ranking approaches are proposed for extractive summarization through the years. More recently, the application of graph-based ranking algorithms

Y. Hou et al. (Eds.): AIRS 2012, LNCS 7675, pp. 376–385, 2012.

has attracted much concern. The basic assumption of graph-based sentence ranking approaches is that the importance scores of two textual units (such as words, sentences or documents) are related to the degree of their relevance. Many graph-based ranking algorithms are examined in the context of document summarization, including the HITS-style mutual reinforcement algorithms, the PageRank-style random walk algorithms and their variations.

As for graph-based sentence ranking approaches, two pivotal issues are the definitions of the nodes and edges of the text graph. In most existing approaches, nodes are usually defined by words, sentences, documents etc. Edges are then defined by the relationship between these textual units. Though various graph-based ranking algorithms based on different textual units are well studied, the definitions of the nodes in most studies are confined to these original textual units in the documents. To overcome this limitation, some recent studies focus on incorporating high-level information beyond the original textual units into the ranking algorithms. Typically, they adopt clustering algorithms to achieve the objective. The sentences in the given documents are first clustered into sentence clusters that are regarded as meaningful topics. The sentence clusters are then integrated into the text graph as extra nodes. Based on this, cluster-based graph models are developed to rank the sentences through a mutual reinforcement process between clusters and sentences, using the cosine similarity metric to construct the edges between sentences and clusters. What we argue here is that it may not be very appropriate to define and manage the topics in this way. In topic detection carried out by K-means clustering or hierarchical clustering, each sentence is assumed to belong to one cluster only. However, the mutual reinforcement process needs to formulate the relevance between every sentence and every cluster, which in fact postulates the hypothesis that a sentence is related to more than one cluster. Therefore, a conflict exists here indeed and it makes the cluster-based reinforcement algorithm less reasonable.

In view of this, we try to define topics in another way so that its definition can conform to the hypothesis of the reinforcement scheme. A powerful topic model, namely the Latent Dirichlet Allocation (LDA), is adopted in our study to form the topics with such characteristics. For the documents to be summarized, the LDA model is first estimated on them to discover a set of LDA topics. Then the estimated model is inferred on each sentence to obtain the topic-sentence relations. Afterwards a bipartite graph is constructed according to the obtained statistics and a reinforcement algorithm is developed to calculate the sentence ranking scores. We believe that LDA is more suitable for topic detection than clustering for the following reasons. First of all, each topic in LDA is defined as a distribution over the words and each sentence is viewed as a mixture of the topics. Therefore, the LDA-based topics can well satisfy the reinforcement hypothesis that a sentence is supposed to be related to more than one topic. Secondly, the edges of the bipartite graph are defined by the conditional probability under the LDA model, which are well consistent with the definition of the nodes.

The remainder of the paper is organized as follows. Section 2 introduces the related work. Section 3 provides a brief introduction of LDA-based topic formation and Section 4 details the topic-sentence reinforcement ranking algorithm. Section 5 presents the experimental results. Section 6 concludes the whole paper.

2   Related Work

Document summarization is to produce a short summary for one or more documents which can convey the most important information in the original documents. Technically, document summarization can be divided into abstractive and extractive. The abstractive technique involves paraphrasing sections of the source documents; while the extractive technique just selects indicative text fragments from the original documents to generate the summary. Up to now, most summarization approaches are still built on the extractive approach which is easy to implement, and we focus on the extractive method as well in this paper.

The most essential issue in extractive method is the ranking of text fragment. The centroid-based MEAD, implemented by [15], ranks sentences by features like sentence length, cluster centroids, position in contexts. In NeATS [12], they add theme information in sentence ranking. Recently graph-based summarization approaches are mainly inspirited by the analysis of the link structure of the World Wide Web, such as Google's PageRank [9] and Kleinberg's HITs [10]. In [8], Zha considered the mutual reinforcement principle that a term which appears in many salient sentences should have a high saliency score and so does a sentence which contains many salient words. Based on this principle, a bipartite graph is built upon the pair-wise relevance between each sentence and each word and then a HITS-style ranking algorithm is applied to calculate the saliency scores of both sentences and words through a mutual reinforcement process. Later, other graph-based models are also investigated for the sentence ranking problem. Erkan and Radev [11] developed an undirected graph to represent the documents by regarding sentences as vertices and calculating the cosine similarity between sentences for establishing the links. The saliency scores of the sentences are then calculated by an algorithm called LexRank, which is derived from PageRank. TextRank [13, 14] is another representative approach motivated by Page-Rank that shares similar ideas. Later on, Otterbacher et al. [6] adopted a weighted undirected graphs to represent the given document sets by the cosine similarity. Then, a PageRank-style random walk model with edge weights is proposed to calculate sentence ranking scores based on the weighted graph. Besides the above works that mainly focus on generic summarization, graph-based ranking algorithms are also employed in query-oriented summarization which is the leading research topic in the area recently.

The usage of clusters in graph-based ranking models was introduced by Wan and Yang [7]. In their method, a set of automatically identified sentence clusters are regarded as the meaningful topics for the given document set. These sentence clusters are then used as the extra nodes in the document graph. It is reported that the cluster-level information is able to improve the sentence ranking result through a mutual reinforcement process between clusters and sentences. In order to better combine the clustering process and the ranking process, Cai et al. [16] conducted a further research in which a reinforcement approach that tightly integrates the graph-based ranking model and the clustering scheme was proposed.

3   Topic Formation by LDA

Latent Dirichlet Allocation (LDA) [1] is a generative model that allows sets of observations to be explained by unobserved groups which explain why some parts of the data are similar. Given a set of documents $D=\{d_1, d_2, ..., d_N\}$ and its vocabulary $W = \{w_1, w_2, ..., w_M\}$, LDA assumes that there are a set of latent topics $Z=\{z_1, z_2, ..., z_K\}$ (K is pre-specified). Each document d_j is viewed as a mixture of the topics in Z and each topic z_k is a distribution over the word vocabulary W. Two kinds of distributions, i.e. the per-document topic distributions $p(Z|d_j)$ and the per-topic word distributions $p(W|z_k)$, are both modeled by multinomial distributions with Dirichlet priors. Different from the well-known bag-of-word model, words in LDA are assumed independent of each other given the topics instead of given the documents, i.e., $P(w_i) = \Sigma_i P(w_i|z_k) P(z_k)$. Compared to other topic models such as pLSA, the use of the Dirichlet priors in LDA results in more reasonable topic mixtures. The smoothed generative model of LDA is illustrated in Fig. 1, in which α and β denote the parameters of the Dirichlet priors for $p(Z|d_j)$ and $p(W|z_k)$ respectively, θ and Φ denote the parameters of the multinomial per-document topic distribution and the per-topic word distributions respectively.

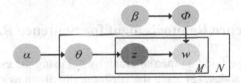

Fig. 1. The generative model of LDA

The LDA model is too complex to be solved by exact inference. In the original work by Blei, the Variational Bayesian method is used to approximate the posterior distribution for inference. Griffiths and Steyvers [4] propose an inference method based on Gibbs sampling that calculates the topics without explicitly representing the parameters. This method is competitive in speed and performance. In the method, θ and Φ are estimated by sampling from the posterior distribution $P(Z|D, \alpha, \beta)$ which is calculated by

$$P(Z \mid D, \alpha, \beta) = \frac{P(D, Z \mid \alpha, \beta)}{P(D \mid \alpha, \beta)}$$

In the above equation, $P(Z, D|\alpha, \beta)$ is calculated by assuming symmetric Dirichlet priors for α and β and integrating θ and Φ out from $P(Z, D, \theta, \Phi|\alpha, \beta)$. On the other hand, $P(D|\alpha, \beta)$ is approximated by a Gibbs Sampling process.

Provided with a set of samples from the posterior distribution $P(Z|D)$, the estimations of θ and Φ can be derived from the number of times a word w_i has been assigned to a topic z_k (denoted by $n(w_i|z_k)$) and the number of times a word from d_j has been assigned to a topic z_k (denoted by $n(d_j|z_k)$ in the samples. Here we can write the estimations as

$$P(z_k \mid d_j) = \frac{n(d_j \mid z_k) + \beta}{\sum_k n(d_j \mid z_k) + K\beta} \quad \text{and} \quad P(w_i \mid z_k) = \frac{n(w_i \mid z_k) + \beta}{\sum_i n(w_i \mid z_k) + M\beta}$$

and illustrate them in Fig. 2 below.

Given the estimated topics Z, we also view a sentence in the document set as a mixture of the topics. Using the similar Gibbs sampling processes, the estimated LDA topics can be inferred on each sentence to obtain its per-topic distribution, i.e. the probabilities $\{P(z_k \mid s_l)\}$ (we denote the sentence set as $S = \{s_1, s_2, \ldots, s_L\}$). We then use these probabilities to obtain the relations between sentences and topics in the bipartite graph, as detailed next.

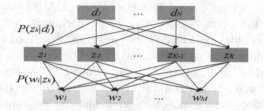

Fig. 2. The estimated probabilities from the Gibbs sampling process

4   Topic-Sentence Reinforcement for Sentence Ranking

For a document set, we build a bipartite graph by (1) using the estimated LDA topics and the sentences as nodes; (2) using the per-topic distributions of the sentences and the per-word distributions of topics to create the bipartite edges, as illustrated in Fig. 3 below.

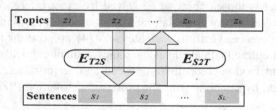

Fig. 3. The topic-sentence bipartite graph

Formally, the constructed graph is denoted as $G = (V, E)$, where the nodes set $V = V_S \cup V_Z$, V_S consists of nodes of sentences in S and V_Z consists of the nodes of topics in Z. As in the mutual reinforcement algorithm, the weighted edges indicate the importance scores passed between nodes. Therefore, it is natural to use the probability $P(Z \mid S)$ inferred by LDA as the weights of the edges from sentences to topics. It reflects the degree of a specific sentence being dominated by a specific topic. On the other hand, the weights of the edges from topics to sentences cannot be simply defined by this probability. These weights reflect how much importance a topic should give to a sentence, or how dominative the sentence is to the topic. A sentence that is much dominated by the topic may still not be dominative since it may only consist of some less indicative words of the topic. Therefore, here we need to re-define the weighted edges from sentences to topics, different from the ones from topics to sentences.   To calculate the weights, we assume that a sentence is supposed to be more dominative to a given topic if the words it contains are dominative to this topic.

In practice, E_{S2T} is used to denote the edges from sentence nodes to topic nodes, and the inferred per-topic distribution of sentences $P(Z|S)$ just acts as the weights for them. On the other side, E_{T2S} denotes the edges from topic nodes to sentence nodes and we use the word distribution $P(T|Z)$ to calculate the average word probability of the sentence as the weight, i.e.,

$$E_{T2S}(k, l) = \frac{\sum_{t \in s_l} P(t|Z_k)}{|S_l|}$$

where t and $|S_l|$ denote a word in the sentence S_l and the total word number of l respectively, $P(t|z_k)$ is from the estimated word proportion of topic Z_k. In most existing approaches, the edges are defined by the cosine similarity metric and thus are symmetric. However, the weighted edges defined in our graph are asymmetric. In fact, in the original HITS model for hyperlink analysis, the edges indicating the linking information between web pages are also asymmetric because the two edges with opposite front and tail ends indeed indicate different relations. Note that the asymmetric weighted edges do not affect the correctness of the mutual reinforcement algorithm.

In the mutual reinforcement algorithm, the authority scores (denoted as Authority$(\cdot)$) of sentences and the hub scores (denoted as Hub$(\cdot)$) of topics are iteratively calculated. The scores in the $(n+1)^{th}$ iteration are calculated based on the scores in the nth iteration, i.e.,

$$Authority^{n+1}(s_l) = \Sigma_k E_{T2S}(k, l) * Hub^n(z_k)$$

$$HubS^{n+1}(z_k) = \Sigma_l E_{S2T}(l, k) * Authority^n(s_l)$$

Denoting the score vectors of sentences and topics at the nth iteration with V_s^n and V_z^n, respectively, the matrix form of the reinforcement process can be given as

$$V^{n+1}{}_z = E_{S2T} V^n{}_s \quad \text{and} \quad V^{n+1}{}_s = E_{T2S} V^n{}_z$$

The above process continues until the convergence is achieved [10]. It can be proved that the hub vector V_Z will converge to the dominant eigenvector of the hub matrix $E_{S2T}E_{T2S}$ and the authority vector V_S will converge to the dominant eigenvector of the authority matrix $E_{T2S}E_{S2T}$.

For the numerical calculation, we set the initial scores of all the sentences and the topics to 1. After each iteration of reinforcement, both the hub vector V_Z and the authority vector V_S are normalized, i.e.

$$Authority^n(s_l) = Authority^n(s_l) / |V^n{}_s|$$

$$Hub^n(z_k) = Hub^n(z_k) / |V^n{}_z|$$

The termination condition is then set as when the maximum gap between the sentence scores in two successive iterations is smaller than a pre-given threshold (say 0.00001 here), i.e.,

$$Max_l |Authority^{n+1}(s_l) - Authority^n(s_l)| < 0.00001$$

5   Experiments

We conduct the experiments on a generic multi-document summarization data set from DUC 2004 [3]. The data set contains 45 document sets, with each set consisting of 10 newswire documents. According to the task definition, a summary is required for each document set and the length of the summary is strictly limited to 665 bytes. The ROUGE[1] [5] toolkit is used for evaluation. ROUGE is a widely recognized automatic summarization evaluation method which evaluates system-generated summaries by matching them to reference summaries. In this study we report two common ROUGE scores, ROUGE-1 and ROUGE-2, which are based on Uni-gram match and Bi-gram match respectively.

In pre-processing, stop-word removal and word stemming are performed by using an NLP toolkit GATE[2]. After the topic-sentence reinforcement process, all the sentences in a document set are ranked by their authority scores and successively extracted into the summary until the length limit exceeds. The MMR algorithm [2] is adopted to control redundancy in the constructed summaries. For each sentence to be extracted, it is compared against the sentences that are already selected in the summary. The sentence is extracted only when it is considered not significantly overlapping any previously selected sentence.

To demonstrate the advantage of LDA topics in the topic-sentence reinforcement scheme, we also implement two other graph-based approaches for comparison. Both of them use the proposed reinforcement algorithm to rank sentences and construct summaries, but they use different topic definitions when constructing the bipartite graph. The first one simply deems each original document as a single approximate topic and the other one adopts the clustering-based method as introduced in [7], regarding a sentence cluster as a topic. Two clustering algorithms, K-means and Spectral Clustering, are implemented and examined in this experiment. In these approaches, the weight of an edge connecting a document/cluster node and a sentence node is calculated by the cosine similarity between the document/cluster and the sentence. For fair comparisons, the MMR algorithm is also implemented on these systems.

Table 1. Results of the graph-based systems

System	ROUGE-1	ROUGE-2
LDA	0.36368 (0.35247 - 0.37505)	0.07678 (0.06933 - 0.08415)
Document	0.31908 (0.30453 - 0.33329)	0.05435 (0.04661 - 0.06184)
Spectral	0.34600 (0.32532 - 0.36700)	0.06679 (0.05821 - 0.07424)
K-means	0.34352 (0.33236 - 0.35197)	0.06501 (0.05991 - 0.08264)

---

[1] We run ROUGE-1.5.5 with parameters "-x -m -n 2 -2 4 -u -c 95 -p 0.5 -t 0 –b 665"
[2] Publicly available at http://gate.ac.uk/

Table 1 shows the average ROUGE-1, ROUGE-2 scores along with the 95% confidential intervals over the 45 document sets for all the three approaches (denoted as LDA, Spectral, K-means, and Document respectively). For the approaches, the topic number and cluster number are both set to the square root of the total sentence number, an experimental value that follows the same strategy in [7].

As demonstrated, the clustering-based approaches and the LDA-based approach indeed outperform the approach that simply uses the original documents as unprocessed natural topics, which shows the benefit of introducing extra nodes into the graph-based sentence ranking model. Moreover, the LDA-based approach can even significantly outperform both clustering-based approaches. This clearly shows that LDA is better at discovering meaningful topics than clustering methods. This result again demonstrates the importance of an appropriate topic definition in topic-sentence reinforcement. As discussed in the introduction section, the assumption implied in the clustering algorithms that a sentence is related to only one cluster conflicts with the relevance assumption of the reinforcement algorithm. This conflict makes clusters less effective in improving sentence ranking results.

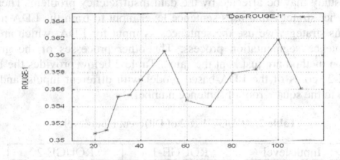

Fig. 4. Topic number K versus ROUGE-1

LDA requires a pre-determined topic number K. Obviously, the estimated topics are dependent on K, and so does the subsequent graph-based ranking results. In the next experiments, we attempt to investigate the influence of K on the performance of the LDA-based approach in turn. Fig. 4 and Fig. 5 illustrate the ROUGE-1 and ROUGE-2 scores of the LDA-based approach with different Ks.

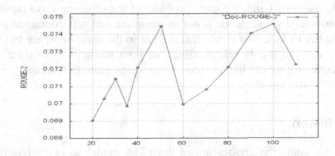

Fig. 5. Topic number K versus ROUGE-2

As illustrated in the above figures, the topic number K indeed influences the performance of the LDA-based summarization method. First of all, the performance drops for very small Ks. Moreover, the performance curve shows obvious fluctuations when K is increased. We think that the fluctuations here may reflect the real topic distributions in the given documents. The peaks of the curve just `indicate the proper topic numbers on different granularities. For example, while using 50 topics has achieved a proper set of topics and a peak performance, further increasing the topic number just causes deterioration of the result since some rational topics may be wrongly split. The deterioration may continue until a new proper set of topics (100 topics in our case) is achieved, which can be viewed as a proper topic set on a more detailed level. When we continuously increase the topic number, similar situations may occur to the fluctuations of the performance. On the other hand, compared to the results of fixed topic numbers, the experimental strategy that uses the square root of sentence number shows its advantage and reasonability because the result of the LDA-based method reported in Table 1 is indeed better than the best result in Fig. 4 and Fig. 5.

In the DUC 2004 data set, a document set only contains about 10 documents. Since LDA is a probabilistic model developed for a large scale corpus, the efficiency of LDA in our study may be affected by the data insufficiency problem. Therefore, we consider another strategy that uses sentence-level input to train the LDA model. According to this strategy, we use the sentences as input for LDA, which are obtained from the sentence segmentation process. The other processes of the graph-based summarization method are just kept the same. Table 2 below provides the ROUGE-1 and ROUGE-2 scores of the LDA-based model with different inputs, and the topic number is set to the square root of sentence number.

Table 2. ROUGE-1 & 2 of Different input-levels

Input-level	ROUGE-1	ROUGE-2
Document-level	0.36368 (0.3525-0.3751)	0.07678 (0.0693-0.0842)
Sentence-level	0.3534 (0.3416-0.3653)	0.0699 (0.0621-0.0770)

From the experimental results, we observe that the document-level input is indeed better than the sentence-level input for the LDA-based ranking algorithm. As mentioned above, the insufficiency of documents is the main problem of the document-level input. On the other hand, the main problem of the sentence-level input is that the information provided by a sentence is not as complete and self-contained as by a document. As a matter of fact, two words that are not in the same sentence but in adjacent sentences may also be highly related. Because both methods have their advantages and disadvantages, we would like to study ways to integrate the advantages of the two strategies in our further studies.

6   Conclusion

In this paper, we study the application of the LDA model to graph-based document summarization. Compared to the existing clustering-based approaches, LDA can discover more meaningful topics and thus can better score and rank the sentences for document summarization. Experimental results on the DUC 2004 data set clearly demonstrate the effectiveness of the proposed approach. In the future work, we would like to explore the role of LDA in graph-based summarization more extensively, including better topic estimation scheme and alternative uses of the graph-based ranking models.

Acknowledgements. The work presented in this paper is supported by a Hong Kong RGC project (No. PolyU 5230/08E).

References

1. Blei, D., Ng, A., Jordan, M.: Latent Dirichlet Allocation. The Journal of Machine Learning Research 3, 993–1022 (2003)
2. Carbonell, J., Goldstein, J.: The use of MMR, diversity based reranking for reordering documents and producing summaries. In: Proceedings of SIGIR 1998, pp. 335–336 (1998)
3. DUC. Document Understanding Conference,
   http://www-nlpir.nist.gov/projects/duc/intro.html
4. Griffiths, T., Steyvers, M.: Finding Scientific Topics. Proceedings of the National Academy of Sciences 101(suppl.1), 5228–5235 (2004)
5. Lin, C.Y., Hovy, E.H.: Automatic evaluation of summaries using n-gram co-occurrence statistics. In: Proceedings of HLT-NAACL 2003, pp. 71–78 (2003)
6. Otterbacher, J., Erkan, G., Radev, D.R., Mihalcea, R.: Using random walks for question-focused sentence retrieval. In: Proceedings of HLT-EMNLP 2005, pp. 915–922 (2005)
7. Wan, X., Yang, J.: Multi-document summarization using cluster-based link analysis. In: Proceedings of the 31st ACM SIGIR, pp. 299–306 (2008)
8. Zha, H.: Generic Summarization and Key Phrase Extraction using Mutual Reinforcement Principle and Sentence Clustering. In: Proceedings of the 25th ACM SIGIR 2002, pp. 113–120 (2002)
9. Brin, S., Page, L.: The anatomy of a large scale hypertextual web search engine. Comput. Netw. ISDN Syst. 30(1-7), 107–117 (1998)
10. Kleinberg, J.M.: Authoritative sources in hyperlinked environment. Journal of ACM 46(5), 604–632 (1999)
11. Erkan, G., Radev, D.R.: LexRank: Graph-based centrality as salience in text summarization. Journal of Artificial Intelligence Research 22, 457–479 (2004)
12. Lin, C., Hovy, E.: From single to multi-document summarization: A prototype system and its evaluation. In: Proceedings of the 40th Annual Meeting on Association for Computational Linguistics (2002)
13. Mihalcea, R.: Graph-based ranking algorithms for sentence extraction, applied to text summarization. In: Proceedings of ACL 2004 (2004)
14. Mihalcea, R.: Language independent extractive summarization. In: Proceedings of ACL (2005)
15. Radev, D.R., Jing, H., Stys, M., Tam, D.: Centroid-based summarization of multiple documents. Information Processing & Management, 919–938 (2004)
16. Cai, X., Li, W., Ouyang, Y., et al.: Simultaneous Ranking and Clustering of Sentences: A Reinforcement Approach to Multi-Document Summarization. In: Proceedings of Coling 2010, pp. 134–142 (2010)

Query Reformulation Based on User Habits for Query-by-Humming Systems

Guanyuan Zhang, Kai Lu, and Bin Wang

Institute of Computing Technology, Chinese Academy of Sciences,
Beijing 100190, China
{zhangguanyuan,lukai,wangbin}@ict.ac.cn

Abstract. Query-by-humming (QBH) systems take human humming audio as input, and use it as a query to retrieve music from a database as accurately as possible. We propose a novel way of query reformulation for QBH search engine by modeling the user's humming habits. Query reformulation technologies have been proven to be very effective in many researches. A critical challenge faced by QBH is that the humming is quite inaccurate. Detailed statistics and analysis on a huge number of humming queries and targets has been done based on the IOACAS dataset. Then we summarized the common humming errors that lead to the poor retrieval effect. We defined a five-tuple to represent errors and proposed a user-humming-model based on Hidden Markov model. The query of user is reformulated by the model. The approach is evaluated on Ict-Muse QBH system, SOSO QBH system and midomi QBH system using the QBSH dataset. The experimental results clearly demonstrate that the approach adopted in this paper can greatly improve the quality of the user humming, which in turn improves the effect of information retrieval.

Keywords: Query reformulation, User-Humming-Model, Query by Humming, Music Information Retrieval.

1 Introduction

Query reformulation is the process of iteratively modifying a query to improve the quality of a search engine result [1]. It achieved great success in the text information retrieval field [2][3][4]. However, in the field of music information retrieval query reformulation is not so popular. We propose a query reformulation approach for QBH system. The main goal of QBH systems is to take human humming audio as input, and use it as a query to retrieve music from a database as accurately as possible. One of the main difficulties in building an effective QBH system is that the users are not always professional singers.

The central idea of our approach is simple and intuitive. Detailed statistics and analysis on a huge number of humming queries and targets has been done based on the IOACAS dataset. Then we summarized the common humming errors that lead to the poor retrieval effect. We defined a five-tuple to represent errors. Before searching, users hum a piece of known music. Then the user humming is corrected on the basis

Y. Hou et al. (Eds.): AIRS 2012, LNCS 7675, pp. 386–395, 2012.

of the standard answer. After that, a user-humming-model is created to optimize the humming query. As soon as a user inputs a humming query, we reformulate the query by using the humming model.

We evaluated our proposed approach on the SOSO QBH system[1], midomi QBH system[2], and our own IctMuse QBH system. The experimental results demonstrate the superiority of our method in its ability to (i) generate reformulated queries with higher retrieval effectiveness, (ii) have sufficient robustness, so it can be used for different humming retrieval systems, and (iii) it can work well in the case of absence of query logs.

The rest of the paper is organized as follows. Section 2 provides a review of the previous work on query reformulation of QBH. In Section 3, we describe the proposed approach in detail. Experiments and results are described in Sections 4. Section 5 concludes the paper and offers directions for future work.

2   Related Works

Query reformulation is an important topic of information retrieval. Several techniques have been proposed based on relevant feedback, query log analysis and distributional similarity.

QBH System: QBH is a Content Based Searching technique used to retrieve relevant songs. Ghias et al. [5] was one of the first to propose QBH in 1995, and coarse melodic contours were used to represent melodic information. Most of the research on QBH is focus on the music representation and matching method [6][7][8][9].

Query Reformulation Using User Relevant Feedback: Rho et al. [10] reformulated the query using user relevant feedback with a genetic algorithm to improve retrieval performance of his MUSEMBLE QBH system. Relevant Feedback is one of the most popular query reformulation methods in the field of information retrieval. Their GA-based feedback scheme needs 20 generations to return result. Their experimental result illuminate that the method can improve the retrieval accuracy up to 20–40%. And the improvement is 15% on the P@5.

Little et al. [11] describes a QBH system that learns from user provided feedback on the search results, letting the system improve while deployed. They propose a parameterized singer error model and a straight-forward genetic algorithm. However their singer error model is used to automatically customize parameters not to reformulate query. They did not introduce and summarize the singer errors and the results of model are not explicable.

Their methods can improve the retrieval accuracy but they need user participation and marked work. They cannot be used for different humming retrieval systems.

Singing Error Model: Meek et al. [12] proposed a model for errors in sing queries. They use the model to solve the problem of identifying the degree of similarity between a query and a potential target in a database of musical works.

---

[1] http://h.soso.com
[2] http://www.midomi.com/

Unal et al. [13] created a large database of humming samples collected from people with various backgrounds. They analyzed the performance of both musically trained and untrained human subjects with statistical techniques. In their experimental result, 94% retrieval accuracy is observed within a test sample of trained subjects, while 72% retrieval accuracy is achieved by a test sample of non-musically trained subjects. However, since the input is totally user-dependent, and includes high rates of variability and uncertainty, the challenge that remains is to achieve robust performance under such conditions.

3   Proposed Architecture

This paper presents a query reformulation framework which is shown in Fig. 1.

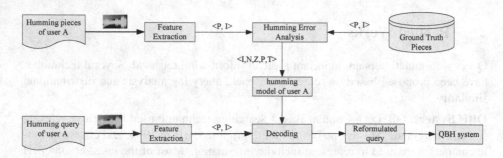

Fig. 1. Query reformulation framework

The humming query is usually between 8 seconds to 20 seconds, which is the average memory length of the user towards an unfamiliar song. We analyze the humming error for the users by using their humming pieces and the ground truth pieces. And then a user-humming-model which is a variant of the Hidden Markov Model (HMM) is trained. The query submitted by the user can be reformulated by his/her humming habit model.

3.1   Users' Humming Habits

We have done the statistics and analysis on the 350 pairs of humming query and target based in the IOACAS_QBH dataset. Humming errors can be divided into five categories. 100 queries in the dataset were selected out randomly. The errors were corrected artificially and then the queries were reformulated by the MIDI music synthesizer. The effect of reformulated query is better than original queries. It can be seen that fixed humming errors in the query can improve the effect.

Common humming errors are summarized as follows:

- Key error: most of the pitches in query are higher or lower than the target.
- Rhythms error: the speed of query is faster or slower than the target.

- Duration error: some measure in the query is not enough or exceed on the time dimension.
- Pitch error: off-pitch.
- Insertions and deletions: adding or removing notes.

The humming error is divided into three categories in our approach. They are global error, local duration error and local pitch error.

The global error includes global pitch error and global rhythms error. Some people cannot remember and reproduce the pitches or temple exactly. The global pitch error means adding a certain value to all pitches. An example is in Fig. 2. The key of query is 3 degrees higher than that of target. The duration of each note is short and the scaling value of tempo is 0.7.

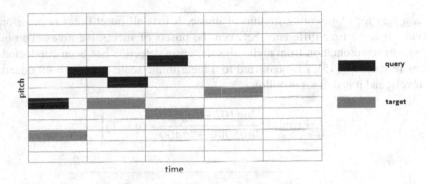

Fig. 2. An example of global error

Local duration error presents the duration error, which means some measure in the query is not enough or exceed on the time dimension.

Local pitch error means some notes in query are off-pitch. An example is in Fig. 3. The insertions and deletions error can be presented by the local duration error and local pitch error.

Fig. 3. An example of local pitch error

3.2 Users Humming Modeling

We proposed a user-humming-model based on Hidden Markov models. Fig .4. is the structure of our model.

The observation of the HMM is the query which is presented by a tuple. It is notated as follows:

$$O_n = <P, I>$$ (1)

Here, P is pitch class which is a representation where all notes are projected into a single octave.

$$P = \text{pitch mod } 12 \quad P \in \{0, 1 .... 11\}$$ (2)

IOI is a way for representing rhythm of music. It lists all onset times or inter-onset intervals. It is the time difference between the onsets of successive notes. To obtain this type of representation from audio signals, onset detection has been conducted by various researchers [15]. I is quantized to a logarithmic scale, using q = 49 quantization levels, and it is defines as follows:

$$I = \text{round} \left[\frac{\log IOI - \log 50}{\log 5000 - \log 50} * (q - 1) \right]$$ (3)

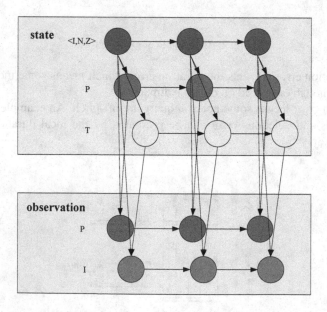

Fig. 4. The structure of user-humming-model

The state the HMM is the users humming error which is presents by a five-tuple. It is notated as follows:

$$S_n = < I, N, Z, P, T > \tag{4}$$

Here, I is the index of notes in target and N is the number of notes affected. Z presents the terminative note in target. P and T present the pitch and temp respectively. We set $P \in \{-5, -4, \ldots, +6\}$ because of the modulus-12 relationship to pitch and $T \in (0.2, 5)$. The <I, N, Z> is defined as follows:

$$\begin{cases} < pitch[n], IOI[n] > = < pitch[I] + P, IOI[I] * T > & if\ N = 0 \\ < pitch[n], IOI[n] > = < pitch[I] + P, \sum_{j=I}^{I+N} IOI[I] > & if\ N! = 0, Z = 0 \\ < pitch[n], \sum_{j=n}^{n+N-1} IOI[j] > = < pitch[I] + p, IOI[I] * T > & if\ N! = 0, Z! = 0 \end{cases} \tag{5}$$

3.3 Query Reformulation

The user's test query is represented as the tuple <P, I>. The state sequence can be decoded out from the model and the observation sequence. The state sequence is an error sequence, from which we can get the reformulated query. MIDI music synthesizer was used to transform the reformulated query to sound.

4 Experiment

4.1 Dataset

We evaluated our proposed query reformulation mechanism using QBSH Corpus database[3]. The corpus has been used for the QBSH task at MIREX (Music Information Retrieval Evaluation eXchange) 2006~2009. This dataset contains 48 ground-truth MIDI files and 4431 singing/humming clips from about 195 subjects.

We selected 100 subjects randomly from 195 subjects in the database. Every subject supply 16 humming clips to the training set and 4 humming clips to the test set. Table 1 shows the details of the training set and testing set.

Table 1. The training set and test set in the experiment

Dataset	Subjects number	Query number	Query number for each subject	Ground truth number
training set	100	1600	16	35
testing set	100	400	4	10

---

[3] Jyh-Shing Roger Jang, "MIR-QBSH Corpus", MIR Lab, CS Dept, Tsing Hua Univ, Taiwan. Available at the "MIR-QBSH Corpus" link at http://www.cs.nthu.edu.tw/~jang

4.2   Evaluation Metric

We measure improvement using the mean reciprocal rank (MRR) of a set of n queries and the Top-X hit rate.

Humming retrieval is different from text retrieval, because in the former, the user has clear search goals and there is only one correct search result. In addition, the humming search engine often returns a fixed number of search results. For example, SOSO only returned 5 results. Therefore, the experiment used the Top-X hit rate, which defined as follows.

$$\text{Top} - \text{X Hit Rate} = \frac{Number\ of\ Query\ That\ \text{The Answer in The First X Position}}{Total\ Number\ of\ Query} \tag{6}$$

MRR emphasizes the importance of placing correct target songs near the top of the list while still rewarding improved rankings lower down on the returned list of songs. MRR is defined as follows.

$$\text{MRR} = \frac{\sum_{i=1}^{n} \frac{1}{r_i}}{n} \tag{7}$$

We define the rank of the *i-th* query, r_i as the rank returned by the search engine for the correct song in the database. Values for MRR range from 1 to 0, with higher numbers indicating better performance. Thus, MRR = 0.5 roughly corresponds to the correct answer being in the top 2 songs returned by the search engine, MRR = 0.25 indicates the right answer is in top 4, and so on.

4.3   Experimental Settings

The experiment is divided into four steps. The specific steps are as follows:

- Training model. We get 100 user-humming-models from our proposed approach and humming clips of these 100 subjects in training set.
- Baseline model. A comprehensive model has been trained out by 1600 query clips in training set. It is trained using our proposed approach, but it is not a personal model.
- Query reformulation. The humming clips of every subject in testing set were reformulated by personal humming model and by comprehensive model separately.
- Evaluation on private system. The original queries and two reformulated queries were input our own QBH system IctMuse separately.
- Evaluation on public systems. In order to prove our approach has sufficient robustness for systems, we carry out experiment not only on our own system but also on another two public QBH system, SOSO and midomi. We ensure three systems contain the 10 ground truth of the testing set.

4.4   Results and Discussion

The top-1 hit rate, top-3 hit rate and top-5 hit rate for both original query and reformulated query on three QBH systems are shown in Table2~4.

Fig. 5 is the original query search result on IctMuse system. We also show the reformulated query search result in Fig. 6. We can observe that the proposed method improves the retrieval effect. Here it is important to note from Fig. 6 that the performance of reformulated query that comes from those 50% top-5 original queries is not descended.

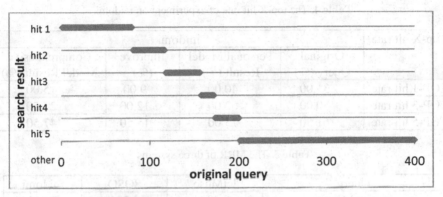

Fig. 5. The original query search result on IctMuse

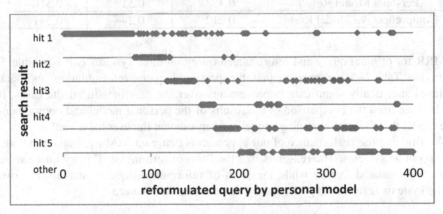

Fig. 6. The reformulated query by personal model search result on IctMuse

Table 2. The top-X hit rate of experiment on IctMuse

Top-X hit rate	IctMuse			
	Original Query (%)	Personal Model Result (%)	Improve (%)	Comprehensive Model Result (%)
TOP-1 hit rate	20.50	32.00	11.50	13.50
TOP-3 hit rate	39.00	56.25	17.25	20.00
TOP-5 hit rate	50.00	69.00	19.00	44.50

Table 3. The top-X hit rate of experiment on SOSO

Top-X hit rate	SOSO			
	Original Query (%)	Personal Model Result (%)	Improve (%)	Comprehensive Model Result (%)
TOP-1 hit rate	29.00	37.00	8.00	16.75
TOP-3 hit rate	46.00	60.00	14.00	22.75
TOP-5 hit rate	59.75	77.00	17.25	46.00

Table 4. The top-X hit rate of experiment on midomi

Top-X hit rate	midomi			
	Original Query (%)	Personal Model Result (%)	Improve (%)	Comprehensive Model Result (%)
TOP-1 hit rate	31.00	40.00	9.00	15.00
TOP-3 hit rate	50.00	62.00	12.00	22.50
TOP-5 hit rate	61.50	75.00	15.50	47.50

Table 5. The MRR of three systems

Query	IctMuse	SOSO	midomi
Original Query	0.303	0.391	0.415
Personal Model Result	0.433	0.512	0.510
Comprehensive Model Result	0.212	0.244	0.241

MRR for original query and reformulated query on three systems is list in table 5. From the Table2~5 we observe that the proposed query reformulation approach achieves statistically significant improvements over the baselines in all the cases. It can be seen from the comparison of the results of the personal model and comprehensive model that the user-humming-model can represented the user habit well.

We find that the performance of our approach is better on IctMuse than on the other two public systems. The reason is that the form of reformulated query on IctMuse is the reformulated feature while the form of reformulated query on the other two public systems is the audio generated by the reformulated feature.

5   Conclusions

In this paper, we presented a query reformulation algorithm for QBH system. Experimental results on three QBH systems are encouraging and our proposed approach has achieved statistically significant improvements. We show that our approach increases the MRR from 0.303 to 0.433 on our IctMuse system.

However, several aspects are out of the scope of this work and offer interesting future research directions. More common humming error discovery and further summary should be done. In fact, the optimized humming error representation could bring significant improvements in retrieval effectiveness. This means that there is still much room to improve the retrieval performance. We have proved that the user-humming-habit indeed helps to improve the effect but the stability of the users' habits needs further validation. Maybe the stability of some users is strong while that of others is weak, and special circumstances maybe change the users' habit. We leave these limitations as our future work.

References

1. Holmqvist, M., Ciaramita, M., Mahler, D.: Generalized Syntactic and Semantic Models of Query Reformulation. In: ACM SIGIR 2010, pp. 283–290 (2010)
2. Riezler, S., Liu, Y., Vasserman, A.: Translating queries into snippets for improved query expansion. In: COLING, pp. 737–744 (2008)
3. Zhang, Z., Nasraoui, O.: Mining search engine query logs for query recommendation. In: ACM WWW, pp. 1039–1040 (2006)
4. Xu, Y., Gareth, J.F., Jones, B.: Query Dependent Pseudo-Relevance Feedback based on Wikipedia. In: ACM SIGIR 2009 (2009)
5. Ghias, A., Logan, J., Chamberlin, D., Smith, B.C.: Query by humming: musical information retrieval in an audio database. In: ACM Multimedia Conferenece 1995 (1995)
6. Ryynanen, M., KJapuri, A.: Query by humming of MIDI and audio using locality sensitive hashing. In: ICASSP 2008, pp. 2249–2252 (2008)
7. Serra, J., Kantz, H., Serra, X., Andrzejak, R.G.: Predictability of music descriptor time series and its application to cover song detection. IEEE Transaction on Audio, Speech and Language Processing 20(2), 514–525 (2012)
8. Salamon, J., Gomez, E.: Melody extraction from polyphonic music signals using pitch contour characteristics. IEEE Transaction on Audio, Speech and Language Processing (2012)
9. Ravuri, S., Ellis, D.: Cover song detection: From high scores to general classification. In: IEEE ICASSP 2010, pp. 65–68 (2010)
10. Rho, S., Han, B., Hwang, E., Minkoo, K.: MUSEMBLE: A novel music retrieval system with automatic voice query transcription and reformulation. The Journal of Systems and Software 81, 1065–1080 (2008)
11. Little, D., Raffensperger, D., Pardo, B.: A query by humming system that learns from experience. In: ICMIR (2007)
12. Meek, C., Birmingham, W.: Johnny Can't Sing: A Comprehensive Error Model for Sung Music Queries. In: IRCAM 2002, pp.124–132 (2002)
13. Unal, E., Narayanan, S.S., Shih, H.-H., Chew, E., Kuo, C.C.J.: Creating data resources for designing user-centric frontends for Query by Humming systems. In: ACM Multimedia Information Retrieval Conference 2003 (2003)
14. Gouyon, F., Dixon, S.: A review of automatic rhythm description systems. Computer Music Journal 29(1), 34–54 (2005)

Query Expansion Based on User Quality in Folksonomy

Qing Guo, Wenfei Liu, Yuan Lin, and Hongfei Lin

School of Computer Science and Technology, Dalian University of Technology
No. 2 LingGong Road GanJingZi District DaLian, 116023, China
{guoqing,wenfei.liu,yuanlin}@mail.dlut.edu.cn,
hflin@dlut.edu.cn

Abstract. With the development of folksonomy, social annotations are widely used in various fields. As one of the applications of social annotations, query expansion has been proved meaningful and effective. Considering that user quality influences the relevance between annotations and resources, we infer that it can affect the relevance between expansion terms and the corresponding original queries. Therefore, we believe that annotations used by high quality users are more appropriate to serve as expansion terms. In this paper, we propose a novel algorithm to obtain user quality, and then explore two methods to integrate user quality into the query expansion procedure: one is to filter the resources returned from initial retrieval using user quality; the other is to increase the weight of the expansion terms which are annotations given by high quality users. Our experiments on Del.icio.us dataset show that user quality is helpful for selecting good expansion terms.

Keywords: User Quality, Social Annotations, Query Expansion.

1 Introduction

As the online social networks develop (e.g., Flikr, Del.icio.us, Bibsonomy, etc [1-2]), the so-called folksonomy [3-4] which allows users to share, store, organize and retrieve online web resources through tagging behaviors, then rises. In folksonomy, the online web resources are defined as resources, the tags attached resources are defined as annotations, and the people who carry out the tagging behaviors are defined as users. With the popularity of folksonomy, social annotations attract the attention of many researchers on various fields.

At the same time, query expansion has been widely used during search to meet user's real information need. Pseudo-relevance feedback (PRF) as one of the effective methods of query expansion, is based on the assumption that the top-ranked documents in the initial retrieval results are relevant and can help describe information needs better [5-6]. Based on PRF, Lin et al. [7] has proved that social annotations could improve the performance for query expansion through mining better expansion terms. However, annotations are always informal and irregular in folksonomy. In other words, not all the annotations can be treated as good expansion terms. So it is essential to achieve more appropriate annotations as good expansion terms.

Y. Hou et al. (Eds.): AIRS 2012, LNCS 7675, pp. 396–405, 2012.

To solve the problem mentioned above, we propose a novel query expansion method based on social annotations with users' quality information added. We attain this goal by mining high quality users in folksonomy systems, because we believe that the high quality users are more likely to use accurate terms as annotations to tag resources, which have a larger possibility to be good expansion terms when working for query expansion. Therefore, annotations after the filter of high quality users would contribute a lot to query expansion.

The contributions of this paper can be summarized as follows: 1) we propose a novel algorithm to give each user a reasonable quality score according to the user's tagging behavior and the mutual reinforcement relationship; 2) we explore two methods to apply user quality to query expansion in folksonomy through extensive experiments, which help select better expansion terms.

The rest of the paper is organized as follows. Section 2 discusses some related work. Section 3 describes the proposed algorithm of receiving user quality and the query expansion approach with user information added. Then the details of our experiment and the performance of the new approach are reported in section 4. Finally, the conclusion and some future works are presented in Section 5.

2   Related Work

Query expansion technique has been relatively mature in information retrieval [8-9]. PRF is an effective one among all methods of query expansion. There are a variety of retrieval models attaching to PRF, such as vector space model [10], probabilistic model [11], relevance model [12], mixture model [13], etc.

With the development of the social networks, more and more researches shift their perspective towards social annotations application. Recommendation [14], semantic Web [15], Web search [1, 16, 17] as different kinds of social annotation applications have already reached very high achievement. Lambiotte et al. [3] first brought the tripartite network model of folksonomy to discover the relationship between annotations, users, and resources. The mutual reinforcement relationship, which benefit the research on folksonomies a lot, can be interpreted as high quality annotations, users, resources are always appear associated with each other. When analyzing different entities in the social network, the mutual reinforce relationship is useful and effective.

Afterwards, Hotho et al. [16] propose FolkRank algorithm, which is based on PageRank [18] and random walk model. So does Social PageRank, proposed by Bao et al. [17]. Both of the two algorithms build an undirected tri-partite graph with users, annotations, and resources to analyze the relationship of every two entities. Noll et al. [19] propose a graph-based algorithm, named Spamming-resistant Expertise Analysis and Ranking (SPEAR), which meets the view that the user who tags a resource first has more chance to be an expert, which is inherited in this paper to mining user quality.

Recently Lin et al. [7] indicate that annotations are good resources for query expansion. Following the theory, we propose a novel query expansion approach based on social annotations with users' quality information added, and explore how well the user quality affect the expansion terms chosen from annotations for query expansion.

3   Query Expansion Based on High Quality Users Approach

The query expansion based on high quality users approach focuses on mining user quality to achieve better expansion terms, thus improving the performance of query expansion. Our approach consists of two phases: the former, calculates user quality, by means of a novel user quality computation metric proposed in this paper based on the FolkRank and the SocialPageRank algorithm; the later, selects appropriate annotations as expansion terms for further retrieval, which is mainly based on the well-known co-occurrence method with user quality integrated.

3.1   User Quality Computation

In this paper, we propose a novel algorithm to obtain user quality. User quality presents how precisely tags the user submit describe the resources, and high quality users refer to the users who always tag resources using appropriate annotations. Our method is implemented in the iterative way on the basis of the mutual reinforcement principle:

$$\begin{cases} U^k = R^{k-1} M_{RU} \\ R^k = U^k M_{UR} \end{cases} \tag{1}$$

in the iterative metric: (i) U^k (resp., R^k) denotes the user (resp., the resource) quality score at the kth iteration; (ii) M_{RU} (resp., M_{UR}) is the transition matrix from resource to user (resp., user to resource). The iterative process will not stop until results get convergent. The basic framework of the algorithm is similar with SocialPageRank and FolkRank. They are all based on the random walk model and the classic PageRank algorithm[18]. The convergence of them can be proved in a similar way. In our experiment, the algorithm converges within 10 times of iterations in our experiments.

In addition, we inherit the demonstrated idea that the earlier a user annotating a resource, the more likely the user being the high quality one. So the calculation for user quality is mixed with tagging time. We sort users for specific resource and annotation by tagging time and assign integral numbers in order to various users from 1 to n according to the sequencing, where n is the number of users who annotate the given resource. Finally, the earliest user who annotated the resource is assigned n for the time score, and the last one is assigned 1.

If we utilize the time score which is linear distribution for various users directly, users who tagged high quality resources earliest will receive rather high quality scores even if they might not have contributed any other resources thereafter. In order to avoid the situation, we control the effect of the time score by receiving the square root of the time score through a number of experiments to make the first user of a resource score more appropriate to be different from other users instead of too high.

Assume that there are N_u users and N_r resources. Let M_{ur} be the $N_u \times N_r$ transition matrix from users to resources and M_{RU} be the $N_r \times N_u$ transition matrix from resources to users. Elements $M_{UR}(u_i, r_j)$ and $M_{RU}(u_i, r_j)$ are assigned as follows:

$$M_{UR}(u_i,r_j) = \frac{\sum_{a_k \in A(u_i,r_j)} \frac{1}{2^k} \cdot p(a_k \mid r_j)}{\sum_{u \in U(r_j)} \sum_{a_k \in A(u,r_j)} \frac{1}{2^k} p(a_k \mid r_j)} , \quad M_{RU}(u_i,r_j) = \frac{\sum_{a_k \in A(u_i,r_j)} \frac{1}{2^k} \cdot p(a_k \mid u_i)}{\sum_{r \in R(u_i)} \sum_{a_k \in A(u_i,r)} \frac{1}{2^k} p(a_k \mid u_i)} \quad (2)$$

where $p(a|r)$ is the generative probability for annotation a from resource r which represents how exactly the annotation a generalizes the resource r, the specific computation of which as equation (3); $p(a|u)$ is the generative probability for annotation a from user u which represents u's interest which is computed as equation (4).

$$p(a \mid r) = \frac{N_u(a,r_j) \times T(u,r,a)}{\sum_{a_x \in A(r_j)} N_u(a_x,r_j)} \quad (3)$$

$$p(a \mid u) = \frac{\sum_{r \in R(u)} p(a \mid r)}{\sum_{a_x \in A(u)} \sum_{r_i \in R(u)} p(a_x \mid r_i)} \quad (4)$$

where $N_u(a,r)$ is the number of users who annotate resource r using annotation a, Then the tagging time is attached to the iterative calculation of the user quality.

After iteration, each user is weighted by a score which measures the contribution of user's tagging behavior, and we call it Quality Score (QS for short) in this paper. The QS is then integrated into the query expansion procedure, thus increasing the ratio of good expanded terms.

3.2   Expansion Terms Selection

The term co-occurrence is usually used to measure how often terms appear together in the text window. In our experiment, it helps measure the frequency of annotations and query terms appearing together, and the whole content of a resource works as the text window. In the initial retrieval, some resources are returned for the query Q which compose the resource set R_j. Then, expansion terms are selected from R_j. QS is applied in expansion terms selected procedure, and there are two means to achieve this goal.

Filtering Resources through User Quality. The first method to apply QS is to re-rank the resources returned from initial retrieval, from which we extract resources tagged by high quality users. Each resource is tagged by many users using different annotations, and each user is weighted by QS for specific annotation. Therefore, each resource can be measured as equation (5), which is called Resource Score (RS for short):

$$RS(r) = \sum_{a \in A(r)} \sum_{u \in U(r,a)} QS(u,a) \quad (4)$$

where $U(r, a)$ is the user set for users who annotate resource r using annotation a; $A(r)$ is the annotation set for annotations which are used to tag the given resource r.

Hence, the returned resources from the initial retrieval are re-ranked by RS. The top N resources are filtered from the re-ranked resources to conduct the further experiment according to the traditional co-occurrence metric (Lin et al. [7]).

Adding User Quality into Co-occurrence Metric. By means of integrating QS to modify the traditional co-occurrence metric, we apply user quality selecting better expansion terms, which is the second way to use user quality. We have tried a number of metric to observe the result, and the best one resulted from the metric below:

$$coo(a_i, q_j \mid R_j) = \sum\nolimits_{r \in R_j} log(tf(a_i, r) \times (1.0 + e^{2QS(u,a_i)}) + 1.0) \times log(tf(q_j, r) + 1.0) \Big/ log(n) \ (5)$$

where $QS(u, a)$ aims at weighting the annotation used by high quality user higher than the one used by other users when we select expansion terms.

Then, based on the assumption that terms in the document are independent from each other, the expansion terms are selected according to the equation below.

$$coof(a_i, Q \mid R_j) = \sum_{q_j \in Q} idf(q_j, R_j) idf(a_i, R_j) log(coo(a_i, q_j \mid R_j) + 1.0) \tag{6}$$

where $idf(*, R_j)$ shares the similar meaning with the inverse document frequency in traditional information retrieval model. Finally, top K expansion terms selected according to equation (7) compose the new expanded query Q_{exp} with the corresponding original query Q. The proportional distribution of the expansion terms and the corresponding original query Q is control by a parameter α.

4   Experiments

4.1   Data Description

In our experiment, the dataset is collected from Del.icio.us and consists of 4414 users, 41204 resources and 28733 annotations in total, which is the same with [20]. The data is processed in two steps: first, stem all of the annotations, making the annotations which share the same stem get together; second, some annotations are made up by several words, such as "*java/programming*", "*java_programming*", and so on. We divide these annotations into several words with the help of the delimiters.

Table 1. Statistics of Evaluation Datasets

Collection	Documents	Topics
AP	242,198	151-200
WSJ	173,252	151-200
Robust 2004	528,155	351-450

The expansion procedure is conducted on the delicious data set, and retrieval procedure is conducted on three standard TREC collections, AP88-90(Associated Press), WSJ87-90 (Wall St. Journal) and Robust 2004 (the dataset of TREC Robust Track started in 2003) are also needed in our experiment. The detail information of the three collections we conduct experiment on in our paper lists in Table 1.

4.2   Experimental Settings

The retrieval procedure is conducted based on the Indri 2.6 search engine which serves as the basic retrieval system. The performance is measured mainly by Mean Average Precision (MAP) for top 1000 documents retrieved from the three TREC collections, meanwhile, P@n also assists to evaluate the retrieval results which means the precision in top n result lists for retrieval.

Table 2. parameter setting of the second method to add the Quality Score(SATUSER1) on AP, WSJ, Robust 2004 collections

Collection	N	M	K	α
AP	30	25	30	0.48
WSJ	25	20	50	0.50
Robust 2004	15	10	45	0.50

Table 3. parameter setting of the first method to add the Quality Score(SATUSER2) on AP, WSJ, Robust 2004 collections

Collection	M	K	α
AP	35	50	0.10
WSJ	15	50	0.50
Robust 2004	10	45	0.30

For the first method to add user quality, we get M returned resources for the initial retrieval from the delicious dataset, and calculate RS by equation (5). Then the M resources are re-ranked according to RS, from which we pick up N topmost resources as the final returned resources to select the candidate expansion terms. The top K annotations are selected according to traditional co-occurrence metric as the final expansion terms. In this paper, the parameters are setting as Table 2.

On the other hand, for the second method adding user quality, the top K annotations are selected from the N resources in the initial retrieval according to the equation (7) and then compose the expanded query Q_{exp}. The specific parameter settings of this paper are shown in Table 3.

Parameters in our experiment are tuned through simple variable method. Figure 1 and 2 show the performance varies with parameters α and M on three TREC collections in the condition of other parameters fixed. Figure 1 displays the MAP on different weights for original query, when α is larger than 0.5, the performance drops with

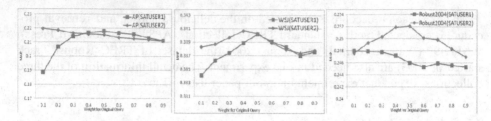

Fig. 1. Performance on different weights of original query for SATUSER1 and SATUSER2 on AP, WSJ, Robust 2004 collections, M=10, K=50

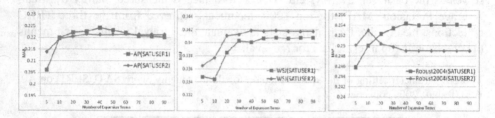

Fig. 2. Sensitivity of SATUSER 1 and SATUSER 2 to the number of expansion terms on AP, WSJ, Robust 2004 collections, $\alpha = 0.5$

α increasing. Figure 2 donates performance changing with the number of expansion terms K. With increasing of K, MAP rises fast to the top, and then remains steady afterwards. For efficiency, we choose K when MAP reaches the top at the first time.

4.3   Evaluation for Query Expansion Based on User Quality

Evaluation on Mining User Quality Algorithm. We compare two methods of gaining user's quality score (QS): SPEAR and our algorithm described in section 3. Because of the difficulty and subjectivity of evaluating user quality, we rely on query expansion results to estimate and evaluate the user quality mining algorithm. Table 4 lists MAP on the three TREC collections based on the QS obtained from SPEAR algorithm and our method.

Table 4. MAP of query expansion for different user quality mining method on AP, WSJ, Robust 2004 collections

Method	AP	WSJ	Robust 2004
SATUSER1(SPEAR)	0.2176	0.3405	0.2526
SATUSER1(our method)	0.2246	0.3419	0.2543
SATUSER2 (SPEAR)	0.2211	0.3402	0.2519
SATUSER2 (our method)	0.2221	0.3411	0.2531

Table 4 shows that our method achieves better performance for mining user quality though a little. The reason for this is that we have considered not only the users' tagging behavior as the SPEAR algorithm does, but also the mutual reinforcement relationship between the users and resources, which represent users' features better.

Evaluation on Query Expansion Based on Our User Quality Mining Approach.
Table 5-7 display MAP, P@5, P@10 and P@20 of retrieving 1000 documents by
various approaches, among which *QL* denotes the Query Likelihood Model retrieval
using original query; *PRFSA* is the retrieval after the query expansion based on social
annotations using PRF method, which serves as our baseline; *SATUSER1* and
SATUSER2 are the retrieval with the query which is expended from social annotations
and then added user quality and time information in two methods respectively.

Table 5. Performance comparisons of different methods on AP collection

Method	P@5	P@10	P@20	MAP
QL	0.2840	0.3000	0.2770	0.2010
PRFSA	0.3400	0.3180	0.2930	0.2142
SATUSER1	0.3680	0.3420	0.3150	0.2246
SATUSER2	0.3600	0.3240	0.3190	0.2221

Table 6. Performance comparisons of different methods on WSJ collection

Method	P@5	P@10	P@20	MAP
QL	0.4760	0.4500	0.4100	0.3037
PRFSA	0.5120	0.4840	0.4360	0.3255
SATUSER1	0.5760	0.5080	0.4610	0.3419
SATUSER2	0.5680	0.5000	0.4590	0.3411

Table 7. Performance comparisons of different methods on Robust 2004 collection

Method	P@5	P@10	P@20	MAP
QL	0.4800	0.4520	0.3930	0.2466
PRFSA	0.4960	0.4580	0.4020	0.2510
SATUSER1	0.5120	0.4560	0.4030	0.2543
SATUSER2	0.5000	0.4560	0.4000	0.2531

As we have seen, *PRFSA* gets much improvement comparing with the performance
of retrieval with original query. Obviously, through integrating user quality into query
expansion procedure, the expansion achieves much better performance. In addition,
SATUSER1 shows better results than SATUSER2. To analyze the reason, we find
that through modifying co-occurrence metric, a large amount of attempt is needed to
discover the best compound mode for the metric and Quality Score. However, due to
the limit of conditions, we cannot test every combinatory formula. While by means of
filtering high quality resources, the user quality contributes to query expansion more
directly. As a result, the performance is better.

SATUSER1 and SATUSER2 show relatively obvious increasing in MAP of retrieval on AP and WSJ collections, however, on Robust 2004 collection, the increasing is small. Then we carefully studied the content of the three TREC collections and

the Del.icio.us dataset. The finding is that Robust 2004 collection contains various kinds of format documents. Because of the irregular document formats, some noise cannot be eliminated during retrieval. As we all see, the baseline doesn't achieve significant improvement, either. So the performance of query expansion using user quality doesn't show very prominent improvement.

Through analyzing the expansion terms, we find the way the user quality operates on query expansion, which is through adding user quality information to increasing the possibility of the relevant annotations being selected as expansion terms. Our experiments show the improvement in query expansion with user quality added, and then we can conclude that user quality is indeed helpful for query expansion.

5   Conclusions

In this paper, we explore the feasibility of the user quality applied on query expansion. We propose an algorithm to obtain user quality in folksonomy, and further demonstrate that the algorithm is valid and works well. Besides, high quality users are proved beneficial to select annotations which are more relevant to the corresponding original query as expansion terms, thus improving query expansion performance. The way it works can be summarized as two methods: the first is to re-rank and filter the returned resources from the initial retrieval by user quality; the second is to integrate user quality into the co-occurrence metric to weight expansion terms by user quality.

In the future, we will test more datasets and methods to observe the effect of user quality in query expansion. What's more, community can be introduced to perfect the user representation, and the relationship between users can be added into query expansion procedure. In addition, we will explore more applications of user information.

Acknowledgment. This work is partially supported by grant from the Natural Science Foundation of China (No.60673039, 60973068, 61277370), the National High Tech Research and Development Plan of China (No.2006AA01Z151), Natural Science Foundation of Liaoning Province, China (No.201202031), State Education Ministry and The Research Fund for the Doctoral Program of Higher Education (No.20090041110002).

References

1. Xu, S., Bao, S., Fei, B., Su, Z., Yu, Y.: Exploring folksonomy for personalized search. In: SIGIR 2008: Proceedings of the 31st Annual International ACM SIGIR Conference on Research and Development in Information Retrieval, pp. 155–162. ACM (2008)
2. Heymann, P., Koutrika, G., Molina, H.G.: Can social bookmarking improve web search. In: Proceedings of the International Conference on Web Search and Data Mining, pp. 195–205 (2008)
3. Lambiotte, R., Ausloos, M.: Collaborative tagging as a tripartite network. In: Proceedings of the International Conference on Computational Science, pp. 1114–1117 (2006)
4. Mathes, A.: Folksonomies - cooperative classification and communication through shared metadata (2004)

5. Lee, K.S., Croft, W.B., Allan, J.: A cluster-based resampling method for pseudo-relevance feedback. In: SIGIR 2008: Proceedings of the 31st Annual International ACM SIGIR Conference on Research and Development in Information Retrieval, pp. 235–242. ACM (2008)

6. Tao, T., Zhai, C.: Regularized estimation of mixture models for robust pseudo-relevance feedback. In: SIGIR 2006: Proceedings of the 29th Annual International ACM SIGIR Conference on Research and Development in Information Retrieval, pp. 162–169. ACM (2006)

7. Lin, Y., Lin, H., Jin, S., Ye, Z.: Social Annotation in Query Expansion: a MachineLearning Approach. In: SIGIR 2011: Proceedings of the 34th Annual International ACM SIGIR Conference on Research and Development in Information Retrieval, pp. 405–414. ACM (2011)

8. Metzler, D., Croft, W.B.: Latent concept expansion using markov random fields. In: SIGIR 2007: Proceedings of the 30th Annual International ACM SIGIR Conference on Research and Development in Information Retrieval, pp. 311–318. ACM (2007)

9. Xu, J., Croft, W.: Query expansion using local and global document analysis. In: SIGIR 1996: Proceedings of the 19th Annual International ACM SIGIR Conference on Research and Development in Information Retrieval, pp. 4–11. ACM (1996)

10. Rocchio, J.: Relevance feedback in information retrieval. The SMART Retrieval System: Experiments in Automatic Document Processing, 313–323 (1971)

11. Robertson, S.E., Walker, S., Beaulieu, M., Gatford, M., Payne, A.: Okapi at trec-4. In: The 4th Text Retrieval Conference, TREC 1996 (1996)

12. Lavrenko, V., Croft, W.B.: Relevance based language models. In: SIGIR 2001: Proceedings of the 24th Annual International ACM SIGIR Conference on Research and Development in Information Retrieval, pp. 120–127. ACM (2001)

13. Zhai, C., Lafferty, J.: Model-based feedback in the language modeling approach to information retrieval. In: CIKM 2001: Proceeding of the 10th ACM Conference on Information and Knowledge Management, pp. 403–410. ACM (2001)

14. Song, Y., Zhuang, Z., Li, H., Zhao, Q., Li, J., Lee, W.C., Giles, C.L.: Real-time automatic tag recommendation. In: SIGIR 2008: Proceedings of the 31st Annual International ACM SIGIR Conference on Research and Development in Information Retrieval, pp. 515–522. ACM (2008)

15. Wu, X., Zhang, L., Yu, Y.: Relevance based language models. In: WWW 2006: Proceedings of the 15th International Conference on World Wide Web, pp. 417–426. ACM (2006)

16. Hotho, A., Jäschke, R., Schmitz, C., Stumme, G.: Information Retrieval in Folksonomies: Search and Ranking. In: Sure, Y., Domingue, J. (eds.) ESWC 2006. LNCS, vol. 4011, pp. 411–426. Springer, Heidelberg (2006)

17. Bao, S., Xue, G., Wu, X., Yu, Y., Fei, B., Su, Z.: Optimizing web search using social annotations. In: WWW 2007: Proceedings of the 16th International Conference on World Wide Web, pp. 501–510. ACM (2007)

18. Brin, S., Page, L.: The anatomy of a Large-Scale Hypertextual Web Search Engine. Computer Networks and ISDN Systems 30(1-7), 107–117 (1998)

19. Noll, M.G., Yeung, C.A., Gibbins, N.C., Meinel, S.N.: Telling Experts from Spammers: Expertise Ranking in Folksonomies. In: SIGIR 2009: Proceedings of the 32nd Annual International ACM SIGIR Conference on Research and Development in Information Retrieval, pp. 612–619. ACM (2009)

20. Lu, C., Hu, X., Chen, X., Park, J.R.: The Topic-Perspective Model for Folksonomy systems. In: Proceeding of the 16th ACM SIGKDD International Conference on Knowledge Discovery and Data Mining, pp. 683–691 (2010)

Semantically Enriched Clustered User Interest Profile Built from Users' Tweets

Harshit Kumar* and Hong-Gee Kim

Seoul National University, Yeongun Campus, South Korea
hkumar.arora@gmail.com, hgkim@snu.ac.kr

Abstract. Existing works in user profiling suffers from two well known problems in IR: polysemy and synonymy. Enriching semantics to terms that represent user interests disambiguate it's context, polysemous topics, and synonyms. One way of enriching semantics to terms is by grouping related terms together into clusters. This work exploits users' tweets to build a Contextualized User Interest Profile($CUIP$) that consist of clusters of (semantically) related terms and their term-weights. We propose two approaches to build the $CUIP$: $svdCUIP$ based on Singular Value Decomposition (SVD); and, $modsvdCUIP$ based on modded SVD (modSVD). We run experiments to determine the appropriate value of various parameters required for building $CUIP$, and also run experiments to compare the two proposed approaches in terms of clustering accuracy and clustering tendency. Results show that the clustering tendency and accuracy of the cluster structure $modsvdCUIP$ is superior than the $svdCUIP$.

Keywords: Factorization, Clustering, User Profiling, Semantics.

1 Introduction

The success of any web portal or social network service (SNS) lies with the number of users registered with that service. Registration with a SNS requires a user to explicitly input his/her personal information which is termed as user profile. The information contained in a user profile is more or less static. Other than the user's personal information, the systems are more interested in users' preferences or users' interests that they can use for recommendation of their services. Search engines and e-commerce portals mine user logs to generate user preferences or a profile of users' interests. We define such profile as the User Interest Profile (UIP); it consists of terms or topics that represent users' interests. The information in a UIP, unlike user profile, is dynamic; it keeps changing with time.

Current research studies[1,2,3,5,6,7] highlight the use of folksonomies for constructing and managing UIP. A very simplistic approach, to build a UIP, could be to collect all the tags that the user input to annotate the resources of interest.

* This research was supported by the KCC(Korea Communications Commission), Korea, under the R&D program supervised by the KCA(Korea Communications Agency)"(KCA-2012-(12-911-05-004)).

Y. Hou et al. (Eds.): AIRS 2012, LNCS 7675, pp. 406–416, 2012.

The frequency of tag usage, its tag weight, indicates the importance of a tag in the UIP. A UIP is defined as a set of tags and their tag-weights. Undoubtedly, a UIP is immensely useful to the web portals for product recommendation, tag suggestion, search results re-ranking, query expansion, etc. Nonetheless, there is still some scope of improvement that we discuss next from Noll et al.[1] that uses folksonomies for building the UIP and use it for personalized search. Authors mention that the search result URL of US Security and Administration is promoted, by re-ranking based on the terms in the UIP, even though it is not related to the input query keyword *security*. There exists no established reasoning that can explain the quantitative effect of the terms, in the UIP, on the re-ranking of the search results, i.e., why does the re-ranking process, based on the UIP, promotes a particular URL more than the other URL? We offer the following reasoning. Some terms, in the UIP, even though not related to the user query *security*, but because they are present in the UIP, contributes to the re-ranking score of the search results. The term, in this case *insurance* in the UIP, has a false positive effect on the re-ranking of the URL of US Security and Administration; this is because of the incapability of the system to judge the context of user query. Note that the terms in the UIP can have, uncalled for, a false positive effect or a false negative effect on the re-ranking of the search results.

In our endeavour to build the $CUIP$, we propose two approaches: $svdCUIP$ based on Singular Value Decomposition (SVD); $modsvdCUIP$ based on a variation of SVD termed as modSvd. A $CUIP$ is defined as a collection of clusters, where each cluster consists of related terms and their term weights. Determining the right contextual cluster from user's $CUIP$ for the given user query and using that cluster for re-ranking search results, or recommending products would result in more user satisfaction. It will also help in disambiguating the context of user query by finding related terms to the user query from the clusters in user's $CUIP$. We run experiments to evaluate which of the two methods have the acceptable or strong clustering tendency and best clustering accuracy. The results obtained in our experiments show that $modsvdCUIP$ clearly outperforms $svdCUIP$.

The rest of the paper is organized as follows. In the next section, we discuss about the current research work that uses folksonomy for building the $CUIP$. Section 3 presents the 2 proposed approaches and their evaluation in Section 4. Finally, we conclude this work in Section 5 followed by references.

2   Related Work

Two key papers exist in the literature that are based on building term cluster clusters for modelling user profile. Shepstein et al.[6] clusters the entire tag space of a folksonomy system to obtain sets of semantically related tags. It models users and resources as vectors over the set of tags. The proposed modified Hierarchical algorithm selects a subset of potential clusters related to user's current navigational activity and use that cluster for recommendation. This way, clusters serve as an intermediate between users and resources. This algorithm has high time complexity, because it calculates the user's interest in each cluster, and each resource's closest clusters- both computations are based on a variation of Jaccard coefficient. Note that the number of clusters could be in millions. Our proposed approaches are centred on user; we propose to discover clusters in user's UIP . Second difference between our work and their work is the use of Latent Semantic Analysis (LSA) for generating similarity matrix. Authors in their future work mentioned that they would like to analyse other approaches like LSA. They also mentioned that they would like to cluster tags in user profiles. This is exactly what we are presenting in this work.

Simpson et al.[7] approach to build CUIP uses cosine similarity measure for discovering term-term similarity matrix. For clustering,authors have used graph based approaches, such as centrality and betweenness. We are using Hierarchical Clustering Algorithm (HCA) that works best when the number of clusters are unknown *a priori*. Moreover, no experiments appear to have been carried out that could verify clustering accuracy, rather they provided some example visualization of cluster structure. We report a more detailed set of experiments that measures the clustering tendency and clustering accuracy of $CUIP$.

3   Building Clustered User Interest Profile

This section describes two methods for generating Clustered User Interest Profile ($CUIP$).

3.1   User's Context and Term Weighting

This work uses Twitter[1] as a source for obtaining terms of user's interest. Tweets inputted by a user can be thought of as texts of user's interest. The first step involves processing tweets to obtain terms by removing stop words, stemming, and extracting nouns[2].

Let Tw be a set of tweets inputted by a user, and let T be a list of terms obtained after processing Tw. A user's context Tw,T,R is defined as a 3-tuple, where R is a binary relation between Tw and T. In order to express that a tweet $tw_1 \in Tw$ is in a relationship with a term $t \in T$, we write $t \ I \ tw_1$ or $(t, tw_1) \in I$, which can be read as "the term t is a topic of tweet tw_1". Focusing only on the relations between tweets and terms in Table 1, based on a dummy example, a user's context in Table 2 is derivable. Note that, in a real scenario, the actual term-values may be greater than 1. This section describes how the terms and their term weights in a UIP are combined, aggregated, and normalized.Each term t in a tweet tw_i has some weight $w(t, tw_i)$ defined as term weight. For ex: the term-weight of term *java* in tweet tw_1 is 1. Term weight of a term in the $CUIP$ is aggregated from from multiple tweets; it is very much possible that the same term may originate from multiple tweets, each with a potentially different weight. We use the standard result set fusion technique to aggregate the weight of term t from tweets $N(tw)$ to calculate the term weight $w(t) = \sum_{i=1}^{N(tw)} w(t, tw_i)$,

---

[1] http://www.twitter.com

[2] http://nlp.stanford.edu/software/tagger.shtml

Table 1. Filtered terms extracted after processing user tweets

Tweet	Terms
tw_1	java, application
tw_2	java
tw_3	travel
tw_4	iphone, game
tw_5	iphone, application

Table 2. User's context derivable from Table 1

	tw_1	tw_2	tw_3	tw_4	tw_5
java	1	1	0	0	0
game	0	0	0	1	0
application	1	0	0	0	1
travel	0	0	1	0	0
iphone	0	0	0	1	1

where $w(t, tw_i)$ is the term-weight of the term t retrieved from the tweet tw_i. For example, the UIP for the user context in Table 2 would be {*java:2, game:1, application:2, travel:1, iPhone:2*}. Using raw frequency of term-weight is problematic; extremely popular terms are used many times, so their term weights are really high. To circumvent this limitation, we propose to normalize the term-weights of terms using the following equation $nw(t_i, up) = \dfrac{w(t_i, up)}{\max_{k=1}^{|up|} w(t_k, up)}$.

3.2   Latent Semantics in User Interest Profile

Latent semantics means those hidden relationship between terms that exists but is not explicitly visible. We propose to build a system that could discover related terms, even though the terms are not identical or do not belong to the same tweet. The approaches to establishing latent structures in a UIP are based on the assumption that more similar terms are more closely related. In this section, we propose two similarity measures to calculate the similarity between terms, and using them for the clustering algorithm.

Term-Term Similarity -   Co-occurrence similarity derives similarity between two or more terms that belong to the same tweet. The degree of their relationship is calculated using the frequency of term co-occurrence, called as 1st order co-occurrence similarity. Another type of co-occurrence similarity is 2nd order co-occurrence similarity that derives the similarity between two terms that do not belong to the same tweet, but they have a common term that relates to both of them. For example, refer Table 2 the term *application* is related to term *iphone* in tweet tw_5, and the same term *application* is also related to term *java* in tweet tw_1; this suggests an indirect relationship or 2nd order co-occurrence relationship between terms *iphone* and *java*. It is analogous to finding friend of a friend and also quantifying the degree of friendship relationship.

The proposed methods employ matrix factorization on user's context or term-Tweet matrix to discover 1st order and 2nd co-occurrence similarity between terms. Latent Semantic Analysis (LSA)[4] uses a matrix factorization technique, Singular Value Decomposition (SVD), to find hidden relationship between terms which is not apparent, otherwise. In the preliminary step, the user context in Table 2 is represented as the term by Tweet matrix, let it be A. The SVD technique decomposes the term-tweet matrix, A, into three matrices, $A = USV^T$,: a term by dimension matrix, U; a diagonal matrix of singular values, S; and, a tweet by dimension matrix V. SVD translates the term and tweet vectors into a space determined by the rank r of matrix A. When we select the k largest singular values from S and their corresponding singular vectors from U and V, we get the rank k approximation A_k, $A_k = U_k S_k V_k^T$. To compute the term-term similarity matrix, we would like to compute, U_k, a low-rank approximation of U matrix. The term-term similarity matrix, Sim_k, is computed using $Sim_k = U_k S_k (U_k S_k)^T = U_k S_k S_k^T U_k^T = U_k S^2{}_k U_k^T$. Determining the value of k is an essential part of generating a good similarity matrix and hence a good cluster structure; we show, in the experiment section, how to determine the appropriate value of k that has the strongest or acceptable clustering tendency. The example below shows the term-term similarity matrix sim, generated using SVD, for the user context in Table 2.

$$Sim = U_2 S_2{}^2 U_2^T = \begin{array}{r} java \\ game \\ application \\ travel \\ iphone \end{array} \begin{matrix} java & game & application & travel & iPhone \\ 1.66 & -0.32 & \mathbf{1.35} & 0.00 & -0.02 \\ -0.32 & 0.61 & 0.30 & 0.00 & \mathbf{1.05} \\ 1.35 & 0.30 & 1.64 & 0.00 & \mathbf{1.03} \\ 0.00 & 0.00 & 0.00 & 0.00 & 0.00 \\ -0.02 & 1.05 & 1.03 & 0.00 & 1.96 \end{matrix}$$

This example is based on a small dummy corpus, therefore the sparseness is low (approx. 40%). In real scenarios, sparseness could be as high as 99% which seriously effects the ability of SVD to calculate the similarity between terms. Our experiments vouch for this statement. To rightly capture the similarity matrix, we go a step further by calculating the cosine similarity between term vectors of similarity matrix sim produced from the SVD process. This method is called as modded Svd (modSvd). Each term vector represents the projection of a term in the term space. For instance, each term t_i in similarity matrix sim has an entry for each term t_j that co-occurs with it.

$$modSim(t_1, t_2) = \frac{\sum_{i=1, j=1}^{n} t_{1i} t_{2j}}{\sqrt{\sum_{i=1}^{n} t_{1i}^2 \sum_{i=1}^{n} t_{2i}^2}}. \tag{1}$$

The similarity matrix $modSim$ calculates the similarity between all pair of term vectors in the entire term space of user; this results in discovering 2nd order co-occurrence similarity between terms. The following example shows the $modSim$ matrix for the matrix Sim, illustrated above, calculated using equation 1.

$$modSim = \begin{array}{r} java \\ game \\ application \\ travel \\ iphone \end{array} \begin{matrix} java & game & application & travel & iPhone \\ 1 & -0.12 & \mathbf{0.84} & 0.00 & 0.19 \\ -0.12 & 1 & \mathbf{0.43} & 0.00 & \mathbf{0.95} \\ 0.84 & 0.430 & 1 & 0.00 & \mathbf{0.69} \\ 0.00 & 0.00 & 0.00 & 1 & 0.00 \\ 0.19 & 0.95 & 0.69 & 0.00 & 1 \end{matrix}$$

The higher value of cosine signifies greater overlap between two vectors across n dimensions, where n is the number of terms in the term space of the user.

Clustering - The input to the clustering algorithm is the similarity matrix calculated in the present section. The output of the clustering algorithm is the clusters of semantically related terms. Since, the number of clusters are unknown before hand, we use Hierarchical Clustering Algorithm (HAC) which works best when the number of clusters are unknown, *apriori*. At the outset, HAC treats each term as a singleton cluster and then successively merge pair of clusters until all clusters have been merged into a single cluster that contain all the terms. The output of HAC is a hierarchy of clusters. We don't need a single hierarchy of cluster, refer fig 1, but l number of clusters. Note that, the value of l is not specified *a priori*. This requires cutting the single cluster; We use distinctness, d, as a measure to cut the single hierarchy of clusters to obtain l clusters. For instance, Table 3 shows the clusters in *svdCUIP* and *modsvdCUIP* generated using HAC for *sim* and *modSim* matrix illustrated above. *svdCUIP* has 4 clusters, it fails to identify that the terms *iphone* and *game* should belong to one cluster, whereas *modsvdCUIP* didn't make that mistake. It is very important to choose the right value of d that could generate appropriate clusters matching user's perspective. Fig. 1 shows the role of distinctness parameter d; if $d >= 1.4$, it would result in only 1 cluster that consist of complete hierarchy of terms. Instead, if $d = 0.4$, then there would be 3 clusters. For value of $d < 0.3$, there would be a plain list of terms. The experiment section shows how to determine the right value of d, to generate crisp clusters, without compromising on the clustering accuracy.

The *CUIP* that results from the application of HAC on *Sim* matrix obtained by applying SVD on term-tweet matrix is called as *svdCUIP*(read as, SVD based CUIP). Whereas, the *CUIP* that results from the application of HAC on *modSim* matrix obtained by calculating the cosine similarity of every pair of term vectors in *Sim* matrix is called as *modsvdCUIP* (read as, modSVD based CUIP).

Table 3. Clusters generated by applying HAC on matrix *sim* and *modSim* for $k = 2$ and $d = 0.35$

Method	Clusters
svdCUIP	[[*iphon*], [*java*, *applic*], [*game*], [*travel*]]
modsvdCUIP	[[*java*, *applic*], [*iphon*, *game*], [*travel*]]

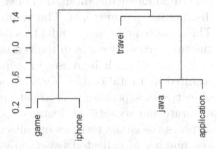

Fig. 1. dendrogram output

4   Evaluation of Clustered User Interest Profile

The experiments in this section examine the clustering accuracy of *svdCUIP* and *modsvdCUIP*. In particular, it addresses the following issues: How does the dimensionality reduction parameter k and distinctness parameter d effects clustering tendency and clustering accuracy?; Which of the two proposed methods have better clustering accuracy? The data set is created from the tweets posted by

Table 4. Term Tweet distribution and characteristics of term-term similarity matrix *sim* and *modSim*

$User_{id}$	$\|TW\|$	$\|T\|$	≤ 0		>0 and ≤ 0.5		>0.5 and ≤ 0.75		>0.75	
			Sim	modSim	Sim	modSim	Sim	modSim	Sim	modSim
$User_1$	143	708	67.38%	45.03%	32.31%	41.7%	0.09%	3.9%	0.2%	9.35%
$User_2$	138	539	59.31%	44.26%	40.17%	44.72%	0.13%	3.1%	0.38%	7.9%
$User_3$	147	679	48%	39.62%	51.57%	50.94%	0.13%	4.64%	0.27%	4.7%
$User_4$	105	430	86.16%	76.21%	12.99%	17.77%	0.22%	0.92%	0.61%	5.09%
$User_5$	144	540	71.1%	47.49%	28.45%	38.28%	0.13%	2.78%	0.32%	11.44%
$User_6$	146	654	55.88%	37.52%	43.76%	49.22%	0.09%	5.37%	0.24%	7.87%
$User_7$	123	563	55.25%	46.64%	44.32%	44.9%	0.15%	3.76%	0.35%	4.68%
$User_8$	162	628	70.84%	49.54%	28.87%	33.84%	0.06%	5.32%	0.21%	11.2%
$User_9$	142	599	63.82%	48.85%	36.30%	41.15%	0.14%	4.1%	0.32%	5.8%
$User_{10}$	156	602	50.71%	40.2%	48.77%	47.36%	0.15%	4.25%	0.35%	8.17%

10 volunteers during the period July to September 2010, refer Table 4. Unlike other evaluations, it is hard to to measure the accuracy of clusters because for clustering no body knows what the correct clusters are. The users themselves are the best judge. To establish the ground truth, we asked each user to group related terms that were extracted from his/her tweets. Generating ground truth manually for evaluation is the normal procedure used in many research papers[9]. Since this process is subjective, we take the average of the scores from all the users as the final score of the clustering. The manual procedure to generate ground truth is obviously a labour intensive and time consuming task; this is the primary reason that forced us to only experiment with a small set of users. For each user, 2 sets of several *CUIPs* were generated. In each set, a *CUIP* is generated for each combination of value of dimension reduction parameter k and distinctness parameter d. The value of k varies from 10 to 110, in steps of 10. This results in 11 sim_k and 11 $modSim_k$ similarity matrices. Similarly, the distinctness parameter d is initialized to 0.03, and it increases in steps of 0.02 until 0.13, after which it increases in steps of 0.1 until 0.93 (total of 14 values). This resulted in a total of 154 *svdCUIPs* and the same number of *modsvdCUIPs*. The objective of experiments is to determine the value of k and d that has the most accurate and acceptable cluster structure generated using each method. Table 4 displays the characteristics of collected tweets and similarity matrix. The average number of collected tweets are 134, and the average number of terms are 621. On average, 62.8% of values in the similarity matrix *sim* are less than or equal to 0, whereas, the average number of values less than or equal to 0 in the similarity matrix *modSim* are 47.53%. Also, the average number of values greater than 0.75 in the similarity matrix, *sim*, are 0.335% compared to 7.2% in the similarity matrix *modSim*. Overall, the number of values less than or equal to 0.5 in the similarity matrix *sim* are 99.6%. On the other hand, the number of values less than or equal to 0.5 in the similarity matrix, *modSim*, is 88.5%. These results suggest that the similarity matrix, *modSim*, is less sparse than the similarity matrix, *Sim*. We choose *silhouettecoefficient* (unsupervised evaluation) [8]

to judge the cluster tendency,and *FScore* (supervised evaluation) to compare the clustering accuracy. A very useful overall quality measure of a given clustering is its average silhouette coefficient, $\bar{s}$. Kaufman and Rousseeuw [8] provide an interpretation of average silhouette coefficient, $\bar{s}$, as a measure of evidence in support of cluster structure: the value of $\bar{s}$ between]0.7, 1.0] suggests strong evidence, between]0.5 ,0.7] suggests reasonable evidence,]0.25, 0.5] suggests weak evidence, and between [-1,0.25] suggests no evidence.

4.1   Evaluating Clustering Tendency

The experiments in this section determine, for both *svdCUIP* and *modsvdCUIP*, the various value(s) of dimensionality reduction parameter k and distinctness parameter d that show(s) strong clustering tendency. The assessment of presence of clusters in a data set is an important step in cluster analysis. The plot of average silhouette coefficient vs. number of clusters in Fig. 2 helps in visualizing cluster tendency, if any. Fig. 2 shows that that *modSvdCUIP* has stronger evidence of cluster tendency, whereas *svdCUIP* exhibits weak or no evidence of clustering tendency. We observed that the clustering tendency in a user's *CUIP* is affected by the ratio of number of zero values to the number of positive values; the lower the better. The average ratio for *modsvdCUIP* is 0.9, and for *svdCUP* is 1.68. The maximum and minimum ratio for *modsvdCUIP* is 3.2 and 0.6, respectively. Whereas, the maximum and minimum ratio for *svdCUIP* is 6.2 and 1.0, respectively. These numbers explain the reason behind the lack of cluster tendency in *svdCUIP*.

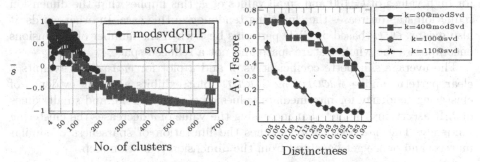

No. of clusters   Distinctness

Fig. 2.   Clustering   Tendency   of   **Fig. 3.** Clustering accuracy of modSvd
modSvdCUIP and *svdCUIP*   (k=30,40) and svd (k=100,110)

The plot also depicts that the value of average silhouette coefficient ($\bar{s}$) decreases as the number of clusters increases beyond 50. This suggests that the best cluster structure is obtained when the number of clusters are less than 50. This is acceptable because the average number of terms is 594, which could possibly result in 30-50 clusters. However, what is surprising is that even when the number of clusters is less than 10, the plot shows strong clustering tendency. To try to find the natural number of clusters in a user's *CUIP*, one should look for

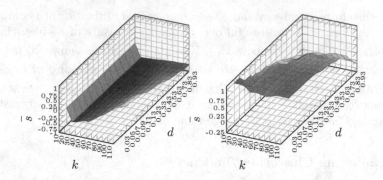

Fig. 4. Determining the value of k and d,the first plot is for $svdCUIP$, and the second plot represents $modsvdCUIP$

a knee, a peak, or dip in the plot of average silhouette coefficient vs. number of clusters [10]. The plot for $modsvdCUIP$ clearly shows a rise and then a dip, and hence a peak, occurring around when the number of clusters are in the range of 30-40. However, it is difficult to find a peak for $svdCUIP$, because there are several such peaks.

Fig. 4 presents 3-dim plot that shows how the value of average silhouette coefficient changes as the value of k and d changes. The average silhouette coefficient vs. k and d plot for $svdCUIP$ exhibits a clear pattern; for low values of k and no matter what the value of d is, there is no evidence of clustering tendency. For high values of k, close to 100 and 110, and low values of d, there is a reasonable evidence of clustering tendency. $svdCUIP$ shows reasonable clustering tendency for high values of $k=100$ and small values of d, this implies that the dimension reduction step decreases the clustering tendency in this case. In other words, it also means, $CUIP$ based on SVD performs best when the number of dimensions encompasses the whole feature space and not a reduced space.

The average silhouette coefficient vs. k and d plot for $modsvdUIP$ exhibits a clear pattern; unlike $svdCUIP$, $modsvdCUIP$ plot exhibits a strong evidence of clustering tendency for intermediate values of $k = 30$ and 40, and small values of d. It ascertains the fact that increasing the value of d decreases the clustering tendency. The $modSvdCUIP$ overcomes the limitation of sparseness of similar matrix, and hence get benefited from the dimension reduction step.

4.2   Clustering Accuracy

This experiment aids in determining the appropriate value of d for the cluster structure that has the highest accuracy. Fig. 3 shows the clustering accuracy for each method. It can be clearly seen that $modsvdCUIP$ has better clustering accuracy than $svdCUIP$. The average clustering accuracy for $modsvdCUIP$ and $svdCUIP$ cluster structure is 0.58 and 0.16, respectively; there is a 244% increase in average clustering accuracy. This indicates that the cluster structure produced by $modsvdCUIP$ is more accurate than the cluster structure produced by $svdCUIP$. When the method is $modSvd$, the dimension reduction parameter

k=30 has better clustering accuracy than k=40. Also, the difference in clustering accuracy between k=30 and k=40 is very small. Moreover, both the curves follow the same pattern, which means the clustering accuracy estimated by $modSvd$ for k=30 and k =40 is nearly identical with a slightly better performance for k=30. The highest clustering accuracy for $modsvdCUIP$ is 0.75, which is obtained when k=30 and distinctness parameter d=0.07. The highest clustering accuracy for $svdCUIP$ is 0.55 which is obtained when k=110 and distinctness parameter d=0.03.

5   Conclusions

We proposed 2 approaches to build Clustered User Interest Profile (CUIP) from user's tweets: SVD based CUIP, ($svdCUIP$), is generated by applying Hierarchical Agglomerative Clustering (HAC) on the term-term similarity matrix computed using SVD; modded Svd based CUIP, $modsvdCUIP$, is generated by applying HAC to the second order co-occurrence matrix $modSim$ obtained from computing the cosine similarity between the term-vectors in the sim matrix. The experiments show that the $svdCUIP$ fails to cluster semantically related terms, however it can, to some extent, cluster co-located terms. To circumvent this limitation, we proposed to generate the cluster structure $modsvdCUIP$. Experimental results suggest that the cluster structure $modsvdCUIP$ has higher clustering tendency and clustering accuracy as compared to the clusters structure $svdCUIP$. The poor clustering tendency of the cluster structure $svdCUIP$ is due to the sparsity(92%) of similarity matrix sim. The best clustering accuracy(0.75) for the cluster structure $modsvdCUIP$ is achieved for dimension reduction parameter k = 30 and distinctness parameter d = 0.07. In contract, the cluster structure $svdCUIP$ showed best clustering accuracy(0.57), though lower than $modsvdCUIP$, for k = 110 and d = 0.03. In the future, we would like to put the $CUIP$ into practice with a search system to develop personalized search.

References

1. Noll, M.G., Meinel, C.: Web Search Personalization Via Social Bookmarking and Tagging. In: Aberer, K., Choi, K.-S., Noy, N., Allemang, D., Lee, K.-I., Nixon, L.J.B., Golbeck, J., Mika, P., Maynard, D., Mizoguchi, R., Schreiber, G., Cudré-Mauroux, P. (eds.) ISWC/ASWC 2007. LNCS, vol. 4825, pp. 367–380. Springer, Heidelberg (2007)
2. Vallet, D., Cantador, I., Jose, J.M.: Personalizing Web Search with Folksonomy-Based User and Document Profiles. In: Gurrin, C., He, Y., Kazai, G., Kruschwitz, U., Little, S., Roelleke, T., Rüger, S., van Rijsbergen, K. (eds.) ECIR 2010. LNCS, vol. 5993, pp. 420–431. Springer, Heidelberg (2010)
3. Xu, S., Bao, S., Fei, B., Su, Z., Yu, Y.: Exploring folksonomy for personalized search. In: SIGIR, pp. 155–162 (2008)
4. Deerwester, S., Dumais, S.T., Furnas, G.W., Landauer, T.K., Harshman, R.: Indexing by latent semantic analysis. JASIS 41(6), 391–407 (1990)

5. Kumar, H., Kim, H.-G.: Using Folksonomies for Building User Interest Profile. In: Konstan, J.A., Conejo, R., Marzo, J.L., Oliver, N. (eds.) UMAP 2011. LNCS, vol. 6787, pp. 438–441. Springer, Heidelberg (2011)

6. Shepitsen, A., Gemmell, J., Mobasher, B., Buke, R.: Personalization in Folksonomies Based on Tag Clustering. In: AAAI 2008, pp. 37–48 (2008)

7. Simpson, E., Butler, M.H.: Analysing Communal Tag Relationships for Enhanced Navigation and User Modeling, pp. 43–64. IGI Global (2009)

8. Kaufman, L., Rousseeuw, P.J.: Introduction, in Finding Groups in Data: An Introduction to Cluster Analysis. John Wiley & Sons, Inc., Hoboken (2008)

9. Liu, B.: Web Data Mining, 2nd edn. (2009)

10. Tan, P.N., Steinback, M., Kumar, V.: Introduction to Data Mining (2005)

Adaptive Weighting Approach to Context-Sensitive Retrieval Model

Xiaochun Wang[1,*], Muyun Yang[1], Haoliang Qi[2], Sheng Li[1], and Tiejun Zhao[1]

[1] School of Computer Science and Technology, Harbin Institute of Technology, Harbin, China
[2] Computer Science and Technology Department, Heilongjiang Institute of Technology, China
xcwang@mtlab.hit.edu.cn

Abstract. To best exploit the context information for meaningful hints to the user's intent, this paper proposes an adaptive weighting approach to improve the current context-sensitive retrieval model. The *potential for adaptability* is first investigated as the performance gap between the current context-sensitive models with a fixed form weight and those with adaptive weights for contextual information. Then the proper context weight is predicated according to the relation strength between the query and its context. The experimental results on a public available dataset indicate that the proposed approach outperforms three baseline methods.

Keywords: Context, information retrieval, interaction data.

1 Introduction

The majority of queries submitted to the search engine are short and ambiguous, and Web users may have different search needs even when they submit the same query [11][12][14][15]. The "one size fits all" approach fails to be optimized for each individual's specific information need. Personalized search appeared as a promising solution to provide different search results according to an individual preference, and it has been revealed that the contextual approach may be a good choice [13].

So far, there exist two frameworks of context-sensitive retrieval. One is the classical statistics language model approach (e.g., [1][7][12]), and the other is the machine learning mechanism, esp. the learning-to-rank techniques (e.g., [3-6][11][16]). An essential issue among these works is how to exploit context information in the retrieval model. Since all the existing studies assume in the paradigm that the context is closely related to current query, a linear combination of current query and context is often adopted in above mentioned studies.

In fact, context may not always help. Several previous works [8][9] have discovered that the current query can also be unrelated to its preceding queries in the same session. For example, in cases of the query specialization and query generalization within a session, documents related to context should be promoted; while in the case

---

* Corresponding author.

Y. Hou et al. (Eds.): AIRS 2012, LNCS 7675, pp. 417–426, 2012.

of query reformulation, documents related to context should be demoted [2]. In other words, context is not always related to current query need, if any, may not be same closeness.

To address this issue, we propose an adaptive weighting approach to the existing context-sensitive retrieval model. We first reveal the potential for adaptability as the performance gap between the current context-sensitive models with a fixed form weight and those with adaptive weight for contextual information. Instead of relying on intuitive heuristics or insights, then build a specific model to predict the relationship strength between current query and context. The results are combined into the retrieval model to best exploit the context information.

The rest of the paper is organized as follows. Section 2 explores Potential for adaptability. Section 3 describes our adaptive model. Section 4 depicts experiment. The paper is concluded in Section 5.

2   Potential for Adaptability

2.1   Context-sensitive Retrieval Model

In a search session, a user may interact with the search system several times. The context here refers to queries and Web pages viewed prior to the current query in the same session. Context-sensitive retrieval problem involves how to exploit the current query together with its context to model a user's search need.

Statistical language models are often used to model a user's information need. One advantage is that the query context can be incorporated naturally as additional evidence into the query model. Language models provide a formal but straightforward method of combining all the pieces of evidence that contribute to user search need.

Context-sensitive retrieval task is to summarize the query history by an n-gram language model and the clickthrough history by another n-gram language model, then linearly interpolate them with the current query model. A typical context-sensitive retrieval framework can be viewed as following [1].

$$p(\omega \mid \theta_k) = \alpha_k p(\omega \mid Q_k) + (1 - \alpha_k)[\beta_k p(\omega \mid H_C) + (1 - \beta_k) p(\omega \mid H_Q)] \qquad (1)$$

In which $\beta_k \in [0,1]$ is a fixed parameter tuning the weight on each history model, and $\alpha_k \in [0,1]$ is a fixed parameter tuning the weight on the current query and its contexts. The search need model $p(\omega|\theta_k)$, is the linear interpolation of the current query model $p(\omega|Q_k)$, history query model $p(\omega|H_Q)$ and history click model $p(\omega|H_C)$ with fixed coefficient.

The key to accurately estimate search need model is to choose parameters appropriate to combine current query model and context model. Different forms of estimation result from specifying the values of. A simplest choice to estimate α_k, β_k set is to set them as constants across all queries, treating the histories equally regardless of the query. A better estimation comes from values dependent on the document length.

Assumed to update the query model over time during search interaction, decaying weights is assigned to the previous query so as to trust a recent query. Known as *Dirichlet* smoothing, Bayesian estimation is generally applied to this process.

2.2   Ideal Performance

In order to calculate the ideal performance of current context-sensitive model, we choose to use the same datasets as described in [1], which is publicly available. The details of the datasets are described in Table 1.

Table 1. Descriptive Statistics about Datasets

#total queries	#docs in collection	#clicked results	#session
120	242,918	91	30

To summarize the quality of a ranked list of results, we use two popular performance measures: (1) *Mean Average Precision* (MAP), a standard non-interpolated average precision and serves as a good measure of the overall ranking accuracy. (2) *Precision at top 20 documents* (pr@20), reflecting the utility of the top 20 documents.

Under the framework described in Section 2.1, we examine the ideal performance of existing context-sensitive retrieval model. Two concepts are defined as follows.

Definition 1. Ideal Parameter. Given a set of candidate parameters A, ideal parameters $\alpha^* \in A$, $\beta^* \in A$ are picked so as to maximize retrieval performance max $E(R(f(q_u, h_c, h_q)), c)$, where q_u is the query submitted by a particular user u, h_q and h_c are queries submitted and pages viewed by u during a search session, respectively; c is the set of clicks on results returned for q_u; $f(\cdot)$ is language model, $R(\cdot)$ is a ranking functi, and $E(\cdot)$ is evaluation function.

Definition 2. Ideal Performance. Given a set of candidate parameters A, a given query q_u, history click h_c, history query h_q, ideal parameters $\alpha^* \in A, \beta^* \in A$, the ideal performance is attained when combine q_u, h_c and h_q using ideal parameters.

The ideal parameter represents an ideal situation in which if each query intention is accurately expressed, there are corresponding α and β in the query intention model. To simplify this issue, we assume that the parameters α, β are independent of each other. Then we use linear search algorithm (step size is 0.1) to try each possible combination. The best combination is picked when retrieval performance achieves its highest value. Thus achieved performance is now called *ideal performance*, and corresponding parameters are called *ideal weights*.

We compare fixed parameters against ideal parameters in Fig.1 and Fig.2. The horizontal axis represents the ID of the queries and the vertical axis represents their corresponding ideal weights. The common fixed parameters of the current model are also included for reference in the figure. Fig.1 shows the distribution of ideal-α, in which is different for each query. In contrast to the fixed parameters (marked as fixed-α). Similar case occurs in Fig.2. In fact, parameters in the current model do not always equal to the ideal ones.

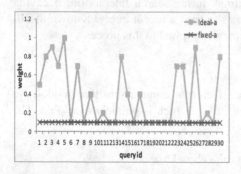

Fig. 1. Weight α distribution

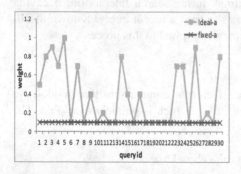

Fig. 2. Weight β distribution

2.3   Potential for Adaptability

Under the same framework as in Section 2.1, we compare the performance gap between the current context-sensitive models with fixed weights and those with adaptive weights for contextual information.

Definition 3. Potential for Adaptability. For a current context-sensitive retrieval model, its potential for adaptability is the actual retrieval performance deviations from the ideal performance.

Fig.3 shows MAP for the ideal context-sensitive ranking with dynamically weighting (dark bar), and current context-sensitive rankings with unified weighting (light bar) for each query. This indicates that dynamic-weighting achieves a 26.58% and a 23.70% improvement over fixed-weighting in terms of MAP and pr@20, respectively. Therefore, there is a substantial retrieval performance gap between current context-sensitive models with unified weights for all queries and those with adaptive weights. We call this gap as the *potential for adaptability*, standing for the potential gain by dynamic context-sensitive search system.

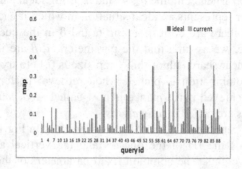

Fig. 3. Potential for single query

To sum up, this section has revealed that parameters in the model are of vital importance to the final retrieval results because it decides whether the user's search intention is properly expressed. In addition, the retrieval model has substantial room for improvement if the context weight could be properly assigned.

3   Adaptive Context-Sensitive Model

3.1   Regression-Based Context Weight Prediction Model

To demonstrate the baseline for the improvement of a dynamic context-sensitive model, we adopt the same framework as described in [1] without any change.

Our chief work is to address the proper assignment of the parameters: α and β, which allows the retrieval system to adapt to changing conditions, changing user behavior patterns and different search setting. We attempt to learn an adaptive strategy automatically instead of relying on heuristics or insights.

We apply machine learning technique to predict suitable parameter according to the search behavior. In practice, there are two extra prediction models in our adaptive context-sensitive retrieval models. One is used for generating proper parameter α, the other is used for generating proper parameter β.

Considering that α and β are weightings from 0 to 1, the prediction task can be viewed as a regression issue naturally. Here we choose Support Vector Machines (SVMs), which performs well for most tasks. Specifically, we choose to use Support Vector Regression (SVR) to demonstrate our approach. The solution to a support vector regression problem is a function that accepts a data point and returns a continuous target value. The support vector regression problem also allows for a zone-of-insensitivity defined by a parameter ε.

We apply ε-SVR, which only cares the errors as long as deviations are larger than ε, as the parameter prediction model. The training data have been taken as $\{(x_1, y_1)...,(x_l, y_l)\} \subset \mathbf{X} \times \mathbf{R}$, where $\mathbf{X}$ denotes the space of the input patterns (e.g. $\mathbf{X} = \mathbf{R}^d$). In our application, x_i presents 39 interactive search behavior features and y_i presents ideal parameters.

3.2   Features

Satisfying users' search needs involves a thorough understanding of their context explicitly through submitted queries and implicitly through viewed Web pages. For this purpose, we design a rich set of features grouped into: *HistoryClick*, *HistoryQuery*, *CurrentQuery&HistoryClick*, and *CurrentQuery&HistoryQuery*, as summarized in Table 2. The broad range of features used enables us to capture many aspects of search activity.

HistoryClick Features: These features are used to capture characteristics of user click behavior, including the total and average length of the viewed page and the total and average number of viewed pages corresponding previous query and all the previous queries respectively.

Table 2. Features Used in Predicting Optimal Parameters

Group	Feature	Feature description
History click features	TotalClick	Number of clicks for this query
	HisClickLen	Total length of history clicks
	DocAveLen	Average length of history clicks for this query
	ClickAve	Average length of one history click
	PreDocLen	Total length of the previous history click
	PreClickNum*	Number of clicks for the previous history click
	PreClick	Average length of the previous click
Current query & History click features	CurrentLen	Length of current query
	SimAverage	Similarity of current query and concatenated history clicks
	ClickMaxSimi	Max similarity of current query and previous history clicks
	ClickPreSimi	Similarity of current query and previous history click C_{k-1}
	ClickAddDup*	Duplication of new query term and previous click C_{k-1}
	ClickAddHc*	Number of new query term occurred in previous click C_{k-1}
	ClickAddHc	Average number of new query term occurred in history clicks
	ClickComDup	Duplication of query terms common and previous click C_{k-1}
	ClickComN	Number of common terms
	ClickComHc	Number of common terms in previous click C_{k-1}
	ClickComHc	Average number of common terms in previous clicks
	ClickSubDup*	Duplication of query term deleted and previous click C_{k-1}
	ClickSubN	Number of query term deleted
	ClickSubHc	Number of query term deleted in previous click C_{k-1}
	CSubHcAve	Average number of query term deleted in previous clicks
History query features	PqrySum	Total length of previous queries
	PqryAve	Average length of previous queries
	PqryNum*	Number of previous queries
Current query & history query features	**caddN**	Number of new query terms
	CqCom*	Fraction of number of common query terms and current query
	SimiAve	Average similarity of current query and previous queries
	MaxSimi	Max similarity of current query and previous queries
	PreSimi	Similarity of current query and previous queries
	AddDup*	Duplication of new query term and the current query
	AddNum*	Number of new query terms
	AddCurAve	Average number of new query term
	SubDup	Duplication of query term deleted and previous query Q_{k-1}
	SubHqPer	Number of query term deleted in previous query Q_{k-1}
	SubHqAve*	Average number of query term deleted in previous queries
	ComDup*	Duplication of common query term and previous query Q_{k-1}
	ComHqN*	Number of common query term in previous query Q_{k-1}
	ComHqAve	Average of common query term in previous queries

CurrentQuery&HistoryClick Features: These features appear to be important in our predicting. They are used to characterize interactions with pages beyond the results page and current query.

HistoryQuery Features: We can capture the pre-query interaction behavior from history queries. To model this aspect of user experience we defined features to characterize the nature of history query from the aspects of query length and the number of previous queries.

CurrentQuery&HistoryQuery Features: Users have their own habits of expressing search intent and there may be common ground between previous queries and current query. These features are also important in predicting parameters and mainly capture aspects of current query and its relation to previous queries, such as overlap between current query terms and previous queries, as well as the fraction of query term deleted in previous queries.

This rich representation of user behavior is similar in many respects to the recent work [6][7]. However, our features are not identical: their features only derived from reformulations between successive history queries whereas ours are also derived from current query and the viewed documents.

Besides these features similar to existing research, we also take new features into account. The new features are distinguished by bold in Table 2. Our major hypothesis is that the terms which are preserved or amended in user's search interactive behavior suggest the strength of the relation between the current query and corresponding context. One advantage of our prediction model is that it can capture the basic rules when context is helpful or harmful to current search need, and assign adaptive weighting accordingly.

4   Experiments

In order to compare the retrieval performances, we use two performance measures MAP and pr@20. For evaluation we use 90 test queries that have contextual information. We adopt SVM-Regression to train the map from interaction features to ideal weights. The others are default settings.

4.1   Baseline Methods

Specifically, we compare the following methods as described in [1]:

FixInt: The information need model is just a fixed coefficient interpolation of three models: current query model, history query model and history click model.

BayesInt: The only difference between BayesInt and FixInt is that, the interpolation coefficients in BayesInt are adaptive to query length. By treating history query model and history click model as Dirichlet priors and current query as the observed data. When viewing BayesInt as FixInt, we see that $\alpha=\frac{|Q_k|}{|Q_k|+\mu+\nu}$, $\beta=\frac{\nu}{\mu+\nu}$, where ν and μ are both prior sample size. ν value is 5.0, and μ value is 0.2.

BatchUp: Once collecting new click or new query, information need model is updated. It introduces a decaying factor to history queries. BatchUp is described as follows:

$$p(\omega|\theta_k) = p(\omega|\varphi_k) \tag{2}$$

$$p(\omega|\phi_k) = \frac{c(\omega,Q_k)+\mu p(\omega|\phi_{k-1})}{|Q_k|+\mu} \tag{3}$$

AdaptiveEW: Our refinements of FixInt and BayesInt take advantages of predicted models based on interaction behavior features to obtain optimal weights, which lead to optimal combination of history queries and history clicks (β_k), as well as optimal combination of current query and context (α_k).

AdaptiveDW: Our refinement of BatchUp takes advantage of a prediction model based on interaction behavior features to obtain proper decaying factor.

FixInt, **BayesInt** and **AdaptiveEW** described above have something significant in common that history information, including history queries and history clicks, are all treated equally, while **BatchUp** and **AdaptiveDW** are context-sensitive models which assign decaying weights on previous queries, and update information need model as soon as new query or new click is obtained.

4.2   Results

There are totally 90 queries that have contextual information. We use 2/3 of these queries for training with the rest 1/3 for testing. Specifically, we classify these queries into three groups according to different size of contextual information. 1/3 of them are queries containing only one preceding query(hence denoted as Q2), 1/3 containing two preceding queries(Q3), and the rest (Q4) containing three preceding queries. For example, when Q2 is selected as testing data, Q3 and Q4 become the training data. It is the default setting for adaptive models, such as AdaptiveEW and AdaptieDW.

From the Figures 4 to 5, we can see that our adaptive model AdaptiveEW consistently outperforms baseline models in terms of all evaluation measures, while AdaptiveDW shows modest retrieval performance. In addition, It is clear that the retrieval performance is substantially improved with more history information available. Gains reach the climaxatwhen the number of preceding queries is 3. Modest improvements occur when the number of preceding queries is 2or 1 This fact also illustrates that the accurate weights predicted by the model depend heavily on rich search behaviors.

In order to analyze feature contribution, we calculate the deviation between ideal weights and actual weights, and obtain top 20 features in prediction model(marked with "*" in Table 2). It reaveals that user interactions behavior is an important indicator when assigning weight between current query and context properly.

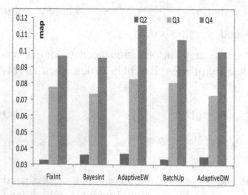
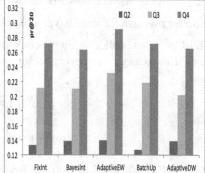

Fig. 4. Performance comparison on MAP   **Fig. 5.** Performance comparison on PR@20

5   Conclusions

In this paper, we address the issue of proper context weight for context-sensitive retrieval model. It is revealed that the existing context-sensitive models with fixed weight (e.g. FixInt method) are not at their best to augment current query with context information. We propose to employ interaction behavior features by Support Vector Regression algorithm to build a weight-prediction model, which enables a more flexible combination of current query and its context. The experimental results show that the proposed model outperforms current static models.

Acknowledgments. The work of this paper is supported by the key project of National High Technology Research and Development Program of China (863 Program, No. 2011AA01A207), the project of National Natural Science Foundation of China (Grant No. 61105072 & 61272384) and the Fundamental Research Funds for the Central Universities.

References

1. Shen, X., Tan, B., Zhai, C.: Context-Sensitive Information Retrieval Using Implicit Feedback. In: 28th Annual International ACM SIGIR Conference on Research and Development in Information Retrieval, pp. 43–50. ACM Press, New York (2005)
2. Xiang, B., Jiang, D., Pei, J., Sun, X., Chen, E., Li, H.: Context-Aware Ranking in Web Search. In: 33rd International ACM SIGIR Conference on Research and Development in Information Retrieval, pp. 451–458. ACM Press, New York (2010)
3. Cao, H., Hu, D.H., Shen, D., Jiang, D., Sun, J.T., Chen, E., Yang, Q.: Context-Aware Query Classification. In: 32nd International ACM SIGIR Conference on Research and Development in Information Retrieval, pp. 3–10. ACM Press, New York (2009)
4. Cao, H., Jiang, D., Pei, J., Chen, E., Li, H.: Towards Context-Aware Search by Learning a Very Large Variable Length Hidden Markov Model from Search Logs. In: 18th International Conference on World Wide Web, pp. 191–200. ACM Press, New York (2009)
5. Cao, H., Jiang, D., Pei, J., He, Q., Liao, Z., Chen, E., Li, H.: Context-Aware Query Suggestion by Mining Clickthrough and Session Data. In: 14th ACM SIGKDD International Conference on Knowledge Discovery and Data Mining, pp. 875–883. ACM Press, New York (2008)
6. Kotov, A., Bennett, P.N., White, R.W., Dumains, S., Teevan, J.: Modeling and Analysis of Cross-Session Search Task. In: 34th International ACM SIGIR Conference on Research and Development in Information Retrieval, pp. 5–14. ACM Press, New York (2011)
7. Luxenburger, J., Elbassuoni, S., Weikum, G.: Matching Task Profiles and User Needs in Personalized Web Search. In: 17th ACM Conference on Information and Knowledge Management, pp. 689–698. ACM Press, New York (2008)
8. Teevan, J., Dumais, S., Horvitz, E.: Personalizing Search via Automated Analysis of Interests and Activities. In: 28th Annual International ACM SIGIR Conference on Research and Development in Information Retrieval, pp. 449–456. ACM Press, New York (2005)
9. Teevan, J., Dumais, S., Horvitz, E.: Potential for Personalization. Transactions on Computer-Human Interaction 17(1) (2010)

10. Teevan, J., Dumais, S., Liebling, D.: To Personalized or Not to Personalize: Modeling Queries with Variation in User Intent. In: 31st Annual International ACM SIGIR Conference on Research and Development in Information Retrieval, pp. 163–170. ACM Press, New York (2008)

11. Pitkow, J., Schütze, H., Cass, T., Cooley, R., Turnbull, D., Edmonds, A., Adar, E., Breuel, T.: Personalized Search. Communications of the ACM 45(9), 50–55 (2002)

12. Shen, X., Tan, B., Zhai, C.: Implicit User Modeling for Personalized Search. In: 14th ACM International Conference on Information and Knowledge Management, pp. 824–831. ACM Press, New York (2005)

13. Pitkow, J., Schütze, H., Cass, T., Cooley, R., Turnbull, D., Edmonds, A., Adar, E., Breuel, T.: Personalized Search. Communications of the ACM 45(9), 50–55 (2002)

14. Teevan, J., Dumais, S., Horvitz, E.: Personalizing Search via Automated Analysis of Interests and Activities. In: 28th Annual International ACM SIGIR Conference on Research and Development in Information Retrieval, pp. 449–456. ACM Press, New York (2005)

15. Teevan, J., Dumais, S., Horvitz, E.: Potential for Personalization. Transactions on Computer-Human Interaction 17(1) (2010)

16. Xiang, B., Jiang, D., Pei, J., Sun, X., Chen, E., Li, H.: Context-Aware Ranking in Web Search. In: 33rd International ACM SIGIR Conference on Research and Development in Information Retrieval, pp. 451–458. ACM Press, New York (2010)

Effective Keyword-Based XML Retrieval Using the Data-Centric and Document-Centric Features

Tsubasa Tanabe, Toshiyuki Shimizu, and Masatoshi Yoshikawa

Department of Social Informatics, Graduate School of Informatics, Kyoto University,
Yoshida-Honmachi, Sakyo-ku, Kyoto 606-8501, Japan
tanabe@db.soc.i.kyoto-u.ac.jp, {tshimizu,yoshikawa}@i.kyoto-u.ac.jp

Abstract. Extensible Markup Language (XML) is used for not only describing structured documents but also for describing data just for generating XML from relational data. The former is called document-centric XML, and the latter is called data-centric XML. From studies on retrieving data-centric XML by using keyword searches, methods based on LCA have been proposed, while from studies on retrieving document-centric XML, methods based on information retrieval that focus on the granularity of XML elements have been proposed. However, documents generally have both data-centric and document-centric elements, so there are cases in which desired results cannot be returned by using existing research. We propose a method for constructing suitable search results for XML documents that include both data-centric and document-centric elements by considering a user's query intention and element features (data-centric or document-centric). Our experiments show that both data-centric and document-centric elements need to be considered for actual XML documents.

1 Introduction

There are many structured documents, and Extensible Markup Language (XML) is used as a typical description format for such documents and structured data. XML is also used as the format for data exchange; therefore, it is more important to be able to retrieve XML effectively. As one of the method for retrieving XML, there is a method called XML keyword search which extracts only part of XML documents including query keywords like web searches. XML keyword search does not require users to know the structure of XML documents and knowledge about the query, so it is easy to retrieve related information by using XML keyword search.

It is important to be aware of whether the target XML is data-centric or document-centric. For example, there are DBLP[2] as data-centric XML, and there are scientific articles in XML from IEEE Computer Society that have been used in INEX[11],the Wikipedia XML Corpus[3], as document-centric XML. Pradhan et al. [1] proposed a method that acquires effective subtrees in document-centric XML, and they believe that the issue of retrieval unit needs to be taken

Y. Hou et al. (Eds.): AIRS 2012, LNCS 7675, pp. 427–436, 2012.

into acccount in document-centric XML, unlike in data-centric XML. We thought that it is important for us to be aware of whether target XML is data-centric or document-centric from this study too.

In the keyword search techniques over XML documents, many methods have been proposed that use the Lowest Common Ancestor (LCA) [5,6] or use Information Retrieval (IR) techniques [9,10]. The former is mainly the methods for data-centric XML, and the latter is mainly the methods for document-centric XML. However, XML documents generally have both data-centric and document-centric parts, so current methods cannot return effective results in some cases. For example, we consider that scientific papers in XML[11] are the ones which include both data-centric and document-centric parts. Paragraph elements are considered as document-centric parts, and elements, such as author and title, are data-centric parts. Therefore, we should focus on element features and construct search results by changing the treatment whether they are data-centric or document-centric parts. In the rest of this paper, we call elements which correspond to data-centric parts as data-centric elements, and elements which correspond document-centric parts as document-centric elements. Also, we call XML keyword search using LCA semantics as the LCA approach, and that using IR techniques as the IR approach.

The LCA approach identifies nodes as search results. However, there are many combinations that include query keywords in identifying nodes, and we specifically need to consider the connection between data-centric and document-centric elements in XML documents that have both elements. By contrast, the IR approach acquires search results by focusing on retrieval granularity; however, data-centric elements are elements that are used as filtering conditions, and such elements should be treated as not having issues concerning retrieval granularity. We need to focus on element features in order to construct effective search results.

We first divide elements into data-centric and document-centric elements and consider the connection between these elements. We propose a method for XML keyword search that is applicable in cases of including both data-centric and document-centric elements.

2   Related Work

The LCA approach returns a node called the LCA which includes all query keywords and is the deepest element in the common ancestor. In research on LCA, Smallest LCA (SLCA) which returns only specific nodes in the LCA, has been proposed [4]. SLCA is an intuitively minimum node that does not include a subtree that has query words at descendant once or more. Fig. 1 shows the use of the Dewey ID, which has been used in previous research [5], and shows elements by element name and the Dewey ID. For example, article(0.0) is the article node whose Dewey ID is 0.0. In Fig. 1, the LCA that includes both "generate" and "snippet" that is a query is sec(0.0.1.1), and the LCA based on "generate" included in title(0.0.1.1.0) and "snippet" included in p(0.0.1.0.2.2) is bdy(0.0.1), and the LCA based on "generate" included in title(0.0.1.1.0) and "snippet" included in p(0.1.1.0.1) is journal(0). In this example, the number of LCA is three and the SLCA is sec(0.0.1.1) from the characteristic.

Bao et al. [6] proposed a ranking method by defining term frequency (TF) and document frequency (DF) which are applicable to XML, especially by focusing on SLCA, and identifying user's intention. Liu et al. [5] estimated a node which a user asks for from a query. Supasitthimethee et al. [8] addressed the issues of using different words to acquire the same results, and which node to return by considering how much information should be included.

When we apply the LCA approach to target XML documents, nodes are identified as search results from a user's query; however, we should consider about not only identifying the node but also combinations in order to acquire effective results. XSemantic [8] or XSeek [5], is discussed which part should be returned under the identified node; however, these methods focus on data-centric XML, so there is the issue that it doesn't consider about including both data-centric and document-centric elements. Therefore, users cannot be satisfied with search results by the application of only the LCA approach in XML documents that include data-centric and document-centric elements. We assume research papers structured as XML documents that include data-centric elements, such as authors' name or title that correspond to article information, and document-centric elements, such as text. These elements are related and their connections need to be considered.

On the other hand, the IR approach focuses on the granularity of XML elements. Hatano et al. [9] automatically determined the suitable granularity of search results by analyzing the content and structure of XML documents. Fujimoto et al. [10] automatically selected document-centric elements by focusing on the element name, rate of period, and the number of different words, and effectively searched XML documents by eliminating unsuitable data-centric elements.

When we apply the IR approach to target XML documents, we focus on the granularity of XML elements, and the element is basically returned as search results. However, this approach is not considered for XML documents that include both data-centric and document-centric elements, which is also true with the LCA approach. Therefore, for the XML documents shown in Fig. 1, the whole article may become a result for the query "Yi XML". However, a user actually asks for information about "XML" which is written by "Yi", so we need to construct search results by focusing on element features that match keywords.

In fact, what kind of elements are typically data-centric or document-centric elements has not been formally defined; however, elements that include text are basically considered as document-centric elements, and others as data-centric elements. Previous research implicitly focused on XML keyword searches for XML documents that include only data-centric or only document-centric elements; however, documents generally include both these elements. Therefore, it is difficult to retrieve results which a user asks for only by using previous research.

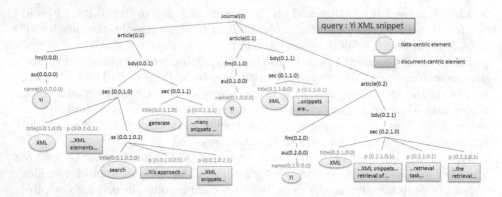

Fig. 1. XML tree of article

We need to divide into data-centric elements and document-centric elements in order to acquire appropriate results. In this paper, we argue that it is preliminarily identified data-centric and document-centric elements by classifying manually.

3 Presumption of Query Intention

There are generally two issues to address when constructing search results. The first issue is how to presume a user's query intention. More effective search results can be obtained if we understand which part query languages should match. The second issue is how to construct search results in XML documents that include data-centric and document-centric elements.

In this Section, we discuss the presumption of query intention and describe our method for constructing search results in Section 4. In Fig. 1, consider that a user's query is "Yi XML snippet". "Yi" appears at name(0.0.0.0.0) and name(0.1.0.0.0), which denote the author's name, and at p(0.0.1.0.2.1), which denotes the paragraph. "XML" and "snippet" also appear at various elements. Therefore, there are various matches by only focusing on a query.

We applied Motomura et al.'s [7] method to presume query intention, and assumed that a user can select the query intention corresponding to a user's intention in a number of interpretations which show user. Motomura et al.'s method mainly focuses on data-centric XML, specifically element name, and acquires a number of interpretations by clustering depending on which element name the query matches. However, we focus on XML documents that include both data-centric and document-centric elements, so we need to consider the existence of document-centric elements. In data-centric elements, for example, au denotes the author's name and is meaningful; however, in document-centric elements, for example, it does not matter which node having element name such as p or sec we should acquire. Therefore, if we use only the method that focuses on the element name, issues arise in document-centric elements. Because of this, we consider that it enables us to show user results estimated query intention by adding being of document-centric elements which do not care about which part should be matched to Motomura's clustering method, and we let these presumptions be interpretations of queries.

Regarding query interpretations, we use the description called Content And Structure(CAS) which is used in INEX by extension. CAS is written like XPath, for example, it is described as //article//sec[about(., mutual information criterion)]. This is a query which asks for sec elements which describes information of "mutual information criterion" in descendant of the sec element or self in article element. We use this CAS description by extension. We propose include and $*_{doc}$. For example, include(.//title, link analysis) means that keywords of "link" and "analysis" are included, and $*_{doc}$ denotes any document-centric elements.

If "web retrieval link analysis" is given, for example, it is considered as an interpretation of //article[include(.//title, link analysis)]//$*_{doc}$[about(., web retrieval)]. This shows that a user asks for information about the part of the description of "web" and "retrieval" in the article that describes "link" and "analysis". CAS, for example, returns fixed results, such as the sec element; however, the user is not concerned with the elements if he/she can acquire desired results, so we propose a description that is not fixed to a specific element by describing $*_{doc}$. The description of include(.//title, link analysis) means that both keywords "link" and "analysis" are included and plays a different role from things corresponding fully to strings written like contains which is described by XPath.

Regarding presumption of query intention, it is preferable to automatically estimate the query intention and show it to users; however, we do not discuss automatic presumption in this paper since we can reduce a user's burden by showing candidates of the query intention and let the user select the presumption of his/her query intention in the results described. We suggest that excessive results are eliminated when constructing search results by considering the extent of influence of data-centric elements based on the selected query.

4   Construction of Search Results

We need to construct search results by taking into account issues due to the inclusion of data-centric and document-centric elements. In this section, we discuss the extent of influence of data-centric elements and the retrieval algorithm based on this.

4.1   The Extent of Influence of Data-Centric Elements

There is a connection between data-centric and document-centric elements when both are included in target XML documents. We focus on this connection by considering the extent of influence of data-centric elements. About the extent of influence of data-centric elements, if a data-centric element that matches a certain path p is unique in a certain subtree T, the data-centric element affects

(a)   (b)

Fig. 2. Connection of elements

the document-centric elements in T. In short, if we focus on the subtree and the path that matches data-centric element is unique, the data-centric element can be estimated information about subtree, so the data-centric element affects document-centric elements in the subtree.

For example, in Fig. 1, if we focus on a subtree that makes article(0.0) a root node, the data-centric element that matches the path described as //article/fm/au/name is unique in article(0.0), so name(0.0.0.0) affects the document-centric elements in article(0.0). However, the data-centric elements that match the path described as //article/bdy/sec/title are title(0.0.1.0.0) and title(0.0.1.1.0) and are not unique, so these elements do not affect the document-centric elements in article(0.0). However, if we focus on a subtree that makes sec(0.0.1.1) the root node, the data-centric element that matches the path described as //sec/title is unique in sec(0.0.1.1), so title(0.0.1.1.0) affects the document-centric elements in sec(0.0.1.1).

Thus, we can find the connection between data-centric and document-centric elements by considering a certain subtree and path, and construct search results.

4.2   Connection of Elements

In Fig. 1, if a user's query is //article[include(.//name, Yi) and include(.//title, generate)]//*$_{doc}$[about(., snippet)], suppose that the result in Fig. 2(a) is returned. The data-centric element that matches the path described as //article/fm/au/name is name(0.0.0.0) and unique in article(0.0), so name(0.0.0.0) affects the document-centric elements in article(0.0). The data-centric element that matches the path described as //sec/title is title(0.0.1.1.0) and unique in sec(0.0.1.1), so title(0.0.1.1.0) affects the document-centric elements in sec(0.0.1.1). Therefore, p(0.0.1.1.1) is affected by both name(0.0.0.0) and title(0.0.1.1.0), as shown in Fig. 2(a). Thus, the more document-centric elements are affected by data-centric elements, the more related results we acquire.

On the other hand, in Fig. 1, if a user's query is //article[include(.//name, Yi) and include(.//title, XML)]//*$_{doc}$[about(., snippet)], suppose that the result in Fig. 2(b) is returned. The data-centric element that matches the path described as //article/fm/au/name is name(0.0.0.0) and unique in article(0.0), so

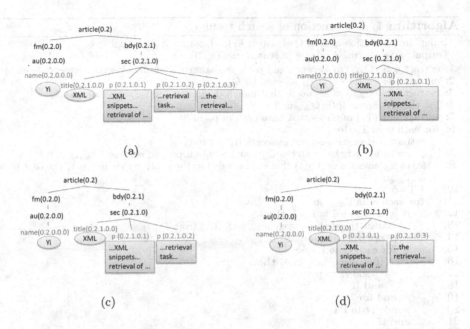

Fig. 3. Construction of search results

name(0.0.0.0.0) affects the document-centric elements in article(0.0). The data-centric element that matches the path described as //sec/title is title(0.0.1.0.0) and unique in sec(0.0.1.0), so title(0.0.1.0.0) affects the document-centric elements in sec(0.0.1.0). However, the data-centric elements that match the path described as //bdy/sec/title are not unique in bdy(0.0.1), so title(0.0.1.0.0) does not affect all the document-centric elements in bdy(0.0.1).

Therefore, in Fig. 2(a) the document-centric element whose value is "snippet" in p(0.0.1.1.1) is the word affected by the data-centric element whose value is "Yi" in name(0.0.0.0.0) and whose value is "generate" in title(0.0.1.1.0). On the other hand, in Fig. 2(b) the document-centric element whose value is "snippet" in p(0.0.1.1.1) is the word affected only by the data-centric element whose value is "Yi" in name(0.0.0.0.0). Thus, the search results in Fig. 2(a) are more effective than those in Fig. 2(b).

Therefore, we can acquire more related and valid results by considering the extent of influence of data-centric elements.

4.3 Algorithm for Constructing Search Results

We construct search results by considering the extent of influence of data-centric elements and the connection between data-centric and document-centric elements based on interpretations. The algorithm for constructing search results is Algorithm 1. For example, suppose that a user's query is //article[include(.//au, Yi) and include(.//title, XML)]//*$_{doc}$[about(., snippet)]. Since "Yi" and "XML"

Algorithm 1. Construction of search results

Input: interpretation queries Q=$(p_r,(p_1, k_1),...)$, target XML tree T
Output: sets of pruned subtree in search results PT

1: extract sets (Q_{dat}) of keyword and path pairs that match data-centric elements from Q
2: extract element sets (E_{infl}) affected by Q_{dat}
3: extract sets (Q_{doc}) of keywords that match document-centric elements from Q
4: use the IR approach by Q_{doc} in T
5: extract sets (T_r) of nodes that have path of p_r in T
6: **for** each t_i in T_r **do**
7:     extract part of data-centric elements $(t_i{}^{dat})$ in t_i
8:     extract the part that matches Q_{dat} and create a pruned subtree (pt_{dat}) in $t_i{}^{dat}$
9:     extract element sets (E_{doc}) that are included in the subtree that makes t_i the root node
       in T_{doc}
10:     PT=ϕ
11:     **for** each e_i in E_{doc} **do**
12:         **if** e_i is included in E_{infl} **then**
13:             extract the shallowest element sets (E_s) from E_{infl}
14:             pt=pt_{dat}
15:             **for** each e_j in E_s **do**
16:                 **if** e_i is included in e_j **then**
17:                     add e_i to pt
18:                 **end if**
19:             **end for**
20:             add pt to PT
21:         **end if**
22:     **end for**
23: **end for**
24: return PT

---

match data-centric elements each other, we extract the part that matches "Yi" and "XML" only in data-centric elements (line 7,8). On the other hand, since "snippet" matches the document-centric elements, we use the IR approach (line 4). In this case, we focus on both data-centric and document-centric elements because there is the case described about "snippet" without using the term in the document-centric elements, however appears in data-centric elements.

We construct search results from acquired results and consider the extent of influence of data-centric elements by constructing these search results. In Fig. 1, if a user's query is //article[include(.//name, Yi) and include(.//title, XML)]//*$_{doc}$[about(., snippet retrieval)], for example, the results in Fig. 3(a)-(d) can be returned. However, in this case, there is not only the possibility for generating a number of similar search results but also of generating a combinatorial explosion. Therefore, about construction of search results we make one result by considering the extent of influence of data-centric elements. Document-centric elements affected by title(0.2.1.0.0) are returned as results, as shown in Fig. 3(a)-(d); thus, it is considered that they treat the same topic. Therefore, a user can effectively acquire search results by summarizing them into one result, and we can also prevent a combinatorial explosion from occurring.

Table 1. Queries used for experiments

query ID	query
Q_1	//article[include(.//title, clustering)]//*$_{doc}$[about(., evaluation measure)]
Q_2	//article[include(.//title, link analysis)]//*$_{doc}$[about(., web retrieval)]
Q_3	//article[include(.//title, data embedding)]//*$_{doc}$[about(., watermarking)]
Q_4	//article[include(.//title, copyright law)]//*$_{doc}$[about(., intellectual property)]
Q_5	//article[include(.//title, gesture recognition)]//*$_{doc}$[about(., hidden markov model)]
Q_6	//article[include(.//title, machine learning)]//*$_{doc}$[about(., mutual information criterion)]
Q_7	//article[include(.//title, digital libraries)]//*$_{doc}$[about(., information retrieval)]
Q_8	//article[include(.//title, operating system)]//*$_{doc}$[about(., thread implementation)]
Q_9	//article[include(.//title, data mining)]//*$_{doc}$[about(., frequent itemsets)]
Q_{10}	//article[include(.//title, frequent itemsets)]//*$_{doc}$[about(., data mining)]
Q_{11}	//article[include(.//title, interconnected networks)]//*$_{doc}$[about(., crossbar networks)]

Table 2. Rate of elements affected by data-centric elements

query ID	the number of article	average rate
Q_1	50	0.370
Q_2	122	0.175
Q_3	19	0.188
Q_4	14	0.145
Q_5	73	0.642
Q_6	22	0.220
Q_7	74	0.509
Q_8	397	0.358
Q_9	397	0.080
Q_{10}	20	0.224
Q_{11}	113	0.737

5 Experiments

We used the IEEE CS data set of INEX [11] as experimental data. The data were 17000 files and the data size was about 740 MB. We made queries by improving CAS which was used as an evaluation in INEX 2005. For example, in //article[about(.//p, data embedding)]//p[about(.,watermarking)] which is CAS, "data embedding" acts as a filter, so we use them as a filter that matches the title. Therefore, we rewrite //article[include(.//title, data embedding)]//*$_{doc}$[about(., watermarking)]. [1]

We consider the extent of influence of data-centric elements, so we examined how many eliminated results are included in actual data by calculating the average rate of affected elements based on the number of elements per article. Table 1 lists the queries used for our experiments, and Table 2 lists the average ratio of elements affected by data-centric elements per article. The ratio is calculated by dividing (the number of elements affected by data-centric elements) by (the number of elements that match the query). In Table 2, we see that there are many elements that are not good for correct answer in actual data. Thus, we

---

[1] In the INEX IEEE CS XML, the atl tag includes the article title, and st tags include section titles. In this experiments, .//title indicates .//atl and .//st.

can acquire more effective search results by using our method. Note that, though the ratios in Table 2 are based on the number of elements, similar results were observed based on the number of words.

6   Conclusion

We proposed a method for constructing search results in XML documents that include both data-centric and document-centric elements by considering the extent of influence of data-centric elements. Our experiments showed that data-centric and document-centric elements need tobe considered for actual XML documents; however, we must check that elements which have been eliminated by considering the extent of influence of data-centric elements are at the top by using the IR approach in the future.

References

1. Pradhan, S.: An algebraic query model for effective and efficient retrieval of XML fragments. In: VLDB, pp. 295–306 (2006)
2. Ley, M.: DBLP - Some Lessons Learned. PVLDB, 1493–1500 (2009)
3. Denoyer, L., Gallinari, P.: The Wikipedia XML Corpus. SIGIR Forum, 64–69 (2006)
4. Xu, Y., Papakonstantinou, Y.: Efficient Keyword Search for Smallest LCAs in XML Databases. In: SIGMOD Conference, pp. 527–538 (2005)
5. Liu, Z., Chen, Y.: Identifying Meaningful Return Information for XML Keyword Search. SIGMOD, 329–340 (2007)
6. Bao, Z., Ling, T.W., Bo, C., Jiaheng, L.: Effective XML Keyword Search with Relevance Oriented Ranking. In: ICDE, pp. 517–528 (2009)
7. Motomura, T., Shimizu, T., Yoshikawa, M.: Alternative Query Generation for XML Keyword Search and Its Optimization. In: Hameurlain, A., Liddle, S.W., Schewe, K.-D., Zhou, X. (eds.) DEXA 2011, Part I. LNCS, vol. 6860, pp. 410–424. Springer, Heidelberg (2011)
8. Supasitthimethee, U., Shimizu, T., Yoshikawa, M., Porkaew, K.: XSemantic: An Extension of LCA Based XML Semantic Search. IEICE Transactions on Information and Systems, 1079–1092 (2009)
9. Hatano, K., Kinutani, H., Yoshikawa, M., Uemura, S.: Information Retrieval System for XML Documents. In: Hameurlain, A., Cicchetti, R., Traunmüller, R. (eds.) DEXA 2002. LNCS, vol. 2453, pp. 758–767. Springer, Heidelberg (2002)
10. Fujimoto, K., Shimizu, T., Terada, N., Hatano, K., Suzuki, Y., Amagasa, T., Kinutani, H., Yoshikawa, M.: Implementation of a High-Speed and High-Precision XML Information Retrieval System on Relational Databases. In: Fuhr, N., Lalmas, M., Malik, S., Kazai, G. (eds.) INEX 2005. LNCS, vol. 3977, pp. 254–267. Springer, Heidelberg (2006)
11. Malik, S., Kazai, G., Lalmas, M., Fuhr, N.: Overview of INEX 2005. In: Fuhr, N., Lalmas, M., Malik, S., Kazai, G. (eds.) INEX 2005. LNCS, vol. 3977, pp. 1–15. Springer, Heidelberg (2006)

Unsupervised Extraction of Popular Product Attributes from Web Sites[*]

Lidong Bing[1], Tak-Lam Wong[2], and Wai Lam[1]

[1] Department of Systems Engineering and Engineering Management
The Chinese University of Hong Kong
{ldbing,wlam}@se.cuhk.edu.hk
[2] Department of Mathematics and Information Technology
The Hong Kong Institute of Education
tlwong@ied.edu.hk

Abstract. We develop an unsupervised learning framework for extracting popular product attributes from different Web product description pages. Unlike existing systems which do not differentiate the popularity of the attributes, we propose a framework which is able not only to detect concerned popular features of a product from a collection of customer reviews, but also to map these popular features to the related product attributes, and at the same time to extract these attributes from description pages. To tackle the technical challenges, we develop a discriminative graphical model based on hidden Conditional Random Fields. We have conducted experiments on several product domains. The empirical results show that our framework is effective.

Keywords: Information Extraction, Conditional Random Fields.

1 Introduction

For developing intelligent E-business systems, an important building block is to automatically extract attribute information from product description pages from different Web sites. For example, a product description page may contain a number of product attributes such as resolution and ISO of a digital camera in the digital camera domain. Existing automatic Web information extraction techniques including wrappers aim at extracting the product attributes from Web pages [1,11,17]. The product attributes of interest are normally specified by users, requiring a substantial amount of domain knowledge and manual effort. On the other hand, users usually are interested in a subset of the product attributes that are relevant to some features for making purchasing decision.

---

[*] The work described in this paper is substantially supported by grants from the Research Grant Council of the Hong Kong Special Administrative Region, China (Project Code: CUHK413510) and the Direct Grant of the Faculty of Engineering, CUHK (Project Codes: 2050476 and 2050522). This work is also affiliated with the CUHK MoE-Microsoft Key Laboratory of Human-centric Computing and Interface Technologies.

Y. Hou et al. (Eds.): AIRS 2012, LNCS 7675, pp. 437–446, 2012.

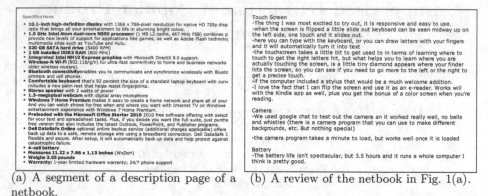

(a) A segment of a description page of a netbook.

(b) A review of the netbook in Fig. 1(a).

Fig. 1. Examples of product description page and user review

For example, in the digital camera domain, users mainly consider the feature "picture quality", which is related to several product attributes including "resolution", "ISO", etc. Some other features such as "supported operating systems" may constitute tiny influence on users' decision making. Existing information extraction methods cannot automatically identify the product attributes that are of the users' interest. Moreover, these kinds of attributes are usually unknown in different domains. As a result, it raises the need for a method that can automatically identify the product attributes that are of users' interests and extract these attributes from Web pages.

We develop an unsupervised learning framework for extracting popular product attributes from different Web product description pages. Unlike existing systems which do not differentiate the popularity of the attributes, we propose a framework which is able not only to detect concerned popular features of a product from a collection of customer reviews, but also to map these popular features to the related product attributes, and at the same time to extract these attributes from description pages. We explain the rationale of our framework using the following example. Fig. 1(a) shows a Web page about a netbook product. This page contains a list of description such as the text fragments "10.1-inch high-definition display ...", "1.5 GHz Intel Atom dual-core N550 processor ...", and "2 GB installed DDR3 RAM ..." showing different attribute information of the netbook. However, not all of them are of interests to most customers and influence users' decision. We wish to extract those attributes which are important for customers to make decision. To achieve this goal, we make use of a collection of online customer reviews available from Web forums or retailer Web sites as exemplified in Fig. 1(b) to automatically derive the popular features. Note that the concerned product from the Web page does not necessarily appear in the collection of reviews. Each popular feature is represented by a set of terms with weights, capturing the association terms related to that popular features. For example, terms like "screen" and " color" are automatically identified to be related to the popular feature "display" of a netbook by analyzing their frequency and co-occurrence information in the customer reviews. Next these terms can help extract the text fragment "10.1-inch high-definition display ..." because it contains the terms "screen" and "color". Our framework can then reinforce that terms like "resolution" and "high-definition", which are contained in the text fragment are also related to the popular feature "display". These newly identified terms can be utilized to extract other attributes related to "display". On the other hand, some other attributes such as "keyboard" are not mentioned in most reviews. Hence, the text fragment "Comfortable keyboard ..." will not be extracted.

2   Problem Definition and Framework Overview

In a particular domain, let $A = \{A_1, A_2, \ldots\}$ be the set of product attributes characterizing the products in this domain. For example, the set of product attributes of the netbook domain includes "screen", "multi-media", etc. Given a Web page W about a certain product in the given domain. W can be treated as a sequence of tokens $(tok_1, \ldots, tok_{N(W)})$ where $N(W)$ refers to the number of tokens in W. We also define $tok_{l,k}$ as a text fragment composed of consecutive tokens between tok_l and tok_k in W, where $1 \leq l \leq k \leq N(W)$. Let $L(tok_{l,k})$ and $C(tok_{l,k})$ be the layout features and the content features of the text fragment $tok_{l,k}$ respectively. We denote $V(tok_{l,k}) = A_j$ if the text fragment $tok_{l,k}$ is related to the attribute A_j.

We denote $A_{POP} \subseteq A$ as the set of popular product attributes. Recall that A_{POP} is related to the popular features, namely, $C(R)$, discovered from a collection of customer reviews, namely, R, about some products in the same domain. Our popular attribute extraction problem can be defined as follows: Given a Web page W of a certain product in a domain and a set of customer reviews R in the same domain. The concerned product in the Web page W does not necessarily appear in R. We aim at automatically identifying all the possible text fragments $tok_{l,k}$ where $1 \leq l \leq k \leq N(W)$ in W such that $V(tok_{l,k}) = A_j$ and $A_j \in A_{POP}$ by considering $L(tok_{l,k})$, $C(tok_{l,k})$, and the popular features $C(R)$. Note that A_{POP} are automatically derived from $C(R)$ beforehand and does not need to be pre-specified in advance.

Our proposed framework is composed of two major components. The first component is the popular attribute extraction component, which aims at extracting text fragments corresponding to the popular attributes from the product description Web pages. Web pages are regarded as a kind of semi-structured text documents containing a mix of structured content such as HTML tags and free texts which may be ungrammatical or just composed of short phrases. Given a Web page W about a certain product in the given domain, W can be treated as a sequence of tokens $(tok_1, \ldots, tok_{N(W)})$. Our goal is to identify all text fragments $tok_{l,k}$ such that $V(tok_{l,k}) = A_j$ and $A_j \in A_{POP}$ where $A_{POP} \subseteq A$. This task can be formulated as a sequence labeling problem. Precisely, we label each token in $(tok_1, \ldots, tok_{N(W)})$ with two sets of labels. The first set of labels

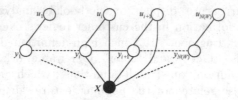

Fig. 2. The graphical model for popular attribute extraction. (Note that all u and y are connected to X in the model. Some obvious links are not shown for clarity.)

contains the labels "B", "I", and "O" denoting the beginning of an attribute, inside an attribute, and outside an attribute respectively. The second set of labels is $A_j \in \boldsymbol{A_{POP}}$, i.e. the type of popular attributes. Conditional Random Fields (CRFs) have been adopted as the state-of-the-art model to deal with sequence labeling problems. However, existing standard CRF models are inadequate to handle this task due to several reasons. The first reason is that each token will be labeled by two kinds of labels simultaneously, whereas standard CRF considers only one kind of labels. The second reason is that the popular attributes are related to the hidden concepts derived from the customer reviews by the second component and are unknown in advance. This leads to the fact that supervised training adopted in standard CRF cannot be employed. To tackle this problem, we have developed a graphical model based on hidden CRF. The proposed graphical model can exploit the derived hidden concepts, as well as the clues from layout features and text content features. An unsupervised learning algorithm is also developed to extract the popular attributes.

The second component aims at automatically derive $\boldsymbol{A_{POP}}$ from a collection of customer reviews $\boldsymbol{R}$. This component generates a set of derived documents from $\boldsymbol{R}$. We develop a method for selecting important terms for constructing the derived documents. Latent Dirichlet Allocation (LDA) is then employed to discover latent concepts, which essentially refer to the popular features of the products $\boldsymbol{C(R)}$, from the derived documents [2]. Each $c \in \boldsymbol{C(R)}$ is essentially represented by a multinomial distribution of terms. For example, one popular feature is more likely to generate the terms "display", "resolution", "screen", etc., while another popular feature is more likely to generate the terms "camera", "speaker", etc. By making use of this term information, our graphical model can extract the text fragments related to the popular attributes.

3   Description of Our Framework

3.1   Our Model

Fig. 2 shows the graphical model capturing the inter-dependency among the essential elements in the extraction problem. Each node and edge of the graphical model represent a random variable and the dependence between two connected nodes. Recall that given a Web page, we can conduct some simple preprocessing by analyzing the DOM structure. The text content in the Web page can be decomposed into a sequence of tokens $(tok_1, \ldots, tok_{N(W)})$. A random variable X refers to the observation from the sequence. For example, it can be the orthographical information of the tokens, or the layout format of the Web page. Another set of random variables denoted as $Y = (y_1, \ldots, y_{N(W)})$ ranging over a finite set of label alphabet $\mathcal{Y}$ refer to the class labels of each token. Recall that each $tok_{l,k}$ corresponds to a contiguous text fragment between tok_l and tok_k. Hence, each y_i can be equal to "B", "I", or "O" denoting the beginning of an attribute, inside an attribute, and out of an attribute respectively. In order to incorporate the information of the derived hidden concepts, which represent the popular product attributes discovered from the customer reviews, we design another set of random variables $U = (u_1, \ldots, u_{N(W)})$ ranging over $A_{POP} \cup \{\bar{A}\}$ where $\bar{A}$ is a special symbol denoting "not-a-popular-attribute". Essentially, each u_i represents the popular attribute that tok_i belongs to. We use V, E^Y, and E^U to denote the set of all vertices, the set of all edges connecting two adjacent ys, and the set of all edges connecting a particular y and a particular u respectively.

Our model is in the form of a linear chain. Hence, the joint distribution $P_\theta(Y = y, U = u | X = x)$ over the class label sequence y and the popular attribute labels u given the observation x and the set of parameters θ can be expressed as follows by the Hammersley-Clifford theorem:

$$P_\theta(Y = y, U = u | X = x) = \frac{1}{Z(x)} \exp\{ \sum_{e \in E^Y, k} \lambda_k f_k(e, y|_e, x)$$
$$+ \sum_{v \in V, k} \mu_k g_k(v, y|_v, x) + \sum_{e \in E^U, k} \gamma_k h_k(e, y|_e, u|_e, x) \}, \quad (1)$$

where $f_k(e, y|_e, x)$ refers to the feature function related to x, the nodes ys connected by the edge $e \in E^Y$. Referring to the text fragments "10.1-inch high-definition display ...", where $tok_1 = $ "10.1-inch", $tok_2 = $ "high-definition", ..., we may design a feature function $f_k(e, y|_e, x) = 1$ if $y_1 = B$, $y_2 = I$, x_1 contains a number, and $f_k(e, y|_e, x) = 0$ otherwise. $g_k(v, y|_v, x)$ refers to the feature function related to x, the node v represented by the vertex $v \in V$; Similarly, we may design a feature function $g_k(v, y|_v, x) = 1$ if $y_i = I$ and x_i is the word "high-definition", and $g_k(v, y|_v, x) = 0$ otherwise. $h_k(e, y|_e, u|_e, x)$ refers to the feature function related to x, the nodes u and y connected by the edge $e \in E^U$. For example, we may design a feature function $h_k(e, y|_e, u|_e, x) = 1$ if $y_i = I$, u_i refers to the popular product attribute "display" and x_i is the word "high-definition", and $h_k(e, y|_e, u|_e, x) = 0$ otherwise. λ_k, μ_k, γ_k are the parameters associated with $f_k(e, y|_e, x)$, $g_k(v, y|_v, x)$, and $h_k(e, y|_e, u|_e, x)$ respectively; $Z(x)$, which is a function of x, is the normalization factor. As a result, the goal of our popular attribute text fragment extraction is to find the labeling of y and u given the sequence x and the model parameter θ which includes all the λ_k, μ_k, and γ_k, such that $P_\theta(Y = y, U = u | X = x)$ is maximized.

3.2 Inference

For simplicity, we use $P_\theta(y, u | x)$ to replace $P_\theta(Y = y, U = u | X = x)$ when the context is clear. Moreover, we will follow the notation in [7] to describe our method. We add the special label "start" and "end" for y_0 and $y_{N(W)+1}$ for easy illustration of our method. Recall that our goal is to compute $\arg\max_{y,u} P_\theta(y, u|x)$ which can be expressed as Equation 1. For each token tok_i in a sequence, we define the following $|Y| \times |U|$ matrix:

$$\Lambda_i^U(y, u|x) = \sum_k \gamma_k h_k(e_i, y|_{e_i}, u|_{e_i}, x). \tag{2}$$

We then can define the following $|Y| \times |Y|$ matrices:

$$\Lambda_i^Y(y', y|x) = \sum_k \lambda_k f_k(e_i, y|_{e_i}, x) + \sum_k \mu_k f_k(e_i, y|_{e_i}, x) + \sum_{u'} \Lambda_i^U(y, u'), \tag{3}$$

$$M_i(y', y|x) = \exp\left(\Lambda_i^Y(y', y)\right). \tag{4}$$

Given the above matrices, $P_\theta(y, u|x)$ for a particular y and u given x can then be computed as follows:

$$P_\theta(y, u|x) = \frac{\prod_{i=1}^{N(W)+1} M_i(y', y|x)\left[\frac{\exp\left(\Lambda_i^U(y, u|x)\right)}{\exp\left(\sum_{u'} \Lambda_i^U(y, u')\right)}\right]}{Z(x)}, \tag{5}$$

where $Z(x) = (\prod_{i=1}^{N(W)+1} M_i(y', y|x))_{\text{start,end}}$ is the normalization factor. Given the sequence of tokens and its observation, we can compute the optimal labeling of y and u using dynamic programming.

3.3 Unsupervised Learning

We have developed an unsupervised method for learning our hidden CRF model. Given a set of M unlabeled data $\mathcal{D}$, in which the observation X of each sequence is known, but the labels y and u for each token are unknown. In principle, discriminative learning is impossible for unlabeled data. To address this problem, we make use of the customer reviews to discover a set of hidden concepts and predict a derived label for u of each token. Note that the derived labels are just used in the learning and they are not used in the final prediction for the unlabeled data. As a result, we can exploit the derived label u and the observation X in learning the model. The approach of discovering hidden concepts will be described in the next section.

Since the class label y of each token is unknown, we aim at maximizing the following log-likelihood function in our learning method:

$$\begin{aligned}\mathcal{L}_\theta &= \sum_{m=1}^M \log P(u^{(m)}|x^{(m)}; \theta) \\ &= \sum_{m=1}^M \log \sum_{y' \in y} P(y', u^{(m)}|x^{(m)}; \theta).\end{aligned} \tag{6}$$

Because of maximizing this log-likelihood function is intractable, we derive its lower bound according to Jensen's inequality and the concavity of the logarithm function. Then the efficient limited memory BFGS optimization algorithm is employed to compute the optimal parameter sets.

3.4 Hidden Concept Discovery

We observe that most customer reviews are organized in paragraphs as exemplified by the reviews in Fig. 1(b). To facilitate the discovery of high quality hidden concepts, we treat each paragraph as a processing document unit. For each review R, we first detect sentence boundaries in R using a sentence segmentator, R becomes a set of sentences denoted as $R = \{S_1, S_2, \ldots\}$. Let $S_i = (\eta_1^i, \eta_2^i, \ldots)$ denote a sequence of tokens in the sentence. Then linguistic parsing is invoked for each S_i to construct a parse tree, in which the constituents of S_i are organized in a hierarchical structure. We extract all the noun phrases located in the leaf nodes of the parse tree. Let the sequence of noun phrases of S_i be represented by $(N_{i1}, N_{i2}, \ldots)$. For each N_{ij}, we construct the context that is useful for latent topic discovery by considering the surrounding terms within a window size in S_i. Then we define ξ_{ij} as the derived content of N_{ij}, and it is composed of all terms in N_{ij} and the context terms. For example, consider a sentence S_i as (It's, almost, soundless, on, the, low, setting, which, is, what, we, used, in, his, old, ,, tiny, bedroom), extracted from the review of an air purifier. The first noun phrase N_{i1} in this sentence is (the, low, setting). If the context window size is 2, the derived content ξ_{i1} for this noun phrase is (soundless, on, the, low, setting, which, is). We remove all stop words from ξ_{ij} and conduct lemmatization for the remaining terms. The derived content representation ξ_i of S_i is obtained by $\xi_i = \bigcup_j \xi_{ij}$.

The derived document Υ for R can be obtained by gathering the derived content of all sentences. Therefore $\Upsilon = (\xi_1, \xi_2, \ldots)$. A collection of derived documents for the review collection $\boldsymbol{R}$ can be obtained as above. We employ Latent Dirichlet Allocation (LDA) to discover the hidden concepts, which essentially refer to the popular features, for a domain.

4 Experimental Results

We have conducted the experiment to evaluate the performance of our framework with product description pages from over 20 different online retailer Web sites covering 7 different domains as depicted in Table 1. In addition, we have collected more than 500 customer reviews in each domain similar to the ones shown in Fig. 1(b) from retailer Web sites. These reviews were fed into the hidden concept discovery algorithm and the number of latent topics was set to 30 for each domain.

Two annotators were hired to identify the popular product attribute text fragments from the product description pages for evaluation purpose. The annotators first read through the reviews. Then they discussed and identified popular features as well as some sample popular product attributes for the domain. After that, such information is used to guide the annotation work of the corresponding domain. Text fragments corresponding to popular attributes were manually

Table 1. The details of the data collected for the experiments

Domain Label	Domain Name	# of products	# of description pages
D1	baby car seat	15	36
D2	carpet cleaner	12	24
D3	disc player	11	22
D4	GPS device	12	22
D5	netbook	10	30
D6	printer	10	21
D7	purifier	12	21

identified from product description pages. The manually extracted popular attribute text fragments were treated as the gold standard for evaluation. The agreement of popular attribute text fragments between the two annotators was about 91%. The others were eliminated from the gold standard.

Since there are no existing methods that directly extract popular product attributes from Web pages and take into account customers' interest revealed in the collection of reviews of the same domain in an unsupervised manner, we implemented a comparative method based on integration of some existing methods. The first comparative method is called "VIPS-Bayes", which consists of two steps. For the first step, we first conduct unsupervised Web data extraction based on VIPS [3]. Since we have observed that almost all of the popular product attribute values (text fragments) are noun phrases, we apply the openNLP[1] package to conduct noun phrase extraction from the text in the product description blocks. The identified noun phrases become the popular attribute value candidates. In the second step, we determine the popular attribute values as follows. We discover the derived hidden concepts for a domain using LDA from the customer reviews. Note that each hidden concept is represented by a set of terms with probabilities. The probability refers how likely that a term is generated from a particular derived hidden concept. Next, each popular attribute value candidate is scored using Bayes theorem. In essence, the score refers to the conditional probability that the candidate comes from a particular derived hidden concept. Those candidates with scores greater than a certain threshold will be considered as popular attribute values. The threshold value is determined by a tuning process so that for each domain the best performance can be obtained.

Table 2 depicts the extraction performance of each domain and the average extraction performance among all domains. It can be observed that our approach achieves the best performance. The average F1-measure of our approach is 0.672, while the average F1-measure value of "VIPS-Bayes" is 0.547. In addition, the paired t-test (with $P < 0.001$) shows that the performance of our approach is significantly better. It illustrates that our approach can leverage the clues to make coherent decision in both product attribute extraction task and popular attribute classification task, leading to a better performance. In our approach, hidden concepts, represented by a distribution of terms, can effectively capture

---

[1] http://opennlp.sourceforge.net/

Table 2. The popular attribute extraction performance of our approach and the comparative method. P, R, and F1 refer to the precision, recall, and F1-measure respectively.

Domain	Our Approach			VIPS-Bayes		
	P	R	F1	P	R	F1
D1	0.681	0.769	0.713	0.523	0.778	0.612
D2	0.630	0.868	0.718	0.482	0.945	0.620
D3	0.548	0.872	0.665	0.388	0.836	0.519
D4	0.606	0.808	0.684	0.354	0.791	0.474
D5	0.776	0.813	0.789	0.722	0.814	0.757
D6	0.479	0.779	0.563	0.372	0.834	0.495
D7	0.589	0.611	0.571	0.558	0.317	0.350
Avg.	0.616	0.789	0.672	0.486	0.759	0.547

the terms related to a popular attribute. As the hidden concept information, together with the content information and the layout information of each token are utilized, our hidden CRF model can effectively extract the popular attribute text fragments from description pages.

5 Related Work

Some information extraction approaches for Web pages rely on wrappers which can be automatically constructed via wrapper induction. For example, Zhu et al. developed a model known as Dynamic Hierarchical Markov Random Fields which is derived from Hierarchical CRFs (HCRF) [17]. Zheng et al. proposed a method for extracting records and identifying the internal semantics at the same time [16]. Yang et al. developed a model combing HCRF and Semi-CRF that can leverage the Web page structure and handle free texts for information extraction [14]. Luo et al. studied the mutual dependencies between Web page classification and data extraction, and proposed a CRF-based method to tackle the problem [9]. Some common disadvantages of the above supervised methods are that human effort is needed to prepare training examples and the attributes to be extracted are pre-defined.

Some existing methods have been developed for information extraction of product attributes based on text mining. Ghani et al. proposed to employ a classification method for extracting attributes from product description texts [5]. Probst et al. proposed a semi-supervised algorithm to extract attribute value pairs from text description [11]. Their approach aims at handling free text descriptions by making use of natural language processing techniques. Hence, it cannot be applied to Web documents which are composed of a mix of HTML tags and free texts. The goal of extracting popular product attributes from product description Web pages is different from opinion mining or sentiment detection research as exemplified in [4,6,8,10,12,13,15]. These methods typically discover and extract all product attributes as well as opinions directly appeared in customer reviews. In contrast, our goal is to discover popular product attributes from description Web pages.

6   Conclusions

We have developed an unsupervised learning framework for extracting precise popular product attribute text fragments from description pages originated from different Web sites. The set of popular product attributes is unknown in advance, yet they can be extracted considering the interest of customers through an automatic identification of hidden concepts derived from a collection of customer reviews.

References

1. Alfonseca, E., Pasca, M., Robledo-Arnuncio, E.: Acquisition of instance attributes via labeled and related instances. In: SIGIR, pp. 58–65 (2010)
2. Blei, D.M., Ng, A.Y., Jordan, M.I.: Latent dirichlet allocation. JMLR 3, 993–1022 (2003)
3. Cai, D., Yu, S., Wen, J.R., Ma, W.Y.: Block-based web search. In: SIGIR, pp. 456–463 (2004)
4. Ding, X., Liu, B., Zhang, L.: Entity discovery and assignment for opinion mining applications. In: KDD, pp. 1125–1134 (2009)
5. Ghani, R., Probst, K., Liu, Y., Krema, M., Fano, A.: Text mining for product attribute extraction. SIGKDD Explorations 8(1), 41–48 (2006)
6. Kobayashi, N., Inui, K., Matsumoto, Y., Tateishi, K., Fukushima, T.: Collecting evaluative expressions for opinion extraction. In: IJCNLP, pp. 584–589 (2004)
7. Lafferty, J., McCallum, A., Pereira, F.: Conditional random fields: Probabilistic models for segmenting and labeling sequence data. In: ICML, pp. 282–289 (2001)
8. Liu, B., Hu, M., Cheng, J.: Opinion observer: analyzing and comparing opinions on the web. In: WWW, pp. 342–351 (2005)
9. Luo, P., Lin, F., Xiong, Y., Zhao, Y., Shi, Z.: Towards combining web classification and web information extraction: a case study. In: KDD, pp. 1235–1244 (2009)
10. Popescu, A.M., Etzioni, O.: Extracting product features and opinions from reviews. In: HLT/EMNLP, pp. 339–346 (2005)
11. Probst, K., Ghai, R., Krema, M., Fano, A., Liu, Y.: Semi-supervised learning of attribute-value pairs from product descriptions. In: IJCAI, pp. 2838–2843 (2007)
12. Tang, H., Tan, S., Cheng, X.: A survey on sentiment detection of reviews. Expert Systems with Applications 36(7), 10760–10773 (2009)
13. Turney, P.D.: Thumbs up or thumbs down?: semantic orientation applied to unsupervised classification of reviews. In: ACL, pp. 417–424 (2002)
14. Yang, C., Cao, Y., Nie, Z., Zhou, J., Wen, J.-R.: Closing the loop in webpage understanding. TKDE 22(5), 639–650 (2010)
15. Zhang, L., Liu, B., Lim, S.H., O'Brien-Strain, E.: Extracting and ranking product features in opinion documents. In: Coling: Posters, pp. 1462–1470 (2010)
16. Zheng, S., Song, R., Wen, J.R., Giles, C.L.: Efficient record-level wrapper induction. In: CIKM, pp. 47–56 (2009)
17. Zhu, J., Nie, Z., Zhang, B., Wen, J.-R.: Dynamic hierarchical markov random fields for integrated web data extraction. JMLR 9, 1583–1614 (2008)

Aggregating Opinions on Hot Topics
from Microblog Responses

Han-Chih Liu and Jenq-Haur Wang

National Taipei University of Technology, Taiwan
joepxu.6@gmail.com, jhwang@csie.ntut.edu.tw

Abstract. Huge volumes of very short microblog messages usually contain diverse contents that make it difficult to detect interesting topics. In this paper, we
propose an opinion aggregation approach based on message influence and hot
topic detection in microblogs. First, message popularity is estimated from the
content features and structural statistics. Then, hot topics are identified from
popular messages and opinion orientations are accumulated from the corresponding responses. In our evaluation on Plurk, the aggregated opinions on 2012
Taiwan Presidential Election showed a high accuracy of 98.74%. This shows
the effectiveness of our proposed approach. Further investigation is needed for
applying the proposed approach to other domains.

Keywords: microblogs, influence analysis, hot topic detection, opinion
aggregation, social media mining.

1 Introduction

With the increasing popularity of microblogging services such as Twitter[1], Plurk[2], and
Tumblr[3], huge volume of very short messages can be generated in realtime that reflect
the latest activity updates or personal thoughts on recent events. However, there are
some issues in microblogs. First, huge volumes of messages with diverse topics make
it difficult for users to track events of interests. Second, since each microblog message
is limited to 140 characters in length, inaccurate, fragmental, and opinionated information often makes it difficult to get the major points. Thus, we need a way to identify hot topics and analyze opinions from huge amount of short messages.

Existing methods for tracking latest updates include event detection, trend detection, and text summarization. These methods usually target at ordinary documents,
news articles, or Web pages which have sufficient content features for applying conventional classification and clustering techniques. However, there are several
challenges in social media. First, social media contain far less content features than
ordinary documents. Conventional features such as TF-IDF weights might not be
directly used in similarity estimation. Second, structural relations such as replies,

---

[1] http://www.twitter.com/
[2] http://www.plurk.com/
[3] http://www.tumblr.com/

Y. Hou et al. (Eds.): AIRS 2012, LNCS 7675, pp. 447–456, 2012.
© Springer-Verlag Berlin Heidelberg 2012

comments, and forwards might imply endorsements from readers. It would be helpful if we can utilize them to compensate the insufficiency in content features. Finally, huge volume of short texts with rich emotional information might be helpful if we can summarize the overall opinion on interesting topics.

In this paper, we propose an opinion aggregation approach based on message influence estimation and hot topic detection in microblogs. In order not to violate user privacy, we do not exploit the social relations among users. Instead, our idea is to incorporate structural statistics among messages such as replies, forwards, and *likes* to help compensate the insufficiency of content features. First, we estimate the message influence by combining content features with simple structural statistics including *user participation* by the number of *replies* and *likes*, and *user propagation* by the number of *forwards* and URIs (Uniform Resource Indicators). Then, we further integrate the *influence persistence* by the duration of responses to give the overall influence score. Finally, daily hot topics are identified from popular messages, and opinion orientations are accumulated from the corresponding responses.

In our experiments, Plurk is used as our major source of microblog data since it is among the top-50 sites, and one of the most popular microblogging sites in Taiwan[4]. During our data collection period, more than 320k plurks with more than 1.8 million responses were collected. Influence score of each message is used to classify its popularity. Then, the aggregated opinions on popular messages are compared with the final result of 2012 Taiwan Presidential Election. Specifically, the experimental results showed a high accuracy of 98.74% in aggregated opinions. Further investigation is needed to verify potential applications in opinion aggregation for other domains.

The remainder of this paper is organized as follows. In Section 2, we review related works. The proposed approach is described in Section 3. We detailed the experimental setup and results in Section 4. Discussions and observations are given in Section 5. Finally, in Section 6, we list our conclusions.

2   Related Work

There has been much research on microblogs especially for Twitter. Kwak et al. [5] analyzed structures in Twitter communities to observe various types of social relations among users. They ranked users according to retweets based on PageRank. Sankaranarayanan et al. [9] developed a message classification system called TwitterStand by content analysis on Twitter. Suh et al. [11] investigated the retweeting behaviors, and monitored the information diffusion on Twitter. Ho et al. [3] analyzed the degree of information propagation in microblogs from three aspects: number of influenced people, time posted, and geographic locations.

To help users browse the huge amount of microblog messages, Weng et al. [16] developed an automatic summarization system called IMASS by analyzing Plurk messages. Yang and Kao [17] performed hot topic analysis on Plurk with modified TF-IDF weighting and association mining. While most researches investigate

---

[4] http://www.alexa.com/topsites/countries/

structural features for users, we focus on combining structural statistics and content features in estimating influence of messages, instead of users.

There's much research effort in analyzing realtime characteristics on Twitter [6, 8, 15]. Sakaki et al. [8] proposed to detect earthquakes on Twitter and to issue early preventive warnings in realtime. Weng and Lee [15] proposed a module called ED-CoW to detect emergent events based on huge amount of realtime messages. Li et al. [6] designed a system called TEDAS to detect the importance and locations of events. In this paper, we performed hot topic detection and response opinion aggregation which combines simple structural statistics with short contents in estimating the popularity. We verified the effectiveness with the final result of 2012 Taiwan Presidential Election. In fact, there have been similar researches in election prediction with Twitter [2, 10, 14]. Most predictions are based on simple frequency counts in tweets. Gayo-Avello et al. [2] suggested that microblogs are not suitable for long-term event detection, while Skoric et al. [10] concluded that election predictions on Twitter are influenced by core political users, celebrity, and media.

Although microblog messages are limited in length, opinion mining and sentiment analysis techniques have been applied to microblogs [12, 13]. Sun et al. [12] utilized opinion lexicon and machine learning methods to classify messages. Tang and Chen [13] used support vector machines for sentiment classification, and verified the effectiveness of non-linguistic features when combined with pure linguistic features. Pak and Paroubek [7] performed sentiment analysis on Twitter using Naïve Bayes (NB) classification on frequencies of adjectives. In this paper, we applied NB classification to classify popular messages by influence scores. By matching opinion terms in microblog responses with two opinion lexicons NTUSD [4] and HowNet [1], we aggregate user opinion orientations on popular topics.

3   The Proposed Approach

There are three major modules in the proposed approach: message influence estimation, hot topic identification, and response opinion aggregation.

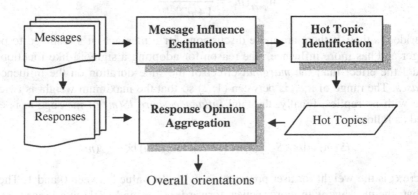

Overall orientations

Fig. 1. System architecture of our proposed approach

As shown in Fig. 1, microblog messages as well as responses such as replies, forwards, and *likes* are first crawled since there's no service readily available for retrieving microblogs during any given period of time. Standard stopword removal and Chinese word segmentation are then conducted, which are not the main focus of this paper. Next, *message influence estimation* module calculates an influence score based on the structural and content features of each message. Then, daily hot topics are identified based on the influence score. Finally, opinion orientations are aggregated from responses of popular messages.

3.1   Message Influence Estimation

Influence estimation of a social message consists of three parts: *user participation*, *user propagation* and *influence persistence*. The most important characteristic of microblogging is to allow user participation in any public message, including *replies* to give comments and *likes* to give recommendations. Given a short message m, the degree of user participation is defined as follows:

$$S_{part}(m) = \log_{10}(res_m + 1) + \log_{10}(fav_m + 1)$$ (1)

Microblog users are also allowed to forward the messages and embed a link to external pages. These actions are defined as user propagation since they allow more chances for the message to be exposed to the public. The degree of user propagation can be similarly defined as follows:

$$S_{prop}(m) = \log_{10}(for_m + 1) + \log_{10}(uri_m + 1)$$ (2)

Each message might have varying impact strengths depending on how long it has been posted and responded. By measuring the time duration $\Delta t(m)$ between the last response of a message m and the time it was initially posted, we can estimate its *influence persistence weight* as follows:

$$w_{pers}(m) = \frac{2}{1 + e^{-\Delta t(m)}}$$ (3)

The idea is that: If there are more discussions for a message, it's assumed to persist longer and has more influence. The reason for adopting a sigmoid-like function is to model the effect that: the *marginal* effect of the time duration on the influence *decreases*. The range of w_{pers} is between (1, 2) so that the maximum weight is twice the case with no replies. Finally, the total *influence score IS(m)* of message m is calculated as follows:

$$IS(m) = (\alpha \times S_{prop}(m) + (1-\alpha) \times S_{part}(m)) \times w_{pers}(m)$$ (4)

where α is the weight for user propagation, with the value between 0 and 1. The idea is that: the message with more replies, forwards, *likes*, and URIs in a longer time period has higher influence.

3.2   Hot Topic Identification

To identify hot topics from short texts, we use the estimated influence score to classify messages into two categories: popular and personal. Since personal messages are usually private thoughts, feelings, or gossips, only limited close friends will be involved in the communication loop. Thus, hot topics are assumed to come from higher influence messages with more participation and propagation among communities.

Since people usually care about the latest events on a *daily* basis, we modified the TF-IDF weights according to the term distribution in a day. The idea is that: terms that occur more often on a particular day than other days are assumed to be more representative of the events on that day. The TF-IDayF weight W_{ij} is defined as:

$$W_{ij} = \left(1 + \log_{10} tf_{ij}\right) \times \log_{10} \frac{N_{day}}{dayf_i}$$

(5)

where tf_{ij} is the term frequency of term t_i in day_j, $dayf_i$ is the "day" frequency or the number of days that contain term t_i, while N_{day} is the total number of days in the collection. Finally, the top-ranked k terms with the highest TF-IDayF weights are selected as the hot topics for day_j.

3.3   Response Opinion Aggregation

Since microblog user opinionss are usually expressed by responding to a message, instead of directly analyzing opinions in the message, we accumulate the opinion orientations from the corresponding *responses* as the overall rating of the message. From the message collection D, the query-relevant messages for query q can be filtered as D_q. For each day_n, the hot topics are represented in a term vector $\hat{h}_n$ and the cosine similarity between each message d_j in D_q and $\hat{h}_n$ is then calculated to verify the relevance. Finally, the top-ranked messages d_j are used as the query-relevant popular messages $M_q(day_n)$.

For each popular message m, we retrieve the corresponding set of responses $Res(m)$ for opinion aggregation. Each response will be assigned an integer score between -1 and 1 depending on its opinion orientation. Specifically, each term in the response is matched against the positive and negative opinion terms in the opinion lexicons. Here we simply consider the relative number of positive and negative opinion terms in calculating the *opinion score* of message m as:

$$OS(m) = \sum_{r \in Res(m)} Sgn\left(tf_{pos}(r) - tf_{neg}(r)\right)$$

(6)

where $Res(m)$ is the set of responses for message m, and $tf_{pos}(r)$ and $tf_{neg}(r)$ are the number of positive and negative opinion terms in response r, respectively. The idea is that: we only need to accumulate the "net" opinion orientation from responses of message m to reflect its overall rating. For huge number of responses, this can be calculated in an efficient way.

Since the opinion score *OS(m)* cannot reflect the relative opinion strengths in responses, we further take its influence score *IS(m)* into account as follows:

$$OIS(m) = \sum_{r \in Res(m)} Sgn(tf_{pos}(r) - tf_{neg}(r)) \times IS(m) \tag{7}$$

The idea is that: the opinion influence score of message *m* is proportional to the accumulated *net* opinion orientations of all responses to message *m*, and the influence score of message *m*. If the message has higher influence scores, the corresponding opinions are more likely to generate more profound influences.

4   Experiments

Although there's standard microblogging dataset such as Tweets2011[5], the duration is limited to two weeks (Jan. 23-Feb. 8, 2011). Thus, Plurk API was used for collecting test data. First, we collected a list of core users including the top-100 officially authorized political users and 10 media users on Plurk. Then, additional 1,640 general users who actively participated in the discussions are included to form 1,750 users. During 2011/10/01 to 2012/01/14, about 350,000 messages and 2,800,000 responses are collected as the raw data. For preprocessing, we used Yahoo Content Analysis API[6] for Chinese word segmentation, and removed meaningless symbols and stop words. Also, Plurk spam such as messages automatically generated by robot accounts, and non-content messages with no text content are filtered. Finally, the remaining 324,512 plurks and 1,863,890 replies are stored as our dataset.

To evaluate the performance of our proposed approach, we conducted two sets of experiments. First, we applied NB classification to classify popular messages by influence score. Second, we aggregated opinions on hot topics related to the 2012 Taiwan Presidential Election, and compared with the final election result.

4.1   Experiment on Influence Estimation for Message Popularity

To evaluate the performance of our proposed influence model, we use the influence score as a threshold to classify messages into popular and personal categories, respectively. The results are evaluated in terms of precision, recall and F-measure. We take the baseline as follows: messages posted by news media and celebrity users are assumed to be popular; while messages without any response are assumed to be personal. From the distribution of influence scores for our dataset, more than 80% of plurks have influence scores of less than 1. This shows an unbalanced distribution, which might greatly bias the classification results. Here, we use simple random sampling to solve the issue. Specifically, given a particular threshold for influence scores, we randomly down-sample the larger set to make a balanced distribution. In evaluating short text classification, we use 3-fold cross validation for training and testing. The experimental results are shown as follows:

---

[5] http://trec.nist.gov/data/tweets/
[6] http://tw.developer.yahoo.com/cas/

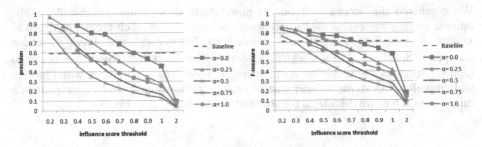

Fig. 2. The precision and F-measure of popular category at different thresholds and α

As shown in Fig. 2, we can find that precision and F-measure of popular category both increase as α decreases in different thresholds. This shows higher impact for user participation on the message popularity than user propagation. Comparing with the precision of 0.59 and the F-measure of 0.714 for the baseline, our method obtains the best F-measure of 0.86 when α=0.25 and threshold=0.2. This shows that popular messages can be more accurately obtained by choosing suitable thresholds for influence scores than simply watching the updates from media or celebrity users.

4.2   Experiment on Opinion Aggregation

To evaluate the performance of our proposed opinion aggregation approach, we use the candidates in 2012 Taiwan Presidential Election as our major queries, which correspond to the three major political parties in Taiwan: Democratic Progressive Party (DPP), Kuomintang (KMT), and People First Party (PFP). Then, the corresponding relevant messages and responses are collected, as shown in Table 1.

Table 1. The (a) query terms and (b) statistics on relevant messages for the three groups of candidates in 2012 Taiwan Presidential Election

Party	Query Terms	# plurks	# responses
DPP	Tsai Ing-wen (蔡英文), Su Jia-chyuan (蘇嘉全), Democratic Progressive Party (民主進步黨, 民進黨)	8,593	54,851
KMT	Ma Ying-jeou (馬英九), Wu Den-yih (吳敦義), Kuomintang (中國國民黨, 國民黨)	13,182	66,450
PFP	James Soong (宋楚瑜), Lin Ruey-shiung (林瑞雄), People First Party (親民黨)	900	5,027

On each day, the top-10 hot topics are identified, and the aggregated opinions of corresponding responses are evaluated by the final election result in terms of mean average error (MAE) and accuracy as follows:

$$MAE = \frac{1}{n}\sum_{i=1}^{n}|f_i - y_i|$$

(8)

$$Accuracy = 1 - MAE$$

(9)

We compared the accuracy of our proposed method with the nation-wide public opinion polling by several media and organizations[7] and the Web predictions such as xfuture ("未來事件交易所", as denoted by FuturePre)[8] and Plurk "Election Campaign Soapbox" ("選戰肥皂箱", as denoted by PlurkPre)[9]. Since we observed high variations in the weekly scores of aggregated opinions as compared with PlurkPre and FuturePre, this shows a rather unstable nature in short-term opinion aggregation on microblogs. Thus, we calculated the accumulated accuracy as in Fig. 3.

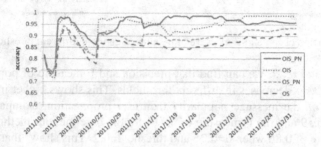

Fig. 3. The accumulated accuracy of OIS_PN, OIS, OS_PN, and OS

Note that OS_PN and OIS_PN scores are calculated as in Eqs.(6) and (7) except that only nouns and proper nouns are considered in hot topics. As shown in Fig. 4, all scores gradually become stable as messages accumulate, and OIS outperforms other scores in the long run. Finally, we checked the performance for the last public opinion polling before presidential election on Jan. 2, 2012. Note that the undecided opinions in public opinion polling are ignored in our comparison, with the remaining percentages normalized to make the total 100%.

Table 2. The comparison of MAE and accuracy in the last public opinion polling before presidential election on 2012/01/02

Organization	Prediction	MAE	Accuracy
China Times	KMT	2.89%	97.11%
Liberty Times	KMT	6.18%	93.82%
UDN	KMT	3.52%	96.48%
TVBS	KMT	2.7%	97.3%
Apple Daily	KMT	6.1%	93.9%
FuturePre	DPP	6.46%	93.54%
PlurkPre	DPP	18.38%	81.62%
OIS	KMT	**1.26%**	**98.74%**

---

[7] http://en.wikipedia.org/wiki/Opinion_polling_for_the_Republic_of_China_presidential_election,_2012
[8] http://xfuture.org/
[9] http://www.plurk.com/Taiwan2012/

As shown in Table 2, the proposed opinion influence score (OIS) obtains the best accuracy of 98.74%, which is higher than the accuracy by TVBS (97.3%) among all public opinion polling predictions from major media. This verifies the effectiveness of our proposed approach in estimating the overall opinion orientations from microblog responses.

5   Discussions

Although the proposed method is simple, our experimental results showed a high accuracy in the aggregated opinions. This shows the effectiveness of the proposed opinion aggregation approach by accumulated orientations from responses of popular messages for hot topics.

From our observation, a relatively low percentage of relevant messages and responses are classified as popular. This is due to the fewer number of query terms in hot topic detection as shown in Table 1. Specifically, the relevant messages and responses are collected by the names of candidates and parties, as used in news reports and editorials. However, informal terms such as abbreviations and nicknames are more common in social media. For example, instead of Ma Ying-jeou (馬英九) and Tsai Ing-wen (蔡英文) in microblogs, nicknames (such as "小馬" and "小英") or even the last name Ma ("馬") and Tsai ("蔡") might be rather used. The issues of abbreviations and nicknames for personal name disambiguation would be further investigated in future work.

6   Conclusion

In this paper, we have proposed an opinion aggregation approach based on message influence and hot topic detection for microblogs. Our approach is novel in opinion aggregation from microblog responses. Empirical evaluations on Plurk showed a very high accuracy of 98.74% in the case of the 2012 Taiwan Presidential Election. Further investigation on the effectiveness for detecting other types of events is needed.

Acknowledgement. We would like to thank the support from National Science Council, Taiwan under the grant number NSC101-2219-E-027-005.

References

1. Dong, Z., Dong, Q.: HowNet and the Computation of Meaning, World Scientific Publishing Co. Ltd. (2006), Dictionary available at http://www.keeenage.com/
2. Gayo-Avello, D., Metaxas, P., Mustafaraj, E.: Limits of Electoral Predictions Using Social Media Data. In: 5th International Conference on Weblogs and Social Media (ICWSM 2011), pp. 178–185. AAAI (2011)
3. Ho, C.T., Li, C.T., Lin, S.D.: Modeling and Visualizing Information Propagation in a Micro-blogging Platform. In: 2011 International Conference on Advances in Social Networks Analysis and Mining (ASONAM 2011), pp. 328-335. IEEE Press (2011)

4. Ku, L., Chen, H.: Mining Opinions from the Web: Beyond Relevance Retrieval. Journal of the American Society for Information Science and Technology (JASIST), Special Issue on Mining Web Resources for Enhancing Information Retrieval 58(12), 1838–1850 (2007), Dictionary available at:
http://nlg18.csie.ntu.edu.tw:8080/opinion/index.html
5. Kwak, H., Lee, C., Park, H., Moon, S.: What is Twitter, a Social Network or a News Media? In: 19th International Conference on World Wide Web (WWW 2010), pp. 591–600. ACM (2010)
6. Li, R., Lei, K., Khadiwala, R., Chang, K.: TEDAS: a Twitter based Event Detection and Analysis System. In: 28th IEEE International Conference on Data Engineering (ICDE 2012), pp. 1273–1276. IEEE Press (2012)
7. Pak, A., Paroubek, P.: Twitter as a Corpus for Sentiment Analysis and Opinion Mining. In: International Conference on Language Resources and Evaluation (LREC 2010), pp. 1320–1326. ELRA (2010)
8. Sakaki, T., Okazaki, M., Matsuo, Y.: Earthquake Shakes Twitter Users: Real-time Event Detection by Social Sensors. In: 19th International Conference on World Wide Web (WWW 2010), pp. 851–860. ACM (2010)
9. Sankaranarayanan, J., Samet, H., Teitler, B.E., Lieberman, M.D., Sperling, J.: TwitterStand: News in Tweets. In: 17th ACM SIGSPATIAL International Conference on Advances in Geographic Information Systems (GIS 2009), pp. 42–51. ACM (2009)
10. Skoric, M., Poor, N., Achananuparp, P., Lim, E., Jiang, J.: Tweets and Votes: a Study of the 2011 Singapore General Election. In: 45th Hawaii International Conference on System Sciences (HICSS), pp. 2583–2591. IEEE (2012)
11. Suh, B., Hong, L., Pirolli, P., Chi, E.H.: Want to be Retweeted? Large Scale Analytics on Factors Impacting Retweet in Twitter Network. In: 2010 IEEE Second International Conference on Social Computing (SocialCom 2010), pp. 177–184. IEEE (2010)
12. Sun, Y., Chen, C., Liu, C., Liu, C., Soo, V.: Sentiment Classification of Short Chinese Sentences. In: 22nd Conference on Computational Linguistics and Speech Processing (ROCLING 2010), pp. 184–198. ACL (2010) (in Chinese)
13. Tang, Y., Chen, H.: Emotion Modeling from Writer/Reader Perspective Using a Microblog Dataset. In: Workshop on Sentiment Analysis where AI meets Psychology (SAAIP 2011), pp. 11–19. ACL (2011)
14. Tumasjan, A., Sprenger, T., Sandner, P., Welpe, I.: Predicting Elections with Twitter: What 140 Characters Reveal about Political Sentiment. In: 4th International AAAI Conference on Weblogs and Social Media (ICWSM 2010), pp. 178–185. AAAI (2010)
15. Weng, J., Lee, B.: Event Detection in Twitter. In: Fifth International Conference on Weblogs and Social Media (ICWSM 2011), pp. 401–408. ACM (2011)
16. Weng, J.Y., Yang, C.L., Chen, B.N., Wang, Y.K., Lin, S.D.: IMASS: An Intelligent Microblog Analysis and Summarization System. In: 49th Annual Meeting of the Association for Computational Linguistics (ACL 2011 System Demonstrations), pp. 133–138. ACL (2011)
17. Yang, Z.H., Kao, H.Y.: The Clustering of Hot Topics on Plurk. Journal of Computers 22(2), 3–10 (2011)

Topic Sequence Kernel

Jian Xu, Qin Lu, Zhengzhong Liu, and Junyi Chai

Department of Computing, The Hong Kong Polytechnic University
{csjxu,csluqin,hector.liu,csjchai}@comp.polyu.edu.hk

Abstract. This paper addresses the problem of classifying documents using the kernel approaches based on topic sequences. Previously, the string kernel uses the ordered subsequence of characters as features and the word sequence kernel is proposed to use words as the subsequences. However, they both face the problem of computational complexity because of the large amount of symbols (characters or words). This paper, therefore, proposes to use sequences of topics rather than characters or words to reduce the number of symbols, thus increasing the computational efficiency. Documents that exhibit similar posterior topic proportions are expected to have similar topic sequence and then should be classified into the same category. Experiments conducted on the Reuters-21578 datasets have proven this hypothesis.

Keywords: Topic sequence, string kernel, classification.

1 Introduction

The Support Vector Machine (SVM) has been widely applied in [1-3] and it is well known for using kernel methods to handle non-separable data points by the hypeplane in the kernel space. The commonly used kernels are linear, polynomial, and RBF kernels. [4] proposed the string kernel which took the ordered subsequence of characters for document representation. This kernel considers the sequential order between characters in a document. However, measuring similarity between two sequences requires a lot of computational resources. To resolve this problem, [4] proposed a dynamic programming technique to promote computation efficiency. To further reduce the computational complexity, [5] used the words instead of characters as the sequences for document representation.

In this paper, we extend the basic idea of string kernel to represent documents as sequences of topics instead of words or characters. As topics are a summary of documents, they can better capture the document semantics. One document might be about crude oil (0.6), ship (0.3) and trade (0.1). Another document may be about trade (0.5), crude oil (0.3) and ship (0.2). Two documents both have three topics but with different topic proportions. For the first document, crude oil has the greatest proportion among the three topics, ship the second greatest proportion and trade is the least. Intuitively, most important topic will be first expressed, and less important topic will be conveyed in succession and the least important topic will be delivered in the last. In so doing, a document can be represented in a sequence of topics according to the topic

Y. Hou et al. (Eds.): AIRS 2012, LNCS 7675, pp. 457–466, 2012.

proportion. It is reasonable to assume that if two documents are similar, they are expected to have not only similar topics, but also the topic sequences. Based on this assumption, we try to classify texts based on the kernel approach using the sequence of topics instead of words or characters. This could greatly reduce the computational complexity of string kernel as there is a small number of topics compared to the words in the whole document collection.

The rest of this paper is organized as follows. Section 2 describes the related works of kernel methods and topic modeling approaches. Section 3 presents the approach of generating topic sequences using topic modeling technique and introduces the string kernel using the topic sequences. Section 4 gives the performance evaluation. Section 5 is the conclusion.

2   Related Works

Kernel functions are computational shortcuts that are able to represent linear patterns in high-dimensional space [6]. They are used to compute pairwise inner products between mapping examples in the feature space [4]. A kernel is valid only when it meets the Mercer's conditions: symmetry and positive semi-definiteness [7].

Currently, there are various kinds of kernels used in SVM, including the polynomial kernel, radio basis function (RBF), and so on. Different from previous kernels that are dependent on the word frequencies, the string kernel takes into account the relative positional information of characters in documents. It compares two documents by enumerating the substrings they contain: the more substring they share, the more similar they are [4].

In this paper, we extend the basic idea of string kernel using the sequence of topics. Hence, topic modeling is vital to generate topic sequences. Methods to discover the semantic structure of a document collection using the probabilistic model include latent semantic indexing (LSI), probabilistic LSI, latent Direchlet allocation (LDA) [8-10]. Besides, two extensions to LDA has been proposed: the correlated topic model and the dynamic topic model [11-12]. To find the posterior distribution of the latent topics given the document collection, various approximate approaches have been used, including mean-field variational inference [8], expectation propagation [13] and Gibbs sampling [14] and collapsed variational inference [15].

3   Methodology

3.1   Topic Modeling and Sequence Generating

Documents that have similar topics are expected to exhibit similar topic proportions. Important topics will have large proportions in a document. Suppose there are four documents: *Doc1*, *Doc2*, *Doc3* and *Doc4*, and three topics: *trade*, *crude oil*, and *ship*. The topic proportions in these four documents are:

Doc1: trade (0.3), crude oil (0.5), ship (0.2)   *Doc2:* trade (0.1), crude oil (0.3), ship (0.6)
Doc3: trade (0.3), crude Oil (0.6), ship (0.1)   *Doc4:* trade (0.2), crude oil (0.3), ship (0.5)

Obviously, **Doc1** and **Doc3** have similar topic proportions. They both have *crude oil* as the most important topic, *trade* as the secondary important topic and *ship* as the least important topic. Similarly, **Doc2** and **Doc4** have a sequential order of *ship, crude oil* and *trade* according to the topic proportions. In this sense, **Doc1** is similar to **Doc3** and **Doc2** to **Doc4**. Besides, **Doc1** and **Doc3** are similar because they have the same sequential topic order. Therefore, the topics in four documents can be re-arranged in a sequential order according to their proportions in each document.

Doc1: *crude oil (0.5), trade (0.3), ship (0.2)*   **Doc2:** *ship (0.6), crude oil (0.3), trade (0.1)*
Doc3: *crude oil (0.6), trade (0.3), ship (0.1)*   **Doc4:** *ship (0.5), crude oil (0.3), trade (0.2)*

Next is to obtain these topic distributions. In this paper, topics are modeled through the LDA [10]. LDA is a generative model which is based on probabilistic sampling techniques investigating how words in documents are generated with the hidden variables [14]. Its main idea is to model documents in terms of topics where a topic is defined as a distribution over a fixed vocabulary of words. In this model, words in documents are observable and topics are latent variables hidden in these documents. Its graphical representation is given in Fig. 1.

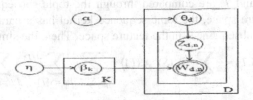

Fig. 1. LDA Graphical Representation

In the Fig.1, each node denotes a random variable and the edge between nodes represents dependency relations between nodes. The double circles around the random variable denote an observable node (evidence node). The plate surrounding the nodes indicates N i.i.d samples. D and K refer to the number of documents and the number of topics, respectively. α and η are hyper-parameters on the mixture proportions for topics and documents. θ_d refers to the multinomial topic distributions for document d and β_k is multinomial word distributions for topic k. $Z_{d,n}$ denotes a topic from which the nth word in document d is drawn and $W_{d,n}$ indicates the observable n^{th} word in d.

In LDA model, for a document d, a vector of topic distributions $\vec{\theta_d}$ is drawn from a Dirchlet distribution; topic assignment for n^{th} word $Z_{d,n}$ follows from a multinomial distribution; and the nth word $W_{d,n}$ in document d is sampled from multinomial distribution. To generate topic sequences, $p(z|w)$ must be obtained for the hyper-parameters α and η. Since exact inference of this distribution is intractable [10], Gibbs sampler is used. When $p(z|w)$ is obtained, the topic distributions θ for each document can be estimated. These topic distributions are used to generate topic sequences.

3.2   Topic Sequence Kernel

The topic sequences are generated in Section 3.1. Then, the subsequences of the topic sequences are extracted as features in the topic sequence kernel model.

Given Σ as a finite topic set, let $S=z_1z_2...z_{|S|}$ be a sequence of topics for a document, $z_i \in \Sigma$ and $1 \leq i \leq |S|$. A subsequence of S, denoted by u, the feature used in the string kernel model, is defined by a index sequence $I=(i_1,...,i_n)$ of S such that $1 \leq i_1 < i_2 < i_n \leq |S|$ and $u=S[I]$, where n is the length of u, the number of topics of the subsequence u. The span of $S[I]$, denoted by $l(I)$, is the distance of the first topic and the last topic of u in S, calculated by $i_n - i_1 + 1$. For example, if S is the topic sequence of $z_1z_2z_4z_3$ and $u = z_1z_4$, then the index set, $I=[1,3]$ such that $u=S[1,3]$, and the span of $S[1,3]$ is 3-1+1 =3. The feature matching of u for a given topic sequence S, denoted by ϕ, is:

$$\phi_u(S) = \sum_{I:u=S[I]} \lambda^{l(I)}$$

where λ is the decay factor, in the range of [0,1], that penalizes the longer span $l(I)$ of subsequences. Based on topic sequence, any two documents, represented by their topic sequences S, and T, are compared through the topic subsequences as features. To control the feature space, the topic sequence kernel has a parameter n which denotes the length of subsequences in the feature space. Then, the similarities are:

$$K_n(S,T) = \sum_{u \in \sum^n} \langle \phi_u(S) \cdot \phi_u(T) \rangle = \sum_{u \in \sum^n} \sum_{I:u=S[I]} \lambda^{l(I)} \sum_{J:u=T[J]} \lambda^{l(J)}$$

$$= \sum_{u \in \sum^n} \sum_{I:u=S[I]J:u=T[J]} \lambda^{l(I)+l(J)}$$

where $\sum^n$ is the set of all topic subsequences of length n. $S[I]$ and $T[J]$ are the subsequences in S and T. $l(I)$ and $l(J)$ are the spans of the subsequences in S and T.

In fact, each topic sequence has the unique topics. This means that the subsequences will occur only once in a topic sequence. Therefore, the feature matching of u in the topic sequences S and T will be changed to,

$$\phi_u(S) = \lambda^{l(I)} \quad \text{and} \quad \phi_u(T) = \lambda^{l(J)}$$

And the kernel function $K_n(S,T)$ will be changed to,

$$K_n(S,T) = \sum_{u \in \sum^n} \langle \phi_u(S) \cdot \phi_u(T) \rangle = \sum_{u \in \sum^n} \sum_{I:u=S[I]} \lambda^{l(I)} \sum_{J:u=T[J]} \lambda^{l(J)} = \sum_{u \in \sum^n} \lambda^{l(I)+l(J)}$$

In this sense, the computational cost will be reduced due to the cancellation of summation procedure in each topic sequence. To avoid enumeration of all subsequences for similarity measurement, dynamic programming, similar to the method in [4] is used here for similarity calculation.

4   Performance Evaluation

Experiments are conducted on the Reuters-21578 dataset, from which we used Modified Apte ("ModeApte") split. Due to the concern of computational complexity of the string kernel, [4] drew a subset of 470 documents with 380 documents for training and 90 documents for testing. [5] proposed to used word sequence kernel on the ten frequent categories. This word sequence kernel, however, is still resource demanding as they claimed. In this paper, the topic sequence kernel will not suffer from this problem since the number of topics is much less than that of characters or words.

In the following experiments, the values for the hyper-parameters α and β are 50/K and 0.01 [14] and the number of iterations is set to 500. Note that the training and testing documents are placed together to obtain the posterior topic distribution. Based on topic distributions in each document, a topic sequence is created for the document.

In terms of classifier, the LIBSVM[1] tool with the one-versus-one strategy is used and default parameters are kept. Since the training sets for the ten categories are unbalanced in favor of negative examples, we weigh the relative importance of positive and negative examples by the ratio between negative and positive examples [5].

For evaluation, we used the precision (**p**), recall (**r**) and F-measure (**F**). In order to have a general overview of performance on the ten categories, the micro-averaged and macro-averaged performances are used. To control the effectiveness of the topic sequence kernel, three parameters need to be tuned manually: length of a subsequence l and the decaying factor λ and the number of topics K. Table 1 gives the overall best performance when $\lambda=0.55$ and $l=2$ and $K=10$. In the following experiments, *TSK* is used to denote the topic sequence kernel.

Table 1. Best Performance when $\lambda=0.55$, $l=2$ and $K=10$

	Micro-average			Macro-average		
	p	r	F	p	r	F
TSK	87.06	86.69	86.88	76.48	75.91	75.98

Table 1 shows that the best performance is achieved when the number of topics K is 10. And $K=10$ is the number of categories in this experimental dataset. This means that if the number of topics is known beforehand, the LDA model can well capture the topic structure in documents. The detailed classification results for each category are listed in Table 2.

From the Table 2, we found that the *earn* category gives the best performance because of its lowest negative-to-positive ratio and the *acq* category gives the second best performance. It is interesting to note that although the *grain* category has a higher negative-to-positive ratio than the *money-fx* category does, it attained a better performance than the *money-fx* category. However, higher negative-to-positive ratio will naturally produce worse classification results. The *corn* and *ship* categories have

---

[1] http://www.csie.ntu.edu.tw/~cjlin/libsvm/

Table 2. Detailed Performance of each Category when λ=0.55, *l=2* and *K=10*

Category	p	r	F
corn	58.93	58.93	58.93
ship	73.91	76.40	75.14
wheat	66.67	78.87	72.26
acq	**93.28**	**92.63**	**92.95**
crude	78.98	73.54	76.16
earn	**95.07**	**95.77**	**95.42**
grain	88.89	85.91	87.37
interest	64.08	50.38	56.41
money-fx	66.18	76.54	70.98
trade	78.85	70.09	74.21

well illustrated this. In the following sections, we will study the effectiveness of the topic sequence kernel (*TSK*) by varying the length of a subsequence *l*, the decaying factor λ and the number of topics *K*.

4.1   Effectiveness of Varying the Number of Topics

The number of topics is crucial to the topic sequence kernel. It determines the computational complexity of the *TSK*. Therefore, for this set of experiments, we kept the values of the parameters the subsequence length *l=2* and the decaying factor *λ=0.55* fixed and observed how the performance is influenced by the number of topics *K* from 5 to 20.

Table 3. Performance of Varying the Topic Number when *λ=0.55* and *l=2*

K	Micro-average			Macro-average		
	p	r	F	p	r	F
5	84.62	73.23	78.52	40.2	42.13	41.14
10	**87.06**	**86.69**	**86.88**	**76.48**	**75.91**	**75.98**
15	84.76	84.82	84.79	70.43	71.01	70.24
20	82.37	82.96	82.66	64.67	66.47	65.47

Table 3 shows that when the topic number *K* is 10, the system gives the best performance, implying that latent topics in the document collection are well captured by *K=10* and other configurations cannot detect the topic structures properly if the *K* is too large or too small. Moreover, *K=10* is the number of categories of document collections. On the other hand, if the number of topics is known beforehand, the latent topics can be well modeled out of the document collections by LDA. It is worth noting the case of *K=5* in which the micro-average score is high, but the macro-average score is rather low. This is because the *corn, ship* and *interest* categories all get zero classification precision and recall values. However, the *acq* and *earn* categories get high precision and recall and these two categories have a large number of testing documents, thus contributing to the overall higher micro-average score.

4.2   Effectiveness of Varying the Subsequence Length

In this set of experiments, the values of the number of topics $K=10$ and the decaying factor $\lambda=0.55$ are fixed and we analyze the effect of varying the subsequence length from 2 to 6.

Table 4. Performance of Varying Subsequence Length when $\lambda=0.55$ and $K=10$

l	Micro-average			Macro-average		
	p	r	F	p	r	F
2	**87.06**	**86.69**	**86.88**	**76.48**	**75.91**	**75.98**
3	74.93	78.4	76.63	58.75	63.83	61
4	78.14	75.53	76.81	63.89	57.45	60.24
5	81.7	73.52	77.4	65	50.34	55.81
6	84.03	66.81	74.44	66.33	38.47	45.65

Table 4 shows that the *TSK* can be more effective for smaller subsequence as compared to larger subsequences since the smaller topic subsequences are able to capture the document semantics than the longer ones. Besides, the longer subsequences have a strict requirement over the matching unit of the two sequences. For example, the topic sequence $S=z_1z_2z_3z_4$ and $T=z_2z_1z_4z_3$, if the subsequence length is set to 3, we will have a set of subsequences $\{z_1z_2z_3, z_1z_3z_4, z_2z_3z_4\}$ from S and another set of subsequences $\{z_2z_1z_3, z_2z_1z_4, z_1z_4z_3\}$ from T. Clearly, we will find no intersections between the two subsequence sets. If the subsequence length is set to 2, we will find an intersection set $\{z_2z_3, z_2z_4, z_1z_3, z_1z_4\}$ between S and T. Hence, we obtained the best micro-average and macro-average scores when the subsequence length is set to 2.

4.3   Effectiveness of Varying the Decaying Factor

The decaying factor λ controls how many gaps are allowed in the matching subsequences of the two sequences. If $\lambda=1$, the gaps between the subsequences are not penalized. If $0<\lambda<1$, the larger the gaps are, the more penalty will be placed on the subsequence. For this set of experiments, we kept $K=10$ and $l=2$ fixed and studied the effects of varying the decaying factor λ.

In Table 5, the highest micro-precision is achieved at $\lambda=0.1$ while all other highest micro-average and macro-average scores are obtained at $\lambda=0.55$. The higher values of λ will place more weights to contiguous topic subsequences. In other words, this is the parameter that penalizes the topic subsequences with large interior gaps.

Table 5. Performance of Varying Decaying Factor when $l=2$ and $K=10$

λ	Micro-average			Macro-average		
	p	r	F	p	r	F
0.1	**87.91**	82.45	85.1	74.01	66.1	67.56
0.15	86.03	82.85	84.41	74.61	72.06	72.87
0.2	86.33	80.7	83.42	65.09	61.96	63.31
0.25	85.85	82.92	84.36	73.15	70.53	71.39
0.3	87.22	85.22	86.21	74.1	72.5	72.96
0.35	87.84	84.54	86.16	72.89	66.8	68.4
0.4	86.28	83.96	85.11	73.62	70.05	70.94
0.45	85.52	81.81	83.62	72.97	65.41	68.49
0.5	86.18	85.04	85.61	75.1	73.51	73.71
0.55	87.06	**86.69**	**86.88**	**76.48**	**75.91**	**75.98**
0.6	84.28	80.8	82.51	69.47	62.88	64.51
0.65	84.86	81.66	83.23	71.38	64.96	66.86
0.7	85.84	83.49	84.65	70.73	64.7	66.49
0.75	84.55	86	85.27	71.57	72.17	70.67
0.8	84.65	85.11	84.88	72.08	72.43	71.93
0.85	84.92	84.64	84.78	72.16	69.49	69.79
0.9	85.37	85	85.19	71.76	70.34	70.09
0.95	86.68	85.22	85.94	74.94	72.47	73.08

4.4   Computational Complexity

To derive an effective computation of this kernel, [4] proposed to use the dynamic programming technique to reduce the complexity of computation to $O(n||S||T|)$. n is the subsequence length and $|S|$ and $|T|$ refer to the number of symbols (words/characters/topics) in S and T. Therefore, the number of symbols in a sequence determines the efficiency of the string kernel. The problem is that the semantic structures of documents cannot be discovered when the number of symbols is greatly reduced. Differently, the topics are a summary of document and can detect the semantic structure of a document.

To further reduce the computational cost, [5] proposed to use the sequence of words instead of characters on the Reuters dataset. The average number of words per document is **141** before removing the stop words and this number drops to **77** after the stop words are removed. In our experiments, the average number of topics is **10**. [5] claimed that if the average sequence length is reduced by about **50%**, the kernel computation time would be reduced by **75%**. By comparison to **77**, the average sequence length is reduced by about **87%** since the topic number of topics is **10** in this paper. Hence, the computational complexity would be greatly reduced.

4.5   Does the Topic Sequential Order Matter?

In this section, we will investigate whether the sequential order of topics matter in text classification. We compared the topic sequence kernel with other kernels including linear, polynomial, RBF kernel and sigmoid kernel. For these kernels, we do not rearrange the topics in a sequential order. We simply use the topic proportions as the weight of each topic. Then each document is represented by a feature vector with topic proportions. Experiments are conducted by varying the number of topics (K) from 5 to 20. From the experimental results, we found that no matter what kind of kernels we used, the performance remains the same for each kernel if the number of topics remains unchanged. Therefore, we name other kernels as *ALL* plus the number of topics. *ALL_5*, for example, indicates the linear, polynomial, RBF or sigmoid kernel with the number of topics being 5. Similarly, *ALL_10*, *ALL_15* and *ALL_20* are either of these kernels with the number of topics being 10, 15 and 20, respectively.

Table 6. Comparison between *TSK* and other Kernels

	Micro-average			Macro-average		
	p	r	F	p	r	F
ALL_5	93.03	61.79	74.26	32.66	25.47	28.11
ALL_10	92.38	74.78	**82.65**	74.71	52.27	59.88
ALL_15	95.61	70.25	80.99	71.96	41.31	49.1
ALL_20	96.84	63.69	76.84	61.31	28.8	35.59
TSK	87.06	**86.69**	**86.88**	**76.48**	**75.91**	**75.98**

Table 6 shows that the topic sequence kernel (**TSK**) gives the best micro-average and macro-average F-scores when compared to other kernels. This testifies our hypothesis that the sequential order of topics really matters in text classification using topics. As the Table 6 shows, among the other kernels, *ALL_10* gives the best micro-average and macro-average F-scores when the number of topics is 10.

5   Conclusions and Future Works

Similar to the string kernel and word sequence kernel, topic sequence kernel considers the sequential structure between the symbols (characters/words/topics). In this paper, we first tried to use the topic sequence kernel operating at the topic level to greatly reduce the computational time cost. Initially, the LDA algorithm is used to extract posterior topic distributions in each document and generate the topic sequence based on these topic distributions. Our observations suggest that the optimal result is obtained when the number of topics is equal to the number of categories and the topic stability might be damaged if a document belongs to more than one category.

We focused on topic sequence kernels which are based on the topics. One advantage is that topics are summaries of documents and they can well capture the semantics of documents. The other advantage is that number of topics per document is much less than the number of words in a document. This can bring the string kernel into practical usage, since string kernel computation is rather resource demanding. Topic sequence kernel is an extension of string kernel. Our work contributes to the structural document representation using topics instead of words or characters and to the reduction of computational runtime cost.

References

1. Joachims, T.: Text Categorization with Support Vector Machines. Technical report, LS VIII NO. 23. University of Dortmund (1997)
2. Osuna, E., Freund, R., Girosi, F.: Training support vector machines: an application to face detection. In: IEEE Conference on Computer Vision and Pattern Recognition, pp. 130–136 (1997)
3. Wang, J.Y.: Application of Support Vector Machines in Bioinformatics. Master's thesis, Dept. Computer Sci. Info. Eng., National Taiwan University (2002)
4. Lodhi, H., Saunders, C., Shawe-Taylor, J., Cristianini, N., Watkins, C.: Text classification using string kernels. The Journal of Machine Learning Research 2, 419–444 (2002)
5. Cancedda, N., Gaussier, E., Goutte, C., Renders, J.M.: Word sequence kernels. Journal of Machine Learning Research 3, 1059–1082 (2003)
6. Shawe-Taylor, J., Cristianini, N.: Kernel Methods for Pattern Analysis. Cambridge University Press (2004)
7. Mercer, J.: Functions of positive and negative type and their connection with the theory of integral equations. Philosophical Transactions of the Royal Society London (A) 209, 415–446 (1909)
8. Deerwester, S., Dumais, S., Landauer, T., Furnas, G., Harshman, R.: Indexing by latent semantic analysis. Journal of the American Society of Information Science 41(6), 391–407 (1990)
9. Hofmann, T.: Probabilistic latent semantic indexing. In: Research and Development in Information Retrieval, pp. 50–57 (1999)
10. Blei, D., Ng, A., Jordan, M.: Latent dirichlet allocation. Journal of Machine Learning Research 3, 993–1022 (2003)
11. Blei, D., Lafferty, J.: Dynamic topic models. In: Proceedings of the 23rd International Conference on Machine Learning, pp. 113–120 (2006)
12. Blei, D., Lafferty, J.: A correlated topic model of science. Annals of Applied Statistics 1(1), 17–35 (2007)
13. Minka, T., Lafferty, J.: Expectation-propagation for the generative aspect model. In: Uncertainty in Artificial Intelligence, UAI (2002)
14. Steyvers, M., Griffiths, T.: Probabilistic topic models. In: Landauer, T., McNamara, D., Dennis, S., Kintsch, W. (eds.) Latent Semantic Analysis: A Road to Meaning. Lawrence Erlbaum (2006)
15. Teh, Y., Newman, D., Welling, M.: A collapsed variational Bayesian inference algorithm for latent Dirichlet allocation. In: Neural Information Processing Systems (2006)

Opening Machine Translation Black Box
for Cross-Language Information Retrieval

Yanjun Ma[1], Jian-Yun Nie[2], Hua Wu[1], and Haifeng Wang[1]

Baidu Inc., Beijing, China
{yma,wu_hua,wanghaifeng}@baidu.com
University of Montreal, Canada
nie@iro.umontreal.ca

Abstract. State-of-the-art Statistical Machine Translation (SMT) systems are widely used for query translation in Cross-Language Information Retrieval (CLIR), but usually as a black box. A strong limitation is that only one-best translation is retained. It is known that CLIR can benefit much from using multiple translation alternatives which produces a query expansion effect. It is then desirable to extend the one-best translation output to a richer translation including more translation alternatives. In fact, translation alternatives are available in SMT before the final best output is selected. A natural way is to open the black box of SMT to access the internal search graph in order to select more translation alternatives. In this paper, we consider the translation alternatives included in the N-best translation outputs. Using our approach for CLIR, we report up to 40% improvement in Mean Average Precision over baseline query translation using an SMT system as a black box on TREC-9 English-Chinese CLIR task, and 10% on NTCIR-5 English-Chinese task. This study demonstrates the usefulness of opening the black box of SMT to produce more adequate query translations for CLIR.

Keywords: Cross-language Information Retrieval, Statistical Machine Translation.

1 Introduction

Cross-Language Information Retrieval (CLIR) aims to retrieve documents in a language different from that of the query. A CLIR system has the potential to extend the searchable information space from a single language to multiple languages. The recent developments in Machine Translation (MT) make it possible for a user to understand the essential contents of a retrieved document in a foreign language. This also makes it easier to implement CLIR by adding an MT system for query translation on top of a general IR system. Intuitively, one would believe that with the help of a good MT system, CLIR can be easily implemented as: Query translation by MT + Monolingual IR + Document translation by MT. This is indeed the approach used in some commercial search engines such as Google, which offers seamless query translation and document translation by its MT system.

However, despite its apparent resemblance to MT, query translation is different from MT from several perspectives [7]. In particular, MT usually produces one single translation for an input sentence, while in CLIR, it is useful to include several translation

Y. Hou et al. (Eds.): AIRS 2012, LNCS 7675, pp. 467–476, 2012.

alternatives of the query in order to obtain a desirable query expansion effect. For example, while the Chinese term "艾滋" is a good translation of the query "AIDS", it is highly useful to also include the alternative "爱滋" in the query translation because both Chinese terms are frequently used in Chinese documents about AIDS. Although this problem has been observed for a long time, to our knowledge, there has not been much extensive work on how to extend the simple use of MT systems for query translation. This situation is largely due to the fact that commercial MT systems such as Systran or Google translate did not offer the possibility to obtain several translation alternatives. [1] In general, several possible translation alternatives for each word or phrase are stored in an internal search graph before the final best translation is selected. It is possible to open the MT black box to gain access to the internal search graph to select more translation alternatives.

In this paper, we attempt to take advantage of structural information in such a translation search graph. Specifically, we leverage portions of the search graphs that can potentially lead to high-quality translations, i.e. the search graphs corresponding to the N-best translations. These translation alternatives are converted into a structured search query. We carried out experiments on English-Chinese CLIR using collections from TREC and NTCIR. The experiments show that our approach can significantly outperform the one that uses a state-of-the-art MT system as a black box. This result clearly demonstrates the importance of exploiting an SMT system more than just using the one-best translation output.

The reminder of the paper is organized as follows. Section 2 will review some related work. Section 3 describes one of the state-of-the-art SMT models - Phrase-Based SMT model. In section 4, we put forward our method of using structural SMT output for CLIR, and report the experimental results in section 5. Finally, we will draw conclusions and point out avenues for future research in section 6.

2   Related Work

Extensive research has been carried out on CLIR for the last 15 years. One can find a complete survey in [7]. Three main families of approaches to query translation have been proposed: using an MT system, using a bilingual dictionary or using a parallel or comparable corpus. Each family of approaches has its advantages and weaknesses. In particular, query translation using MT often can suggest an appropriate translation, but is limited by the fact that only one possible translation is obtained. In many cases in IR, a query is just one possible expression of an information need, and there may be many others. In query translation, the selection of one translation artificially confines one to only one possible expression of the information need (even though it is appropriate), leading to low recall. Nevertheless, compared to the other approaches to query translation, MT-based approaches benefit much from the translation selection process undertaken for producing the best translation output, which is based on a target language model. This selection process takes into account whether different translation terms may occur together in the target language, thereby the translation candidates that do not fit well the query context are discarded.

---

[1] A recent version of Google translate offers the possibility to see several translation alternatives.

Bilingual dictionary is often seen as an inexpensive alternative to MT. It also provides multiple translations for each term. Including multiple translations for a query usually yields higher recall ratio. However, a bilingual dictionary contains the translations of different meanings of a term. Without an effective disambiguation or selection process, dictionary-based translation will include many inappropriate translation terms, leading to a possible topic drift. In order to select appropriate translations from a bilingual dictionary, several approaches have been proposed to select translation terms using a measure of cohesion [2,5,11]. However, we observe that still only one-best translation is selected for each query term in the above approaches. These approaches are thus limited in the same way as with MT systems. As the expansion effect during query translation has proven to be beneficial, one can legitimately question about the reasonableness of such a limitation to one-best translation.

Parallel corpora have been used to train translation models (usually IBM model 1) for query translation. However, one usually performs a much less strict translation selection than for SMT: for each query term, a certain number of best translation alternatives, or those whose translation probability is higher than a threshold, are included in the translated query [4]. Such an approach can indeed include many possible translation alternatives, but also many inappropriate ones. Compared to SMT, we observe that one missing element is the language model that helps selecting appropriate translation candidates.

The previous experiments using different approaches draw a clear picture of what is needed for query translation: On the one hand, it is useful to include multiple translation alternatives. On the other hand, it is useful to perform some selection of translation terms. A possible way to achieve this goal is to open the black box of an MT system so that multiple translation alternatives are kept, while some selection on translation alternatives is still performed using a language model. This is the approach that we investigate in this paper.

Notice that there have been some attempts to open the MT black box for CLIR. For example, Sakai [10] used alternative translations from a rule-based MT system. We are however not aware of similar studies using an SMT system. Compared to a rule-based MT system, an SMT system has the advantage of also providing a probability for each translation output, which allows us to weigh the translation in the final translated query. In a recent study Magdy and Jones [6] tried to construct an SMT system specifically for query translation by performing preprocessing for stop-word removal and word stemming before the training phase. However, still one-best translation is used in this study. The purpose of this attempt is different from ours in this paper.

3   Phrase-Based Machine Translation

As our investigation is based on an SMT system that uses phrase-based translation, let us first briefly describe the principle used in our SMT system.

Given a source sentence $s_1^I = s_1, ..., s_j, ...s_I$, which is to be translated into a target sentence $t_1^J = t_1, ..., t_i, ..., t_J$, among all the possible target sentences, we will choose the sentence with the highest probability as in (1):

$$\hat{t}_1^J = \underset{t_1^J}{\operatorname{argmax}}\{Pr(t_1^J|s_1^I)\} = \underset{t_1^J}{\operatorname{argmax}}\{\sum_{m=1}^{M} \lambda_m h_m(t_1^J, s_1^I)\} \tag{1}$$

In this log-linear framework [8], we have a set of M feature functions $h_m(t_1^J, s_1^I)$, $m = 1, ..., M$, and the translation and language models can be incorporated into this framework as features.

In state-of-the-art Phrase-Based SMT [3], phrase translations are normally induced using word alignment techniques [1]. The word alignment information are preserved in our translation model so that the alignment information is available during translation.

4 Exploiting MT for Query Translation

In language models for IR, the score of a document D is determined as follows:

$$score(D, Q) = \sum_{t \in Q} P(t|Q)log P(t|D) \tag{2}$$

in which the document language model $P(t|D)$ should be smoothed. A common method for smoothing is the following Dirichlet smoothing [13]:

$$P(t|D) = \frac{tf_{t,D} + \mu P(t|C)}{|D| + \mu} \tag{3}$$

where C denotes the collection, and μ is tunable parameter.

In CLIR, the purpose of query translation is to define a language model $P(t|Q_s)$ where t is in the target language while Q_s is in the source language. Using an SMT system, if only one-best translation is used, the generated query translation is used to define it as follows:

$$P(t|Q_s) = P(t|Q_t) \tag{4}$$

where Q_t is the translation retained for Q_s. This is the traditional approach using MT as a black box, which we use as our baseline. Assume that the query "environmental protection laws" is translated to "环境保护法", which can be segmented into "环境" and "保护法", the latter two terms are used as query terms to retrieve documents in Chinese. Using Indri retrieval system [12], the corresponding query is

```
#combine(环境  保护法)
```

4.1 Structured Query Generation

As we mentioned earlier, an SMT will generate multiple translation candidates for each term in an internal search graph. which will be used by SMT to select the best translation sequence. Figure 1 is an example of MT search graph. Each arc in the graph represents a possible translation for the word/phrase it covers. The graph contains translations not only for words, but also for phrases. For a given term (a word or a phrase), there are

Fig. 1. An example of MT search graph

often several possible translation candidates. For example, the word "environmental" can be translated as "环境", "环保", etc. The one-best translation selected as the final SMT output corresponds to the best path that covers the whole input word sequence. To take advantage of the other translation alternatives, a possible solution is to extend the one-best translation by including other translation alternatives for each term. The translation alternatives are considered as synonyms. In Indri system, a set of synonyms can be combined using #syn operator. Assume that the path in bold in Figure 1, with the underlined translations, corresponds to the best one in the search graph. Then we also consider the translation alternatives for the terms "environmental" and "protection laws". The structured query that we generate is as follows:

```
#combine(#syn(环境 环保 ...) #syn(保护法 保护法规 ...))
```

The internal search graph may contain many translation alternatives, including those that do not fit well the given query context. As we discussed earlier, the selection made in SMT is crucial for removing translation candidates that are inappropriate for the given query. Therefore, we do not retain all the translation alternatives for each term in the search graph, but only those that appear in the N-best translations, which all went through a translation selection process.

4.2 Combining Multi-path Translation Results

In addition to the multiple translation alternatives for each source term, the source sentence can also be segmented in different ways, either as single words or as phrases. In the example shown in Figure 1, we can see that "environmental protection laws" can be segmented into either a single phrase, a phrase + a word, or three words. Different segmentations will result in different translations. To take into account the multiple ways to segment the input query, we will allow the SMT system to generate N-best translations which covers the most probable ways to segment the source query. For each translation output, we can proceed in the same way as for the best translation by including all the translation alternatives as synonyms, or we can just use the N-best translations without extending them by synonyms. We will test both options in our experiments.

With both options, we have to run N retrieval processes, resulting in N sets of documents $\{R_1, R_2, ..., R_N\}$ where $R_n = D_1, D_2, ..., D_M$ contains M documents, each of which is associated with a relevance score $S_{rel}(R_n, D_m)$. The scores for each document are linearly combined with equal weights. To punish the documents contained in fewer sets of retrieval results, the score for document D_m can be calculated as follows:

$$S_{intp}(D_m) = \frac{N}{\sum_{n=1}^{N} \delta(R_n, D_m)} \sum_{n=1}^{N} S_{rel}(R_n, D_m)\delta(R_n, D_m) \qquad (5)$$

where $\delta(R_n, D_m)$ is a Kronecker function with value 1 if $D_m \in R_n$ and 0 otherwise, and $\frac{N}{\sum_{n=1}^{N} \delta(R_n, D_m)}$ is used to punish documents less frequently retrieved according to the N-round retrieval results. [2]

5   Experiments

5.1   Data

Our experiments are conducted on two CLIR collections, i.e. TREC-9 and NTCIR-5 with 25 and 50 topics respectively. Current experiments focus on English-Chinese CLIR, i.e. to retrieve Chinese documents using English queries. To accomplish this task, we translate the English titles into Chinese using our in-house SMT system. The stop words in the translation output are removed.

5.2   Systems

Query translation are carried out using Baidu MT system.[3] The retrieval experiments are conducted using an state-of-the-art IR system - Indri[4] with support to popular structured query operators. To evaluate our approach, we use Mean Average Precision (MAP) as our main measure. To show the impact on the top retrieval results, we also include P@5 and P@10.

5.3   Using Synonyms and Interpolation

We first run monolingual IR (C-C-T) and CLIR with the baseline query translation (E-C-T). The results are reported in Table 1. Both our monolingual and baseline cross-language retrieval performance is comparable to top systems that participated in TREC-9 and NTCIR-5 evaluation campaigns and those reported in previous studies. We observe that query translation does not lead to a large loss in retrieval effectiveness on TREC-9 collection, while the loss is much larger on NTCIR-5. This difference is due to the differences in the collections. It turns out that the queries in NTCIR-5 are much harder to translate by SMT than those in TREC-9, because there are many named entities, whose translation may be problematic. Table 1 also shows our experimental results using top 50 best translations for synonym derivation (+syn@50). Using synonyms leads to 11% relative improvement in MAP over baseline CLIR system (E-C-T) on TREC-9 collections, even outperforming the monolingual (C-C-T) retrieval results. On NTCIR-5 collection, using synonyms only lead to modest improvement in MAP.

---

[2] Note that the relevance score S_{rel} is a negative value so that a smaller $\delta(R_n, D_m)$ will decrease the final document score.
[3] http://fanyi.baidu.com
[4] http://www.lemurproject.org/indri.php

Table 1. Using synonyms derived from N-best translations (An exponent n on MAP means a statistical difference ($p < 0.05$) with the line n)

		TREC-9			NTCIR-5		
		MAP	P@5	P@10	MAP	P@5	P@10
1	C-C-T	0.2181	0.3040	0.2400	0.3785	0.6440	0.5460
2	E-C-T	0.2077	0.2720	0.2080	0.2644	0.4640	0.4220
3	+syn@50	0.2280^2	0.3040	0.2240	0.2650	0.4680	0.4000
4	+intp@50	$0.2615^{1,2,3}$	0.3040	0.2240	$0.2817^{2,3}$	0.4800	0.4340
5	+intp+syn@50	$0.2725^{1,2,3}$	0.3120	0.2280	$0.2896^{2,3}$	0.4880	0.4320

Table 2. Examples of structured queries that improve CLIR

Input Query	stealth technology in Asia
one-best translation	隐身 技术 亚洲
+Synonyms	#syn(隐形 隐身) 技术 亚洲
Additional translations	#syn(亚 #1(亚洲 区) 亚洲 亚太 #1(亚洲 地区) 东亚 东南亚) 隐身 技术
Input Query	Daya Wan nuclear power plant
one-best translation	广东 核电厂
+Synonyms	广东 核电厂
Additional translations	大亚湾 #syn(湾 万) 核电厂

From Table 1, we can also observe that interpolating multi-path retrieval results (+intp@50) consistently improve the performance of the retrieval system. When the multi-path retrieval system is further enriched with synonyms in the search query (+intp@50 +syn@50), the performance can be further improved, especially on TREC-9. The above observations also apply to P@5 and P@10. This means that the enriched translations does not increase recall to the detriment of precision. On the contrary, precision on top results is also improved.

To understand why our approach can improve retrieval effectiveness, we show several examples in Table 2. These examples show that using synonyms can alleviate the errors or term variances introduced by the one-best translations. In the first example, the addition of synonym "隐形" to "隐身" (both are legitimate translations of "stealth") can improve the retrieval performance because the first term is more frequently used for "stealth technology". Using additional translations in N-best list could produce more translation alternative due to different segmentations over the source input query. For this example, however, the additional translations are not really helpful - "Asia" is also translated into "亚太" (Asia-Pacific), "东亚" (East Asia) and "东南亚" (Southeast Asia), which may not be appropriate for this query. In the second example, N-best translations contains the correct translation "大亚湾" for English word "Daya Wan" (the name of a place), which is not correctly translated in the first best translation.

5.4 The Impact of N-Best Size

The previous experiments used 50 best translations. The evaluate the impact of the N-best size, we conducted retrieval experiments using different values for N.

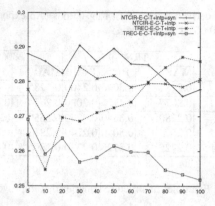

Fig. 2. The Impact of N-best Size on MAP

Figure 2 depicts the curve of MAP when N is set to different values. It can be observed that the optimal values are different for TREC-9 and NTCIR-5 collections. It can also be observed that using synonyms on TREC-9 collection consistently improves MAP as N increases, while the system without synonyms fluctuates and starts to dip when N is set to 80. On NTCIR-5 collection, the system using synonyms also yields more stable performance. Moreover, a larger N-best list is more beneficial to the shorter TREC-9 queries where query expansion more useful in improving retrieval performance.

5.5   Dynamic Selection of N-Best Translations

Given the fact that translation search graphs of the input queries vary from one to another, using the same number of N-best translations for all queries may not be optimal. For instance, a longer input query normally implies a larger number of vertices and edges in the search graph and respectively more translation candidates compared to a shorter query. Meanwhile, even for queries with equal number of translation candidates, the distribution of posterior scores may vary. It is therefore necessary to dynamically determine the number of N-best translations for each query.

N-best translations are associated with their translation scores and can be ranked according to them. Intuitively, a sharp decrease in translation score generally indicates an appropriate cut-off point. Given an input source language query Q_s and its N-best translations $Q_{t_1}, Q_{t_2}, ..., Q_{t_N}$, we use $TS(Q_{t_n})$ to denote the translation score of Q_{t_n}. An cut-off point can be established where the difference $Dist$ between consecutive N-best translations Q_{t_n} and $Q_{t_{n-1}}$ reach a threshold α:

$$\text{Dist} = \frac{TS(Q_{t_n}) - TS(Q_{t_{n-1}})}{TS(Q_{t_n})} \tag{6}$$

We further assume that the optimal α for a particular query depends on the length of the query, and the distance between consecutive N-best translations of longer input queries tend to be smaller than those of shorter input queries. We can define α as a exponential function of input query length l: $\alpha = \frac{j}{lk}$, where $j \in (0, \infty)$ and $k \in (0, \infty)$ are constants. Albeit j and k can grow without bound in theory, the differences in translation scores for two consecutive N-best translations is normally bounded. Here we empirically set $j \in [0.5, 5]$ and $k \in [0.5, 5]$. We then perform grid search for optimal j and k with a step size 0.5 using NTCIR-6 topics as development set. We found the optimal value for j and k are respectively 2 and 2. As can be seen from Table 3, to dynami-

Table 3. Dynamic selection of N-best translations

	TREC-9			NTCIR-5		
	MAP	P@5	P@10	MAP	P@5	P@10
E-C-T	0.2077	0.2720	0.2080	0.2644	0.4640	0.4220
+intp+syn-opt	0.2871	0.3040	0.2720	0.2896	0.4880	0.4320
dynamic	0.2903	0.3360	0.2840	0.2883	0.4960	0.4420

cally determine optimal number of N-best translations yields comparable performance in MAP compared to when N is set set at optimal numbers - 0.2871 for TREC-9 and 0.2896 for NTCIR-5. This result confirms our intuition that different queries require different N depending on how many translation alternatives there are and how likely they are. It also shows that our method to automatically determine N is quite effective. Notice, however, that N shown in Figure 2 is set for the whole collection. If N were set to its best for each query, the effectiveness on a collection could be even higher. This means that the best effectiveness shown in Figure 2 is not truly optimal and there is still much room for improving the dynamic search for the optimal N for each query.

6   Conclusions and Future Work

In this paper, we presented an approach to exploit more translation alternatives in an SMT system in English–Chinese CLIR. More specifically, we considered N-best translations from an SMT system, which are then converted to structured queries. Two different types of structural information can be induced from the N-best translations: (1) synonyms, i.e. different translations of one source word/phrase, and (2) different ways to segment the input query. Incorporating such information into input query composition yields gains in MAP on both TREC-9 and NTCIR-5 collections. Our experimental results clearly showed that it is beneficial to open the MT black box to obtain more translation alternatives.

The methods we explored in this paper can be further improved from several perspectives. (1) We did not consider translation probabilities when synonyms are combined. Intuitively, it would be reasonable to combine synonyms according to their respective translation probability from the source term. Technically, this is feasible with the #wsyn operator in Indri. (2) The creation of the final structured query requires the SMT system to generate N-best translations. This explicit generation of N final translations can be avoided and we can indeed create the structured query directly from the internal search graph. This would correspond to a different decoding process than for MT. (3) The final N-best translations are selected using a traditional statistical language model. As shown in several previous studies, query terms often do not form a sequence of words as in a general language, so are their translations. In [2,5,11], co-occurrence statistics are used instead of traditional language models for translation selection. Such an approach may be more appropriate for queries. The solutions to all these problems may lead to even larger improvement in CLIR effectiveness. We will investigate them in our future research.

Acknowledgments. This work is supported by 863 State Key Project (Grant No. 2011AA01A207). The authors would like to thank the reviewers for their insightful comments.

References

1. Brown, P.F., Della-Pietra, S.A., Della-Pietra, V.J., Mercer, R.L.: The mathematics of Statistical Machine Translation: Parameter estimation. Computational Linguistics 19(2), 263–311 (1993)
2. Gao, J., Nie, J.Y., Xun, E., Zhang, J., Zhou, M., Huang, C.: Improving query translation for cross-language information retrieval using statistical models. In: Proceedings of SIGIR 2001, pp. 96–104 (2001)
3. Koehn, P., Och, F., Marcu, D.: Statistical Phrase-Based Translation. In: Proceedings of the 2003 Human Language Technology Conference and the North American Chapter of the Association for Computational Linguistics, Edmonton, AB, Canada, pp. 48–54 (2003)
4. Kraaij, W., Nie, J.Y., Simard, M.: Embedding Web-Based Statistical Translation Models in Cross-Language Information Retrieval. Computational Linguistics 29(3), 381–420 (2003)
5. Liu, Y., Jin, R., Chai, J.Y.: A maximum coference model for dictionary-based cross-language information retrieval. In: Proceedings of SIGIR 2005, pp. 536–543 (2005)
6. Magdy, W., Jones, G.J.F.: Should MT Systems Be Used as Black Boxes in CLIR? In: Clough, P., Foley, C., Gurrin, C., Jones, G.J.F., Kraaij, W., Lee, H., Mudoch, V. (eds.) ECIR 2011. LNCS, vol. 6611, pp. 683–686. Springer, Heidelberg (2011)
7. Nie, J.: Cross-Language Information Retrieval. Morgan & Claypool Publishers (2010)
8. Och, F., Ney, H.: Discriminative training and maximum entropy models for Statistical Machine Translation. In: Proceedings of the 40th Annual Meeting of the Association for Computational Linguistics, Philadelphia, PA, pp. 295–302 (2002)
9. Pirkola, A.: The effects of query structure and dictionary-setups in dictionary-based cross-language information retrieval. In: Proceedings of SIGIR 1998, pp. 55–63 (1998)
10. Sakai, T.: Advanced Technologies for Information Access. International Journal of Computer Processing of Oriental Languages 18(2), 95–113 (2005)
11. Seo, H.-C., Kim, S.-B., Rim, H.-C., Myaeng, S.-H.: Improving query translation in English-Korean cross-language information retrieval. Information Processing and Management 41, 507–522 (2005)
12. Strohman, T., Metzler, D., Turtle, H., Croft, W.B.: Indri: A language model-based search engine for complex queries. In: Proceedings of the International Conference on Intelligence Analysis, May 2-6, McLean, VA (2005)
13. Zhai, C., Lafferty, J.: A study of smoothing methods for language models applied to information retrieval. ACM Trans. Inf. Syst. 22(2), 179–214 (2004)

Topic Based Author Ranking
with Full-Text Citation Analysis

Jinsong Zhang[1], Chun Guo[2], and Xiaozhong Liu[2]

[1] Dalian Maritime University, Dalian, China
zjs.dlmu@gmail.com
[2] School of Library and Information Science, Indiana University Blooming, IN, USA
{chunguo,liu237}@indiana.edu

Abstract. Author metadata provide significant scientific publication characterization, which often represents important domain knowledge. Publications from existing or potential reputable authors motivate further research as "stand on the shoulder of giants". This paper addresses author ranking problem for information retrieval and recommendation, and the contributions of this research are four-fold. First of all, we employed full-text citation analysis (citation context) to enhance the classical author citation network. Second, supervised topic modeling method is used to determine the contribution of a specific author (as a vertex) or a citation (as an edge). Third, PageRank with prior and transitioning topical probability distributions measured the importance of authors (in the graph) based on each scientific topic. Last but not least, we proposed a novel evaluation method to compare the result of PageRank with prior with classical ranking methods, i.e., BM25, TFIDF and Language Model, and PageRank. The result shows that our ranking method with full-text citation analysis significantly (p<0.001) outperforms than the other ranking methods.

Keywords: Author ranking, Labeled-LDA, Full-text citation extraction, PageRank with prior.

1 Introduction

Currently, most studies of Bibliometrics have mainly focused on the quantitative analysis of scientific and technological literature, i.e., publications, authors, venues, topics and other related metadata by a set of methods, i.e., direct citation, bibliographic coupling, and co-authorship analysis. As we known, author metadata provide significant scientific publication characterization, which often represents important domain knowledge, so publications from reputable authors motivate further research, as "stand on the shoulder of giants"[1].

In fact, author modeling and author ranking tasks have been studied for a long time in previous research, and made fruitful achievements. Among them, the most simple, easy, but effective ranking indicator is citation counts [2], which mean that the importance of publications/authors is higher than others with more citation counts. However, the limitation of this method is caused by a fundamental problem of bibliometrics:

Y. Hou et al. (Eds.): AIRS 2012, LNCS 7675, pp. 477–485, 2012.
© Springer-Verlag Berlin Heidelberg 2012

the citation relationship between papers and authors are oversimplified, namely, we credit all the citations in a very similar way, while ignoring the important citation topical information.

To overcome this limitation, we proposed a novel author ranking method based on full-text citation analysis, where each citation is represented by a citation context window along with its topic distribution. The remainder of this paper is organized as follows: Session 2 reviews related work for author ranking analysis; Session 3 presents the main methodologies of this paper; Session 4 is about the experiment, experimental data and results will be described; In the last session, we will discuss the findings and limitations of this paper and identify subsequent research steps.

2   Related Work

So far, author ranking has long been heavily discussed by fundamental researches, and made fruitful achievements by various methods, which are used to characterize the author network. The most common approach is citation count [2]. Besides, impact factor [3] and h-index [4] are also influential methods in this field. Additionally, some articles measuring the author ranking by hyperlink graph, a common method is Page-Rank algorithm [5], which is used for measuring the relative importance of the link-based graph, webpages, publications or authors graph.

While most of these methods are topic independent, which cannot be used for measuring author ranking based on selected topic. Actually, most traditional researches of citation analysis treated all citations equally. However, in reality, Herlach [6] try to distinguish the contributions of references by citation position or citation content. In fact, topic modeling is also a valuable method for determining the diverse importance of vertices, including ACT, LDA [7], and Labeled-LDA [8] etc. On the other hand, someone tried to optimize PageRank, which is used widely in bibliometrics, i.e., White and Smyth [9] first proposed the priors idea in their formalization of a relative-rank extension to both PageRank and HITS.

In this paper, topic based author ranking will be derived from full-text citation analysis, and topic dependent ranking lists for authors will be measured by an optimized PageRank algorithm, which take into account the initial value of vertices and transitioning probability of edges. The main methods will be introduced in the next session.

3   Proposed Method

3.1   Author Graph Based on Full-Text Extraction

Most previous studies for citation network analysis in bibliometrics are based on one assumption: $paper_1$ and $paper_2$ are connected, when $paper_1$ cites $paper_2$. It also applies to author analysis network: if $author'_1 s\ paper$ cites $author'_2 s\ paper$, then $author_1$ and $author_2$ are related somehow, the relation of which can be characterized on a directed graph with authors as vertices and citations as edges. So the graph can be expressed as a set of Vertices and Edges: $G = (V, E)$.

V as a set of vertices represents all the authors, $V=\{v_1, v_2, v_3, \cdots, v_n\}$. In most cases, one author has multi-publications, so v_i as a single vertex of author is a set of papers, as: $v_i = \{p_1, p_2, p_3, \cdots, p_m\}$.

E as a set of edges means the relationships between authors generated from citation network. One vertex could have more than one edge to the other vertex. For edges of author graph can be seen as a set of any two authors' citation relationship, $E=\{\langle v_1, v_2 \rangle, \langle v_1, v_3 \rangle, \cdots, \langle v_i, v_j \rangle\}$, in which $\langle v_i, v_j \rangle$ means vertex v_i connects to v_j, so each edge is expressed as $\langle v_i, v_j \rangle = \{\langle p_1, p_2 \rangle, \langle p_1, p_3 \rangle, \cdots, \langle p_t, p_k \rangle\}$, only when $p_t \in v_i$ and $p_k \in v_j$. Please note, unlike classical method, as our method employs full text citation extraction, one publication may have more than one citation relation to the other publication. For instance, one publication may cite the other publication more than one time in different locations.

As shown in the following figure, two authors were seen as v_1 and v_2, where $v_1 = \{p_1, p_3\}$ and $v_2 = \{p_2, p_4\}$, the relationship of papers were shown in the left figure. In the right part, the relation between authors can be characterized on the directed graph: e_1 and e_2 represented the edges from v_1 to v_2, in which e_1 is a set of citation, p_1 to p_2, for the twice citing relationship; and e_3 characterized the relation from v_2 to v_1. While both p_1 and $p_3 \in v_1$, the direct self-citation relation of $p_1 \rightarrow p_3$ may be a different kind of citation indicator for research assessments, so we do not characterize it when creating the graph [10].

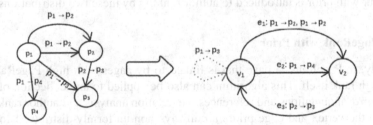

Fig. 1. Simple graph for author network

Currently, most studies of citation analysis are based on the metadata of a paper, i.e., title, abstract and reference in publications. In this research we used regular expression to extract the citation context, left and right n words surround the citation, like "… …[number]… …", "… …[number-number]… …" and "…….[number, number……]". The citation context will be used to represent and character the relation (edge on the graph) between authors.

Meanwhile, we also used the text content to represent each author vertex, where all the full-text papers, published by the target author, are used to generate a word distribution. Later in the next section, both author vertex (papers distribution) and citation edge (citation context distribution) are presented by topic distribution using label LDA algorithm.

3.2   Topic Modeling Based on Labeled-LDA

Labeled-LDA (Labeled Latent Dirichlet Allocation) [8] is a supervised topic model for credit attribution in multi-labeled corpora. Unlike LDA method, Labeled-LDA assumes the availability of topic labels and the characterization of each topic by a multinomial distribution $\beta_{key_{i,}}$ over all vocabulary words. W is a set of words w_i, chosen from the document d in the training text, $W=\{w_1, w_2, \cdots, w_n\}$, the word is picked in proportion to the publication's preference for the associated label $\theta_{paper,label_i}$ and the label's preference for the word $\beta_{label_i,w}$. So two matrices can be constructed by p_{P_i,l_t} and p_{l_t,w_s}, that represent the cooccurrence probability of paper and label, and the cooccurrence probability of label and word, respectively.

$$\theta_{paper,label} = \begin{pmatrix} p_{P_1,l_1} & \cdots & p_{P_1,l_t} \\ \vdots & \ddots & \vdots \\ p_{P_i,l_1} & \cdots & p_{P_i,l_t} \end{pmatrix} \; ; \; p_{p_i,l_t} = \frac{P(l_t|paper_v)}{\sum_{x=1}^{|V|} P(l_t|paper_x)} \tag{1}$$

$$\beta_{label,w} = \begin{pmatrix} p_{l_1,w_1} & \cdots & p_{l_t,w_s} \\ \vdots & \ddots & \vdots \\ p_{l_t,w_1} & \cdots & p_{l_t,w_o} \end{pmatrix} \; ; \; p_{l_t,w_s} = \frac{P(w_s|l_t)}{\sum_{t=1}^{|W|} P(l_t|w_s)} \tag{2}$$

Therefore, the author graph described in session 3.1, author vertex (all the full-text papers) and citation edge (citation context window), can infer the possible topics distribution for each text by using Labeled-LDA modeling, and get two matrices respectively, papers distribution and citation context distribution. Later in the next section, PageRank with prior is introduced to author ranking by these two distributions.

3.3   PageRank with Prior

Generally, in PageRank, a page that is linked to by pages with high PageRank receives a high rank itself. This algorithm can also be applied to any collection of entities with reciprocal quotations and references, i.e. citation analysis and author ranking.

While the vertex and edge priors are always non-uniformly-distributed for author graph, the importance and contribution of vertices and edges are quite different. For this reason, unlike classical PageRank, PageRank with prior algorithm [11] calculates the relative importance of set with the vertex prior probability distribution. In this context, before we calculate the vertex (author) importance, some vertices (authors) are more important than others for certain topic. For instance, for some authors, based on their publication topic distribution, we already know their expertise in some specific topics, and they should be more important than other authors in their areas. At the same time, each edge is characterized by a transitioning topic probability distribution based on the citation context, which means citing author's credits on a topic are more likely to be transferred to a cited author, if the transitioning probability of the this topic between them is high.

Based on the matrix by Labeled-LDA (session 3.2), we can get prior probability score of each vertex (author) based on each topic, as:

$$p_{v,l_t} = \sum p_{P_i,l_t} = \sum_i \frac{P(l_t|paper_v)}{\sum_{x=1}^{|V|} P(l_t|paper_x)} \tag{3}$$

where $p_i \in a$, and the transitioning probability score of edge can be calculated as:

$$p_{e,l_t} = \sum p_{l_t}(v_i|v_j) = \sum \frac{P(l_t|citation_{j,i})}{\sum_{x=1}^{d_{out}(v_j)} P(l_t|citation_{j,x})} \tag{4}$$

where $e = <a_i, a_j>$, $v_i \in a_i$ and $v_j \in a_j$. So the score of author ranking by Page-Rank with prior can be calculated as:

$$\pi_{key_t}(v)^{i+1} = (1 - \beta_b)\left(\sum p_{e,l_t} \pi_{key_t}{}^{i+1}(u)\right) + \beta_b p_{v,l_t} \tag{5}$$

This equation represents a Markov chain for a random surfer who transitions "back" to the root vertexes R with probability β_b at each time-step. For each incoming link (citation) from v the PageRank score is updated with respect to edge (citation) transitioning probability p_{e,l_t}.

In order to interpret the result of PageRank with prior, we will use a very simple instance: $author_1$ citing $author_2$ and $author_3$, where $author_1$ has two topics: key_1 and key_2, and scores are $p_{v_1,l_1} = 0.3$ and $p_{v_1,l_2} = 0.7$, respectively. $e_1 = \langle v_1, v_2 \rangle$, $e_2 = \langle v_1, v_3 \rangle$, the scores of citation based on label are represented as: $p_{e_1,l_1} = 0.4$, $p_{e_1,l_2} = 0.6$ and $p_{e_2,l_1} = 0.2$, $p_{e_2,l_2} = 0.8$.

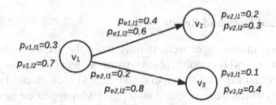

Fig. 2. Sample of PageRank with prior

For key_i, we will calculate the ranking scores for these three nodes:

$$p_{v_2,l_1} = 0.3 * \frac{0.4}{0.4+0.2} = 0.2, p_{v_3,l_1} = 0.3 * \frac{0.2}{0.4+0.2} = 0.1$$
$$p_{v_2,l_2} = 0.7 * \frac{0.6}{0.6+0.8} = 0.3, p_{v_3,l_2} = 0.7 * \frac{0.8}{0.6+0.8} = 0.4$$

Based on these scores, we found that, for topic key_1, $author_2$ is more important than $author_3$; for topic key_2, $author_3$ is more important than $author_2$. We can interpret this exemplar graph in this way: $author_1$'s expertise is more focusing on key_2 ($p_{v_1,l_2} > p_{v_1,l_1}$), so $author_3$ is more significant than $author_2$.

3.4   Evaluation Method

As this research focuses on the method of calculating publication topic importance, we can hardly compare the authority vector with other classical bibliometric indicators, such as h-index or impact factor, which are topic independent. So, in order to get the "ground truth" for author ranking with topic independence, some review or survey papers were collected with one assumption: the authors cited by a review paper are the relatively important authors for a specific topic. Since degree of importance of cited authors may be different, we used the number of citations (by a review paper) to characterize the importance. Thus, if a review paper for keyword key_i cited $author_1$ twice and $author_2$ once, then, $Importance_{key_i}(author_1) = 2$ and $Importance_{key_i}(author_2) = 1$. We also assume that if an author is not cited by the target review paper, then the importance of this author for the target topic is 0. We also assume that if an author is cited 4 or more times by the review paper, then its importance is equal to 4.

Although we recognize the review paper cannot cover all important authors for the target area, the goal of this evaluation is to compare the performance of our approach with a list of baseline algorithms:

Original PageRank: calculate the vertices' importance scores by original PageRank, and the vertex and edge priors are uniformly distributed. Some other methods (**TFIDF, BM25 and Language Model**) based on index file: use the keyword (topic label) as the query to search all the paper content (abstract and full text), and then rank the list based on TFIDF, BM25 or Language Model. **Language Model + Page-Rank**, which combined the language model with PageRank, where PageRank score is used as the prior likelihood.

4   Experiments

4.1   Experimental Data and Result

The experiment data in this paper were mainly from the ACM digital library, we selected 14,222 authors, whose publications were at least two, for measuring. The resulting dataset contained 41,370 papers and 223,811 citations. For training topic modeling, 3911 keywords were selected as topics, who appeared more than 10 times in the selected publications of training data and trained a Labeled-LDA model (mentioned in 3.2) based on 14,232 papers, which was used in the proposed ranking algorithm. As shown in figure 3, the left figure is an example of the topic distribution over a paper and the right figure is the topic distribution for a citation. The node with the highest probability in figure is the most relevant topic for the sample text.

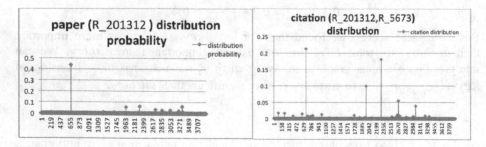

Fig. 3. Example of distribution probability of paper and citation

Based on topic distribution, author graph calculated by PageRank with prior output different ranking lists for different topics.

4.2   Experimental Evaluation

The evaluation method mentioned in session 3.4 will be used in this part, and 104 review or survey papers were chosen as "ground truth" based on 104 topics. The result of PageRank with prior was compared against several traditional methods, i.e. PageRank, TFIDF, BM25, and Language Model on two indicators, MAP (Mean Average precision) and nDCG (normalized Discounted Cumulative Gain) [12].

Table 1. Evaluation result of MAP

	Pagerank	TFIDF	BM25	LM	Pagerank_LM	Prior
MAP@10	0.1393	0.2386	0.1357	0.2517	0.3534	0.4307
MAP@30	0.1545	0.2198	0.1274	0.2326	0.3152	0.3733
MAP@50	0.156	0.204	0.1232	0.2111	0.2939	0.3411
MAP@100	0.1404	0.1802	0.1106	0.181	0.2655	0.3054
MAP@300	0.1249	0.1501	0.0974	0.147	0.2165	0.2464
MAP@500	0.1153	0.1369	0.0923	0.1343	0.1974	0.2228
MAP@1000	0.1005	0.1204	0.0859	0.1187	0.1762	0.194
MAP@3000	0.079	0.0979	0.0717	0.0955	0.1455	0.1583
MAP@5000	0.0698	0.0894	0.0677	0.0878	0.134	0.1451
MAP@@ALL	0.0562	0.052	0.0351	0.0487	0.0842	0.1274

Table 2. Evaluation result of nDCG

	Pagerank	TFIDF	BM25	LM	Pagerank_LM	Prior
nDCG@10	0.0666	0.1137	0.0575	0.1098	0.1782	0.2637
nDCG@30	0.0858	0.1025	0.0558	0.0955	0.1656	0.2274
nDCG@50	0.0862	0.0984	0.0587	0.0927	0.1568	0.2143
nDCG@100	0.0978	0.0968	0.0586	0.0919	0.1515	0.2007
nDCG@300	0.111	0.1075	0.0666	0.1018	0.1598	0.2037
nDCG@500	0.1216	0.1196	0.0751	0.1133	0.1724	0.2226
nDCG@1000	0.1627	0.1488	0.0987	0.1394	0.2009	0.264
nDCG@3000	0.2585	0.2271	0.1647	0.21	0.2737	0.364
nDCG@5000	0.3152	0.2716	0.2062	0.2491	0.3105	0.4144
nDCG@ALL	0.432	0.3338	0.3041	0.3275	0.3623	0.4981

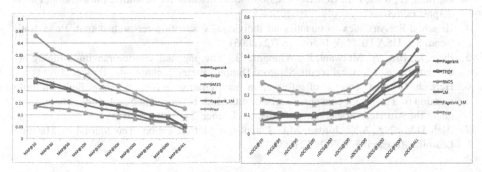

Fig. 4. Evaluation Result of MAP and nDCG

As shown in the tables and figures, we got positive results. PageRank alone (topic independent) performed the worst (as this is topic independent), while PageRank with prior achieved the best result both for MAP and nDCG (with t-test $p<0.001$). For this research, nDCG@n is a more important measure, because it can tell the degree of importance. As show in the right figure above, it's clear that PageRank with prior is always better than all other baseline methods, and the results are significant.

5   Conclusion

In this paper, we first create an author citation network with 14,222 authors as vertices and more than 223,000 citations as edges, in which full-text citation extraction (citation context) improved the performance. Then, Labeled-LDA algorithm is employed to generate topics for the full-text of publications (as a vertex) and the context of citations (as an edge), which determine the topic contribution of a specific author and a citation. After that, the importance of authors is measured by PageRank with prior and transitioning probability distributions based on each scientific topic. Finally, some authors are selected from the review or survey papers as "ground truth" for evaluating our approach compared with some other baselines. Based on the evaluation results of MAP@n and nDCG@n, we found that PageRank with prior can produce reliable high quality topic ranking results.

While there are still limitations in this paper, for instance, the context of citation located by regular expression can hardly handle all the publications, only 67.7% were extracted, because of citation format and encoding problems. So, for future work, we will try to find other methods to extract the citation context more effectively, and also optimize the ranking algorithm.

References

1. Franceschet, M.: PageRank: standing on the shoulders of giants. Communications of the ACM 54(6), 92–101 (2011)
2. Garfield, E.: Citation analysis as a tool in journal evaluation: journals can be ranked by frequency and impact of citations for science policy studies. Science 178, 471–479 (1972)
3. Garfield, E., Sher, I.H.: Genetics Citation Index. Institute for Scientific Information, Philadelphia (1963)
4. Hirsch, J.E.: An index to quantify an individual's scientific research output. Proc. Natl. Acad. Sci. USA 102(46), 16569–16572 (2005)
5. Page, L., Brin, S., Motwani, R., Winograd, T.: The pagerank citation ranking bringing order to the web. Technical report, Stanford Digital Library Technologies Project (1998)
6. Herlach, G.: Can retrieval of information from citation indexes be simplified? Multiple mention of a reference as a characteristic of the link between cited and citing article. Journal of the American Society for Information Science 29(6), 308–310 (1978)
7. Blei, D.M., Ng, A.Y., Jordan, M.I.: Latent dirichlet allocation. The Journal of Machine Learning Research 3, 993–1022 (2003)

8. Ramage, D., Hall, D., Nallapati, R., Manning, C.D.: Labeled LDA: A supervised topic model for credit attribution in multi-labeled corpora. In: EMNLP 2009 Proceedings of the 2009 Conference on Empirical Methods in Natural Language Processing, pp. 248–256. Association for Computational Linguistics (2009)
9. White, S., Smyth, P.: Algorithms for estimating relative importance in networks. In: Proceedings of the Ninth ACM SIGKDD International Conference on Knowledge Discovery and Data Mining, pp. 266–275. ACM (2003)
10. Cheng, A., Friedman, E.: Manipulability of PageRank under Sybil strategies. In: First Workshop on the Economics of Networked Systems, NetEcon 2006 (2006)
11. Rodriguez, M.A., Bollen, J.: Simulating network influence algorithms using particle-swarms: Pagerank and pagerank-priors (2006)
12. Järvelin, K., Kekäläinen, J.: Cumulated gain-based evaluation of IR techniques. ACM Transactions on Information Systems (TOIS) 20(4), 422–446 (2002)

Vertical Classification of Web Pages
for Structured Data Extraction

Long Li, Dandan Song*, and Lejian Liao

Beijing Engineering Research Center of High Volume language Information
Processing and Cloud Computing, Beijing Lab of Intelligent Information Technology
School of Computer Science, Beijing Institute of Technology, Beijing, China
lilongChinese@gmail.com, {sdd,liaolj}@bit.edu.cn

Abstract. We propose a general hierarchical vertical classification framework, which can automatically discover the inherent hierarchical structure of relationships among verticals based on flat datasets, and then build a hierarchical classifier. We conducted a set of comparison experiments to verify the performance of it, such as with flat vs hierarchical structure of relationships, as well as among different feature selection and classification methods. Experimental results show that the hierarchical classifiers built on the basis of the proposed framework make big improvements over the flat classifiers when classifying unseen web pages. Among them, the Support Vector Machine using Odds Ratio to select discriminative features performs best.

Keywords: vertical classification, automatic hierarchy, hierarchical classifiers, structured data extracting.

1 Introduction

Recent years, structured data extraction attracts many interests and shows a promising future [1–3]. However, a common hypothesis behind a great portion of these methods is that the web pages should have been classified into specific verticals (e.g., books, cameras, jobs). Classification techniques can be used for this, and we call this kind of classification task to divide web pages into diverse verticals as *Vertical Classification*.

A lot of hierarchical classifiers [4–6] have emerged. The hierarchy structure of categories used by traditional hierarchical classifiers is constructed manually or given in advance. However, to build the hierarchy structure of categories manually is time-consuming and the results are sensitive to the subjective opinions of different persons. Moreover, most of them aim at classifying the news-like documents, which apply many sentences to convey some information, into different topics or genres. None of them focus on classifying web pages containing structured data into verticals to our knowledge. We will give a comprehensive study on vertical classification methods in this paper.

---

* Corresponding author.

Y. Hou et al. (Eds.): AIRS 2012, LNCS 7675, pp. 486–495, 2012.

There are some specific issues for the vertical classification techniques which make traditional web page classifiers could't be directly applied. First, the general document representation TF-IDF doesn't perform well, owing to that all words in one page should function identically when more discriminative features are selected. In other words, the common assumption for web page classifications that the feature with high term frequency in one page is considered as more important than others [7] is no longer suitable. Second, different verticals may have semantic similarities, which bring troubles to vertical classification. For instances, the four verticals *Cameras, Cellphones, Tablets, Mp3players* have many identical characteristics and are difficult to be distinguished by general web page classification.

These issues will be systematically investigated in this paper, which reports a general hierarchical vertical classification framework. Given flat datasets whose relationship among verticals is unknown or not provided, our framework can firstly makes use of agglomerative hierarchical clustering to exploit the intrinsic hierarchical structure of relationships among verticals automatically. With a metric which is proposed to partition the hierarchical structure into levels, training data is accordingly reconstructed into hierarchy, and a hierarchical classifier is learned.

We conducted a set of comparison experiments for the design of vertical classifiers, such as with flat vs hierarchical structure of relationships, as well as among different feature selection and classification methods. Experimental results show that the hierarchical classifiers built on the basis of the proposed framework make big improvements over the flat classifiers in classifying unseen web pages, with Support Vector Machine using Odd Ratio as the feature selection strategy performs best. Furthermore, The hierarchical classifiers require only a quarter or less number of features of the corresponding flat ones, which would reduce the training time and facilitate the memory utilizing.

2   Related Work

Web page classification has attracted much attention for a long time, and a huge amount of classifiers have been employed into the task, including naive bayes classifier [4], decision trees [8], hidden naive bayes [9], support vector machines [5, 6, 10]. These classifiers are organized either as flat (non-hierarchy) form or hierarchical form, namely flat classifiers or hierarchical classifiers.

Comparing to flat classifiers, the hierarchical ones have many advantages. The main of them is that as the intrinsic hierarchy structure can factually describe the reality, it can be used to improve the performance. Ceci et al. [4] applied the hierarchy of categories into all phase of text classification to construct hierarchical classifiers, and comparison experiments were done on Yahoo, ODP and RCV1 dataset. Cai et al. [6] built a SVM classifiers based on discriminant functions, which were structured in the way that the class hierarchy was mirrored to include the inner-class relationships, on the WIPO-alpha patent collection. Dumais et al. [5] constructed a two-level hierarchical SVM classifier with two

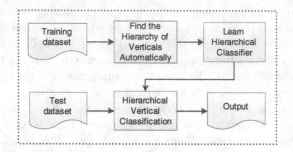

Fig. 1. Overview of the Framework

different combination rules manually. Their experiments on LookSmart showed small advantages over baseline flat models. Gentile et al. [10] proposed to turn the hierarchical SVM classifier into a Bayes classifier and the method outperformed the simple hierarchical SVM on RCV1 and OHSUMED.

In above hierarchical classification methods, the hierarchical structure of categories are manually annotated or simply referred from public directories, which requires amounts of human efforts or may not be suitable for all applications. Weigend et al. [11] employed cluster analysis to suggest an implicit hierarchy structure firstly and human assignments were applied to verify the structure. Its main drawback is that it isn't suitable for building large-scale heterogeneous web page classifiers automatically and can not keep pace with the explosive growth of web pages on the internet.

Moreover, existing web page classification researches are mainly focusing on text classification for news-like pages, and their targets are to label the contents with different topics, taxonomies or genres, not the verticals. Vertical classification, which is assumed as a prerequisite process of structured data extraction methods [1–3], is not well studied.

3   Automatic Hierarchical Classification: the Framework

We propose a framework which can learn a hierarchical classifier from flat non-hierarchy dataset automatically. It is shown in Figure 1. By using agglomerative hierarchical clustering, the intrinsic hierarchy structure of the verticals can be constructed automatically, and then the training dataset is reconstructed into hierarchical levels. Afterwards, a classifier can be learned to distinguish a sub-level vertical from other verticals within the same top vertical.

3.1   Find the Hierarchy of Verticals Automatically

In order to automatically explore the inherent hierarchical structure of verticals, an agglomerative hierarchical clustering process ("from bottom to up") should be performed ahead. In our framework, we employ a vector $\boldsymbol{v_i} = (v_{i1}, v_{i2}, \cdots, v_{iN})$ to represent a special original vertical. The value v_{ij} is the percentage of feature

f_j in vertical v_i and N is the total number of features. When one or more verticals make up a cluster, their vectors are grouped to represent the cluster. In order to figure out the two closest clusters which should be merged together, a measure of dissimilarity between vertical clusters is required. The metric to evaluate the distance between two clusters of verticals in our proposed framework is defined as:

$$Distance(\boldsymbol{A}, \boldsymbol{B}) = \max_{\boldsymbol{v_x} \in A, \boldsymbol{v_y} \in B} \|\boldsymbol{v_x} - \boldsymbol{v_y}\| = \max_{\boldsymbol{v_x} \in A, \boldsymbol{v_y} \in B} \sqrt{\frac{\sum_{j=1}^{N} (v_{xj} - v_{yj})^2}{\sum_{j=1}^{N} \sigma_j}} \quad (1)$$

where the vector $\boldsymbol{v_x}$ and $\boldsymbol{v_y}$ is a vertical of the cluster $\boldsymbol{A}$ and $\boldsymbol{B}$ respectively, and v_{xj} and v_{yj} is a feature of vector $\boldsymbol{v_x}$ and $\boldsymbol{v_y}$. The scalar σ_j is defined as:

$$\sigma_j = \begin{cases} 1, & v_{xj} \neq 0, \ or \ v_{yj} \neq 0 \\ 0, & otherwise \end{cases} \quad (2)$$

The output of most hierarchical clustering algorithms is a nested hierarchy of graphs which can be cut at a desired dissimilarity level and form a partition (clustering) [12]. As there is no versatile approach to find the right number of clusters till now, many heuristic strategies are used in the clustering procedure. Here, we employ a threshold θ to terminate the process of clustering to get a desired hierarchical structure, and the θ is defined as:

$$\theta = \frac{2 \times \sum_{x \neq y}^{M} \|\boldsymbol{v_x} - \boldsymbol{v_y}\|}{M(M-1)} \quad (3)$$

where $\boldsymbol{v_x}$ and $\boldsymbol{v_y}$ represent a pair of different verticals in the original dataset, and M is the total number of verticals, so θ corresponds to the average distance between all pairs of verticals. Consequently, the procedure of finding the hierarchy structure of verticals automatically is shown in Algorithm 1. Given the flat non-hierarchy dataset, the algorithm can iteratively build a hierarchy tree from bottom to up by merging the two closest clusters into a larger one in each iteration, and the process continues until the shortest distance of each cluster is greater than the threshold θ.

All plain texts of each web page are extracted during the preprocessing stage. The title, description and keywords fields of <META> tag and the content within <TITLE> tag provide useful information for classification and are also extracted. After that, all these texts are tokenized into words, and then are lemmatized.

Under the context of vertical classification, all words in one page function equally when selecting discriminative features, because the common assumption for general web page classifications that the feature with high term frequency in one page is considered as more important than others [7] is not suitable. For instance, one of the most discriminative feature "ISBN" of "Books" vertical appears only one or two times, however the word "girl", which is one word of a special book title and is not important, may occur more than forty times in

Algorithm 1. An Agglomerative Hierarchical Clustering Process to Find the Hierarchical Structure of Verticals Automatically

---

1: **input:** a set of flat (non-hierarchical) verticals V_f.
2: Figure out the threshold θ based on V_f;
3: Take each vertical $v_x \in V_f$ as a single cluster c_x and add c_x to the cluster set C;
4: **repeat**
5:     Calculate the minimum distance L_{min} of each two clusters in set C;
6:     **if** $L_{min} \leq \theta$ **then**
7:         Merge the two closest clusters c_A and c_B into one c_{AB};
8:         Put the new cluster c_{AB} into the cluster set C;
9:         Remove c_A and c_B from set C;
10:    **end if**
11: **until** $L_{min} > \theta$
12: **return** the cluster set C.

Amazon.com. And the general representation TF-IDF used by common web page classifiers does not perform well. So in our framework, we regard the features in one page with diverse occurrence as equivalent and a binary vector corresponding to the presence/absence of each feature is adopted.

3.2 Learn Hierarchical Classifier

In this framework, all verticals will be organized as a vertical-relational tree and each vertical node can has at most one parent. Besides, all pages can only be assigned to no more than one leaf vertical. We propose a top-down level-based approach to learn hierarchical classifiers. A classifier is built for each non-leaf node and different classifiers use different document representations. A new web page is classified by searching the vertical-relational tree top-down from root to leaves greedily. The procedure of learning the hierarchical classifier is that:

(1) for each non-leaf vertical-relational node (initially the root), select discriminative features for its sub-level vertical classification.
(2) learn a classifier to distinguish sub-level nodes.
(3) continue the learning process on each child non-leaf node, until a leaf vertical is reached.

4   Classification Techniques

4.1   Feature Selection

A web page is generally represented by a feature vector with thousands of unique terms. However, the high dimensionality could be problematic and expensive for computation. Many feature selection methods can reduce the dimensionality without decreasing and even increasing the accuracy of classification.

Odds Ratio (OR): The idea of OR is that the distribution of features on relevant documents is different from the distribution of features on uncorrelated documents. The features with high OR are considered more important than others. Mladenic et al. [13] showed that the OR method performed best over all others in their experiments.

Information Gain (IG): The IG of a feature is a metric of the expected reduction in entropy when the training data is divided into small sets owing to the feature. The entropy of a training dataset indicates the impurity of it. And, the higher IG of one feature, the more discriminative. Yang et al. [14] declared the IG was the most competitive approach in their work.

4.2   Web Page Classifiers

After more discriminative features have been selected, many machine learning methods can be used to build a function mapping between documents and categories over a set of training data [7].

Naive Bayes (NB): NB is one of most popular classifiers in text classification, due to its easiness to conduct and surprisingly effective [4]. NB is constructed by maximizing the posterior probability $P(c_i|d)$ of category c_i given a document d. It assumes that all features are independent of each other given a category, which makes the posterior probability easy to be estimated.

Hidden Naive Bayes (HNB): Although NB is simple and efficient, its hypothesis of feature conditional independence is always violated for real world data. In order to improve NB classifiers, the HNB [9] introduces a hidden parent for each node to combine the influences from all others. Jiang et al. [9] demonstrated that HNB significantly outperformed NB and other variants of NB, such as naive Bayes tree (NBTree), tree-augmented naive Bayes (TAN), and averaged one-dependence estimators (AODE).

Support Vector Machine (SVM): SVM is used for a wide rang of classification problems and often as the benchmark to evaluate other classifiers, because of its outstanding performance and generalization capability. SVM is originally designed for binary classification, as a linear SVM seeks a maximal margin hyperplane to separate a set of positive examples from a set of negative ones. In order to extend the SVM (with binary form) to multi-class case, many algorithms have been developed. A very efficient method designed by Platt [15] employs the sequential minimal optimization (SMO) to break the large quadratic programming problem into a series of smaller ones, which makes it applicable for the dataset with a large amount of features.

5   Experiments

In this section, we want to know whether the performance of the proposed framework described in Section 3 is better than the corresponding flat ones under the context of vertical classification. Among the six combinations of two feature selection methods (*i.e.*, IG and OR) and three famous classifiers (*i.e.*, HNB, NB and

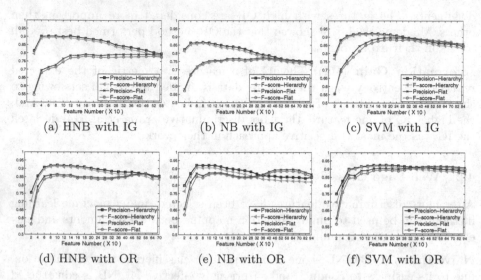

(a) HNB with IG   (b) NB with IG   (c) SVM with IG

(d) HNB with OR   (e) NB with OR   (f) SVM with OR

Fig. 2. The precision and F-score comparison of six combinations of two feature selection methods and three different classifiers with different feature numbers

SVM), we are interested in which one performs best based on the framework. The dataset used in our experiments is partially derived from the structured data extraction method by [3]. The other is collected from some popular websites, and involves 3 semantically similar verticals, including *Cellphones, Tablets, Mp3players*. In the dataset, every vertical has 6 websites with about 100 pages for each site. Four of them will be randomly selected from each vertical as the training dataset, and the rest is regarded as the testing dataset, which is unseen for the learned classifiers. In addition, each page can only belong to one vertical.

5.1 Flat vs Hierarchical Classifiers

On All Verticals: The most important experiment should be done is to investigate the effectiveness of the hierarchical vertical classifiers with respect to flat ones. In order to achieve a fair comparison, all datasets used to train and evaluate the learned hierarchical classifiers are the same as the flat ones. Besides, all experiments are conducted more than 20 times for randomly selecting the training websites and getting a more reliable results.

Figure 2 shows that all hierarchical classifiers, built on the basis of the proposed framework, outperform the flat classifiers significantly. Hierarchical HNB, NB, SVM get about 11%, 9.5%, 2.5% (11%, 9.5%, 3%) improvement respectively over flat non-hierarchy classifiers with IG feature selection method in *precision* (in *F-score*). As to OR method, hierarchical HNB, NB and SVM classifiers achieve a small improvement over the flat ones respectively in both *precision* and *F-score*. Inspiringly, only 80~100 features are needed for hierarchical classifiers to obtain the best performance, no matter which feature selection strategy is adopted. On the contrary, the flat non-hierarchy classifiers demand more than

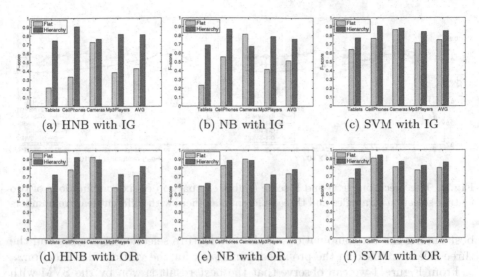

Fig. 3. The F-score comparison of hierarchical and flat classifiers of six combinations of two feature selection strategies and three different classifiers on 4 semantically similar verticals (*i.e. Tablets, Cellphones, Cameras, Mp3players*) and the average of them (AVG)

4~5 times numbers of features to achieve approximate performance. Therefore, another conclusion is drawn that the hierarchical classifiers can obtain excellent results by using a quarter or more less number of features than flat non-hierarchy classifiers, which can effectively reduce the training time and favor a memory-efficient implementation.

On Four Similar Verticals: In order to check the effectiveness of the proposed framework for dealing with semantically similar verticals (e.g., *Cameras, Cellphones, Tablets, Mp3players*), we compare the *F-score* of these verticals to find out the difference between hierarchical and flat classifiers. The results are illustrated by Figure 3.

The results show that these hierarchical classifiers outperform the corresponding flat ones and attain satisfactory results. Among the six combinations (of two feature selection methods and three excellent classifiers), the biggest improvement is gained by the HNB classifier using IG to select features, which almost increase 40%. The second is the NB classifier adopted IG as the feature selection method, which is about 25%. Besides, the hierarchical classifiers also outperform the corresponding flat non-hierarchy ones with the OR method, although the flat classifiers could achieve acceptable results.

5.2   Comparison between Hierarchical Classifiers

In Section 5.1 we can find great advantages of hierarchical classifiers comparing to the corresponding flat ones. In this section, we want to know which one perform

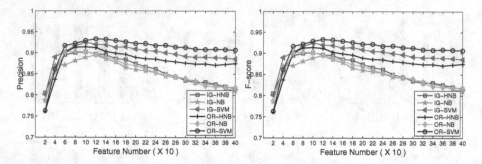

Fig. 4. The precision (left) and F-score (right) comparison of six combinations of two feature selection strategies and three different classifiers with different feature numbers

best among the six combinations of the two feature selection strategies and the three classifiers used in the proposed framework for the vertical classification.

From Figure 4 we can observe that the best result is won by the SVM with OR in both *precision* and *F-score*, when the number of features reaches 120. The SVM with IG gains an approximately good result when the feature number is 100. Consequently, SVM is the best classifier among the three hierarchical classifiers. As regards to HNB and NB, the former performs better than the latter, especially when using OR to select features. Besides, the OR method performs better than the IG method across the three classification algorithms, which is similar to the above conclusions. This phenomenon may be caused by the fact that every vertical has some unique attributes to be distinguished from others and OR method can figure out them by only considering the present features, while the IG approach emphasizes both the attributes belonging to the special vertical and the ones not in part of it when choosing representative features.

6   Conclusions and Future Work

This paper describes a general framework to build hierarchical classifiers for vertical classification automatically from flat datasets, which contains two parts. First, exploit the intrinsic hierarchy structure of relationships among verticals automatically by using an agglomerative hierarchical clustering process. Second, learn a classifier for every non-leaf node in the hierarchy tree and then form a hierarchical classifier. A series of experiments demonstrate that the hierarchical classifiers built on the basis of the proposed framework outperforms the flat non-hierarchy ones, and the internal hierarchical structure of different verticals can be used to improve the effectiveness and efficiency.

In this paper we have not investigated the effectiveness of the framework with a large number of verticals and has a deep-level hierarchy structure in nature. We will expand our framework to this kind of dataset in the future. We also have not known whether the hierarchy structure learned by our framework is better or poorer than the one built by human beings manually due to the lack of this kind of dataset currently, and we will do this in the following work.

Acknowledgment. This work is funded by 973 project (Grant No. 2013CB329605), NSFC (Grant Nos. 60873237 and 61003168), Natural Science Foundation of Beijing (Grant No. 4092037), Outstanding Young Teacher Foundation and Basic Research Foundation of Beijing Institute of Technology, and partially supported by Beijing Key Discipline Program.

References

1. Zhai, Y., Liu, B.: Structured Data Extraction from the Web Based on Partial Tree Alignment. TKDE 18(12), 1614–1628 (2006)
2. Wong, T.L., Lam, W.: Learning to Adapt Web Information Extraction Knowledge and Discovering New Attributes via a Bayesian Approach. TKDE 22(4), 523–536 (2010)
3. Hao, Q., Cai, R., Pang, Y., Zhang, L.: From One Tree to a Forest: a Unified Solution for Structured Web Data Extraction Categories and Subject Descriptors. In: SIGIR, pp. 775–784 (2011)
4. Ceci, M., Malerba, D.: Classifying web documents in a hierarchy of categories: a comprehensive study. JIIS 28(1), 37–78 (2007)
5. Dumais, S., Chen, H.: Hierarchical classification of Web content. In: SIGIR, pp. 256–263. ACM, New York (2000)
6. Cai, L., Hofmann, T.: Hierarchical document categorization with support vector machines. In: CIKM, pp. 78–87. ACM, New York (2004)
7. Ben Choi, Z.Y.: Web Page Classification. In: Chu, W., Lin, T.Y. (eds.) Foundations and Advances in Data Mining. STUDFUZZ, vol. 180, pp. 221–274. Springer, Heidelberg (2005)
8. Finn, A., Kushmerick, N.: Learning to classify documents according to genre: Special Topic Section on Computational Analysis of Style. JASIS 57(11), 1506–1518 (2006)
9. Jiang, L., Zhang, H., Cai, Z.: A Novel Bayes Model: Hidden Naive Bayes. TKDE 21(10), 1361–1371 (2009)
10. Gentile, C., Zaniboni, L.: Hierarchical Classification: Combining Bayes with SVM. In: ICML (2006)
11. Weigend, A.S., Wiener, E.D., Pedersen, J.O.: Exploiting Hierarchy in Text Categorization. IR 1(3), 193–216 (1999)
12. Jain, A.K., Murty, M.N., Flynn, P.J.: Data clustering: a review. CSUR 31(3), 264–323 (1999)
13. Mladenic, D., Grobelnik, M.: Feature Selection for Unbalanced Class Distribution and Naive Bayes. In: ICML, pp. 258–267. Morgan Kaufmann Publishers Inc., San Francisco (1999)
14. Yang, Y., Pedersen, J.O.: A comparative study on feature selection in text categorization. In: ICML, pp. 412–420. Morgan Kaufmann Publishers Inc., San Francisco (1997)
15. Platt, J.C.: Fast training of support vector machines using sequential minimal optimization. In: Schölkopf, B., Burges, C.J.C., Smola, A.J. (eds.) Advances in Kernel Methods, pp. 185–208. MIT Press, Cambridge (1999)

Grid-Based Interaction for Exploratory Search*

Hideo Joho[1] and Tetsuya Sakai[2]

[1] Faculty of Library, Information and Media Science
University of Tsukuba
hideo@slis.tsukuba.ac.jp
[2] Microsoft Research Asia
tetsuyasakai@acm.org

Abstract. This paper presents a grid-based interaction model that is designed to encourage searchers to organize a complex search space by managing $n \times m$ sub spaces. A search interface was developed based on the proposed interaction model, and its performance was evaluated by a user study carried out in the context of the NTCIR-9 VisEx Task. With the proposed interface, there were cases where subjects discovered new knowledge without accessing external resources when compared to a baseline system. The encouraging results from experiments warrant further studies on the model.

1 Introduction

Exploratory search systems are designed to help people with ill-defined information needs. People's information need tends to be ill-defined when they engage in a complex task. Exploratory search systems can be categorized into two groups based on accessibility to an entire document collection. When one has an access to an entire document collection, some level of content analysis can be performed and a set of common properties can be identified. These common properties are then used to organize the collection for users to explore. Examples of this type of exploratory search systems are Flamenco [9] and mSpace [7]. Another type of exploratory search systems does not assume an access to an entire collection. Therefore, support is often given to users dynamically based on their search history and some analysis of search results. Examples of this type of exploratory search systems are AspecTiles [2], Aspectual browser [8], and Slice n' Dicer [3]. This paper proposes a new interaction model for the latter type of exploratory search systems, and studies its effect on task performance, information seeking behavior, and user perceptions.

1.1 Grid-Based Interaction Model

The interaction model proposed in this paper was derived from two lines of research in interactive information retrieval (IIR). One is an instance finding task that has been studied in a series of Interactive Tracks in TREC [1]. In this task, searchers were asked

---

* An early unrefereed version of this paper was presented at the NTCIR-9 workshop [4]. This work was funded by Microsoft Research Asia under the 7th MSR CORE Project.

Y. Hou et al. (Eds.): AIRS 2012, LNCS 7675, pp. 496–505, 2012.

to find instances of events, achievements, countries, technologies that matched a particular condition. The idea was to go beyond a conventional document retrieval which was often not sufficient to study interactive aspects in IR. For example, searchers received no reward to find duplicated instances in the task. A key element of this task was that it used instances as an axis of information seeking process. Instances can be a generic yet powerful property to search, organize, and analyze a given topic and its information space. For example, Barack Obama, David Cameron, and Hu Jintao are an instance of world leaders. Similarly, iOS, Android, and Windows Mobile are an instance of mobile operating systems. Instances tend to co-occur in relevant documents since they are often contextually related. Therefore, we consider that the notion of instances plays an important role in supporting exploratory search tasks.

Another line of research is called aspectual search [2,8]. Aspectual search emphasizes to find aspects to complete an exploratory search task. For example, writing a biography of a world leader requires to find and select what aspects of the leader should be included. Making a summary of a large event like Olympics also depends on exploration of potential angles. These aspects or angles can be relatively simple such as time and location in some cases. However, when a topic becomes complex, determining appropriate aspects is not trivial. Again, a key element of this search was that it used aspects as an axis of information seeking process. Aspects can be seen as a property of instances. For example, age, education, and political agenda are an aspect of world leaders. Similarly, price, required memory, and hardware compatibility might be an aspect of mobile operating systems. If instances are a vertical axis, aspects can be seen as a horizontal axis in exploration and organization of a search space.

As can be seen, these two lines of research are closely related, and thus, can be integrated into a single model. This was our motivation and we call it a *grid-based interaction model*. Existing studies did not make a clear distinction of the two kinds of notion. In this sense, our model was an extension of the research which looked at instance finding and aspect finding. An important consideration here is that searchers are unlikely to find instances and aspects at the beginning of search. Instead, they are more likely to discover these elements of a given topic as they make progress in a search task. Therefore, we need an interaction model that can support exploration of a search space using the notion of aspects and instances.

We consider that an expression of information needs is crucial for supporting exploratory search, and thus, devised a syntax to formulate a grid-based query as shown in Figure 1. Dimensions in a query space are separated by a bar sign (|), which means that this syntax can represent as many dimensions as needed. However, in this paper, we only consider two dimensions. All the terms placed before the bar sign represent the first dimension while those after the bar sign represent the second dimension. Keywords in individual dimensions are separated by a comma (,). When more than one keyword is used between commas, such term will be taken as a phrase. For example, the query, 'black,

```
term [, term...] | term [, term...]
```

Fig. 1. Query syntax for a grid search

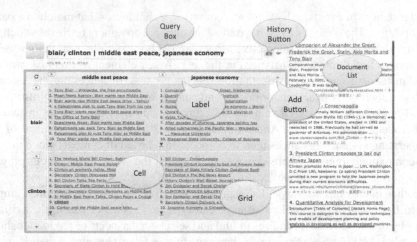

Fig. 2. Grid Search UI (Query: Blair, Clinton | Middle East Peace, Japanese Economy)

white | cat, dog', will represent four search queries such as 'black cat', 'black dog', 'white cat', and 'white dog'. The query, 'david cameron, barack obama | approval rate', will yield two queries such as <"david cameron" "approval rate"> and <"barack obama" "approval rate">. As can be seen, the syntax is simple and intuitive. A user study suggests that participants did not have a major problem to operate a search task with the proposed syntax. To link with the notions of instances and aspects described earlier, the first query had aspects first then instances, while the second query had instances first then aspects. Therefore, the proposed syntax does not explicitly define the order of dimensions expressed in the query. It is up to a designer (or possibly user) of search systems. The next section describes a grid search interface which can be seen as one possible case of implementing the proposed query syntax.

A search interface was developed based on the interaction model discussed above. The interface is composed of three main areas, namely, Query Box (1), Document List (3), and Grid (4), as shown in Figure 2. When a user submits a query in Query Box using the proposed syntax, both Document List and Grid are shown with search results. There is a History button next to Query Box which allows users to revisit and rerun previous queries submitted in a search session.

The Grid area consists of Labels (5) and Cells (6). Labels are derived from individual terms given in Query Box. Given that a user submits a query Blair, Clinton | Middle East Peace, Japanese Economy to the interface, terms on the left part of a query (i.e., Blair and Clinton) are placed as a row label (Green Line), while terms on the right part (i.e., Middle East Peace and Japanese Economy) are placed as a column label (Red Line). Moreover, the terms in Query Box and Labels are synchronized, and users can edit either of them to reformulate the query. Double-clicking a label allows a user to change the keywords. Add Button (7) allows a user to append a new label to the grid. Full-text is shown in a pop-up window when the title of the result is clicked from a cell or document list. Cells (6) show search results of the grid-based queries as shown in Figure 3. Each cell shows the result of a particular combination of terms

Fig. 3. Query and Grid

derived from the query. Taking our earlier example query, Cell 1 shows a search result of a query,<Blair "Middle East Peace ">, Cell 2 shows that of <Blair "Japanese Economy">, and so forth. As a consequence, one grid presents the results of four sub-queries in this case. Grid does not have to be $n \times n$, and can be $n \times m$. Furthermore, the order of rows and columns can be changed by dragging a label.

When a cell is selected, Document List shows search results of a particular sub-query (e.g., <Blair "Japanese Economy">). When multiple cells are selected, Document List shows search results by merging selected individual cells. Currently, the merged ranking is based on redundancy and round-robin. In other words, those documents that are commonly retrieved by sub-queries are ranked higher than those are retrieved once. Tie-documents are ranked in a round-robin manner across the retrieved documents of individual sub-queries. When a query is submitted or reformulated in Query Box or via Labels, Document List shows a set of retrieved documents by merging the results of all sub-queries. This is equivalent to select all cells in Grid. A more effective way to generate the document list is of our future work.

2   Experiment

The evaluation of the proposed search interface was carried out as a user study in the framework of NTCIR-9 VisEx Task[1]. Since the detail of experimental design is given in the overview paper of VisEx Task [5], this section only gives a brief summary of how a user study were carried out. It should be noted that a team who participated in VisEx Task will be referred to as *participants*, while people who participated in the user study as a subject will be referred to as *subjects* in the rest of this paper.

The task required participants to develop a user interface that can support either an event collection task or trend summarization task or both. An example of the event collection task is to find as many instances of airplane crashes that occurred in Asia as

[1] http://must.c.u-tokyo.ac.jp/visex/hiki.cgi?FrontE

Table 1. Number of nuggets found (N=20)

	Event collection		Trend Summary	
	Baseline	Experimental	Baseline	Experimental
Mean	7.6	7.4	7.6	6.9
SD	3.5	2.8	3.1	2.8
Min	2.0	4.0	3.0	3.0
Max	16.0	15.0	14.0	12.0

possible. An example of the trend summarization task is to find as many information that describes a prime minister's approval rate as possible. As can be seen, both tasks were exploratory and required a number of iterations of search to complete them. Subjects were given up to 50 minutes to complete the tasks. A baseline system was provided by the organizer and all systems used the API of a common backend search engine. Subjects were given an instruction (6 slides) of how to use the systems and a training session to familiarize with tasks and behavior of the experimental system. In the system instruction, an emphasis was made to encourage people to organize a problem space using a $n \times m$ notion such as *people × year*, *place × event*, and *things × attributes*.

3   Results

This section presents results of our preliminary analysis on the two exploratory tasks of VisEx. It should be noted that our analysis is intentionally *qualitative* due to the experimental design. Therefore, no statistical test is performed unless otherwise stated. Finally, please note that we use the term *nugget* [6] to refer to a unit of information required to collect in individual tasks, which might be different from traditional use.

Task Performance. Both tasks asked subjects to collect as many relevant nuggets as possible within allocated time. Therefore, an overall task performance can be measured by the number of nuggets found. The results are shown in Table 1. If you look at the third row of Table 1, the performance was comparable between the baseline and experimental systems in the former task, while the difference between the two systems was found to be larger in the latter task. A similar trend was found in the maximum number of nuggets found by subjects. A grand average number of nuggets found by subjects was 7.5 (SD: 3.1) and 7.2 (SD: 2.9) for Event collection task and Trend summarization task, respectively (not shown in the table).

   The next analysis looked at whether or not a particular topic was easy (or difficult), or a particular subject performed really well (or bad). The results are shown in Figure 4. For the topic breakdown of the number of nuggets, the X axis represents Topic 1 to 4 of the two task while the Y axis represents the number of nuggets found. Each topic line has five data points which correspond to five subjects. A horizontal bar is the median value of the five data points. Please note that Topic 1 of Event collection task has nothing to do with Topic 1 of Trend summarization task. Similarly, for the subject

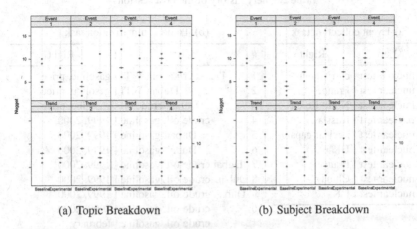

(a) Topic Breakdown   (b) Subject Breakdown

Fig. 4. Breakdown of the number of nuggets found

breakdown, the X axis represents Subject 1 to 5, and data points are the number of nuggets found by each subject across four topics. Again, Subject 1 in Event collection task is a different person from Subject 1 in Trend summarization task. As for the topic breakdown, no obvious pattern was observed from the analysis. Topic 3 of Event collection task seems to have the best performance in both systems, but the data range is large, so this is not necessarily the easiest topic. Subjects with the baseline system appear to struggle in Topic 2 of Event collection task while Topic 1 appears to be the most difficult one by subjects of the experimental system. In Trend summarization task, fewer noticeable pattern was observed. As for the subject breakdown, it appears that subjects' performance varies over the system groups as well as the task groups. All groups seem to have a good performer and poor performer, some had a large difference across topics, while others had a similar performance.

Finally, we looked at whether or not we had order effects on subjects' performance. There were at least two reasons for us to investigate the effect. First, since this was the first time for subjects to carry out a search task with the grid search interface, their performance might increase as they got used to the system. Second, since individual tasks lasted 50 minutes, we suspected that there might be fatigue effect on their performance towards later topics. Pearson's correlation shows a significant correlation ($p \leq .003$) between the number of nuggets and topic order, but coefficient was $r = .44$ which means the contribution ratio is below 20%. Therefore, an order effect appears to be weak.

To summarize the results of task performance, subjects with the experimental system appear to achieve a comparable performance to the baseline system in Event collection task, while there seems to be some factor in the experiment system that caused a performance loss in Trend summarization task. These observations did not seem to be influenced by a particular topic, subject, nor order, although their interaction effect might exist. The following sections look at information seeking behavior and subjective assessments to gain a further insight into these results.

Table 2. Query history of the best sessions

(a) Event collection task				(b) Trend summarization task	
#	Left	|	Right	#	Left | Right
1	nuclear test | carried out			1	Dubai crude oil, WTI | gasoline, price
2	nuclear test | France			2	Dubai, WTI | gasoline, price
3	nuclear test | North Korea			3	Dubai, crude oil | gasoline, price
4	nuclear test | Russia			4	crude oil, gasoline | 1999, 2000
5	nuclear test | nuclear-capable			5	Dubai, gasoline | 1999, 2000
6	nuclear test | UK			6	Dubai, gasoline | 1999, 2000
7	nuclear test | France			7	Dubai crude oil, gasoline | 1999, 2000
8	nuclear test | China			8	Dubai>crude oil, gasoline | 1999, 2000
9	nuclear test | UK			9	Dubai, crude oil, gasoline | 1999, 2000
				10	crude oil, gasoline | 1999, 2000
				11	crude oil, gasoline | february
				12	Dubai, crude oil | gasoline, price
				13	Dubai, crude oil | gasoline, price, 1999
				14	Dubai, crude oil | gasoline, price, 1998
				15	decline | 2000
				16	decline | 2000, Dubai

Information Seeking Behavior. An advantage of the proposed query syntax was that it allowed a user to express a complex search space as they made progress in exploratory search tasks. Therefore, we first looked at how subjects formulated and reformulated their queries during the VisEx tasks. Query history of the *best performing session* (i.e., the session with the largest number of nuggets found) using the experimental system was shown in Table 1(a) and Table 1(b) for Event collection task and Trend summarization task, respectively. Queries are a translation from the original language used in the experiment (i.e., Japanese).

Table 1(a) shows the queries for the topic which asked for event information (e.g., time, location, people) about nuclear tests all over the world. The first query appears to try to retrieve documents that includes a text like "carried out a nuclear test". In fact, the subject found several nuggets from this query. Then, except Query #5, the rest is a combination of "nuclear test" and a country name. In other words, the grid did not really expand as the task developed. The number of nuggets found by those queries was mixed. More importantly, those queries with country names could have been expressed as nuclear test | France, North Korea, Russia, UK, China. Table 1(b) shows the queries for the topic which asked for trend information about the price of crude oil and regular gasoline in Dubai. This topic was more complex than the previous example, since it explicitly asked for three dimensions: time, price, and oil type. Using the proposed syntax, users would have to somehow divide them into a set of two dimensions. Such attempt can be found in Query #4 and #11 where the subject focused on oil type and year. However, it seems that the results were not satisfactory without a location in the query, and thus, other queries contains Dubai in either side of the query.

The last analysis on information seeking behavior examined to what extent subjects accessed external sources to complete tasks. In the VisEx tasks, subjects were allowed

Table 3. Perception of tasks (N=20) (1: Strongly Agreed, 4: Either, 7: Strongly Disagreed)

	Event collection		Trend Summary	
	Base.	Exp.	Base.	Exp.
Satisfaction	3.0	3.0	4.0	3.0
Difficulty	3.5	4.0	4.0	3.5
Time	4.0	3.0	4.0	3.0
Exhaustive	4.0	3.0	4.0	4.0
Resource	2.0	3.0	4.0	3.0

to access external sources to support their tasks, although none of the nuggets found in external sources counted in performance measures. The results are as follows. There were 10 (Baseline) and 5 (Experimental) subjects who answered yes to the question in Event collection task, while there were 3 (Baseline) and 4 (Experimental) in Trend summarization task. As can be seen, there was a noticeable difference between the two systems in Event collection task while the frequency was comparable in Trend summarization task. Given that an overall task performance of Event collection task was comparable between the two systems, subjects with the baseline system appeared to need more support outside the given system than the experimental system. The motivation for accessing external sources varied. The post-search questionnaire established that subjects sought for a detail of a particular event, definition of a technical term, or information needed to judge relevance of nuggets.

To summarize the results of information seeking behavior, we have observed cases where the proposed syntax could be effective to organize a search space. There were cases where subjects organized a query space two-dimensionally (e.g., oil types × year in Table 1(b)). However, the frequency of effective use of the proposed syntax was relative low. As for relevance judgements on document surrogates, the experimental system appeared to increase a chance of finding nuggets in click-through documents in Event collection task. Finally, subjects tended to need an access to external sources in the baseline system when compared to the experimental system.

User perceptions. The last part of our analysis looked at subjects' perceptions on tasks and systems. Subjective assessments were captured by a 7-point Likert scale where subjects indicated a degree of agreement to a given statement. For example, a statement was "The task I performed was complex" and the scale were Strongly Agree (1), to Either (4), to Strongly Disagree (7). Task perceptions were captured after every topics, while system perceptions and interaction perceptions were captured after all topics.

The results of task perceptions are shown in Table 3. The numbers in Table 3 are a median of corresponding data. As can be seen, we did not observe a large difference in any aspects of task perceptions. However, a topic-breakdown of the data (not shown) suggests that the difference in Satisfaction in Trend summarization task is likely to be due to Topic 3 and 4. On the other hand, the difference in the perception of time (You had sufficient time to complete a task) seems to be consistent over all topics in both tasks. The difference in the perception of resources (You found a sufficient amount of news articles to complete a task) is likely to be due to Topic 1 and 3 in Event collection task. Little pattern was observed in the rest of questions.

Another question we asked in the post-search questionnaire was whether or not subjects discovered new knowledge which was somehow unexpected, during the task. The results are as follows. There were 1 (Baseline) and 16 (Experimental) subjects answered yes to the question in Event collection task, while there were 10 (Baseline) and 8 (Experimental) in Trend summarization task. As can be seen, there was a noticeable difference between the two systems in Event collection task, while the number was comparable in Trend summarization task. Examples of discovery reported by subjects include an association between two events, varied amount of information across countries, lack or bias of information in news articles, as well as topics themselves. Finally, the results of subjects' perceptions of the systems will be briefly reported. In Event collection task, subjects appeared to find the baseline system easier to learn and to operate than the experimental system. There was a clear trend in the assessments of functionality. Additional comments suggest that subjects wanted an ability to submit a standard query, to move to next 10 results within a cell, and to sort documents by date. Response speed seems to be acceptable. All subjects seemed to feel some level of frustration during the task, but the variance was much larger in the experimental system than the baseline.

To summarize user perceptions, we did not observe a large difference between the two systems in terms of the perception of tasks although some values varied over topics. However, with the experimental system, subjects tended to encounter new knowledge during the tasks when compared to the baseline system. As for the perception of systems, subjects seemed to find it more familiar to the baseline system than the experimental system. Subjects expressed several features that they would like to have in the experimental system, which can be considered for further development.

4   Concluding Discussion

A new grid-based interaction model was proposed, and the performance of the grid-based search interface was measured by three aspects in our study: task performance, information seeking behavior, and subjective assessments. This section first summarizes the major findings of the experiment and discuss their implications for further research and development of the grid-based interaction model.

The overall task performance of the two systems was comparable in Event collection task. However, we observed some positive signals in the experiment. Based on the frequency of external source access and of discovering new knowledge, the grid-based interface might facilitate the exploration of a document collection through analytical search process during the tasks. It might be possible that the performance of the baseline system was actually due to a frequent access to external sources. We need more analysis to exploit this aspect. We also obtained ideas to improve the current implementation of the grid-based interaction model. The query analysis of the best performing session suggests that there were cases where the proposed query syntax can be effective to represent a complex search space. However, subjects did not appear to take advantage of the syntax. Since our tasks were not simple like home page finding, it is possible that subjects were focusing on finding nuggets than effectively leveraging the potential benefit of the syntax. Thus, more examples should be given in the instruction, and a step-by-step tutorial might be needed in a training session. A comparison of query formulation process between the two systems should give us a better idea of how exactly such instruction should be formed.

The overall task performance of the experimental system appeared to be lower than the baseline system in Trend summarization task. We did not observe a particular topic or subject strongly affected the average performance. Furthermore, several aspects such as the successful click-through rate, frequency of accessing external sources, and frequency of discovering new knowledge seem to be comparable between the two systems. We speculate that information needs often formulated in Trend summarization task require more than two dimensions to express, which was not supported well in the current implementation of the interface. We intentionally limited the dimension size to two, but this could cause an extra effort to subjects to divide a search space to a set of two dimensions. This was exemplified in the query reformulation of the best session in this task. In short, this task was more complex than the current search interface was designed to support. It is technically possible to expand the proposed query syntax to accept more than two dimensions. However, this would require further consideration regarding how to present search results in a way that they make sense to searchers. More fundamentally, we need to study how to guide searchers to divide high-dimensional complex search into sub spaces, and how to support such tactics using the grid-based interaction model. These are all interesting research questions to pursue as future work.

References

1. Dumais, S., Belkin, N.: The TREC interactive tracks: Putting the user into search. In: Voorhees, E.M., Harman, D.M. (eds.) TREC: Experiment and Evaluation in Information Retrieval, pp. 123–153. MIT Press (2005)
2. Iwata, M., Sakai, T., Yamamoto, T., Chen, Y., Liu, Y., Wen, J.-R., Nishio, S.: AspecTiles: Tile-based visualization of diversified web search results. In: Proceedings of ACM SIGIR 2012 (2012)
3. Joho, H., Jose, J.M.: Slicing and dicing the information space using local contexts. In: Proceedings of the First Symposium on IIiX, pp. 111–126 (2006)
4. Joho, H., Sakai, T.: Grid-based interaction for ntcir-9 visex task. In: Proceedings of NTCIR-9, pp. 533–540 (2011)
5. Kato, T., Matsuhita, M., Joho, H.: Overview of the VisEx task at NTCIR-9. In: Proceedings of the Ninth NTCIR Workshop Meeting. NII, Tokyo (2011)
6. Pavlu, V., Rajput, S., Golbus, P.B., Aslam, J.A.: IR system evaluation using nugget-based test collections. In: Proceedings of ACM WSDM 2012 (2012)
7. Schraefel, M.C., Wilson, M., Russell, A., Smith, D.A.: mspace: improving information access to multimedia domains with multimodal exploratory search. The Communication of ACM 49, 47–49 (2006)
8. Villa, R., Cantador, I., Joho, H., Jose, J.M.: An aspectual interface for supporting complex search tasks. In: Proceedings of the 32nd ACM SIGIR Conference, pp. 379–386. ACM (2009)
9. Yee, K.-P., Swearingen, K., Li, K., Hearst, M.: Faceted metadata for image search and browsing. In: Proceedings of the SIGCHI Conference, pp. 401–408. ACM (2003)

Analyzing the Spatiotemporal Effects on Detection of Rain Event Duration

Jyun-Yu Jiang, Yi-Shiang Tzeng, Pei-Ying Huang, and Pu-Jen Cheng

Department of Computer Science and Information Engineering
National Taiwan University, Taiwan
{b98902114,r00944020}@ntu.edu.tw, {d96004,pjcheng}@csie.ntu.edu.tw

Abstract. There has been significant recent interest in using the aggregate information from social media sites to detect and predict real-world phenomena. Temporal and geographic effects are often considered as two possible impact factors on detection of rain event from microblog data. However, the actual contribution of them to rain event detection has yet to be defined. To investigate this issue, one method considering overall effects of time and geography is proposed for detecting the rain event. Our analysis implies that the way people post tweets changes dynamically during a day. The number of tweets grows from the early morning and peak at midnight. Besides, distribution of the population and user responses to the rain event are both not the same in different regions. Our findings therefore suggest that temporal and geographic effects may play an important role in the detection of rain event. We also apply our strategy to forecast the rain events. Our results show that our strategy performs well both in detecting and predicting events of rain. Comparative analysis with existing methods is also presented to demonstrate the effectiveness of our method. Our proposed scheme is therefore practical and feasible to be deployed in the real world.

Keywords: Event detection, Event duration, Microblog, Temporal, Geographical.

1   Introduction

When people plan to travel to somewhere, the first thing comes up to their mind might be how the weather is there (e.g., Is it raining now? Or will it rain in the near future?). The weather forecast for a city is summarized from different regions which belong to this city. It cannot reflect the regional variation individually. For example, in a city, it might rain in some places while it might not rain in other places. Or some places may be located at the boundary between two cities. However, most weather stations only provide weather forecast for cities or famous scenic spots. There are no weather data for local areas which lie in the city. It is therefore infeasible for people to obtain exact weather conditions for local regions.

That is now changing. Data from increasingly popular online social network sites (e.g., Twitter) allow us to study the detection of rain event duration in real

Y. Hou et al. (Eds.): AIRS 2012, LNCS 7675, pp. 506–517, 2012.

time in a way that is both fine-grained and massively global in scale [14]. Twitter, as a popular microblogging service, has become a new information channel for users to receive and share information. As of March 2012, there are more than 140 million active users and over 340 million tweets are created and redistributed a day[10]. Compared to news or blogs articles, Twitter messages(tweets) have 140-character-message limit in length, resulting in a short and ungrammatical textual feature. Twitter users tend to use abbreviations and acronyms (e.g., IC refers to I see, BTW refers to By the way). To detect the rain events by using tweets collected from Twitter is therefore a real challenge due to the heterogeneous and noisy nature of the data. On the contrary, such limitation enables users to update information instantly. With the popularity of mobile device, Twitter users worldwide act as a group of sensors, forming a social sensor network to share what is happening around (e.g., tsunami, rain, personal status), making it possible to real-time report the rain event which happened anywhere at any time.

However, except the limitations of Twitter in nature mentioned above, the rain event detection might still be potentially influenced by many external factors (e.g., geo-location, time, human behavior). By considering the time and location information, we can detect target events and estimate location of target events. As a tweet is often associated with a post time and a geo-location, we can detect when and where a rain event happens. For example, a user might make a tweet such as "Now it is raining" at 7:13pm on December 24. Consequently, if a rain event happens in an oceanic area, it is more difficult to locate it precisely from tweets. It also becomes more difficult to make good estimation in less populated areas. These two cases imply that practicability of detection of rain events mainly relies on the number and spatial dispersion of Twitter users. The way people post tweets changes dynamically during a day. The number of tweets grows from the early morning and peak at midnight. It should be noted that tweets around the time and geographically close to such areas would be considered alternatively as approximate indicators to detect rain events happened in such queried areas at a given time.

As mentioned above, we conjecture that the spatial and temporal features may play the important role in rain event detection. However, to the best of our knowledge, there are no previous research studies targeted to a spatiotemporal issue in detection of rain events. Thus, it remains unclear to what extent and in what way the effects of time and geography would be imposed on the detection of rain event. In this paper, we therefore focus on understanding the influence of time and geographic on the detection of rain event by aggregating Twitter messages (tweets). Our study provides clear evidences that the spatiotemporal feature is an essential factor in detection of rain event duration.

2   Related Work

In the literature, several approaches are proposed to detect events. Allan [1] studied the topic detection and tracking from documents. Allan et al. [3] and

Yang et al. [15] also use documents to do the on-line event detection. They calculate similarity among the existing documents and new coming one by the incremental clustering to determine a new generated event. Besides, there are many previous works studied the event detection on web pages. However, their proposed methods might not be applicable to deal with the detection of the rain events from microblogs. Microblogs are often shorter and more frequent updated than documents and blogs. Therefore, conventional statistic-based term weighting strategies like frequency might not reliable in analyzing microblog contents.

Some researches studied the event detection problem from the view of signal by analyzing the frequency of the time series data. Chen et al. [5] and Jianshu et al. [9] consider words on Twitter and tags of photos on Flickr as the energy, the events are then detected by analyzing the energy distribution with wavelet transform. He et al. [7] analyzed words in both time and frequency domain with Fourier transform. Such methods can detect well while signal alters suddenly. However, the rain events usually have the duration such that we need not only detect the beginning but the ending whose variation is not obvious. Cataldi et al. [4] introduced the aging theory to emerged terms and group them into some topics. Sakaki et al. [12] studied whether users observe the effect of earthquake and locate it in a probability-based approach. The same idea can also be applied to other short length document. Teevan et al. [13] traced the trajectory of storm by the query logs of a search engine.

Various studies have been made of the analysis of microblogging data (e.g., Twitter) from spatial and temporal perspectives. Sakaki et al. [12] introduced a concept of social sensors to detect earthquake event in real time. They examined the time-series data to create a temporal model to calculate the probability of an event occurrence. Spatial models such as Kalman filtering and particle filtering are then proposed to estimate the locations of events. Java et al. [8] considered both geographical and topological properties from twitters to analyze the distribution of users and their tweets. MacEachren et al. [11] used the geographical information of tweets to visualize the location and content of tweets.

Even different from general events, weather events are usually related to regional and local information [13]. Cox and Plale [6] used the Twitter data to improve the weather observation. Sakaki et al. [12] determined whether a user observes the earthquake. But the weather events are always with the duration such that we cannot detect only the happening of an event like first story in topic detection and tracking problem (TDT problem) [2] but all process of the event. Unlike the traditional TDT problem, the duration of rain events are usually shorter. The strength of signal is also weaker than usual events in the later stage in the duration.

3   Modeling Signals

Our goal is using the twitter data to determine whether it rains or not for anywhere and anytime. For any locations we are interested in their weather conditions, we collect tweets nearby them and use corresponding tweets to create their own signals so as to detect the rain events. In the rest of this paper, we use L_j and T_i to represent locations and tweets, respectively.

3.1   Uniform Weighted Signal

As shown in Figure 1, we segment time line into equal size. The value in each time slot is the summation of $score(T_i)$, where T_i is tweets posted in the period and $score(T_i)$ is defined as follows:

$$socre(T_i) = 1 \ \text{if } T_i \text{ is related to rain, otherwise } 0$$

The score function preserves all rain related tweets and views them identical. In the implementation, we construct a classifier trained by support vector machine (SVM) to filter out the tweets not talking about rain. The signal in Figure 1 now reflects the probability of the whether it rains in L_j in each time slot to a certain extent. The higher the value in a time slot, the higher the chance that it rains in L_j at that time.

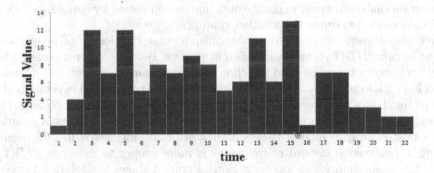

Fig. 1. Signal example

3.2   Temporal and Geographical Weighted Signal

We furthermore take time and geographic factors into consideration.

Temporal Aspect. For the same location but different time, we observe that the number of rain-related tweets in a day changes significantly (Figure 2). Many twitter users like to post tweets in the night and the number of tweets drops dramatically in the early morning. The phenomenon probably makes detecting rain event in those "inactive" period hard. Moreover, the durations of rain events are often less than a day, also highlighting the importance of daily dynamics.

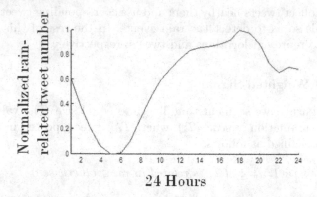

Fig. 2. Distribution of rain-related tweets in a day

Geographical Aspect. For the same time but different locations, we also find the similar situation. Figure 3 is the sampling tweets distribution of several locations. In this example, we find the distributions in these locations are not uniform. Tweets in some sub-regions are quite dense while some are very sparse. The serious unbalance may make the outcome be dominated by dense sub-regions and thus can't determine the weather condition objectively.

We also observe an interesting phenomenon that the degree of people's interest in rain($RID(L_j)$) varies in different regions. Before further discussing the relation between them, we need to define how to measure the "degree of interest" first. For each location L_j, we calculate the number of rain related tweets, divided by the number of the tweets in each raining time slot, and then set $RID(L_j)$ by averaging the values over all of the raining time slots. The measurement assumes users post more rain related tweets if they are more interested in rain events. Using normalization instead of frequency is more proper to measure $RID(L_j)$ since the population varies among regions. Figure 4 shows the relation between $RID(L_j)$ and their corresponding raining frequency. After taking logarithm, we find they have negative correlation with R^2 =0.73. Surprisingly, it quite fits power law. The finding suggests if it seldom rains in a region, the rain events tend to catch one's eye. Thus, more information about weather condition will be shared on microblogging. The experiment demonstrates how geographical factor influences the signals in Figure 1 again.

Weighted Score Function. All of above observations suggest temporal and geographical factors affect signals, we therefore tune the score function $score(T_i)$ to reflect their influences. The updated score function is named weighted score, donated as $WScore(T_i)$, and defined as follow:

$$WScore(T_i) = Tmp(T_i) \times Geo(T_i) \times score(T_i),$$

where $Tmp(T_i)$ and $Geo(T_i)$ are the temporal and geographical weights, respectively.

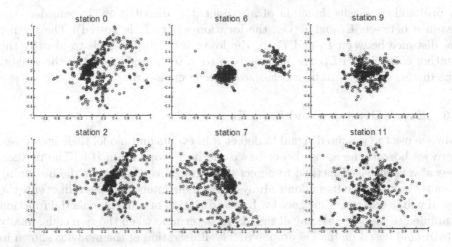

Fig. 3. The distribution of users for six stations. Each station is located at (0,0)

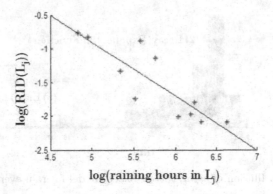

Fig. 4. The relation between the frequency of rain and the degree of users' interest in rain

$$Tmp(T_i) = \frac{1}{\text{the average number of the users who are interested}}$$
$$\text{in rain in the time slot when } T_i \text{ is posted}$$

$$Geo(T_i) = \frac{(1 + dist(T_i, L_j))^\alpha}{\#tweets \text{ in the sub-region where tweet is posted}}, where \; \alpha \leq 0.$$

In our weighting function $Tmp(T_i)$, "the users" who are interested in rain are those users posting rain related tweets, which can be learned from training data. $Geo(T_i)$ contains two parts. In the denominator, we divide location L_j into 4 by 4 sub-regions. Then a tweet T_i is normalized according to the population of a sub-region where it is posted. The numerator is a function proportional to probability density function of a exponential distribution. It considers the distance between L_j and T_i(i.e., the location where T_i is posted). The farther the distance between L_j and T_i is , the lower reliability of T_i is to identify the weather condition of L_j. The parameter α ($\alpha \leq 0$) is used to control the weight. The distance factor will be emphasized if we decrease α.

3.3   Event Detection and Modeling

Now we use the weighted signal to detect rain events and model their life cycles. Here we borrow the aging theory based method proposed in [14]. The method uses a wavelet based method to detect the burstiness of signal as the beginning of rain events. Since users don't always keep their interest in the rain event, the signal will decrease as time goes by. In this duration of rain, exponential function is applied to model the decay of signal. To determine when the rain ends, finally a threshold based method is proposed. The illustration of the model is shown in Figure 5.

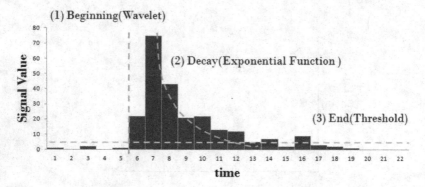

Fig. 5. Illusion of aging theory based model for rain event

4   Experiments

To evaluate the performance of our system, we gather the weather information(rain event) from thirteen weather stations in American and tweets posted around these stations(20 miles × 20 miles) over a period of three months, from November 1, 2011 to January 31, 2012. We measure the performance by three-fold cross validation on four indicators including precision, recall, F1-score and accuracy. The definitions of four indicators are defined as follows:

	Rain(Actually)	No Rain(Actually)
Rain(Reported)	A	B
No Rain(Reported)	C	D

$$precision = \frac{A}{A+B}$$

$$recall = \frac{A}{A+C}$$

$$F1 - score = \frac{2 \times precision \times recall}{precision + recall}$$

$$accuracy = \frac{A+D}{A+B+C+D}$$

We compare our performance with a threshold based method[14]. If the value of uniform weighted signal in any time slot is larger than the threshold, then the method judges the time slot as raining. We hope the system alarms everytime when it rains, but we do not expect many false alarms. Therefore, F1-score is adopted as the main indicator. Results are shown in Table 1. We can see that the performance of T-Signal is improved on recall. When we add the time factor into our model, the more desolate time slot will be more sensitive and more raining slots can be detected. Inversely, G-Signal improves uniform-Signal on precision. It is caused by weighting tweets with the geographical factor, and thus closer users have larger weights. Then the model can detect events more exactly. Overall, our experiments show that the spatiotemporal factor is helpful to detect the duration of a rain event.

Table 1. Performance of rain event detection model. G and T donates temporal and geographical weights respectively

	Threshold-Based	Uniform-Signal	T-Signal	G-Signal	T+G-Signal
Precision	0.4972	0.5610	0.5528	0.6160	0.6036
Recall	0.6253	0.7102	0.7254	0.6741	0.7059
F1-Score	0.5424	0.6239	0.6255	0.6425	0.6507
Accuracy	0.9299	0.9432	0.9424	0.9501	0.9492

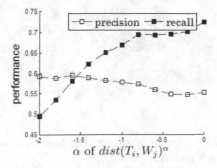

Fig. 6. The performance under various *alpha*'s

We also repeat the experiments under various α's (in $Geo(T_i)$) to see how the region size we use to crawl data affects the performance. The tendency in precision and recall is plotted in Figure 6. Smaller regions lead to lower recall and higher precision. For location L_j , merely using its nearby tweets is better in identifying its weather condition. However, not every location L_j always has sufficient nearby tweets, which will lead to the low recall.

We further categorize rain as heavy, normal and light, and then examine their properties. In our experiment, recall values for three are 0.9173(heavy), 0.7200(normal) and 0.6633(light) using T+G-Signal. The average number of rain related tweets normalized by the number of total tweets during heavy rain is 0.0598, which is greater than 0.0404 and 0.0301 for normal rain and light rain respectively. More tweets are posted during the more serious events, making it easier to detect in these periods.

As we discussed in Section 3.2 (geographical aspect), the location of users is an important factor to measure reliabilities of tweets. It also suggests that the bias of location affects performance a lot. For instance, station 0 and station 2 are near to each other, but their performance is quite different (Table 2). As shown in Figure 3, the users in station 2 are bias to upper right, namely, the direction of station 0. In contrast, the users in station 0 are closer to the center. This example may suggest the performance will be better if the users are closer to the station. In the case of station 7, it is limited by the topography(Figures 3 and 8) but it also has the good performance. The station 9 is another extreme instance. The users of station 9 are very close to the center and it results in a great performance.

Table 2. Performance of each station

station id	0	2	6	7	9	11
F1-score	0.6602	0.2100	0.5062	0.5664	0.7368	0.6399

Fig. 7. The locations of station 0 and station 2. The blue balloon means the center of station 0; the red thumbtack means station 2's.

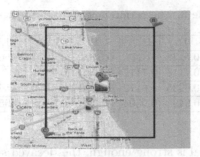

Fig. 8. The topography around station 7. Balloon A is the center, and other balloons and black frame are the boundaries for tweets gethering.

5   Rain Prediction

In the previous section, we have showed that our system is practical for detecting the rain events. Here we further want to understand whether we can predict weather conditions in the next n time slots. In Figure 9 we plot the signal during a rain event. From our observation, we believe there are two possible directions for weather prediction. First, the signal of L_j in this example starts to rise before it starts raining in L_j. A possible explanation is that before it rains in L_j, it rains near L_j so the users nearby post rain-related tweets. We therefore can detect rain events before it occurs. Second, whether it rains or not in consecutive time slots is not independent, making it possible to use previous data to infer the weather conditions in the near future.

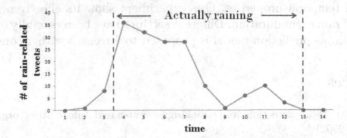

Fig. 9. A rain event example

We extend the aging based model [14] and construct a simple weather prediction system by modeling the life cycles of rain events. For a location L_j, after catching the beginning of a rain event, we start to monitor its signal. The signal generally doesn't reach maximum at once since users need time to respond the new event. To predict if it is still raining in the following time slots, we always assume the signal will decay exponentially(see Figure 5). This is because we have a few hints about future. What we know is how many tweets in the current time

Table 3. The performance of rain prediction

Time(Hour)	0.25	0.5	0.75	1	1.25	1.5	1.75	2
Precision	0.5562	0.5548	0.545	0.5401	0.4974	0.5018	0.4926	0.4868
Recall	0.7050	0.7118	0.7117	0.7089	0.7191	0.7233	0.7268	0.7171
F1-Score	0.6218	0.6235	0.6172	0.6130	0.5880	0.5925	0.5872	0.5799
Accuracy	0.9484	0.9498	0.9504	0.9508	0.9508	0.9516	0.952	0.9519

slot. If the current signal is strong enough, the decayed signal we expect in the next slot will not be lower than a given threshold. Then we predict the rain event will continue; otherwise, it will stop. Table 3 displays the results.

As we expect, the performance drops when the time we'd like to predict is farther from now, but it doesn't drop dramatically. We get 0.579 in F1-score for predicting if it still rains two hours later. One way to improve the performance is considering meteorological knowledge simultaneously. For instance, the length of rain in one location may vary with season and thus we can use an adaptive threshold instead a static one.

6   Conclusions

In this paper, we discuss the influence of spatiotemporal factor on rain events detection and prediction. The number of tweets changes in a day, making it hard to detect events on those "inactive" periods. The different distribution of population also leads to similar problem. Moreover, the degree of users' interest in rain varies with regions. We, therefore, re-weight tweets according to their spatial and temporal properties. Our experiment show its effectiveness in the detection of rain event duration. Different settings have been carefully examined. Finally, a simple prediction model is proposed to forecast weather conditions.

References

1. Allan, J.: Topic detection and tracking: event-based information organization, vol. 12 (2002)
2. Allan, J., Lavrenko, V., Jin, H.: First story detection in tdt is hard. In: Proceedings of the Ninth International Conference on Information and Knowledge Management. pp. 374–381. ACM (2000)
3. Allan, J., Papka, R., Lavrenko, V.: On-line new event detection and tracking. In: Proceedings of the 21st Annual International ACM SIGIR Conference on Research and Development in Information Retrieval (1998)
4. Cataldi, M., Di Caro, L., Schifanella, C.: Emerging topic detection on twitter based on temporal and social terms evaluation. In: Proceedings of the Tenth International Workshop on Multimedia Data Mining (2010)
5. Chen, L., Roy, A.: Event detection from flickr data through wavelet-based spatial analysis. In: Proceedings of the 18th ACM Conference on Information and Knowledge Management (2009)

6. Cox, J., Plale, B.: Improving automatic weather observations with the public twitter stream (2011)

7. He, Q., Chang, K., Lim, E.P.: Analyzing feature trajectories for event detection. In: Proceedings of the 30th Annual International ACM SIGIR Conference on Research and Development in Information Retrieval, SIGIR 2007 (2007)

8. Java, A., Song, X., Finin, T., Tseng, B.: Why we twitter: understanding microblogging usage and communities. In: Proceedings of the 9th WebKDD and 1st SNA-KDD 2007 Workshop on Web Mining and Social Network Analysis, pp. 56–65. ACM (2007)

9. Jianshu, W., Bu-Sung, L.: Event detection in twitter. In: Proceedings of the Fifth International AAAI Conference on Weblogs and Social Media (2011)

10. Li, C., Sun, A., Datta, A.: Twevent: Segment-based event detection from tweets. In: Proceedings of the 21th ACM International Conference on Information and Knowledge Management, CIKM 2012. ACM (2012)

11. MacEachren, A.M., Robinson, A.C., Jaiswal, A., Pezanov, S., Savelyev, A., Blanford, J., Mitra, P.: Geo-Twitter analytics: Application in crisis management. In: 25th International Cartographic Conference (2011)

12. Sakaki, T., Okazaki, M., Matsuo, Y.: Earthquake shakes twitter users: real-time event detection by social sensors. In: Proceedings of the 19th International Conference on World Wide Web (2010)

13. Teevan, J., Ramage, D., Morris, M.R.: #twittersearch: a comparison of microblog search and web search. In: Proceedings of the Fourth ACM International Conference on Web Search and Data Mining, WSDM 2011, pp. 35–44. ACM (2011)

14. Tzeng, Y.S., Jiang, J.Y., Cheng, P.J.: Event duration detection on microblogging. In: Proceedings of the 2012 IEEE/WIC/ACM International Conference on Web Intelligence, WI 2012 (2012)

15. Yang, Y., Pierce, T., Carbonell, J.: A study on retrospective and on-line event detection. In: Proceedings of the 21st Annual International ACM SIGIR Conference on Research and Development in Information Retrieval (1998)

Outpatient Department Recommendation
Based on Medical Summaries

Hen-Hsen Huang, Chia-Chun Lee, and Hsin-Hsi Chen

Department of Computer Science and Information Engineering,
National Taiwan University
#1, Sec. 4, Roosevelt Road, Taipei, 10617 Taiwan
{hhhuang,cclee}@nlg.csie.ntu.edu.tw, hhchen@ntu.edu.tw

Abstract. The family medicine in some regions is not as popular as that in the United States. Most patients choose the outpatient department without professional advice. In this work, we propose a health care aiding system that recommends the outpatient department for a patient according to his/her chief complaint and personal attributes. The recommendation is based on the past medical summaries of a hospital. Three methods including language model, support vector machine, and k-nearest neighbor algorithm along with different features are explored. The experimental results show that the SVM classifier with features selected from chief complaint, as well as personal attributes such as age, gender, and disease information achieves an f-measure of 79.35%.

Keywords: Health Care Aiding, IR Application, Medical Informatics, Recommendation.

1   Introduction

In some regions such as Taiwan, the family medicine is not popular. Rather than ask the professional advice, most patients have to choose one of outpatient departments from a hospital by themselves when they feel ill. The outpatient departments are classified based on the aspect of medical professionalism, rather than on the individual patient's cognition. For this reason, choosing a proper department is often a challenging task, especially for those patients who lack of medical knowledge or whose disease is imperceptible. Mistaking the proper department not only wastes the patients' time and money and increases their suffering, but also is inherent in the risk of disease progression.

Although the outpatient guide is available in some hospitals, it is still not convenient for many patients because the information on the guide tends to be too brief to help patients rapidly find their most proper departments, especially for the large hospitals including dozens of departments.

Fig. 1 shows a screenshot of the patient guide provided by the website of National Taiwan University Hospital (NTUH), the largest hospital in Taiwan. In this patient guide, the information is organized in the style of frequently asked questions, aka FAQ. All the health problems are classified into nine categories including "Children",

Y. Hou et al. (Eds.): AIRS 2012, LNCS 7675, pp. 518–527, 2012.

"Nerves and Sensory", "Maternity and Urinary", "Skin", "Mental and Psychological", "Health check and Others", "Physique", "Beauty and Body", and "Eye, Ear, Nose, and Throat". The first entry in the guide is a pair of questions, such as "Which department should be taken for the children with development delay?" and the corresponding suggestion, "The pediatric rehabilitation and child psychologist in the department of pediatrics". For the category "Children", only 9 questions are listed in the guide. Even in the largest category "Health check and Others", only 19 questions are listed. It is obvious that this guide is too limited to cover thousands of health issues. In addition, the FAQ-styled data is not ideal for most patients to search the needed information. For instance, the question "Which department should be taken for the patient of stroke?" is aligned in the category "Nerves and Sensory", and the suggestion of this question is "department of neurology". In Chinese, "Nerves" and "Neurology" share the same terminology "神經". If a patient knows the stroke is an issue related to the nerves, and s/he is likely to know the department of neurology is the proper department to take. On the other hand, a patient who does not have any knowledge about the stroke and the nerves will never find this entry in the category "Nerves and Sensory" in this guide. The other instance is observed in the category "Mental and Psychological". The suggested department for all the questions in this category is "department of psychiatry", which is already suggested in the category name. This is a case of non-informative organization because the patient will choose the department of psychiatry without any doubt since s/he knows that her/his health trouble is related to psychology.

Fig. 1. Screenshot of the patient guide on the website of NTUH

From the above observations, the limited coverage and non-informative organiza-tion of outpatient guide is useless to many patients, and the patients are still used to ask the staffs in the hospital to get the right outpatient clinic. As a result, how to use the expensive medical resources efficiently becomes an important issue.

Instead of the outpatient guide and the human suggestion, we propose an automatic recommendation system that suggests a patient the most proper outpatient department according to the patient's chief complaint and personal attributes including age, gender, and chronic diseases. The rest of this paper is organized as follows. We intro-duce the experimental dataset in Section 2. The recommendation methods and the features to be explored are presented in Section 3. Section 4 shows and discusses the experimental results. Finally, we conclude this paper in Section 5.

2   Experimental Dataset

In the past, some information retrieval systems use the medical documents to tackle the issues relating to health care [1-2]. Some other work suggests patients useful health information based on learning algorithms [3]. To the best of our knowledge, using the medical documents to suggest patients the proper outpatient department is a new idea.

A set of medical documents from NTUH is employed to develop such a recom-mendation system. A data entry in the collection is a medical summary of a patient's visit. There are three parts in a medical summary, including a chief complaint, a brief history, and a course and treatment. The chief complaint is mostly a short statement that is said by a patient in the outpatient clinic declaring the purpose of his/her visit, which is usually a description of the patient's physical discomfort. The samples (S1), (S2), (S3), and (S4) are some instances of chief complaints in our dataset.

(S1) Epigastralgia for 10 days
(S2) Tarry stool twice since last night
(S3) Left thyroid goiter for about 1 year
(S4) Cough and dyspnea for 3 days

As shown in the instances, a typical chief complaint consists of two elements, i.e., the symptoms found by the patient and the duration of these symptoms. The brief history is a summary about the physical conditions of the patient. The sample (S5) is the first two paragraphs of a brief history in our dataset.

(S5) This 61y/o female was a case of (1) schizophrenia, paranoid type, for 1 year and (2) asthma for more than 30 years, and she has taken medicine as instructed at 三總 hospital.

Several months ago, she began to have abdominal fullness especially after drink. Acid regurgitation sensation was found at times. And sleeplessness was caused by abdominal fullness. Abdominal pain, nausea, dysphagia, melena or heart burn sesation were not noted. However, body weight loss 4kg was noticed in recent 3 months (63kg ->59kg). Therefore, she went to 三總 hospital. On

94/12/30, PES revealed gastric mass on body post and her family was told of early gastric cancer. Pathology showed signet-ring adenocarcinoma.

The brief history is written by physicians in English with mixing a few Chinese terms such as person names and hospital names. As shown in the sample (S5), the word "三總" is the name of another large hospital in Taiwan. The physicians describe the personal information about the patient and the past medical treatment of the patient in brief history as a reference for advanced treatment. In (S5), the personal attributes including the age and the gender of the patient are mentioned. In addition, the chronic diseases of the patient are also recorded in the brief history. For instance, the first part of the brief history in another medicine summary is provided as the sample (S6).

(S6) The 74 year old woman was a patient with diabetes mellitus, hypertension, chronic renal insufficiency, neurogenic bladder under medical control for six years.

The chronic diseases of the patient such as diabetes mellitus and hypertension are described in the brief history as information for the medical diagnosis and treatment.

The third part of a medical summary is the course and treatment. In this part, the treatment processes such as medication administration, inspection, surgery and the treatment outcomes are described in detail. For instance, the first paragraph of a course and treatment is shown as the sample (S7).

(S7) For her uremic symptoms and hyperkalemia, she started to receive regular hemodialysis three times a week at the day on admission. Peritoneal dialysis was chosen for long term dialysis and Tenckhoff catheter implantation was performed on 95/1/23.

The other instance (S8) shows how the medication administration and the outcomes of inspection described in a course and treatment as follows.

(S8) After admission, he received Gemzar 1920mg (1000mg/m2) and Carboplatin 450mg(AUC=6) on 2006/01/09 smoothly. Besides, chest echo on 1/09 showed minimal left pleural effusion and left supravicular lymph node aspiration was done on the same day. The cytology result was pending. No fever or specific discomforts developed. Under the relative stable condition, he was discharged on 1/10 and followed up at OPD.

All the fields of each medicine summary are listed in Table 1. Besides the three major parts, i.e., chief complaint, brief history, and course and treatment, several metadata are available in each medical summary. The patient's name, the physician's name, and the date of the visit are removed in this dataset because of the privacy. Department is the critical information for our work. After the medical summaries from the smaller departments whose medical summaries are no more than 1,000 are excluded, all the rest medical summaries are classified into 14 classes according to their departments. The statistics of our dataset is shown in Table 2.

From the dataset, we can obtain the department and the chief complaint of each medical summary. In addition, the age, the gender, and the chronic diseases of the

Table 1. The fields in a medicine summary in our dataset

Fields	Explanation	Samples
Chief Complaint	The statement said by a patient describing the symptoms of the comfort or the illness	See (S1), (S2), (S3), and (S4)
Brief History	The personal information and the past medicine events of the patient	See (S5) and (S6)
Course and Treatment	The processes and the outcomes of this treatment	See (S7) and (S8)
Department	The outpatient department	department of neurology
Ward	The patient's ward number	13D 1201
Year	The year of the patient's visit	2006

Table 2. Statistics of the dataset

Department	Number of entries
Dental	1,969
Dermatology	1,144
ENT (Ear, Nose, and Throat)	6,400
Internal Medicine	32,160
Neurology	2,544
Obstetrics and Gynecology	8,928
Oncology	5,082
Ophthalmology	4,332
Orthopedics	8,759
Pediatrics	10,904
Rehabilitation	1,575
Psychiatry	1,777
Surgery	27,531
Urology	7,216
Total	120,321

patient are extracted as features as well. Therefore, we can use this dataset as training data to build a recommendation system that ranks the departments according to a patient's input such as the chief complaint and the personal attributes.

3   Methods

We model the recommendation task as a ranking problem, and three ranking approaches and various features are explored in the experiments. The ranking algorithms are introduced as follows.

3.1   Language Model

The idea is to find a department such that the chief complaint of a patient is generated most likely by the model trained by the previous patients' complaints of the department. In the training stage, we train each department as a language model based on the patients' medical summaries in the department. In the test stage, the perplexities between the input and all the departments are calculated to rank the departments.

The lower the perplexity is, the more likely the input is related to the department. As a result, the department with the lowest perplexity is the most proper department for the input. The SRI Language Modeling Toolkit (SRILM)[1] is utilized to training and testing the language models. The depth of language models are set as trigram, which is better than unigram and bigram in the experiments.

3.2   Support Vector Machine

Department recommendation can be modeled as a text classification task. Support vector machine (SVM), a powerful classification algorithm in many applications, is adopted. In the training stage, we train a multi-class SVM classifier with the training set. In the test stage, the SVM classifier not only predicts the most proper class for the given input, but also reports the confidence measures of all classes for the input. Thus, we can rank the departments by their confidence measures for an input instance. The LIBVSM[2] [4] is used in this work, and the kernel is L2-regularized logistic regression. The parameters are optimized with a grid search algorithm.

3.3   k-Nearest Neighbor Algorithm

We can predict the most suitable department based on collective intelligence embedded in medical summaries. The basic idea is that the department selection of patients can be regarded as their votes on the departments. The patients' experiences will give a hint to make decision. The instance-based k-nearest neighbor (kNN) algorithm is explored. We set the k to 5, which is the best value in the preliminary experiments. In contrast to SVM, the kNN algorithm is local sensitive. We will compare the performances of the three different ranking algorithms and show their differences.

3.4   Features

For each medical summary in the dataset, the chief complaint and the personal information extracted from the brief history are encoded as features.

Chief Complaint: A chief complaint is a sentence or a fragment of a sentence written in English. We first convert all the alphabetic characters in chief complaints to lowercase. All the words are segmented and performed with the Porter2[3] algorithm to obtain their root forms. For the language model, each chief complaint is represented as a sequence of words. For SVM and kNN, the bag-of-words representation is applied.

Personal Attributes: The personal attributes of a patient including the age, the gender, and the chronic diseases of the patient are extracted from the brief history. The beginnings of most brief histories are described in a similar pattern like "This X year-old man", "This X years old female", or "This X y/o woman". Therefore, age extraction can be formulated by a pattern matching. Similarly, the gender can also be captured

---

[1] http://www.speech.sri.com/projects/srilm/

[2] http://www.csie.ntu.edu.tw/~cjlin/libsvm/

[3] http://pypi.python.org/pypi/stemming/1.0

with a dictionary where the gender-related nouns including "female", "man", "girl", etc. are collected. Some instances are shown as follows.

(S9) This 70 year-old gentleman had past history of ....

(S10) This 42 y/o female was healthy before ...

(S11) This 64 y/o lady was well before. No weight loss, general malaise, or other subjective complaints was noted recently...

To extract the chronic diseases, we prepare a dictionary in which 21,030 terms about diseases are collected. The simple pattern matching algorithm does not work well due to the noises from the long patient histories. In other words, many diseases that do not really attack a patient may be also mentioned in the brief history as background information. For instance, the patient's family history is frequently described in the brief history, and the suspected diseases that have been ruled out will be also recorded. In order to avoid introducing these noises, we use a position sensitive algorithm to matching the patterns. The idea is that the real chronic diseases of a patient tend to be described in the first two sentences. For this reason, we only match the diseases that appear in the first two sentences.

4   Experiments and Discussion

4.1   Experimental Results

The experimental results of using language model, SVM, and kNN with all the features for each department are shown in Table 3. All the models in the experiments are evaluated by 10-fold cross validation. The reported metrics are precision, recall, and f-measure. The t-test is used for significance testing. P (%), R (%), and F (%) in the header denote precision, recall, and f-measure in percentage, respectively. SVM significantly outperforms the other two models at $p=0.0001$. The departments whose f-measures are more than 90% are dental, obstetrics & gynecology, ophthalmology, and psychiatry. The common property of these four departments is that the health troubles related to them are limited to definite parts of the body or the specific functions of the human being. For instance, the department of dental is related to the troubles of the mouth and teeth, and the department of psychiatry is related to the mental. The department of oncology is the one having the poorest performance. This is understandable because cancer can attack all organs in human body, and various symptoms may point to cancer. Fortunately, the poor performance of the department of oncology is not an issue in practice. In fact, the cases of cancer are usually found in the other specialist outpatients. For this reason, suggesting the patients with cancer potential for a specialist outpatient is not really wrong.

In the second experiment, we explore the performance of individual features used in the best model, the SVM classifier. The results are shown in Table 4. All the numbers in the table are f-measures in percentage. The CC in the header is the abbreviation of Chief Complaint. The additional gender information does not significantly improve the performance at $p=0.05$. One reason may be that many diseases are unrelated to a specific gender. On the other hand, the gender information is also useless

Table 3. Performance of the three algorithms with all the features

Departments	Models								
	Language Model			SVM			kNN		
	P(%)	R(%)	F(%)	P(%)	R(%)	F(%)	P(%)	R(%)	F(%)
Dental	80.72	90.35	85.26	91.44	85.17	88.19	82.38	79.08	80.69
Dermatology	53.62	79.02	63.89	82.53	67.74	74.41	75.07	50.26	60.21
ENT	86.82	83.39	85.07	85.67	86.78	86.22	58.02	81.69	67.85
Internal Medicine	73.11	75.54	74.30	73.67	82.01	77.62	69.90	73.24	71.53
Neurology	47.90	59.28	52.99	66.10	52.12	58.29	53.87	33.14	41.03
Obstetrics & Gynecology	94.34	90.40	92.33	94.84	91.83	93.31	88.10	86.42	87.26
Oncology	52.24	56.75	54.40	67.20	47.36	55.56	60.67	30.83	40.89
Ophthalmology	95.17	95.48	95.32	95.59	94.97	95.28	89.50	88.78	89.14
Orthopedics	84.51	85.27	84.89	85.32	87.27	86.29	68.90	84.72	76.00
Pediatrics	87.02	85.28	86.14	87.50	88.77	88.13	87.33	84.87	86.08
Rehabilitation	49.39	56.57	52.74	73.09	53.46	61.75	53.81	38.98	45.21
Psychiatry	87.45	93.36	90.31	94.38	88.86	91.54	94.77	69.27	80.04
Surgery	69.20	63.66	66.31	70.45	69.48	69.96	60.99	59.05	60.00
Urology	84.79	81.43	83.08	87.33	81.46	84.29	71.01	71.78	71.39
Macro-Averaged	74.73	78.27	76.22	82.51	76.95	79.35	72.45	66.58	68.38

Table 4. F-measures of the SVM classifier with different feature combinations

Departments	Features				
	CC	CC+Age	CC+Gender	CC+Diseases	All
Dental	84.44%	82.61%	84.10%	89.35%	88.19%
Dermatology	65.87%	67.47%	67.56%	72.57%	74.41%
ENT	82.21%	82.60%	82.21%	85.88%	86.22%
Internal Medicine	70.14%	73.81%	70.23%	76.59%	77.62%
Neurology	47.92%	49.18%	48.45%	58.06%	58.29%
Obstetrics & Gynecology	88.00%	88.72%	88.50%	93.39%	93.31%
Oncology	40.38%	41.66%	41.04%	54.45%	55.56%
Ophthalmology	93.34%	93.33%	93.43%	95.34%	95.28%
Orthopedics	84.17%	84.30%	84.13%	86.01%	86.29%
Pediatrics	66.73%	85.58%	67.14%	84.13%	88.13%
Rehabilitation	56.40%	57.34%	57.08%	58.55%	61.75%
Psychiatry	84.54%	85.75%	85.20%	91.01%	91.54%
Surgery	64.17%	64.47%	64.17%	69.75%	69.96%
Urology	80.48%	81.37%	81.01%	84.08%	84.29%
Macro-Averaged	72.06%	74.16%	72.45%	78.51%	79.35%

for the gender-specific departments such as obstetrics & gynecology. As expected, the feature of chronic diseases is very useful, especially for certain departments such as oncology and psychiatry. The overall performance of using the features Chief Complaint and chronic diseases (denoted as CC+Diseases in Table 4) is close to using all the features (denoted as All). However, using all the features still yields a significant improvement at p=0.0001.

The confusion matrix of SVM classification is shown in Table 5. Each row of Table 5 shows the percentages of instances of a department that are classified into the 14 departments. The two largest departments, Internal Medicine and Surgery, are the two departments that most instances are misclassified into. The reason is that these two departments are more general than the specific organ specialist departments. It is a common situation that a patient would be referred to the department of surgery from the department of internal medicine for surgery, and then would be referred to the internal medicine back for postoperative follow. Another common situation is that a patient of a specialist department would be referred to the department of surgery for surgery, and then would be referred to the department of oncology for chemotherapy after the surgery. For this reason, the specialist departments, the department of internal medicine, the department of surgery, and the department of oncology share a number of common patients.

A special case is that 10.22% of Rehabilitation instances are misclassified into Neurology. This is the most frequent misclassification besides the cases that Internal Medicine and Surgery are involved. A major disease of the department of neurology is the stroke (cerebrovascular accident, CVA), which might cause the inability to move limbs. To recover the loss body function of stroke patients, rehabilitation treatment is usually used on those patients. Thus, the patients of Neurology and Rehabilitation are higher overlapped. (S12) is an instance of a stroke patient who accepted the treatment from the department of rehabilitation.

(S12) Sudden onset right limb weakness and speech disturbance on 98/3/6.

Table 5. Confusion matrix of SVM classification

Actual Class	Predicted Class													
	Dent	Derm.	ENT	Med	Neur.	O&G	Onc	Ophth	Ortho	Pedi	Reha	Psyc	Surg	Urol
Dent	85.17	0.10	2.84	1.83	0.05	0.10	0.36	0.05	0.46	0.76	0.00	0.00	7.87	0.41
Derm	0.52	67.74	0.87	16.78	0.17	0.26	0.44	0.26	1.66	1.05	0.00	0.00	9.88	0.35
ENT	0.23	0.05	86.78	3.73	0.06	0.05	0.97	0.19	0.13	0.61	0.00	0.00	7.03	0.17
Med	0.05	0.28	0.44	82.01	0.82	0.39	2.02	0.19	0.71	1.31	0.27	0.13	10.48	0.90
Neur	0.00	0.08	0.94	23.70	52.12	0.16	0.67	0.79	1.02	0.63	3.69	1.26	14.31	0.63
O&G	0.04	0.02	0.08	2.26	0.00	91.83	0.17	0.00	0.21	0.48	0.00	0.01	4.05	0.83
Onc	0.10	0.08	3.92	28.87	0.35	0.61	47.36	0.08	1.38	0.26	0.14	0.02	15.72	1.12
Ophth	0.07	0.02	0.23	1.04	0.07	0.12	0.07	94.97	0.39	0.35	0.00	0.00	2.45	0.23
Ortho	0.17	0.06	0.13	1.78	0.06	0.05	0.25	0.07	87.27	0.53	0.13	0.00	9.37	0.15
Pedi	0.13	0.07	0.29	4.23	0.14	0.22	0.23	0.08	0.34	88.77	0.05	0.06	5.25	0.15
Reha	0.00	0.00	0.32	16.70	10.22	0.19	0.32	0.00	3.62	0.76	53.46	0.06	12.70	1.65
Psyc	0.06	0.11	0.28	5.46	1.07	0.17	0.11	0.00	0.06	0.51	0.11	88.86	2.98	0.23
Surg	0.27	0.15	1.50	18.80	0.68	0.58	1.20	0.24	2.92	2.61	0.33	0.04	69.48	1.17
Urol	0.06	0.03	0.21	6.75	0.03	1.05	0.44	0.10	0.28	0.33	0.15	0.00	9.12	81.46

4.2   Discussion

Our department recommendation system is an aiding tool for the patients seeking the proper outpatient department. Even if the accuracy of the recommendation is not perfect, some incorrect suggestions will not cause serious risk in practice. If a patient takes a wrong department, the physician will find this error and give an appropriate treatment. Such a change can also be a feedback to refine the recommendation system. In general, the proposed system provides decent and efficient suggestions to most patients.

Although the chief complaints are said by patients, all the chief complaints in our dataset are recorded by physicians. The physicians try to record the patients' words as realistic as possible, but the gap between experts and the public cannot be ignored. Different choices of words by the general public may decrease the performance of this system in real world if the inputs are done by patients themselves. How to narrow down the gap is a future work.

Furthermore, all the chief complaints in our datasets are written in English. In other words, the knowledge extracted from the medical summaries is in terms of English. For serving the patients speaking in other languages, e.g., Chinese, a medical summary translation system [5] should be integrated into the recommendation system. We will develop a multilingual outpatient department recommendation system in the future. The chief complaints in Chinese will be translated into English, and submitted to our outpatient department recommendation system. Vocabulary gap and translation accuracy should be tackled at the same time.

5   Conclusion

In this paper, we introduce the collective intelligence embedded in medical summaries to design an outpatient department recommendation system. This system suggests a patient an outpatient department according to the patient's chief complaint and personal attributes. Three different learning models and various feature combinations are proposed and evaluated. In the experiments, the SVM classifier with all features achieves an f-measure of 79.70%. This system aids the patients without medical knowledge to seek the proper outpatient departments. That reduces the risk of misdiagnosis and helps save the expensive medical resources. In the future work, we will recommend outpatient departments of much finer grain, in particular, for the two largest departments, i.e., Internal Medicine and Surgery. Besides, resolving the information gap in practical uses and introducing the multilingual capabilities into the proposed system will be investigated.

Acknowledgment. Research of this paper was partially supported by National Science Council (Taiwan) under the contract NSC 101-2221-E-002-195-MY3.

References

1. Mondal, D., Gangopadhyay, A., Russell, W.: Medical decision making using vector space model. In: ACM SIGHIT 2010, pp. 386–390 (2010)
2. Zhang, Y., Wang, P., Heaton, A., Winkler, H.: Health information searching behavior in MedlinePlus and the impact of tasks. In: ACM SIGHIT 2012, pp. 641–650 (2012)
3. Sondhi, P., Vydiswaran, V.G.V., Zhai, C.: Reliability Prediction of Webpages in the Medical Domain. In: Baeza-Yates, R., de Vries, A.P., Zaragoza, H., Cambazoglu, B.B., Murdock, V., Lempel, R., Silvestri, F. (eds.) ECIR 2012. LNCS, vol. 7224, pp. 219–231. Springer, Heidelberg (2012)
4. Chang, C.-C., Lin, C.-J.: LIBSVM: a library for support vector machines. ACM Transactions on Intelligent Systems and Technology 2, 27:1–27:27 (2011)
5. Chen, H.-B., Huang, H.-H., Tan, C.-T., Tjiu, J., Chen, H.-H.: A statistical medical summary translation system. In: ACM SIGHIT 2012, pp. 101–110 (2012)

Actively Mining Search Logs for Diverse Tags*

Lun Yan, Congrui Huang, and Yan Zhang**

{pkualan,huangcongrui}@gmail.com, zhy@cis.pku.edu.cn

Abstract. Social tagging has become a very important mechanism for organizing information on the Web. Usually, people tag a web page manually, just as what they do on a social bookmarking web site. In this paper, we will demonstrate a brand-new perspective - tagging web pages automatically by mining search logs. In order to keep diversity, we first classify web queries into different categories and then extract tags from queries to depict each category. Thereafter we describe a web page with all queries which are related to this page, and finally we get the recommended tags for each web page after mapping the related queries into corresponding diverse tags. The experiments conducted on a real search log show that our method can dig out accurate and meaningful diverse tags for web pages more effectively.

Keywords: Social tagging, Search log, Recommendation.

1 Introduction

Users tag objects for different purposes. Regardless of the purpose, tags can save users' effort in tracking information they need. Tags can also help search engines retrieve web pages more precisely and can be used to identify user intents. For example, traditional keyword-based retrieval cannot return web pages about 'Microsoft Office' when a user submits 'Microsoft'. Tags will make search engines aware of such potential relations between 'Office' and 'Microsoft'.

Carman et al. [5] find that the terms people use to annotate web sites are similar to the ones they use in the corresponding web search. They show that the vocabulary contains a large amount of overlap and the term frequency distributions are correlated. Usually people use search engine actively. When they submit queries to seek information they need, they often refine the queries to get more precise and accurate results. To some extent, queries have comparable ability with tags in describing a web page. Therefore, if we extract tags for a web page from queries people have posted to view that page, the extracted tags will be of high quality as they are from different queries issued by different users.

The contribution of our work is summarized as follows:

- We extract tags from the query terms for a web page.
- We use online encyclopedia to help to extract as much tags as possible.
- We can rank the tags related to a web page diversely.

---

* Supported by NSFC under Grant No. 61073081.
** Corresponding author.

Y. Hou et al. (Eds.): AIRS 2012, LNCS 7675, pp. 528–538, 2012.
© Springer-Verlag Berlin Heidelberg 2012

2   Related Work

2.1   Tag Generation and Tag Recommendation

Basically, the tags are generated from text or from social sites. Zhou et al. [18] propose a probabilistic generative model for generation of document content as well as associated tags. The community influence on tag selection in Flickr has been studied by Marlow et al. [12]. Tags can be recommended based on their quality, co-occurrence, mutual information and object features [7]. Xu et al. [17] propose a set of criteria for tag quality and present a collaborative tag suggestion algorithm with these criteria to recommend high-quality tags. In [15], authors exploit terms extracted from the object metadata and tag co-occurrence patterns in order to suggest tags for users on Flickr. Liu et al. [11] rank the tags associated with a given image automatically according to their relevance to the image content to select tags for an image. Fabiano et al. [1] exploit not only previously assigned tags, but also terms extracted from other textual features associated with the target object to recommend tags.

2.2   Tag Classification

We have to classify queries first and try to understand queries related to a page. Jansen et al. [9] develop a software application that can automatically classify queries using a Web search engine log. Shen et al. [14] try to enrich queries using WordNet for web-query classification. In our work, we use Baidu Baike (also known as Baidu Encyclopedia) [1] to help tag generating and query classification. Hu et al. [8] map the query into the Wikipedia representation space to capture its semantic interpretation. Bordino et al. [4] project the query-flow graph on a low-dimensional Euclidean space to capture a notion of semantic similarity between queries.

3   Methodology of Tagging Web Pages

In our work, we extract tags from related queries of a web page and finally annotate the web page with ranked tags diversely. The framework of our methodology is shown in Fig. 1.

Tag Generating is a basic part of our work and will be discussed in Section 3.1. The task of Tag Generating is to map queries to the corresponding tags.

As shown in the dotted box on the upper left corner in Fig.1, all queries in search logs will be classified first. In our work, queries are classified with the help of Baidu Baike. Briefly, we can exploit the hierarchical structure of Baidu Baike to build a semantic tree [2,10]. With the assistance of this tree, we can classify all entries in Baidu Baike and then classify the queries.

For each web page d, a query q is directly related to d if and only if a user u posts the query q and in the results returned by the search engine, u clicks d.

---

[1] http://baike.baidu.com/

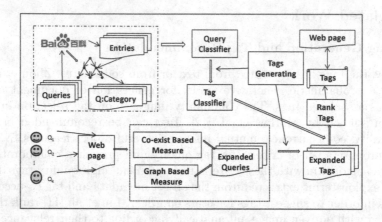

Fig. 1. Frame Overview

The query q is called a DR query of page d. We collect DR queries for a web page first and filter out those ambiguous ones. In order to get semantically related tags for a web page, we will expand DR queries of a web page with Coexistence-based Measurement or Graph-based Measurement which will be both discussed in Section 3.3.

3.1   Tag Generating

In order to map queries to tags, we imitate the language habit of online users. For this purpose, we want to get the frequently-used words of the tagging users. Entries of Baidu Baike can meet our demand. Besides, these entries can be considered as tags. We collect the entries and use an n-gram algorithm to segment each query q into Baidu Baike entries. As an n-gram is both an entry of Baidu Baike and a substring of q, we use formula (1) to compute the appearance probability of this n-gram. $C(\omega_1\omega_2\ldots\omega_n)$ is the frequency of a certain n-gram in the query and N is the appearance of distinct queries of the n-gram in Baidu Baike.

$$P(\omega_1\omega2\ldots\omega_n) = \frac{C(\omega_1\omega_2\ldots\omega_n)}{N} \tag{1}$$

We first select the highest probability of n-gram out as a tag of q, and then remove this substring from q and extract new tags for q repeatedly. For example, this is a query q "who is the writer of A Mid-Summer Night's Dream". In Baidu Baike, we get entries "A Mid-Summer Night's Dream" and "Writer". So query q will be mapped into tags "A Mid-Summer Night's Dream" and "writer". Given a query q, we use T_q to represent the set of probability of each n-gram extracted from query q. p_t represents the probability of tag t, an n-gram of q. The possibility of tag t generated from q could be calculated by formula (2). $|\cdot|$ is the L_1 norm.

$$P(t|q) = \frac{p_t}{|T_q|} \tag{2}$$

3.2 Tag Classification

For some kinds of web pages, people often have similar viewpoints. To avoid such topic bias, we will make the tags of a web page come from as more categories as possible.

In our work, we build a tag classifier. We classify all tags into different categories and refer this classification information when we rank tags for a web page, just as shown in Fig.1. In Section 3.1, we see that tag classification can be realized with query classification. A tag t may be generated from several queries which we call source queries of t. Once classification results of t's source queries are abstained, the category that the majority of source queries belong to will be marked as t's category.

We generate training set by Baidu Baike and consider using the k-nearest neighbor algorithm to conduct query classification. Let Q be the set of queries, D be the set of web pages and U be users that own at least one complete query process: post one query and has clicked a web page in the returned results. For information in the search log, we have existing ternary relation S, and $S \subseteq Q * D * U$. Thus each record $(q, d, u) \in S$ means that a user u has posted query q and clicked page d in the returned results. Each query has a distribution of web pages. The vector of web page occurrence can be used to represent the query. For the query q_k,

$$V_{q_k} = \{n_{q_k,d_1}, n_{q_k,d_2} \ldots n_{q_k,d_m} | \exists u \in U, i \in [1, m] : (q_k, d_i, u) \in S\} \quad (3)$$

Here, n_{q_k,d_i} means the number of times that all users in U have clicked d_i when submit q_k.

$$Sim(q_i, q_j) = \frac{V_{q_i} \cdot V_{q_j}}{\|V_{q_i}\| \times \|V_{q_j}\|} \quad (4)$$

$\| \cdot \|$ is the L_2 norm. Basically we can draw similarity between q_i and q_j with formula (4), however, not every page in V_{q_k} can well reflect the meaning of q_k. Consider the following situation:

- **Spam Page.** Spam page contains lots of hot keywords. It may be attached by many queries without any related information.
- **Exception Page.** Exception page is such a page that a user has clicked accidentally. For instance, a user sends the query 'Harry Potter and the Deathly Hallows' to find the book, but he clicks some pages talking about J. K. Rowling, who is the author.

$$\acute{n}_{q_k,d_i} = \frac{n_{q_k,d_i}}{\sum_j n_{q_k,d_j}} * \log(\frac{N}{|m : n_{q_m,d_i} \in V_{q_m}|}) \quad (5)$$

To combat spam pages, in vector V_{q_k}, we use $\acute{n}_{q_k,d_i}$ in (5) to replace n_{q_k,d_i}. N is the number of unique queries in the search log, while $|m : n_{q_m,d_i} \in V_{q_m}|$ is the total number of queries that users have posted with page d_i clicked in the corresponding returned results.

To reduce the noise of exception pages, we try to measure how likely page d_i can match user's intent. Suppose that a user sends query q_k, and then in the returned results page d_i is clicked. If d_i can well satisfy the user's intent, the user will probably send no more queries and click no other pages in the same session. For that reason, we segment each user's query stream into sessions based on the method mentioned in [16]. We name d_i and q_k as a pair. The matching degree of this pair is inversely proportional to the number of web pages and the number of queries of a session in which this pair has occurred. While this pair can occur in several sessions, in formula (6), $D_{k,session}$ denotes the web page array in a session and $Q_{k,session}$ denotes the query array in a session. Pages and queries that follow the record of q_k in a session will be put into the arrays. Finally, the mode values of sizes of different arrays are used to generate md_{q_k,d_i}.

$$md_{q_k,d_i} = \frac{1}{mode(|D_{k,session}|) \times mode(|Q_{k,session}|)} \tag{6}$$

Thus we get a new array to represent V_{q_k}:

$$N_{q_k,d_i} = \acute{n}_{q_k}, d_i \times md_{q_k,d_i} \tag{7}$$

$$V_{q_k} = \{N_{q_k,d_1}, N_{q_k,d_2}, \ldots N_{q_k,d_m} | \exists u \in U, i \in [1, m] :$$
$$(q_k, d_i, u) \in S\} \tag{8}$$

3.3   Tag Expanding

For a page, we will extract tags from queries users send directly to target the web page which we call directly related queries of the web page, shortened as DR queries. However, many web pages have a very small amount of DR queries. These DR queries can provide limited tags to describe a web page which force us to find out more information related to the web page.

For each web page, we build a matrix $R_{K \times N}$ to denote relevance score between $d's$ directly related query set Q_d ($Q_d = \{q_1, q_2, \ldots q_K\}$) and other queries r_j in the search log. Each element of the matrix will be computed by (8).

$$R[q_i, r_j] = \{rel(q_i, r_j) \quad 0 \le i \le K; 0 \le j \le N; r_j \ne q_i 1 \quad r_j = q_i \tag{9}$$

We will present two methods to compute $rel(q_i, r_j)$, Coexistence-based Measurement and Graph-base Measurement. The former focuses on the queries which have clicked same web pages or have occurred in same sessions historically with DR queries, while the latter is an extension of the former.

Coexistence-Based Measurement. To some extent, queries that have occurred in same sessions or have clicked same web pages with web page $d's$ DR queries (d itself excluded) will be more relevant with page $d's$. We call queries that have occurred in same sessions or have clicked same web pages (d itself excluded) with page d DR queries coexistence queries of d. In Coexistence-based Measurement, for a web page d we only use its coexistence queries to build matrix $R_{K \times N}$. According to the Jaccard coefficient, we can normalize the coexistence of two queries as follows:

$$rel(q_i, r_j) = \frac{|q_i \bigcap r_j|}{|q_i \bigcup r_j|} \tag{10}$$

The coefficient takes the number of times that both q_i and r_j have occurred, divided by the number of times that the two queries have occurred in total. The Jaccard coefficient is known to be useful to measure the relevance between two objects or sets. There are two kinds of coexistence definitions: click same web pages or occur in same sessions. We will discuss both of them in experiments.

Graph-Based Measurement. Coexistence-based Measurement discovers limited queries for a page, especially when the log data is sparse. If we draw edges between coexistence queries and DR queries of web page d, we will get Fig.2.

In Fig.2, we see that Coexistence-based Measurement only considers queries in the first level. If we expand the graph from this level, we will get a DAG, as shown in Fig.2. Although r_{21} does not occur with q_{01}, it is still probably related to q_{01}. An approach to compute the relevance between two queries can be used to compute the maximum information flow between two points. The maximum information flow is influenced by three conditions:

· The ratio of $|q_i \rightarrow r_j|$ to $d_o(q_i)$. $d_o(q_i)$ is the number of routes from q_i to r_j, while $d_o(q_i)$ is the out-degree of vertex q_i
· The ratio of $|q_i \rightarrow r_j|$ to $d_i(r_j)$. $d_i(r_j)$ is the in-degree of vertex r_j
· The number of nodes on the shortest path between q_i and r_j, which is denoted as $min(N_{q_i \rightarrow r_j})$.

Then we have:

$$rel(q_i, r_j) = \frac{|q_i \rightarrow r_j|}{d_o(q_i)} \times \frac{|q_i \rightarrow r_j|}{d_i(r_j)} \times \frac{1}{min(N_{q_i \rightarrow r_j})} \qquad (11)$$

In Graph-based Measurement, we mark $q_i \rightarrow r_j$ as the routes between q_i and r_j and compute the relevance between two queries with formula (10).

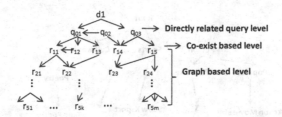

Fig. 2. An example for Graph-based Measurement (dotted line means multi level or multi nodes)

3.4   Tag Ranking

For a web page d_i, we use Q_{d_i} to denote DR queries of d_i. In (11), q_k belongs to Q_{d_i}, and N_{q_k} denotes the number of DR queries of d_i. q represents the query related to page d_i which means $R[q_k, q]$ is greater than zero. $R[q_k, q]$ can be obtained in matrix R (which is discussed in Section 3.3). $p(t|q)$ aims to calculate the probability of tag t with formula (2) mentioned before. The tags are ranked by $p(t|d_i, q)$.

$$p(t|d_i, q) = p(t|q) \times p(d_i|q) = \max_{k=1}^{N_{q_k}} p(t|q) \times R[q_k, q] \quad q_k \in Q_{d_i} \quad (12)$$

D_{q_k} denotes the set of web pages that query q_k is directly related to. EA_{q_k, d_i} denotes the discrimination and is calculated with (13). On average, if d_i has appeared in major sessions that D_{q_k} gets involved, q_k is supposed to be a convincing query of d_i. In (14), we change the method for calculating $p(d_i|q)$ as we think the higher EA_{q_k, d_i} is, the more relevant query q to page d_i will be.

$$EA_{q_k, d_i} = \frac{\sum\limits_{p_j \in D_{q_k}, d_i} \frac{|d_i \cap p_j|}{|d_i|}}{|D_{q_k}|} \quad (13)$$

Thus in (12) we can use (13) to calculate $p(d_i|r_j)$:

$$p(d_i|q) = EA_{q_k, d_i} \times R[q_k, q] \quad (14)$$

For each web page, we generate tags from queries in Section 3.1 and collect all related tags with the method presented in Section 3.3. Finally, we put all related tags into their categories with category information obtained in Section 3.2. For each category, we will select top three ones. All tags selected will be ranked together by their relevance with the web page.

4   Experiments

In experiments, we introduce how to gather training set and demonstrate the evaluation method. We show the performance of the two relevance measurements mentioned in Section 3 and compare our methods with the language-model based relevance measurements. Ranking effect of EA parameter will also be investigated.

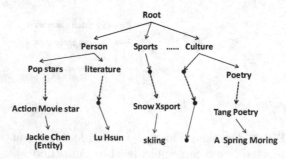

Fig. 3. Baidu Baike Semantic Tree

Besides applying entries of Baidu Baike to generate tags, we convert Baidu Baike into a semantic tree to help gathering training set and evaluating semantic relevance between a tag and its corresponding web page. In Fig.3, the second-layer nodes represent top categories information, and there are 12 in total. Each entry exists as a leaf node of this tree.

4.1   Data Set

We use a part of Sogou search log. The log contains 3,121,019 queries, 8,459,725 web pages and 6,415,029 sessions. We select 23990 web pages randomly.

4.2   The Training Set

To get the training set, we collect related entries of each query with the help of Baidu Baike's search engine. Each query is connected with some entries which are remarkably similar to it.

After getting related entries of a query, we calculate semantic similarity between each query's similar entries and category nodes of the semantic tree with formula proposed in [13]. The most similar category of these entries will be treated as the query's category. With the help of this training set and the method mentioned in Section 3, we get tags for each category. A sample of tags in some categories is shown in Table 1.

Table 1. Tags Category Sample

Category	Tags
Geography	Mount Tai City, Tarim Basin
Economics	Consumption-based value-added tax
History	Capital war, Charles I
Sports	Asian Winter Games, NBA
Person	Lady Gaga, George Bush, Min Yao
Special Category	Download, 2006, Software

4.3   Case Study

We select web pages and illustrate their tags obtained by the two measurements. In Table 2, Cosession and Coclick are two types of coexistence relevance measurement. The web page in Table 2 describes a type of car of Brilliance. In Table 2, Cosession method provides tags about Nimble as users always maintain Nimble and Brilliance in a search task of a session. It is obvious that graph based method has supplied more related tags and give us more understanding of related concepts in the web page.

Table 2. Sample Tags

Web page	Brilliance Auto officially announced the listing of "National amazing price": Nimble 85,800 yuan from the
Directly related Query	Brilliance Nimble
Cosession	Brilliance Nimble,FRV,Brilliance Auto, BMW-Brilliance,price list
Coclick	DPCA,buick excelle,Brilliance Nimble, Sporty Car,Mazda,Dongfeng Citroen
Graph Based	Brilliance Nimble,Brilliance,Brilliance Auto, China Motor,motor corporation, BMW-Brilliance,saloon car,Hyundai Motor

4.4   Comparison and Ranking Evaluation

We use Baidu Baike semantic tree to validate the relevance between a tag and its corresponding web page. A similarity proposed in [13] is employed. For each tag, we make the average score it receives from the current page's words as its final similarity score to the page. Specifically, we use semantic similarity scores as golden standards. To avoid noisy words in the web page, we only employ the words whose similarity scores to tags have exceeded a threshold, which is empirically set as 0.5. As the value range of the similarity formula is between 0 and 1 (larger value means more relevant), we mark a tag as a relevant tag if its similarity score to the web page exceeds 0.8.

Tags can also be extracted from web pages content, hence, we extract keywords from a web page [6] to generate its tags. We use Latent Dirichlet Allocation (abbreviated as LDA) [3] model to generate tags for web pages. In Table 3, we will see the difference between these methods. DR method means we generate tags only from those directly related queries.

Graph-based method gets higher precision than other methods when n is smaller than 5. LDA method also gains satisfactory results, while both keyword-based method and Cosession-based method have poor performance. The reason is, content-based method relies too much on the quality of the web page contents.

Table 3. *Precision@N* of Content-Based Method and Query-Based Method

Precison@N	Content-Based		Query-Based			
	Keyword Based	LDA Based	DR	Cocclick Based	Cosession Based	Graph Based
P@1	0.077	0.106	0.172	0.109	0.084	0.134
P@2	0.079	0.11	0.116	0.095	0.076	0.12
P@3	0.075	0.1	0.084	0.086	0.072	0.11
P@4	0.075	0.096	0.065	0.081	0.068	0.104
P@5	0.075	0.092	0.052	0.075	0.065	0.099
P@6	0.075	0.091	0.044	0.07	0.062	0.095
P@7	0.076	0.091	0.038	0.065	0.06	0.091
P@8	0.074	0.09	0.033	0.06	0.057	0.088
P@9	0.073	0.088	0.029	0.056	0.055	0.085
P@10	0.073	0.088	0.026	0.051	0.052	0.082

Table 4. Ranking Performance of EA-Independent and EA-Dependent

NDCG@N	EA-Dependent			EA-Independent		
	Coclick Based	Cosession Based	Graph Based	Coclick Based	Cosession Based	Graph Based
NDCG@1	0.295	0.191	0.442	0.433	0.533	0.425
NDCG@2	0.234	0.346	0.352	0.474	0.569	0.485
NDCG@3	0.256	0.337	0.386	0.486	0.623	0.519
NDCG@4	0.296	0.383	0.435	0.499	0.648	0.533
NDCG@5	0.285	0.422	0.451	0.502	0.652	0.553
NDCG@6	0.297	0.521	0.456	0.517	0.665	0.557
NDCG@7	0.332	0.524	0.508	0.526	0.679	0.574
NDCG@8	0.345	0.524	0.544	0.532	0.694	0.582
NDCG@9	0.345	0.524	0.544	0.537	0.706	0.594
NDCG@10	0.367	0.524	0.544	0.539	0.711	0.612

In tag ranking part, we have discussed EA parameter which is calculated based on the session relation between queries and web pages. As shown in Table 4, we see that EA definitely promotes the ranking result of different measures.

4.5 Manual Evaluation

To do manual evaluation, five volunteers rate each tag at five levels: 1(Poor), 2(Bad), 3(Acceptable), 4(Good), and 5(Excellent), based on their understanding of the web page. We randomly select 100 web pages. Each page has three groups of tags and every group has at most 5 tags. For every tags of each page, we make the average score as the final score. To measure precision, we mark a tag as a relevant one when its final score exceeds 4. The results of manual evaluation are demonstrated in Table 5.

Table 5. The Results of Manual Evaluation

N	Precision@N			NDCG@N		
	Coclick Based	Cosession Based	Graph Based	Coclick Based	Cosession Based	Graph Based
1	0.133	0.02	0.186	0.605	0.658	0.61
2	0.092	0.03	0.16	0.659	0.664	0.606
3	0.065	0.02	0.131	0.718	0.708	0.65
4	0.056	0.015	0.116	0.776	0.749	0.727
5	0.045	0.012	0.099	0.849	0.84	0.781

Precision results of manual evaluation and semantic tree evaluation are basically consistent. NDCG results become higher when we only keep 5 tags for each page. We emphasize the category of tags, however, for a given web page, there are many improper categories which should not be used to generate tags. To reduce the sacrifice of relevance, most of the web pages only involve 2 or 3 categories.

5 Conclusion

In this paper, we propose a new prospective of mining search logs for tags. We present two approaches to dig out tags related to a web page and rank them properly. In order to keep diversity, we classify web queries into different categories and then extract tags from queries to depict each category. As a future work, we will exploit the extracted tags to improve query recommendation and web search.

References

1. Belém, F.M., Martins, E.F., Almeida, J.M., Gonçalves, M.A., Pappa, G.L.: Exploiting co-occurrence and information quality metrics to recommend tags in web 2.0 applications. In: CIKM (2010)

2. Bing, L., Sun, B., Jiang, S., Zhang, Y., Lam, W.: Learning ontology resolution for document representation and its applications in text mining. In: CIKM (2010)
3. Blei, D.M., Ng, A.Y., Jordan, M.I.: Latent dirichlet allocation. The Journal of Machine Learning Research (3), 993–1022 (2003)
4. Bordino, I., Castillo, C., Donato, D., Gionis, A.: Query similarity by projecting the query-flow graph. In: SIGIR (2010)
5. Carman, M.J., Baillie, M., Gwadera, R., Crestani, F.: A statistical comparison of tag and query logs. In: SIGIR (2009)
6. Chien, L.-F.: Pat-tree-based keyword extraction for chinese information retrieval. In: SIGIR (1997)
7. Gupta, M., Li, R., Yin, Z., Han, J.: Survey on social tagging techniques. In: SIGKDD (2010)
8. Hu, J., Wang, G., Lochovsky, F., Tao Sun, J., Chen, Z.: Understanding user's query intent with wikipedia. In: WWW (2009)
9. Jansen, B.J., Booth, D.L., Spink, A.: Determining the informational, navigational, and transactional intent of web queries. Information Processing and Management 44(3), 1251–1266 (2008)
10. Jiang, S., Bin, L., Sun, B., Zhang, Y., Lam, W.: Ontology enhancement and concept granularity learning: keeping yourself current and adaptive. In: SIGKDD (2011)
11. Liu, D., Hua, X.-S., Yang, L., Wang, M., Zhang, H.-J.: Tag ranking. In: WWW (2009)
12. Marlow, C., Naaman, M., Boyd, D., Davis, M.: Ht06, tagging paper, taxonomy, flickr, academic article, to read. In: HYPERTEXT (2006)
13. Scriver, A.D.: Semantic distance in wordnet: A simplified and improved measure of semantic relatedness. Master Thesis, University of Waterloo, Canada (2006)
14. Shen, D., Pan, R., Sun, J.-T., Pan, J.J., Wu, K., Yin, J., Yang, Q.: Query enrichment for web-query classification. ACM TOIS 24(3), 320–352 (2006)
15. Sigurbjornsson, B., van Zwol, R.: Flickr tag recommendation based on collective knowledge. In: WWW (2008)
16. White, R.W., Bilenko, M., Cucerzan, S.: Studying the use of popular destinations to enhance web search interaction. In: SIGIR (2007)
17. Xu, Z., Fu, Y., Mao, J., Su, D.: Towards the semantic web: Collaborative tag suggestions. In: WWW (2006)
18. Zhou, D., Bian, J., Zheng, S., Zha, H., Giles, C.L.: Exploring social annotations for information retrieval. In: WWW (2009)

Author Index